Electromagnetic Wave Propagation and Transmission: Theory and Concepts

Electromagnetic Wave Propagation and Transmission: Theory and Concepts

Edited by **Edgar Wilson**

Willford Press
New York

Published by Willford Press,
118-35 Queens Blvd., Suite 400,
Forest Hills, NY 11375, USA
www.willfordpress.com

Electromagnetic Wave Propagation and Transmission: Theory and Concepts
Edited by Edgar Wilson

International Standard Book Number: 978-1-68285-249-1 (Hardback)

Printed in China.

Contents

	Preface	**IX**
Chapter 1	**An accelerated hybrid TLM-IE method for the investigation of shielding effectiveness** N. Fichtner and P. Russer	**1**
Chapter 2	**Antenna with optimized pattern for simultaneous reception of terrestrial signals and signals of geostationary satellites** G. Saala, D. Müller, J. Hopf and S. Lindenmeier	**7**
Chapter 3	**Motion compensation of short-range, wide-beam synthetic aperture radar** T. Reichthalhammer and E. Biebl	**13**
Chapter 4	**Leaky wave antenna with amplitude controlled beam steering based on composite right/left-handed transmission lines** M. A. Eberspächer and T. F. Eibert	**19**
Chapter 5	**A cooperative transponder system for improved traffic safety, localizing road users in the 5 GHz band** B. Schaffer, G. Kalverkamp, M. Chaabane and E. M. Biebl	**25**
Chapter 6	**Pedestrian recognition using automotive radar sensors** A. Bartsch, F. Fitzek and R. H. Rasshofer	**31**
Chapter 7	**Design and analysis of an isotropic two-dimensional planar composite Right/Left-handed waveguide structure** M. A. Eberspächer, M. Bauer and T. F. Eibert	**42**
Chapter 8	**Suppression of grating lobes for MMW sparse array setups** S. Bertl, A. Kirschner and J. Detlefsen	**48**
Chapter 9	**Ray tracing with multi-radiation transmitters** R. Brem and T. F. Eibert	**53**
Chapter 10	**Reconfigurable ultra-wideband transmitter for generation of arbitrary impulse shapes and modulation schemes** A. T. Ott, C. J. Eisner, M. D. Blech and T. F. Eibert	**57**
Chapter 11	**Influences of weather phenomena on automotive laser radar systems** R. H. Rasshofer, M. Spies and H. Spies	**63**

Chapter 12	**A fast three-dimensional deterministic ray tracing coverage simulator for a 24 GHz anti-collision radar** H. Azodi, U. Siart and T. F. Eibert	**75**
Chapter 13	**Near-field radiation pattern distortion of antenna attached to wall in through-the-wall radar imaging** B. Cetin, H.-R. Benedickter and P. Leuchtmann	**81**
Chapter 14	**Electromagnetic diffraction and scattering of a complex-source beam by a semi-infinite circular cone** H. Brüns and L. Klinkenbusch	**90**
Chapter 15	**A fully probe corrected near-field far-field transformation technique employing plane-wave synthesis** R. A. M. Mauermayer and T. F. Eibert	**96**
Chapter 16	**Using quasi-guided modes for modeling the transfer behavior of bent dielectric slab waveguides** M. Stallein, B. Bandlow and R. Schuhmann	**104**
Chapter 17	**Phaseless spherical near-field antenna measurements for low and medium gain antennas** C. H. Schmidt, S. F. Razavi, T. F. Eibert and Y. Rahmat-Samii	**112**
Chapter 18	**High resolution time of arrival estimation for a cooperative sensor system** C. Morhart and E. M. Biebl	**118**
Chapter 19	**Sensitivity analysis of waveguide eigenvalue problems** N. Burschäpers, S. Fiege, R. Schuhmann and A. Walther	**124**
Chapter 20	**Automotive radar – investigation of mutual interference mechanisms** M. Goppelt, H.-L. Blöcher and W. Menzel	**129**
Chapter 21	**Improved measurement surface for MEG using magnetic-dipole sources and a spherical-multipole expansion** F. Argin, H. Ahrens and L. Klinkenbusch	**135**
Chapter 22	**Properties of axial surface waves along dielectrically coated conducting cylinders** U. Siart, S. Adrian and T. Eibert	**140**
Chapter 23	**Scattering centers modeling of non-anechoic measurement environments** K. A. Yinusa, C. H. Schmidt and T. F. Eibert	**146**
Chapter 24	**Generation of postured voxel-based human models for the study of step voltage excited by lightning current** J. Gao, I. Munteanu, W. F. O. Müller and T. Weiland	**151**
Chapter 25	**A comparison of minimum norm and MUSIC for a combined MEG/EEG sensor array** H. Ahrens, F. Argin and L. Klinkenbusch	**158**

Chapter 26 **Simulations of magnetocardiographic signals using realistic geometry models of the heart and torso** 164
C. V. Motrescu and L. Klinkenbusch

Chapter 27 **Performance analysis of a low cost wireless indoor positioning system with distributed antennas** 171
A. T. Ott, M. Shalaby, U. Siart, E. Kaliyaperumal, T. F. Eibert, J. Engelbrecht and R. Collmann

Chapter 28 **Modeling of electrical transport in zinc oxide varistors** 177
K. Bavelis, E. Gjonaj and T. Weiland

Chapter 29 **A mixed finite-element formulation for the modal analysis of electromagnetic waveguides featuring improved low-frequency resolution of transmission line modes** 183
R. Baltes, J. Al Ahmar, O. Farle and R. Dyczij-Edlinger

Chapter 30 **Analytical finite element matrix elements and global matrix assembly for hierarchical 3-D vector basis functions within the hybrid finite element boundary integral method** 190
L. Li, K. Wang, H. Li and T. F. Eibert

Chapter 31 **Local microwave heating of sand molds as a means to overcome design limitations in sand mold casting** 201
O. Wiedenmann, R. Ramakrishnan, P. Saal, E. Kılıç, U. Siart, T. F. Eibert and W. Volk

Chapter 32 **Comparison of matching layers for automotive radome design** 209
F. Fitzek, R. H. Rasshofer and E. M. Biebl

Chapter 33 **A reconfigurable software defined ultra-wideband impulse radio transceiver** 215
M. D. Blech, A. T. Ott, P. Neumeier, M. Möller and T. F. Eibert

Chapter 34 **Modeling of field singularities at dielectric edges using grid based methods** 222
C. Classen, E. Gjonaj, U. Römer, R. Schuhmann and T. Weiland

Chapter 35 **Impact of system parameter selection on radar sensor performance in automotive applications** 228
H.-L. Blöecher, M. Andres, C. Fischer, A. Sailer, M. Goppelt and J. Dickmann

Chapter 36 **Eigenmode analysis of the electromagnetic field scattered by an elliptic cone** 233
M. Kijowski and L. Klinkenbusch

Chapter 37 **Eigenvalue extraction from time domain computations** 240
T. Banova, W. Ackermann and T. Weiland

Chapter 38 **The method of lines in the time domain** 247
S. F. Helfert

Chapter 39 **Multi-sensor doppler radar for machine tool collision detection** 254
T. J. Wächter, U. Siart, T. F. Eibert and S. Bonerz

Permissions

List of Contributors

Preface

This book has been a concerted effort by a group of academicians, researchers and scientists, who have contributed their research works for the realization of the book. This book has materialized in the wake of emerging advancements and innovations in this field. Therefore, the need of the hour was to compile all the required researches and disseminate the knowledge to a broad spectrum of people comprising of students, researchers and specialists of the field.

Electromagnetic wave propagation plays a crucial part in telecommunication engineering and networking. This book provides an in-depth analysis of electromagnetic wave propagation and transmission with the help of topics like signal processing, antenna wave propagation, transmitters, etc. The researches and case studies incorporated in this book attempt to highlight the emerging concepts and technological advancements in this field. It will help students and researchers by foregrounding their knowledge in this field.

At the end of the preface, I would like to thank the authors for their brilliant chapters and the publisher for guiding us all-through the making of the book till its final stage. Also, I would like to thank my family for providing the support and encouragement throughout my academic career and research projects.

Editor

An accelerated hybrid TLM-IE method for the investigation of shielding effectiveness

N. Fichtner and P. Russer

Institute for High Frequency Engineering, Technische Universität München, Arcisstr. 21, 80333 München, Germany

Abstract. A hybrid numerical technique combining time-domain integral equations (TD-IE) with the transmission line matrix (TLM) method is presented for the efficient modeling of transient wave phenomena. This hybrid method allows the full-wave modeling of circuits in the time-domain as well as the electromagnetic coupling of remote TLM subdomains using integral equations (IE). By using the integral equations the space between the TLM subdomains is not discretized and consequently doesn't contribute to the computational effort. The cost for the evaluation of the time-domain integral equations (TD-IE) is further reduced using a suitable plane-wave representation of the source terms. The hybrid TD-IE/TLM method is applied in the computation of the shielding effectiveness (SE) of metallic enclosures.

1 Introduction

Shielding effectiveness (SE) is a measure for spurious radiation leaking out of metallic enclosures which contain electronic circuitry and moreover it denotes the prevention from external radiation to penetrate into an enclosure (Paul, 2006). The computation of SEs is usually a very complex task as the SE depends on various phenomena such as geometrical parameters, materials, apertures, contacts, cabling and connections. In order to account for all effects which may influence the SE an electromagnetic full-wave simulation is required which allows to incorporate all structural details and material properties. Further, it is desirable to compute the SE for a broad band of frequencies to identify forbidden or allowed operating frequency bands. Under this consideration the use of full-wave time-domain modeling techniques such as the finite-difference time-domain (FDTD) or the transmission line matrix (TLM) method is advantageous. Using either FDTD or TLM, complex 3-D structures of finite spatial extension can be modeled in the time-domain (TD). For SE measurements and simulations, however, the electromagnetic fields some distance away from the enclosure are required (Paul, 2006). In this case integral equation (IE) based methods are preferable as the space between the source and the observation will not be discretized and hence does not contribute to the computational complexity. A hybrid technique therefore is proposed consisting of the TLM method in conjunction with TD integral equations (Fichtner and Russer, 2009). In the hybrid TD-IE/TLM method the circuitry together with the metallic enclosure and the aperture is modeled with TLM. The electric and magnetic fields leaving the metallic enclosure are probed at any time step and are used as source terms in the TD-IEs. From the TD-IEs, which are a TD version of the Stratton-Chu IEs (Hansen and Yaghjian, 1999; Tai, 2000), the fields at any position exterior of the TLM domain can be computed. In combination with the total-field/scattered-field (TF/SF) technique, spatially remote TLM subdomains can be coupled using TD-IEs (Fichtner and Russer, 2009). Moreover, a plane-wave time-domain (PWTD) algorithm based on the plane-wave representation of the radiated fields is employed to accelerate the TD-IE computation (Ergin et al., 1999; Chew et al., 2001).

2 The hybrid TD-IE/TLM method

2.1 The total-field/scattered-field technique

The hybrid TD-IE/TLM technique requires for the distinction between incident and scattered fields respectively inside a TLM domain. With the total-field/scattered-field (TF/SF) technique a partitioning between the total field, i.e. incident plus scattered field, and the scattered field only is possible. Consider a TLM domain Ω as shown in Fig. 1 in a cross-sectional view. The inner region Ω_1 contains all scatterers and sources and is surrounded by a second region Ω_2 without any structure elements or sources. The boundary between the two regions is denoted as Huygens' surface $\partial\Omega_{12}$ (Fichtner and Russer, 2009; Taflove, 2000). Fields excited and/or scattered within region Ω_1 will cross $\partial\Omega_{12}$ and are only outward propagating waves in region Ω_2 as long as a absorbing boundary at the TLM domain boundary $\partial\Omega$ is used.

Correspondence to: N. Fichtner
(fichtner@tum.de)

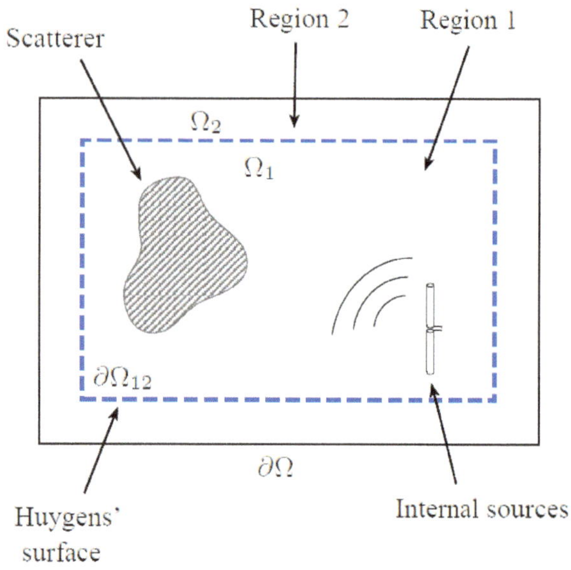

Fig. 1. TLM simulation domain Ω containing scatterers and source elements. Ω is subdivided in a total field region Ω_1 and scattered field region Ω_2. The dashed line indicates the Huygens' surface $\partial\Omega_{12}$.

In all simulations a perfectly matched layer (PML) boundary with 12 layers and a geometrical conductivity profile is used (Fichtner and Russer, 2009; Taflove, 2000). Consequently, the fields in region Ω_2 can be used for the computation of the radiated near-fields and far-fields as well as for the coupling of different TLM domains (Fichtner and Russer, 2009).

Using the TF/SF technique it is also possible to excite arbitrary wave forms within region Ω_1. These waves can either be analytically precomputed, e.g. for the excitation of arbitrary plane-waves (Taflove, 2000) or fields which originate from other remote TLM domains (Lindenmeier et al., 1999; Fichtner and Russer, 2009). For the one-side excitation of incident waves, a modified TLM connection operator has been derived (Fichtner and Russer, 2009). With ${}_k\boldsymbol{E}^{\mathrm{inc}}_{l,m,n}$ and ${}_k\boldsymbol{H}^{\mathrm{inc}}_{l,m,n}$ as the incident electric and magnetic fields, the modified TLM connection operator is written as

$$\begin{aligned}{}_k\boldsymbol{a}_{l,m,n} &= \boldsymbol{\Gamma}\cdot{}_k\boldsymbol{b}_{l,m,n}\\ &+\frac{1}{2}\left(\boldsymbol{\Gamma}_E\cdot{}_k\boldsymbol{E}^{\mathrm{inc}}_{l,m,n}+\boldsymbol{\Gamma}_H\cdot{}_k\boldsymbol{H}^{\mathrm{inc}}_{l,m,n}\right).\end{aligned} \tag{1}$$

The connection matrices for the electric and the magnetic fields are

$$\boldsymbol{\Gamma}_E=\begin{bmatrix}\boldsymbol{A}&\boldsymbol{0}&\boldsymbol{0}\\ \boldsymbol{0}&\boldsymbol{A}&\boldsymbol{0}\\ \boldsymbol{0}&\boldsymbol{0}&\boldsymbol{A}\end{bmatrix},\quad \boldsymbol{A}=\Delta l\begin{bmatrix}1&-1&0&0\\-1&1&0&0\\0&0&1&-1\\0&0&-1&1\end{bmatrix} \tag{2}$$

and

$$\boldsymbol{\Gamma}_H=\begin{bmatrix}\boldsymbol{B}&\boldsymbol{0}&\boldsymbol{0}\\ \boldsymbol{0}&\boldsymbol{B}&\boldsymbol{0}\\ \boldsymbol{0}&\boldsymbol{0}&\boldsymbol{B}\end{bmatrix},\quad \boldsymbol{B}=Z_0\Delta l\begin{bmatrix}1&-1&0&0\\1&-1&0&0\\0&0&-1&1\\0&0&-1&1\end{bmatrix}. \tag{3}$$

With Eqs. (1) to (3) it is possible to excite fields only within region 1. No fields in region 2 will be excited as required for the hybrid TD-IE/TLM method. For vanishing incident fields Eq. (1) reduces to the standard TLM connection operator $\boldsymbol{\Gamma}$ as required.

2.2 Coupling integrals

In numerical field computations the surface equivalence principle can be employed to compute the radiated electromagnetic fields everywhere in space from the knowledge of the field distribution on an arbitrary closed surface ∂V, which encloses all source elements (Balanis, 1997; Taflove, 2000). Consequently, the radiated near-fields or far-fields can be computed for simulation domains of finite extension (Taflove, 2000). The TD version of the Stratton-Chu IE using exterior differential forms is given by

$$\begin{aligned}\mathcal{E}(\boldsymbol{x},t) &= \mu\frac{\partial}{\partial t}\int_{\partial V'}\mathcal{I}\wedge\left(\mathrm{n}\lrcorner\,\mathrm{n}\wedge\frac{\mathcal{H}(\boldsymbol{x}',t')}{4\pi r}\right)\\ &-\int_{\partial V'}\left(\star(\mathcal{R}'\wedge\mathcal{I})\right)\wedge\frac{\mathcal{E}(\boldsymbol{x}',t')}{4\pi r^3}\\ &-\int_{\partial V'}\frac{\partial}{\partial t'}\left(\star(\mathcal{R}'\wedge\mathcal{I})\right)\wedge\frac{\mathcal{E}(\boldsymbol{x}',t')}{4\pi r^2 c}\\ &+\int_{\partial V'}\mathcal{R}\wedge\frac{\star\mathcal{E}(\boldsymbol{x}',t')}{4\pi r^3}\\ &+\int_{\partial V'}\frac{\partial}{\partial t'}\mathcal{R}\wedge\frac{\star\mathcal{E}(\boldsymbol{x}',t')}{4\pi r^2 c}\end{aligned} \tag{4}$$

where $\mathcal{I}$ is the unit double-form

$$\mathcal{I}=\left(\mathrm{d}x\,\mathrm{d}x'+\mathrm{d}y\,\mathrm{d}y'+\mathrm{d}z\,\mathrm{d}z'\right). \tag{5}$$

(Details on exterior differential forms can be found in Russer (2006)). By $\mathcal{R}$ and $\mathcal{R}'$ the distance 1-form $\mathcal{R}=(x-x')\mathrm{d}x+(y-y')\mathrm{d}y+(z-z')\mathrm{d}z$ is denoted whereas primed or unprimed differentials are used respectively. The unit 1-form n in Eq. (4) is orthogonal to the considered plane of the Huygens' surface and is oriented outwards. An expression for the radiated magnetic field analog to Eq. (4) follows from the principle of duality.

A time and space discrete version of Eq. (4) and its dual is obtained by expanding the fields into two-dimensional pulse-functions of edge length Δl and applying point-matching. Δl is identical to the TLM cell length in order to use directly the fields obtained from the TLM computation. After several step, which will not be outlined here, one obtains the compact representation

$$_k\mathcal{E}(\boldsymbol{x}) = \sum_{i=1}^{6}\sum_{j}\Big[\boldsymbol{L}_{\mathcal{E}\mathcal{E}}\cdot {}_{k'}\mathcal{E}(\boldsymbol{x}_j) + \boldsymbol{L}_{\mathcal{E}\mathcal{H}}\cdot {}_{k'}\mathcal{H}(\boldsymbol{x}_j)\Big], \tag{6a}$$

$$_k\mathcal{H}(\boldsymbol{x}) = \sum_{i=1}^{6}\sum_{j}\Big[\boldsymbol{L}_{\mathcal{H}\mathcal{E}}\cdot {}_{k'}\mathcal{E}(\boldsymbol{x}_j) + \boldsymbol{L}_{\mathcal{H}\mathcal{H}}\cdot {}_{k'}\mathcal{H}(\boldsymbol{x}_j)\Big]. \tag{6b}$$

The time index k' is the retarded discrete time given by $k' = k - |\boldsymbol{x} - \boldsymbol{x}_j|/c\Delta t$ where $\boldsymbol{x}_j$ is the source position and $\boldsymbol{x}$ is the observation position. The operators $\boldsymbol{L}$ are computed from Eq. (4) and consist of discrete temporal and spatial derivatives. From Eq. (6) the field at position $\boldsymbol{x}$ is computed from the retarded field values on a closed Huygens' surface ∂V (see Fig. 1). In the hybrid TD-IE/TLM method the outgoing fields are probed on a closed Huygens' surface which is located within region Ω_2. Using the probed fields as source terms, the radiated fields outside of the TLM domain Ω can be computed from Eq. (6). The radiated fields are used as excitation within a second TLM subdomain by using the TF/SF technique and the modified connection operator (1). Consequently, two or more TLM subdomains are electromagnetically coupled without discretizing the empty space between the subdomains.

3 The plane-wave time-domain algorithm

The dominant cost in the hybrid TD-IE/TLM method results from the computation of Eq. (6) which scales with $\mathcal{O}(N_s^2 N_t)$. The number of basis-functions which constitute $\partial\Omega_{12}$ is given by N_s and N_t is the number of considered time-steps. A more efficient computation scheme is obtained if several source elements are first grouped together and in the following the contribution of every group is considered individually. This is basically done in the fast multipole method (FMM) used for the acceleration of frequency domain (FD) moment method (MoM) implementations (Chew et al., 2001). A TD analog of the FMM, which uses a plane-wave expansion for the representation of the outgoing fields, has been developed by Ergin et al. (1999); Chew et al. (2001). The plane-wave time-domain (PWTD) algorithm employs a Whittaker-type field representation of the form

$$\tilde{u}_i(\boldsymbol{x},t) = -\frac{1}{8\pi c}\frac{\partial}{\partial t}\int \mathrm{d}\Omega \int_V \mathrm{d}\boldsymbol{x}' q_i(\boldsymbol{x}',t') \tag{7a}$$

$$= u_i(\boldsymbol{x},t) - \int_V \mathrm{d}\boldsymbol{x}' \frac{q_i(\boldsymbol{x}',t+r/c)}{4\pi r} \tag{7b}$$

where $\int \mathrm{d}\Omega$ is the area integral over the unit sphere. The unit direction vector is given by $\boldsymbol{k} = \boldsymbol{e}_x \sin\theta\cos\phi + \boldsymbol{e}_y \sin\theta\sin\phi + \boldsymbol{e}_z\cos\theta$ and t'. The quantities $\tilde{u}_i(\boldsymbol{x},t)$, $u_i(\boldsymbol{x},t)$ and $q_i(\boldsymbol{x},t)$ are arbitrary field and source components respectively. In Eq. (7b) the second term on the right-hand side is anti-causal and can be eliminated by using proper time-gating techniques as outlined in Ergin et al. (1999); Chew et al. (2001).

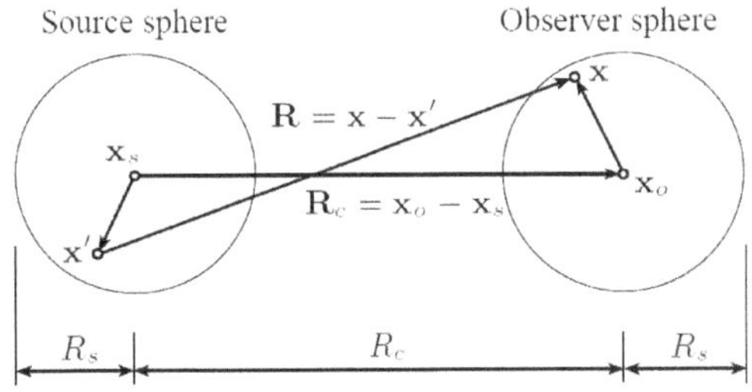

Fig. 2. Grouping of the source and observation positions inside two spheres of finite spatial extension.

Starting with Eq. (7) it is possible to derive a three-stage implementation for the summation in Eq. (6). According to Fig. 2 the distance vector $\boldsymbol{R}$ between the source and observation positions is first rewritten as

$$\boldsymbol{R} = \boldsymbol{x} - \boldsymbol{x}' = (\boldsymbol{x} - \boldsymbol{x}_o) + \boldsymbol{R}_c - (\boldsymbol{x}' - \boldsymbol{x}_s) \tag{8}$$

with $\boldsymbol{R}_c = \boldsymbol{x}_o - \boldsymbol{x}_s$. In the first stage of the PWTD algorithm the source distribution is mapped onto outgoing plane-waves propagating in all directions. This operation is denoted as Slant-stack transform (SST) and is given by

$$q_i^{\mathrm{out}}(\boldsymbol{k},t) = \int_V d\boldsymbol{x}' q_i\left(\boldsymbol{x}', t + \frac{\boldsymbol{k}\cdot(\boldsymbol{x}'-\boldsymbol{x}_s)}{c}\right). \tag{9}$$

In Eq. (9) the time-dependent source distribution $q_i(\boldsymbol{x}',t)$ is represented by outgoing rays in the $\boldsymbol{k}$-th direction. In the second stage the rays leaving the source sphere are translated onto the observation sphere via the convolution operation

$$q_i^{\mathrm{in}}(\boldsymbol{k},t) = \mathcal{T}(\boldsymbol{k},\boldsymbol{x}_c,t) * q_i^{\mathrm{out}}(\boldsymbol{k},t) \tag{10}$$

where $\mathcal{T}(\boldsymbol{k},\boldsymbol{x}_c,t)$ is the diagonal translation operator

$$\mathcal{T}(\boldsymbol{k},\boldsymbol{x}_c,t) = -\frac{1}{8\pi c}\frac{\partial}{\partial t}\delta\left[t - \frac{\boldsymbol{k}\cdot\boldsymbol{R}_c}{c}\right]. \tag{11}$$

The incident rays depend on the direction of arrival $\boldsymbol{k}$ and time t. In the third and last step the incoming rays $q_i^{\mathrm{in}}(\boldsymbol{k},t)$ are projected onto the observation positions using

$$\tilde{u}_i(\boldsymbol{x},t)\int d\Omega q_i^{\mathrm{in}}\left(\boldsymbol{k}, t - \frac{\boldsymbol{k}\cdot(\boldsymbol{x}-\boldsymbol{x}_o)}{c}\right). \tag{12}$$

In summary for a compact set of source and observation positions each with finite spatial extension the evaluation of Eq. (6) is accelerated by concatenation of all source elements, which are in case of the hybrid TD-IE/TLM method the fields $_{k'}\mathcal{E}(\boldsymbol{x}')$ and $_{k'}\mathcal{H}(\boldsymbol{x}')$ on the Huygens' surface. After the translation operation (10) a de-concatenation (12) is

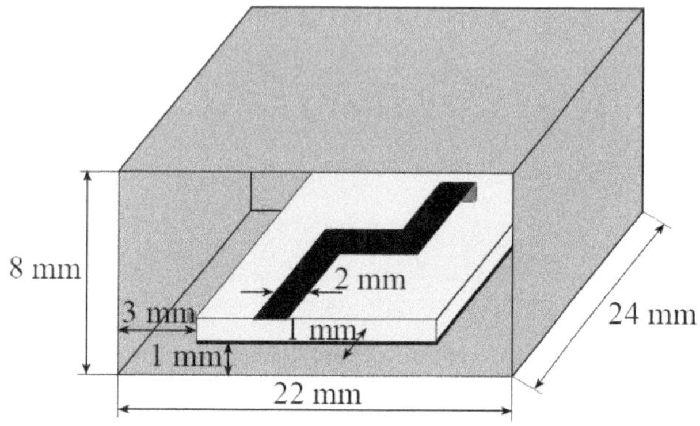

Fig. 3. Dimensions of the microstrip line inside the metallic enclosure.

applied. It can be shown that using the PWTD technique outlined between Eqs. (7) to (12) the computation of the TD-IEs scales with $\mathcal{O}(N_s^{1.5} N_t \log N_s)$. The acceleration is basically achieved by replacing the convolution in Eq. (10) with a multiplication whereas first the terms $\mathcal{T}(\boldsymbol{k}, \boldsymbol{x}', t)$ and $q_i^{\text{out}}(\boldsymbol{k}, t)$ are Fourier-transformed. After multiplication an inverse Fourier transform is applied. More details on the PWTD algorithm can be found in Ergin et al. (1999); Chew et al. (2001).

4 Simulation results

4.1 Shielding effectiveness (1)

The hybrid TD-IE/TLM method provides an efficient tool for the simulation and analysis of shielding problems. A measure for the shielding performance of enclosures and housings is the shielding effectiveness (SE), defined as

$$\text{SE} = 20 \cdot \log_{10} \left| \frac{E_0}{E_S} \right|, \tag{13}$$

where E_0 is the measured or simulated electric field without shield and E_S is the electric field when the shielding is used (Paul, 2006).

In a first test case a microstrip line with two 90° bends and a short at the end is placed inside a metallic enclosure. The microstrip line inside the metallic enclosure is shown in Fig. 3. The 90° bends of the line are not compensated using mitered bends and hence spurious fields are excited at the bends. The microstrip line structure is excited using a waveguide port with a modulated Gaussian pulse defined between the frequencies 5 and 20 GHz. From the short the main part of the incident wave is reflected and leaking out of the circuit.

Additional to the metallic box as shown in Fig. 3 a front panel will be used to reduce field leakage out of the enclosure. Two different configurations for the front panel are investigated, which are shown in Fig. 4 (a) and (b). The front panel in (a) exhibits a large air inlet of 8 mm width and 4 mm height and panel (b) has several smaller air inlets. Both air inlets (a) and (b) may serve as airflow apertures.

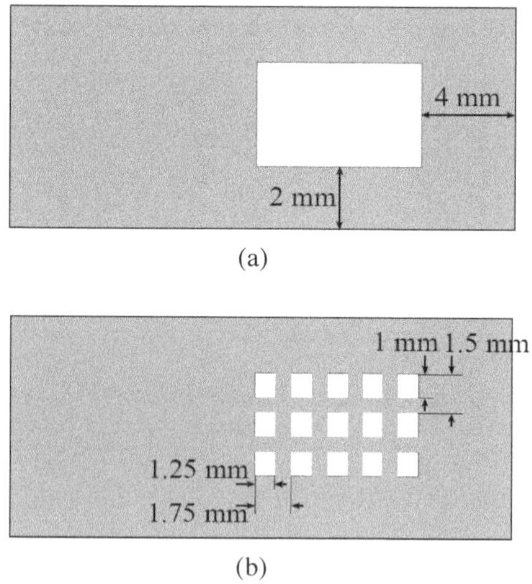

Fig. 4. Different configurations for the front panel of the enclosure. **(a)** exhibits a large air inlet of width 8 mm and height 4 mm. **(b)** consits of several small air inlets.

The electric field is measured 25 cm in front of the panels. In case of the TD-IE/TLM simulation the microstrip line and the enclosure is simulated with TLM whereas a Huygens' surface close to the box is used to probe the electric and magnetic fields leaking out from the enclosure. These fields are used as sources in the TD-IE computation (6). The Huygens' surface ∂V is embedded within the source sphere as shown in Fig. 2. As the radiated fields at one position are required only, the first two steps of the PWTD algorithm are needed.

As a reference a CST simulation of the identical structure has been performed. The shielding effectiveness (SE) from 5 to 20 GHz for the two panel configurations are shown in Fig. 5. For panel configuration (a) SE values between 0 and 60 dB are achievable. At some frequencies, in particular at 16.5 GHz the SE is well below zero. In this case the air inlets even enhance field leakage as they operate as aperture antennas (Paul, 2006). Using panel (b) the SE is about 50 dB higher in comparison to panel (a). The SE characteristic with respect to frequency of panel (b) is almost identical to panel (a) except for a positive 50 dB offset.

The accordance between CST and TD-IE/TLM simulated results is very good for all considered frequencies. As in the TD-IE/TLM method the space between box and sense point is not discretized the hybrid method turns out to be approximately five times faster in comparison to a CST simulation when the same machine is used, where all the space between enclosure and measurement point is meshed.

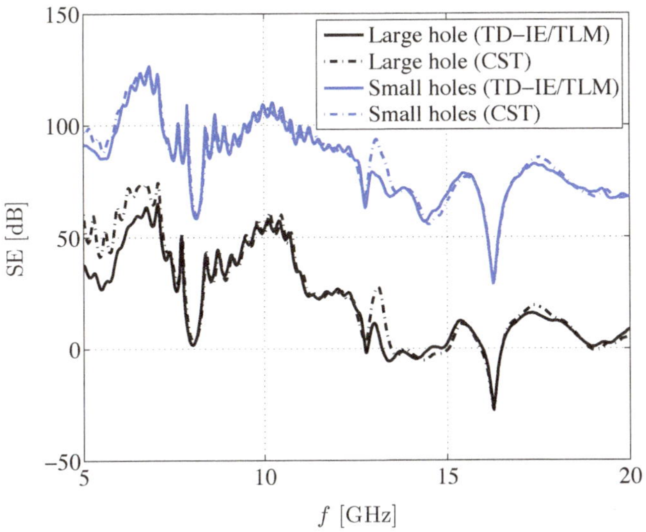

Fig. 5. TD-IE/TLM and CST computed shielding effectiveness (SE) for the two front panel configurations of Fig. 4.

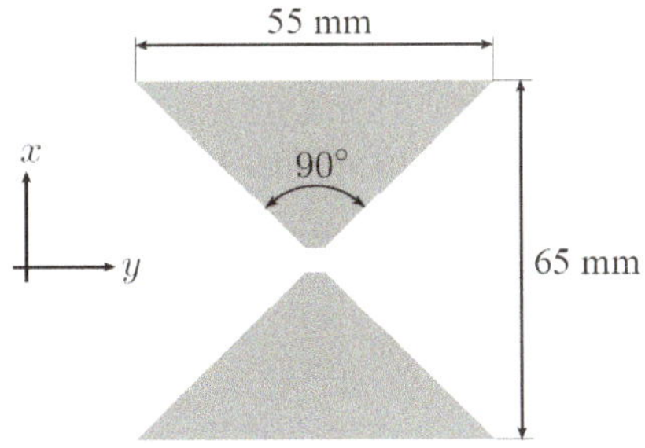

Fig. 6. Geometry of planar bowtie antenna.

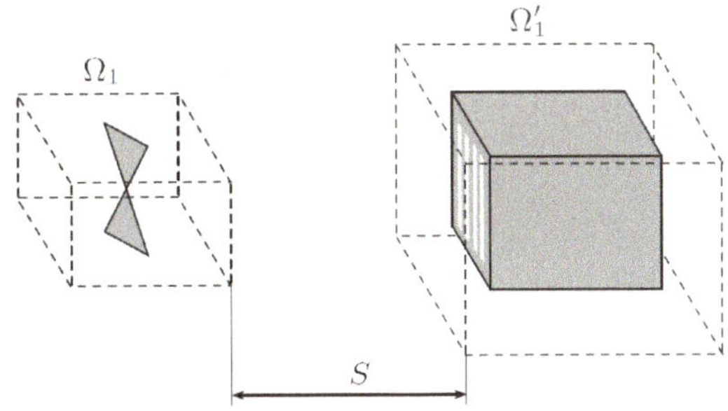

Fig. 7. TLM simulation scenario for the investigation of the shielding effectiveness of a metallic box irradiated by a planar bowtie antenna. The distance between the two TLM subdomain Ω_1 and Ω_1' is $S = 430$ mm.

4.2 Shielding effectiveness (2)

A further example concerning the shielding effectiveness is on the prevention of external radiation to penetrate into a metallic enclosure whereas different air inlet configurations are used. For this purpose a metallic box is considered which is irradiated by a broadband antenna. As antenna a planar bowtie will be used which can be modeled straightforward in TLM. The bowtie antenna geometry is shown in Fig. 6. For the excitation of the antenna an electric field oriented in x-direction is impressed at its center. A modulated Gaussian pulse defined between the frequencies 1 and 5 GHz will be used for the excitation. Usually such investigations are performed in the far-field (Paul, 2006). As the limit between near-field and far-field, the Fraunhofer distance $d = 2D^2/\lambda$ is considered, where D is the maximum overall antenna dimension and λ is the wavelength of interest (Balanis, 1997). For a frequency of $f = 5$ GHz and a maximum antenna dimension of $D = 95$ mm, given from the bowtie diagonal, the Fraunhofer distance d is 300 mm. In order to have some additional space between box and antenna the separation is set to $S = 430$ mm.

The complete simulation scenario is shown in Fig. 7. The TLM subdomain Ω_1 with the dimension $20 \times 20 \times 20$ cells contains the planar bowtie antenna of Fig. 6. The metallic box is modeled within a second TLM subdomain Ω_1' of dimension $41 \times 41 \times 31$ cells. In both TLM subdomains the discretization $\Delta l = 2.5$ mm is used. The box has a square front with a side length of 105 mm and is 130 mm in depth. The two TLM subdomains in Fig. 7 show only the total-field regions. This means that the regions Ω_1 and Ω_1' are in addition surrounded by a scattered-field domain Ω_2 and Ω_2' respectively, whereas each domain is four TLM cells in width. These scattered-field domains are not explicitly shown in Fig. 7. The TLM domains Ω_1 and Ω_1' are embedded within the source and observation sphere respectively of the PWTD algorithm.

For the front plane of the metallic box four different configurations of air inlets are investigated. The front planes are shown in Fig. 8. For configuration (a) the inlets are square and have a width of 15 mm. The separation between the inlets is 10 mm. In configuration (b) the width of the square inlets and the separation is identical to 5 mm. In case of configurations (c) and (d) the horizontal and vertical stripes have a width of 5 mm and a length of 65 mm.

In this analysis the excitation of resonant modes inside the metallic enclosure caused be the fields penetrating into the enclosure, i.e. the type of modes and their amplitudes are of interest. The electric field radiated from the bowtie antenna is polarized in x-direction. Thus, the E^x component inside the box is measured 10 cells behind the front panel. The magnitude spectra of the probed electric field using the four front plane configurations (a) to (d) are shown in Fig. 9. In case of configuration (a) the signal amplitudes are very high because of the strong coupling. The interior of the box is only weakly shielded from external radiation if large air inlets are used. Moreover, the resonating modes exhibit high quality factors as can be seen from the sharp peaks at the resonant frequencies. This indicates that the resonating modes

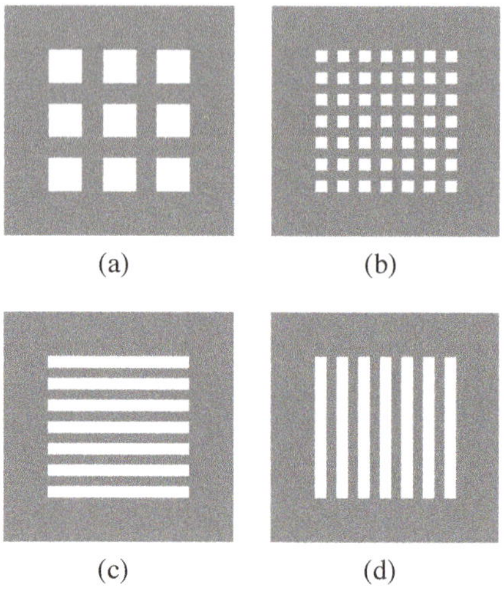

Fig. 8. Air inlet configurations for the shielding effectiveness investigations on the irradiated metallic enclosure.

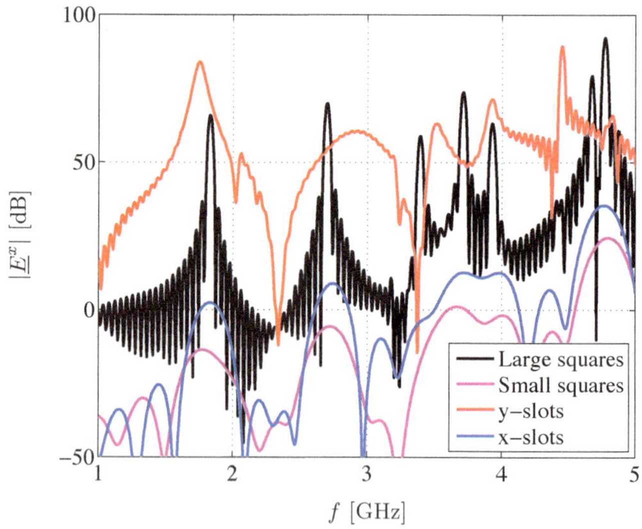

Fig. 9. Magnitude spectrum of the E^x field components measured inside a metallic box for the different front panel configurations of Fig. 8.

are only weakly disturbed by the inlets. From the magnitude spectrum in Fig. 9 it can be deduced that the highest shielding effect can be achieved if either the small inlets of configuration (b) or the x-directed slots of configuration (d) are used. In case of (b) the shielding is so strong that almost no resonances are excited within the box. The slots of configuration (d) behave like a polarization filter and prevent most of the radiation from penetration into the box. The situation is different in case of configuration (c). Here the slots are oriented orthogonal to the polarization of the incoming electric field and thus the fields can easily penetrate into the box. Though the formation of resonating modes is rather weak. It can be shown that the lowest order resonating modes are of TE_{10l}-type with an electric field in x-direction. Due to the orientation of the metallic bars in case of configuration (c) the electric fields of the TE_{10l} modes can easily leak out from the metallic box which explains the low quality factors.

5 Conclusion

A hybrid TD-IE/TLM method using the plane-wave time-domain technique has been presented for the fast and efficient modeling of typical shielding problems. Circuits together with enclosures are modeling with TLM and the radiated fields in the near-as well as in the far-field are obtained from TD integral equations. The computational burden for the evaluation of the integral equations is lowered by using a suitable plane-wave representation. In case of the presented SE simulations the computational effort is lowered by a factor of 3 to 5 if the hybrid TD-IE/TLM method is used compared with a pure TLM simulation. Using the PWTD algorithm further 20% reduction in computational cost are achievable.

Acknowledgements. This work was supported by the Deutsche Forschungsgemeinschaft (DFG).

References

Balanis, C.: Antenna Theory – Analysis and Design, John Wiley & Sons, 2nd edn., 1997.

Chew, W., Michielssen, E., Song, J. M., and Jin, J. M.: Fast and Efficient Algorithms in Computational Electromagnetics, Artech House, Norwood, MA, USA, 2001.

Ergin, A., Shanker, B., and Michielssen, E.: The Plane-Wave Time Domain Algorithm for the Fast Analysis of Transient Wave Phenomena, IEEE Trans. Antennas Propagat. Magazine, 41, 39–52, 1999.

Fichtner, N. and Russer, P.: A Total-Field/Scattered-Field Technique Applied for the TLM-Integral Equation Method, 2009 IEEE MTT-S Int. Microwave Symp. Dig, 325–328, 2009.

Hansen, T. and Yaghjian, A.: Plane-Wave Theory of Time-Domain Fields, IEEE Press, 1999.

Lindenmeier, S., Pierantoni, L., and Russer, R.: Hybrid Space Discretization – Integral Equation Methods for Numerical Modeling of Transient Interference, 1999, IEEE Trans. Electromagn. Compatibility, 41, 425–430, 1999.

Paul, C.: Introduction to Electromagnetic Compatibility, Wiley Series in Microwave and Optical Engineering, John Wiley & Sons, New York, 2nd edn., 2006.

Russer, P.: Electromagnetics, Microwave Circuit and Antenna Design for Communication Engineering, Artech House, 2nd edn., 2006.

Taflove, A.: Computational Electrodynamics – The Finite Difference Time-Domain Method, Artech House, Boston, 2nd edn., 2000.

Tai, C.-T.: Direct Integration of Field Equations, Progress in Electromagn. Research, 28, 339–359, 2000.

2

Antenna with optimized pattern for simultaneous reception of terrestrial signals and signals of geostationary satellites

G. Saala, D. Müller, J. Hopf, and S. Lindenmeier

Institut für Hoch- und Höchstfrequenztechnik, Universität der Bundeswehr München, Werner-Heisenberg-Weg 39, 85577 Neubiberg, Germany

Abstract. The link margin for satellite communication services targeting mobile users is tight. This leads to reception problems in areas with strong fading scenarios like urban areas or forested areas with dense foliage. In order to avoid loss of service in urban areas with week coverage by satellite signals, terrestrial repeaters are used. Therefore in the design of antennas for mobile reception of satellite services, an optimized characteristics for the satellite constellation, as well as an adequate coverage of the terrestrial repeaters has to be accounted for. In this paper an antenna with optimum shaped pattern for the satellite mode and adjustable gain for vertical polarization is investigated. Beside constraints regarding the radiation pattern, measurements concerning the dependence of the antenna on noise temperature and therefore the behavior in real noisy environments are carried out. A comparison with a conventional antenna shows the high performance of the new antenna in combination with a very low volume requirement.

1 Introduction

Services in satellite communications are not only set up for stationary receiver stations any more. During the last decade more and more services are targeting also mobile users not only for telephony but also for broadcasting radio and data services. Satellite systems such as Satellite Digital Audio Radio Services (SDARS) in the USA and Worldspace in Africa, Europe and parts of Asia, are already in operation. Aspects of the coverage of Europe by Worldspace are examined in Sallam et al. (2008).

For each service a different constellation of satellites might be reasonable, depending on area of coverage, data rates and available transmission power of the satellite. For a wide area of coverage the geostationary orbit (GEO) is chosen for several systems, like Worldspace and XM-SDARS, whereas other systems operate in high earth orbits (HEO), such as Sirius-SDARS which is described in Briskman and Prevaux (2004) using several satellites to realize the continuous coverage of the service. An overview of several satellite services is given in Lida et al. (2003).

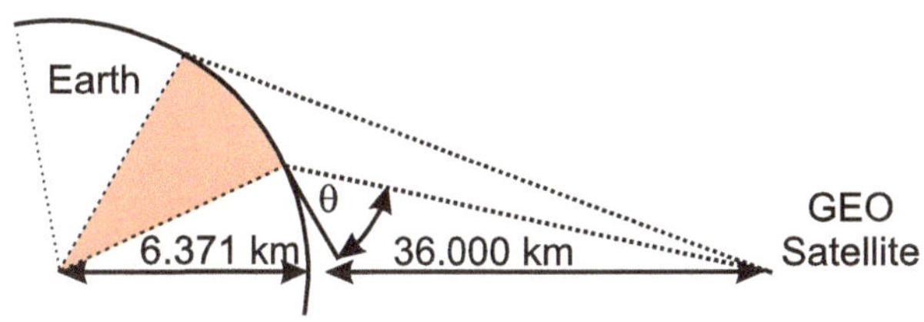

Fig. 1. Sketch of the geostationary satellite constellation for 2-D-approximation of the elevation angle above the horizon, the satellite is seen under from an earth station.

For a coverage of the USA and Europe the minimum latitude is approximately $\mathrm{Lat}_{\min} \approx 26°$ N for the southern point of Texas. The maximum latitude for the north of Denmark is approximately $\mathrm{Lat}_{\max} \approx 57°$ N. From Fig. 1, with an easy two dimensional approximation it can be seen, that geostationary satellites in these areas of Europe and the USA are seen in an elevation angle from 25° to 60° (Morgan and Gordon, 1989). In the following antennas are investigated which are designed for services using geostationary satellites. As the satellites are never seen at zenith or the horizon, respectively, the radiation pattern should be low at these positions for optimum power distribution in ranges of the elevation angles the satellites may appear in. In contrast to stationary receiving stations where the antenna beam can always point towards the satellite in a fixed direction, the azimuth angle under which the satellite occurs is unknown in mobile

Correspondence to: G. Saala
(gerd.saala@unibw.de)

Fig. 2. Complete antenna structure with vertical feed section realized as double sided parallel strip line.

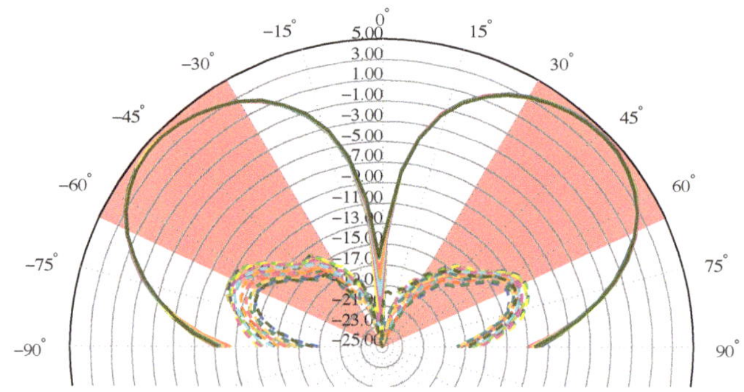

Fig. 3. Basic antenna setup with coaxial feed. 16 elevation cuts for vertical (− −) and horizontal (—) polarization in azimuthal steps of $\Delta\varphi = 11.25°$.

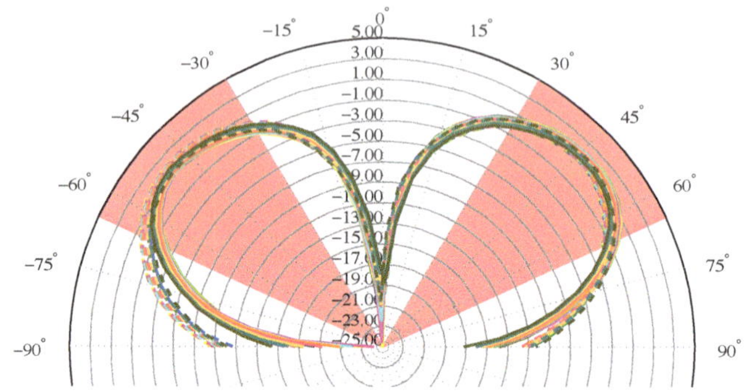

Fig. 4. Basic antenna setup with coaxial feed. 16 elevation cuts for right (—) and left (− −) hand circular polarization in azimuthal steps of $\Delta\varphi = 11.25°$.

applications. Hence, the radiation pattern should be omnidirectional over the azimuth. A further constraint arises from the sufficient coverage of areas with strong fading scenarios or weak signals like urban areas. In those areas the coverage is supported by additional terrestrial repeaters which transmit in vertical polarization. In this paper, an antenna with an adjustable power ratio between vertical and horizontal polarization is investigated. Considering the vertical polarization with respect to the performance of the reception of satellite signals, also the noise behavior of the antenna with better terrestrial reception is analyzed. Finally field tests are presented which compare the performance in signal to noise ratio (SNR) of the new antenna type with a commercially available antenna with respect to different receiving environments.

2 Antenna structure

The principle of a new antenna structure has been presented in Saala et al. (2009). It consists of an electrically large loop structure over a ground plane. This loop structure is segmented in order to achieve a homogeneous current distribution over the complete circumference of the horizontal part of the antenna. With that homogeneous current over the ground plane the main beam is directed to an elevation angle of approximately $\theta \approx 45°$ and a null at the zenith and horizon. In the following several variants of this antenna, shown in Fig. 2, are investigated with respect to the adjustment of the vertical polarized part of the radiation pattern. The antennas are designed for a center frequency of $f_c = 1.486$ GHz.

2.1 Basic operation mode

The basic antenna consists of the horizontal loop structure which is fed by horizontal double sided parallel strip line connected to a vertically oriented feed section, e.g. coaxial line. By proper design an additional matching network can be avoided at the base point of the coaxial line. The vertical part can also be realized by a double sided parallel strip line for easier production. In Figs. 3 and 4 the radiation patterns are depicted for linear and circular polarizations, respectively. The horizontal polarization is perfectly rotational symmetric over the azimuth. The vertical polarization is more than 15 dB lower than the horizontal polarization. The circular polarizations are also equally balanced with a little asymmetry over the azimuth, which is due to the asymmetry of the vertical polarization caused e.g. by effects of the antenna chassis. With this antenna a simultaneous reception of data streams transmitted in right and left hand polarization is possible without necessity of a switch, which might interrupt the data stream when switching from one polarization towards the other one.

2.2 Stronger terrestrial mode

For optimum coverage not only via satellites but also by terrestrial repeaters, the gain in vertical polarization of the antenna has to be adjusted to the needs of the service. Therefore the loop structure of the antenna is enhanced to perform in a mixed mode operation by introducing a capacitor at the base

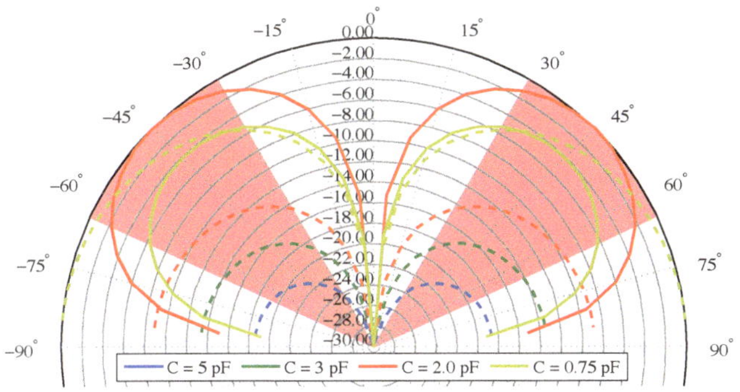

Fig. 5. Simulation: Different values of capacitance introduced at the feed point of the antenna. Vertical (– –) and horizontal (—) polarized radiation pattern, the curves are average values over the azimuth.

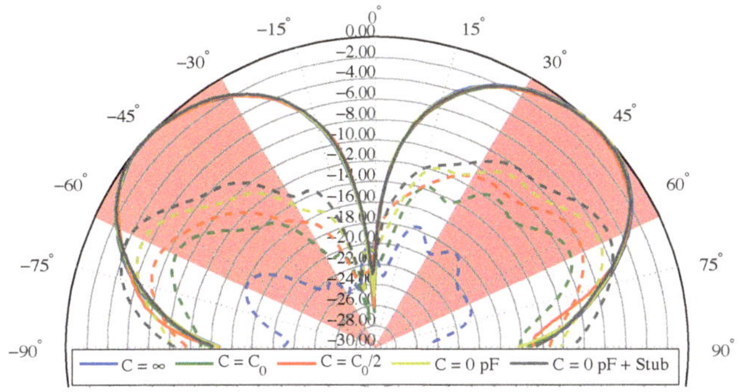

Fig. 7. Measurement: Vertical (– –) and horizontal (—) polarized radiation pattern for different capacitances at the feed point. Each curve is the average over four equally spaced elevation cuts with $\Delta\varphi = 45°$.

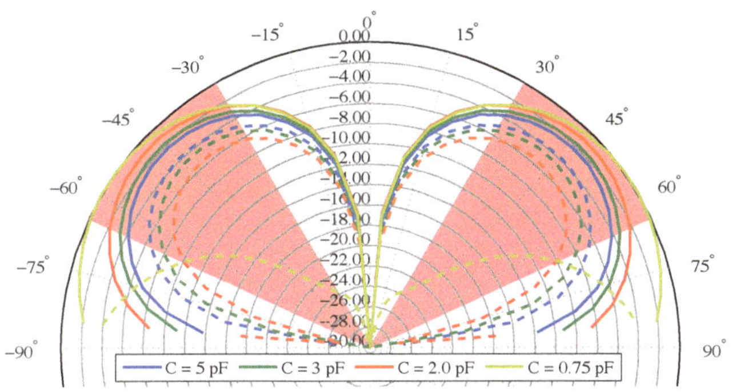

Fig. 6. Simulation: Different values of capacitance introduced at the feed point of the antenna. Right (—) and left (– –) hand circular polarized radiation pattern, the curves are average values over the azimuth.

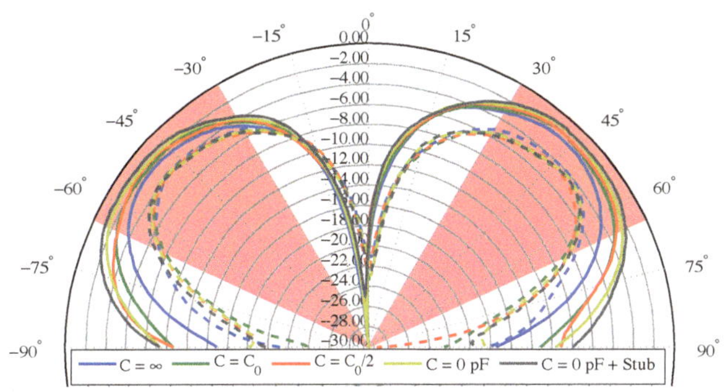

Fig. 8. Measurement: Right (—) and left (– –) hand circular polarized radiation pattern for different capacitances at the feed point. Each curve is the average over four equally spaced elevation cuts with $\Delta\varphi = 45°$.

point of the vertical feed section of the antenna. With this capacitor a common mode current is excited on the vertical section, which leads to a raised radiation in vertical polarization.

Compared to the antenna operating in its basic mode with mainly horizontal polarized radiation, the introduction of a capacitor changes the input impedance of the antenna, which cannot easily be accounted for by change of the loop-structure or the feed line. Hence an additional matching network is required. The power ratio between the horizontal and vertical polarization is not influenced by this matching network, compared to the unmatched operation.

Figures 5 and 6 show simulation results of antennas with different values of capacitors introduced at the feed point of the antenna. In order to show the tendency of the power ratio independent of the matching, the graphs are normalized values of each antenna. For the linear polarizations a fine graduation over the capacitance is achieved. For very small capacitors the vertical polarization even exceeds the horizontal polarization. Because of the excitation of the common mode current at the base point of the antenna the phase for the received signals is not necessarily the same for vertical and horizontal polarization. This leads to an imbalance between the circular polarizations (Fig. 6) which gets greater with rising vertical polarization. This imbalance can be used to account for an imbalance of the power distribution to the two circular polarizations on the satellite.

For the measurements, shown in Figs. 7 and 8, the capacitors at the base point of the antenna and the matching network have been introduced on the rear side of the ground plane. The plots for the linear polarizations in Fig. 7 show the precisely adjustable power ratio H/V between the horizontal and vertical polarization starting from the basic mode operation for the short with $C_{\text{feed}} = \infty$ and ending with a large reactance for the open circuit with $C_{\text{feed}} = 0\,\text{pF}$. Compared to the simulations the power ratio H/V is different e.g. in case of $C = 0\,\text{pF}$. This is due to a stray capacitance between the open conductor and the ground plane.

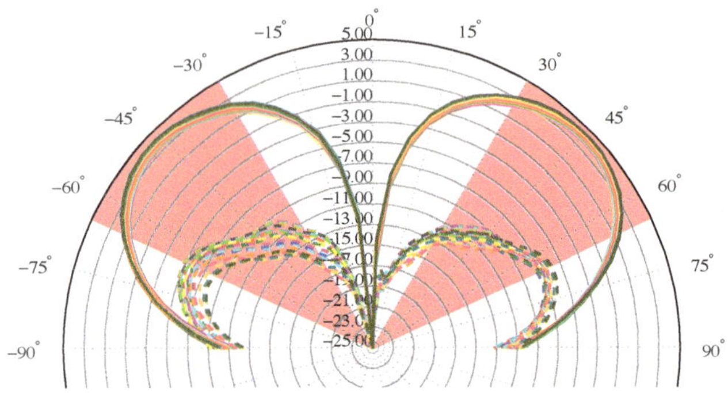

Fig. 9. Capacitor at the base of the vertical feed structure. 16 elevation cuts for vertical (– –) and horizontal (—) polarization in azimuthal steps of $\Delta\varphi = 11.25°$.

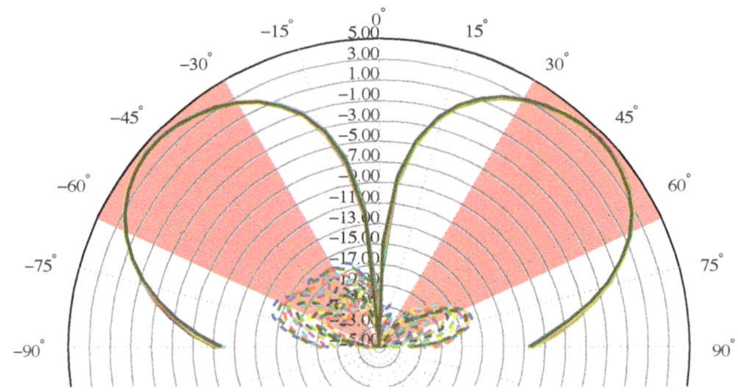

Fig. 11. Capacitors at the top of the vertical feed section. 16 elevation cuts for vertical (– –) and horizontal (—) polarization in azimuthal steps of $\Delta\varphi = 11.25°$.

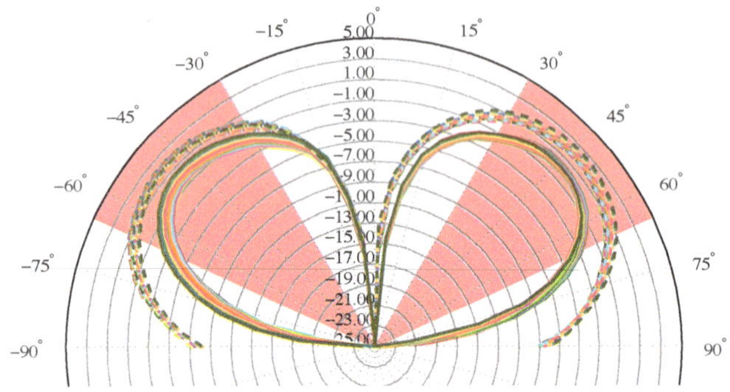

Fig. 10. Capacitor at the base of the vertical feed structure. 16 elevation cuts for right (—) and left (– –) hand polarization in azimuthal steps of $\Delta\varphi = 11.25°$.

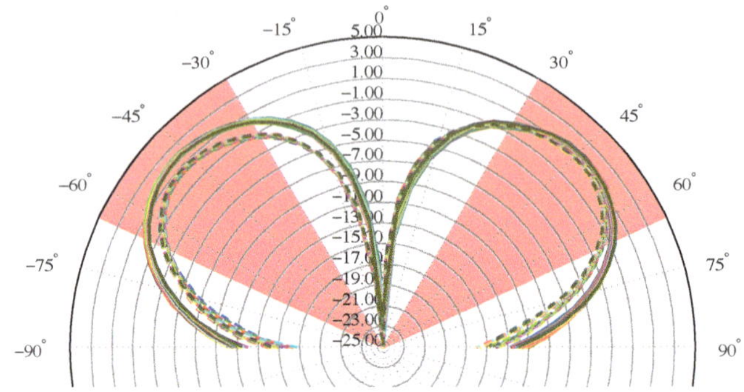

Fig. 12. Capacitors at the top of the vertical feed section. 16 elevation cuts for right (—) and left (– –) hand polarization in azimuthal steps of $\Delta\varphi = 11.25°$.

The radiation pattern of one antenna sample with increased shape of vertical the polarization, matched to $50\,\Omega$ is shown in Fig. 9 for the linear polarization with a power ratio of $H/V \approx 9\,\text{dB}$. The radiation pattern of the circular polarizations in Fig. 10 show an imbalance of up to 2 dB.

2.3 Suppression of the terrestrial mode

The increased gain in vertical polarization, described in Sect. 2.2, is due to a common mode current excited at the base of the vertical feed section. A small fraction of a common mode current is also excited using a coaxial line or parallel strip line, respectively for this feed section. For an additional discrimination of the parasitic vertical polarized radiation a balanced unbalanced transition could be introduced at the feed point of the antenna. Another solution is achieved by introducing two capacitors at the top of the vertical feed section. Thereby a common mode current is excited also at the top section of the vertical feed, which interferes destructively with the parasitic common mode current excited at the base point of the antenna. The resulting radiation pattern for this antenna is shown in Fig. 11 for the linear polarizations with a power ratio of $H/V \approx 18.5\,\text{dB}$. The left hand and right hand circular polarizations in Fig. 12 are almost balanced.

A comparison of the phases of the received signals of the antennas described in the last sections is shown in Fig. 13. Only for the basic mode the signals for vertical and horizontal polarization are in phase or anti-phase respectively, which is the reason for the very good balance between left and right hand circular polarizations in this case.

3 Noise measurements and antenna temperature

The system noise temperature T_S at the antenna terminals considering antenna losses, attenuation on the transmission line and noise of the amplifiers or receiver respectively, which is used to calculate the G/T of a satellite receiving antenna, is given by Kraus and Marhefka (2003) (Eq. 1).

$$T_S = T_{A,\mathrm{rad}} + T_0\left(\frac{1}{\eta_A} - 1\right) + T_0\left(\frac{1}{\eta_L} - 1\right) + \frac{T_R}{\eta_L} \quad (1)$$

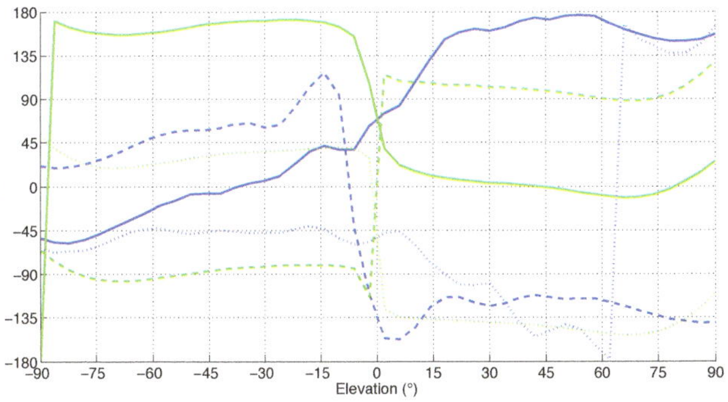

Fig. 13. Transmission phases for vertical (blue) and horizontal (green) polarizations; antenna in basic (—), stronger vertical (– –) and suppressed vertical polarization (···) mode.

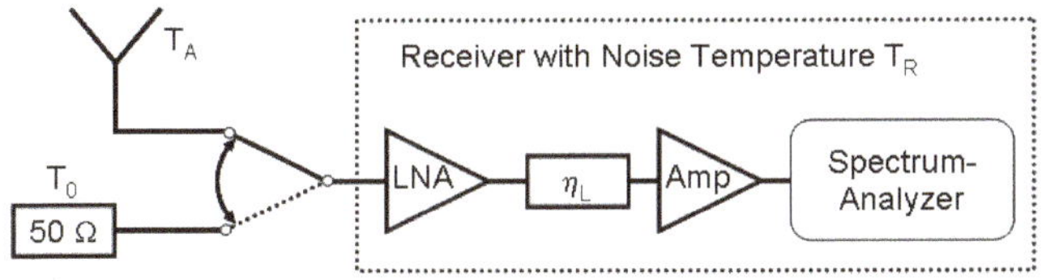

Fig. 14. Measurement setup for determination of the antenna noise temperture.

In order to investigate the impact of the surrounding temperature on the noise characteristics of antennas with varied adjustment of the radiation pattern the three antennas described in Sects. 2.1, 2.2 and 2.3 are compared in noise measurements with respect to the radiation part of the antenna noise temperature $T_{\mathrm{A,rad}}$.

Each of the three antennas are matched to the system impedance of $Z_\mathrm{S} = 50\,\Omega$. In the noise measurement each antenna is compared to the temperature of a reference resistor with a temperature of $T_\mathrm{Ref} = T_0 = 293\,\mathrm{K}$, using the measurement setup depicted in Fig. 14. The noise temperature of the low noise amplifier (LNA), the transmission line and the power amplifier are summarized in the receiver temperature T_R at the input of the LNA. The Antenna Temperature $T_{\mathrm{A,rad}}$ has to be determined. With the reference resistor switched to the low noise amplifier (LNA) the system noise temperature at the input of the receiver is given from Eq. (2). Equation (3) determines the measured noise temperature with the antenna under test switched to the receiver.

$$T_{\mathrm{S,Ref}} = T_0 + T_\mathrm{R} \quad (2)$$

$$T_{\mathrm{S,A}} = T_{\mathrm{A,rad}} \cdot \eta_\mathrm{A} + T_{\mathrm{A,Loss}} + T_\mathrm{R}, \quad (3)$$

where $T_{\mathrm{A,Loss}} = T_0 \cdot (1 - \eta_\mathrm{A})$ is the noise temperature contributed by the losses in the antenna. The efficiency of the antenna is determined by calculations from gain measurements of the radiation pattern. Therefore the measured directional characteristic of the antenna is compared with an ideal isotropic radiator at the same sampling points as in the measurement. With the power ratio

$$\Delta P_\mathrm{S} = \frac{P_{\mathrm{S,A}}}{P_{\mathrm{S,Ref}}} = \frac{T_{\mathrm{S,A}}}{T_{\mathrm{S,Ref}}} \quad (4)$$

the radiation part of the antenna temperature $T_{\mathrm{A,rad}}$ can be determined to

$$T_{\mathrm{A,rad}} = \frac{1}{\eta_\mathrm{A}} \cdot \left[\Delta P_\mathrm{S} \cdot (T_0 + T_\mathrm{R}) - T_\mathrm{R} - T_{\mathrm{A,loss}}\right] \quad (5)$$

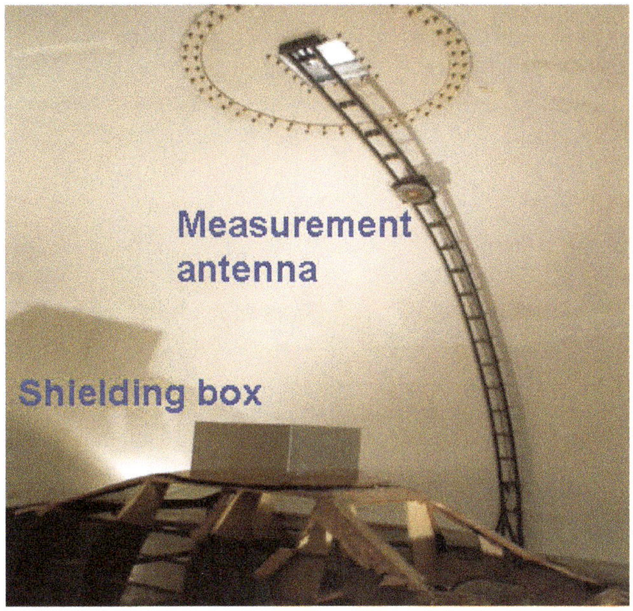

Fig. 15. Facility for antenna noise temperature T_A measurements.

In the presented measurements the radiation part of the antenna temperature is investigated in dependence of the radiation characteristics, especially the power ratio between horizontal and vertical polarization. The measurements have been performed in a T_A measurement facility shown in Fig. 15 which is covered by a radom for weather protection. The antenna unter test is positioned on a car model. Using the measurement antenna, which can be positioned in elevation, also G/T-mesurements can be performed. The results of the measurements are shown in Table 1. The difference within $\Delta f = 2\,\mathrm{MHz}$ is, beside a slight change of the antenna impedances, mainly caused by the edge of the transmitted signal being close to the center frequency. For the two antennas with lower vertical polarization the results are similar. A significant increase of the antenna temperature occurs for the antenna with stronger vertical polarization. The difference in antenna temperature is comparable with the noise contributed by losses in the antenna of more than 0.2 dB to 0.3 dB. Hence, considering noise aspects the vertical polarized radiation of an antenna should be kept as low as possible for satellite applications. Overall the new antenna shows very good results in noise temperature.

Table 1. Results of noise measurements of three antennas with different power ratio (H/V) of vertical and horizontal polarization with center frequency f_c.

Antenna	H/V-ratio	$T_{A,rad}$ @ f_c	$T_{A,rad}$ @ $f_c + 2$ MHz
Reduced terr. pol.	18.5 dB	28.6 K	16.0 K
Basic terr. pol.	15.0 dB	31.1 K	13.3 K
Increased terr. pol.	9.0 dB	57.8 K	36.5 K
$T_{A,loss}$ @ $\eta = -0.2$ dB	n/a	13.2 K	13.2 K

4 Results of field-tests

For performance tests in real receiving scenarios the new antenna has been tested in comparison with a conventional antenna which can be switched between left and right hand circular polarization. The optimum polarization matching due to switching is achieved by a seriously higher effort in the amplifier network. Both antennas have been placed in similar positions on a car in three different scenarios. Hardware effort and volume requirement of the new antenna are less than half the one of the reference antenna. At the open range no surrounding objects have been present. In the second scenario the car has been positioned at the edge of a forest with moderate shading of the satellite signals by the trees. In the third case the test position has been inside a forest with shading by high trees. The signals for the circular polarizations have been imbalanced at the satellite with lower power for LHCP. The results in Table 2 show a comparable performance for the new antenna. With only half the hardware effort and volume requirement, in noisy environments like the forest it even outperforms the reference antenna.

5 Conclusions

A new antenna for mobile reception of signals transmitted via geostationary satellites is presented. With an optimum shaping of the radiation pattern it is possible to receive signals of left and right hand polarization simultaneously. Therefore disruption of the signal due to switching is avoided. For the coverage of terrestrial repeaters in urban areas, a method for an adjustable power ratio between vertical and horizontal polarization is presented. Noise measurements show how important a precisely adjustable vertical polarization is for satellite applications. Finally tests in real receiving scenarios have shown a very good performance of the new antenna comparable to a high end switchable reference antenna which needs seriously higher mounting volume and effort in fabrication.

Table 2. Results of field tests; comparison of the signal to noise ratio between commercially available antenna with the new loop type antenna.

SNR of	Open Range	Edge of Forest	Inside Forest
Ref. Antenna RHCP	9.74 dB	5.77 dB	2.60 dB
New Loop Antenna RHCP	9.68 dB	4.60 dB	4.40 dB
Ref. Antenna LHCP	6.90 dB	1.97 dB	1.00 dB
New Loop Antenna LHCP	4.43 dB	1.60 dB	2.30 dB

References

Briskman, R. D. and Prevaux, R. J.: S-DARS broadcast from inclined, elliptical orbits, Acta Astronaut., 54, 503–518, 2004.

Kraus, J. D. and Marhefka, R. J.: Antennas: for all applications, vol. 3, McGraw-Hill, New York, 2003.

Lida, T., Pelton, J., and Ashford, E.: Satellite Communications in the 21st Century: Trends and Technologies, vol. 202, Prog. Astronaut. Aero., 2003.

Morgan, W. L. and Gordon, G. D.: Communications Satellite Handbook, vol. 1, John Wiley, New York, 1989.

Saala, G., Hopf, J., and Lindenmeier, S.: Small satellite car antenna for simultaneous reception of LHCP and RHCP signals, in: 3rd European Conference on Antennas and Propagation (EuCAP) 2009, vol. 3, 2698–2700, 2009.

Sallam, H., Abdel-Nabi, T., and Soumagne, J.: A GEO satellite System for Broadcast Audio and Multimedia Services Targeting Mobile Users in Europe, in: 4th Advanced Satellite Mobile Systems Conference (ASMS) 2008, vol. 4, 134–139, 2008.

3

Motion compensation of short-range, wide-beam synthetic aperture radar

T. Reichthalhammer and E. Biebl

Fachgebiet Höchstfrequenztechnik, Technische Universität München, Munich, Germany

Abstract. Up to now, SAR systems are a well known possibility for long-range detection. Applying them for short-range applications with wide-beam antennas, of course, does not increase the resolution but the probability to detect hidden targets with an anisotropic radar cross section significantly in comparison to other detection systems. Working with an appropriate wavelength even improves the possiblity to look through natural cover like grass. An application is detecting fawn while pasture mowing. The main issue in such applications is the antenna's motion in range direction as it is carried by cars or traction engines. If motion is not compensated, the phase cannot be reconstructed correctly, the resolution gets poorer and, in worst case, the target even disappears. Conventional methods for motion compensation either fail for wide beam antennas, since for contributions of wide angles the phase reconstruction is incorrect, or is not applicable for realtime data processing, because the processing time due to interpolation or similar steps is very high. We present a method of image reconstruction regarding motion of the antenna as well as wide beamwidth. This method is analyzed concerning processing time in comparison to the conventional image reconstruction. In our system we use a combination of algorithms. There is shown a comparison for different algorithms dependent of the antenna's motion and aperture angle.

Correspondence to:
T. Reichthalhammer
(reichthalhammer@tum.de)

1 Introduction

One of the main characteristics of an SAR system is its independence of the distance between the antenna and the targets concerning the resolution in azimuth direction. That means the resolution stays constant with varying distance. Anyhow a short-range SAR system has advantages, first and foremost a wide-beam system (Gerbl, 2007). Such a system can be used for imaging in automotive applications or in farming to detect wild animals. (Reichthalhammer, 2010) The probability of detection increases if the aperture angle becomes larger since the target is illuminated from different angles. The contributions of e.g. a point scatterer are summed up in proper phase relation. The bigger the aperture angle, the more contributions and the better the contrast in the image.

The quality of an SAR image depends on the accuracy of the phase reconstruction of the single signal contributions of different angles. A consequence of a phase error is a picture with degradation in contrast and resolution. To achieve an exact phase rotation one has to know the exact position of the antenna at each recording point. This issue enlarges if the antenna is not carried by a plane but by a vehicle and the ground is bumby. So motion compensation requires an exact detection of the position and an appropriate signal processing algorithm regarding wide beam contributions.

At short-range systems the special focus concerning motion compensation lays on motion compensation in range direction. The velocity in azimuth can be measured easily due to having exact information of the vehicle's current velocity. But the oscillation of the antenna in range direction is high compared to the distance between target and antenna. In addition the wavelength has a great influence on the quality of the picture as shown later. With existing algorithms for motion compensation in range direction, either contributions of wide angles are not compensated correctly, or realtime processing does not allow them.

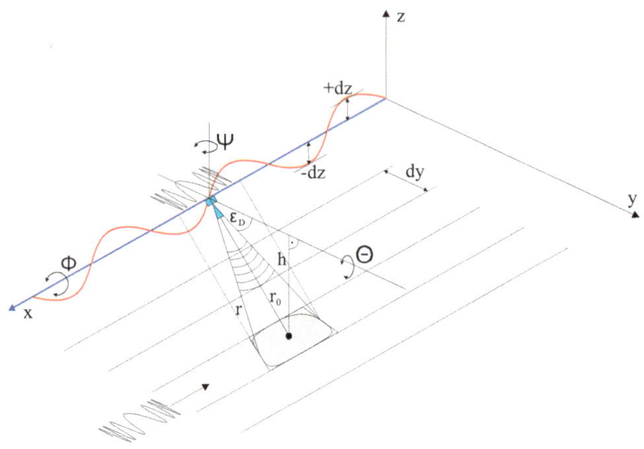

Fig. 1. Coordinate system.

2 System setup

Short-range radar imaging with wide-beam antenna is discussed detailed in (Gerbl, 2007). An applicable signal source therefore was developed in (Reichthalhammer, 2010). The crucial point of wide-beam systems is the fast growing doppler bandwidth that postulates a high sample rate. In turn this means a high data volume and a longer processing time. In addition the doppler bandwidth and the processing time is even enlarged if motion has to be compensated. In Fig. 1 the coordinate system of the setup is shown.

The azimuth direction of the system is x, the range direction r_0 is perpendicular to the azimuth direction. The depression angle ϵ_D is the angle between the x-y-plane and the antenna.

3 SAR imaging

The general way to reconstruct a recorded scene in azimuth direction is a convolution of the recorded signal with the complex conjugate signal of a point scatterer for each line d_y of the picture. Our system works in a stop-and-go manner and the doppler frequency caused by the motion of the vehicle can be neglected. So for a stepped frequency system the equation therefore is

$$x[l] = \sum_{l=-r_0 \tan\alpha}^{+r_0 \tan\alpha} x_1[n] \cdot x_2[l-n], \tag{1}$$

where x_1 is the SAR system response and x_2 the complex conjugate function of x_1.

Because the illuminated area enlarges with growing distance to the antenna, the compensation function changes with varying distance (Klausing, 2000). The real part of the received signal at a frequency of 24 GHz of a point scatterer at

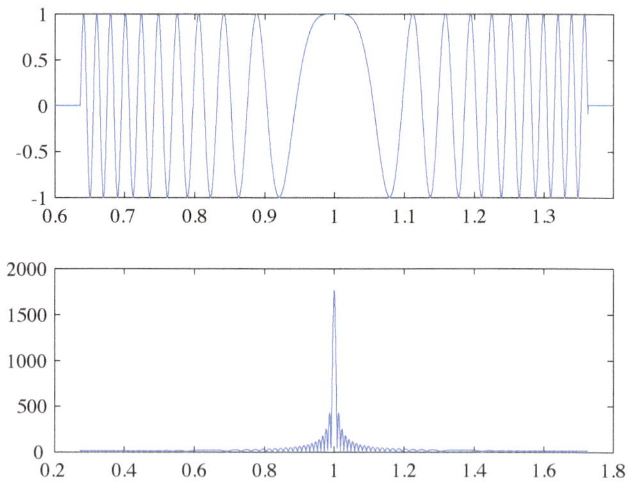

Fig. 2. real part of received signal and magnitude of compressed signal.

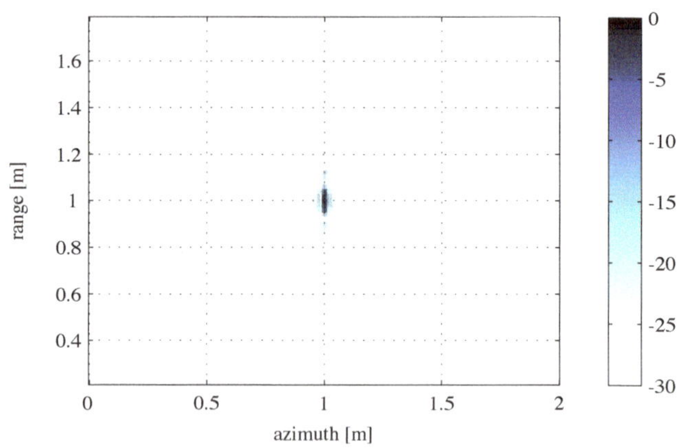

Fig. 3. SAR picture of a point scatterer, 1.8GHz bandwidth, scaled to its maximum.

a distance r_0=1m and an aperture angle of 20° is depicted in Fig. 2 assumed constant directivity of the antenna.

The resulting function Fig. 2 is a compression of the spatially extendend received signal with a peak at the position of the target. Obviously the phase velocity increases with increasing angles. The spatial doppler frequency is zero, where the target is located. To get resolution in range equidistant frequency steps inside a given bandwidth are necessary. This results in an unambiguous range which is treated in (Reichthalhammer, 2010). Conventional SAR imaging uses a two dimensional Fourier transformation of the signal $s(x,t)$ into $s(k_x,\omega)$ as described e.g. in (Soumekh, 1999). Since our data is sampled equidistantly in azimuth direction the raw data format is $s(x,\omega)$. A typical motionless SAR picture of a point scatterer like the sphere in Fig. 1 is depicted in Fig. 3.

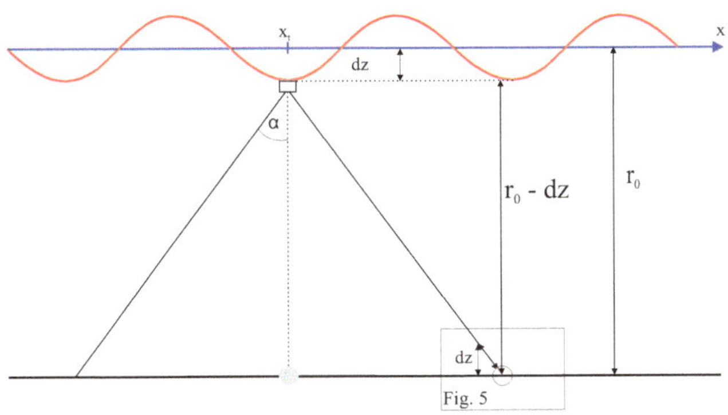

Fig. 4. Occuring compensation error due to antennas movement dependent of aperture angle.

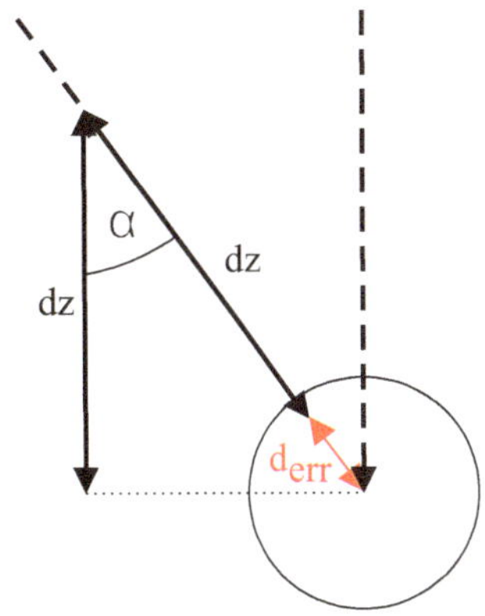

Fig. 5. Magnification Fig. 4.

Usually the wavefront curvature cannot be neglected what enforces a transformation of the data from polar into cartesian coordinates. This step can be circumvented if the range stacking algorithm is used (Soumekh, 1999). At range stacking for each given range bin the azimuth compressed signals for each frequency are added what yields the range resolution. This way even for wide beam antennas an interpolation can be avoided and computational time saved.

4 Motion in short-range SAR systems

In a 3-dimensional space in principle all 6 degrees of freedom have to be considered. A translational motion in x, y and z-direction and a rotatory motion around each axis (pitch, gear, yaw). In our case a few assumptions are done to simplify the problem. Supposed that only a rotation around x exists along the path x, the target would appear at different positions y in the antenna's field of view. In worst case -depending on the aperture angle in range and the rotation- the target would even disappear. To compensate this rotation the antenna pattern has to be multiplicated inverse to the image columns. Rotations around y and z can be neglected, since a huge turn radius of the vehicle and a little curvature of the ground in respect of a short aperture time T_{SA} - the time one target is illuminated by the antenna.

Assumed that the vehicle moves straight ahead the azimuth direction during one synthetic aperture time T_{SA} one can neglect the translational motion in y as well. As described in (Reichthalhammer, 2010) the system records data along its path x equidistantly independent of the velocity. Hence a quasi static system is supposed in azimuth direction and motion compensation is not necessary. So the system's movement is reduced to an oscillation in z and a rotation Φ around x which hase to be considered. So the 6-dimensional problem can be reduced to a 2-dimensional problem. For further calculation and simulation even the rotation is neglected.

5 Motion compensation vs. processing time

To reconstruct a recorded scene correctly, the main postulation to an SAR system is, that its trajectory has to be known. Simplified a rectilinear trajectory can be assumed, but carried by a vehicle an SAR system cannot follow a straight line. A solution is to detect the motion and rotation and compensate it. For this purpose different algorithms are recommended in literature. The majority of these algorithms try to detect the deviation either by inertial measurement supported by GPS or by analyzing the received signal. Afterwards the received data is manipulated, what means that the phase and the amplitude are corrected. This procedure holds for the case of a small aperture angle and is know as the narrow-beam-assumption. For increasing aperture angles these algorithms become defective. The error appears since the reconstruction of the phase is done for each recording point, which corresponds to a column in the SAR picture. If the aperture angle of the antenna becomes bigger each data point can contain information of an object far away from the perpendicular, but a translatory deviation of the antenna always is a term perpendicular to the trajectory. For example the signal of a point scatterer at distance r is

$$s = \mathrm{a} \cdot \exp(-2\mathrm{j}\mathrm{k}r) \tag{2}$$

where a is an amplitude term and $k = \frac{2\pi}{\lambda}$ the wavenumber.

With classical compensation like the 2-D Fourier method the singular data point is treated as it only contains information of the scene perpendicular to the trajectory. For example, the distance r between a sphere and an antenna for r_0=1 m and $\alpha = 20°$ degree as depicted in Fig. 4 varies with a deviation of the antenna at x_n in +z-direction. The magnification of the error made is depicted in Fig. 5.

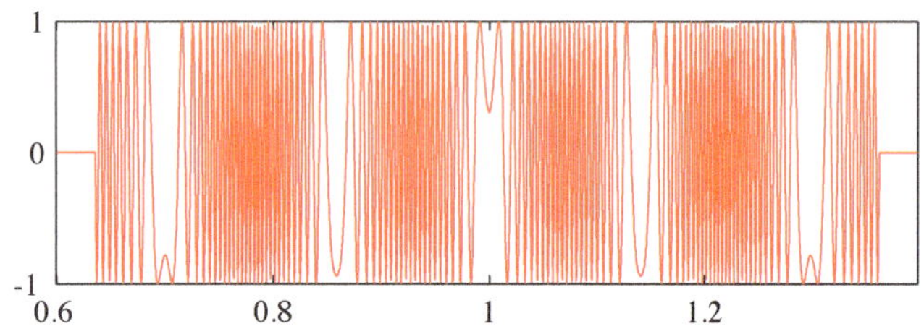

Fig. 6. real part of received signal with motion, f=24GHz, $\alpha = 20°$, 5 Hz/aperture, a=0.1m, $\epsilon_D = 45°$.

To compensate this deviation the data at point x_n is manipulated by a phase term

$$s = a \cdot \exp(-2jkr) \cdot \exp(-2jkd_z \sin\epsilon_D). \tag{3}$$

d_z can be positive or negative. In case of a small beamwidth the variation of d_z over the angle can be neglected. Higher angles cause too less phase subtraction in case of motion in $+d_z$-direction and too much phase addition in case of motion in $-d_z$-direction.

The absolute error can be defined as

$$d_{err} = \frac{|d_r|}{\cos\alpha} - |d_r|, \tag{4}$$

where α is the angle between target and antenna as depicted in Fig. 4 and $|d_r| = |d_z|\sin\epsilon_D$.

If an arbitrary harmonic deviation is introduced to the antenna's trajectory with an amplitude of 10 cm and 5 oscillations per synthetic aperture with ϵ_D=45°, which is a realistic assumption, the real part of the received signal is depicted in Fig. 6. Hence the typical chirp-like signal form of Fig. 2 is lost. In case of no compensation, the SAR picture of a point scatterer is depicted in Fig. 7.

If compensated with the narrow-beam assumption the signal is filtered as described by Eq. (1) and the picture of Fig. 7 turns to Fig. 8. The maximum of the target decreases, sidelobes appear and the resolution in azimuth decreases significantly, but the processing time itself stays constant.

Note that for using the SAR system for nearly real-time image processing, due to actual processor capabilities the biggest issue is to find a way between sufficient contrast and short processing time. Since there are no smooth transitions between and algorithms mainly focus on one of both sides respectively, one cannot do a compromise.

A well known solution that avoids errors appearing for higher aperture angles is the SAR backprojection algorithm, see (Soumekh, 1999). This algorithm works as follows: SAR data firstly is filtered in range. For each pixel its contribution s_{d_r} to the relevant part of the SAR data is reconstructed and vectorial multiplicated with the corresponding SAR data s_{az_n}. The sum of the resulting vector yields the reflectivity

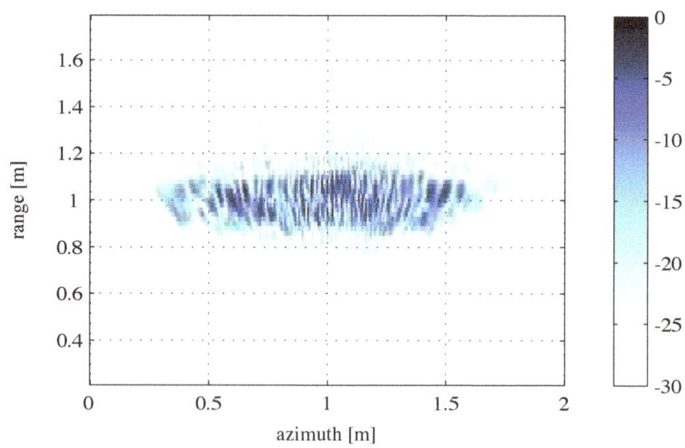

Fig. 7. SAR picture of a point scatterer, 1.8GHz bandwidth, scaled to its maximum, uncompensated motion.

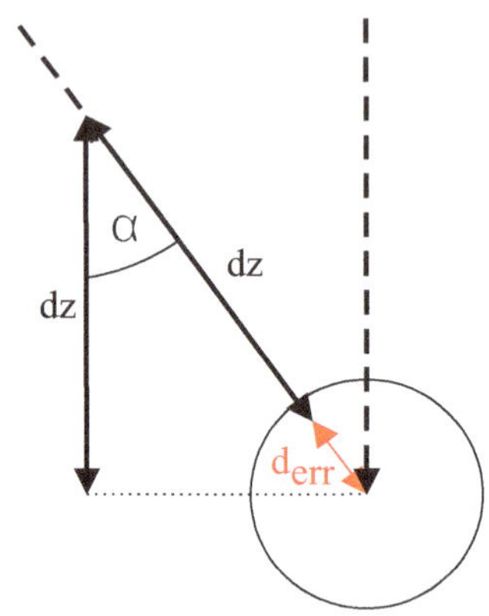

Fig. 8. SAR picture of a point scatterer, 1.8GHz bandwidth, scaled to its maximum, conventional motion compensation.

of the pixel. This way even a translational motion of the antenna can be compensated, but note that the data have to be filtered in range before, what does not fit to the range stacking algorithm, where an azimuth compression of the data in the first step is compellent. Moreover this algorithm causes a high processor load as treated later. The equation for backprojection is

$$x[n] = \sum_{l=n-(r_0-d_z)\tan\alpha}^{n+(r_0-d_z)\tan\alpha} s_{az_n}[l] \cdot s_{d_r}[l], \tag{5}$$

where x_{d_r} is the signal different at each point n.

A combination of range stacking algorithm and back projection algorithm is an effective way of combining both high quality image reconstruction and acceptable processing time, but as treated above for range stacking the data firstly has to be compensated in azimuth.

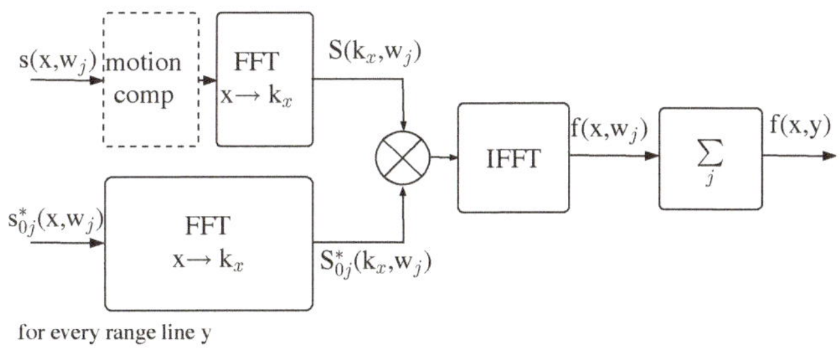

Fig. 9. SAR signal processing with and without motion compensation with narrow beam assumption.

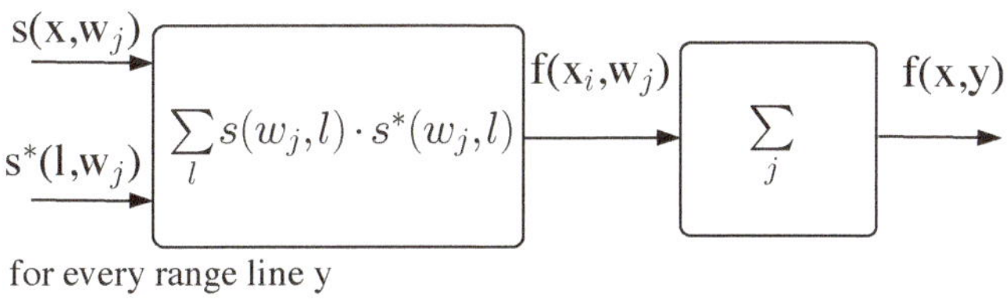

Fig. 10. SAR signal processing with combination of range stacking and backprojection algorithm.

Furthermore the interpolation step of the backprojection algorithm in azimuth direction does not have to be done with our system since the image resolution in azimuth and so r_{ij} inherently are defined by the sampling space Δx.

In Fig. 9 our processing of the SAR signal is shown with and without motion compensation (dashed box) with the narrow beam assumption, which leads to an image as shown in Fig. 8. The processing has to be done for each range line y. To achieve range resolution with range stacking the results for different frequencies are calculated and added each time.

In Fig. 10 the processing is shown if wide beams and other image degrading effects have to be considered. One can see, that nothing of the 2-D Fourier algorithm is left. In both dimensions the processing is substituted by a summation. For each range line, each pixel is processed concerning its reflectivity, as described above. Each range line has to be processed for each frequency. The results for one range line are summed up vectorialy. This way, the picture is reconstructed like the one depicted in Fig. 11.

A comparison for one range line d_r=1 m is made in Fig. 12. In the upper figure the reconstruction of a point scatterer without motion is depicted in comparison to the processing result if a typical motion is not compensated. If not compensated, the image contains no information as depicted in Fig. 7. In the lower figure the results are shown for both, a compensation as described in Fig. 9 and a compensation as treated in Fig. 10.

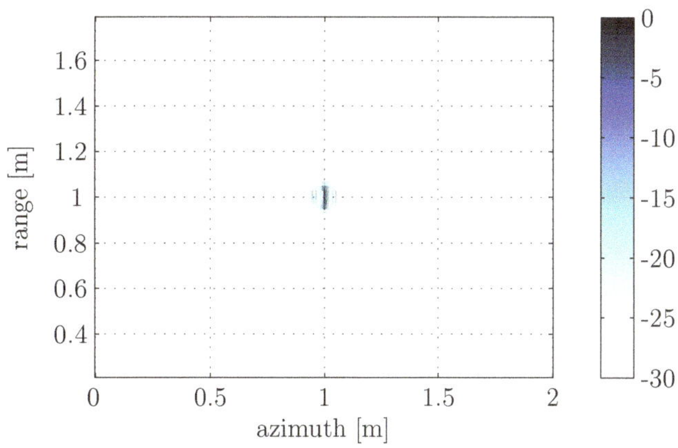

Fig. 11. SAR picture of a point scatterer, 1.8 GHz bandwidth, scaled to its maximum, motion compensation with derivative of backprojection algorithm.

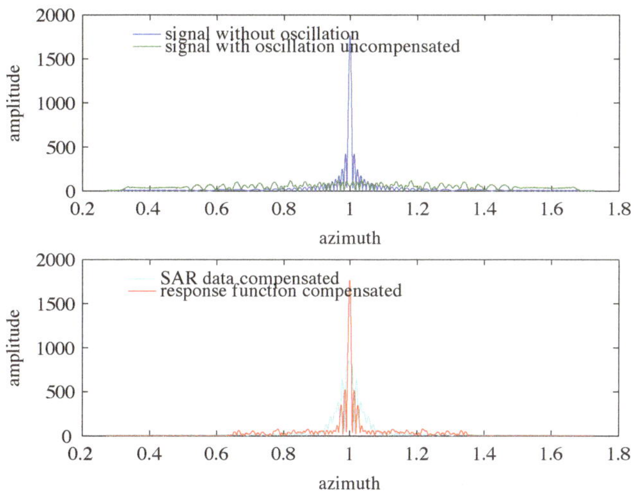

Fig. 12. comparison of SAR signal reconstruction at one range line for no motion, motion without compensation, motion with different compensation methods.

6 Processing time

An estimation of the processing time can be done. Therefore a reference processing time have to be defined. In our case this is the processing time without motion compensation as depicted in Fig. 9, that can be described as a 1-D Fourier algorithm with range stacking.

Let y be the number of lines we want to process, x the number of data points in azimuth and f the number of frequencies used. Than the processing load O without motion compensation is

$$O = y[\underbrace{x \cdot \log_2(x) \cdot f}_{\text{FFT}} + \underbrace{x \cdot f}_{\text{summation}}] =$$

$$= y \cdot x \cdot f[\log_2(x) + 1]. \tag{6}$$

If the SAR raw data is manipulated the processing load of Eq. (6) is extended by the term x and reads as

$$O = y \cdot x \cdot f[\log_2(x) + 1] + x. \tag{7}$$

In the case of motion compensation as depicted in Fig. 10 the processing load O increases to

$$O = y \cdot [\underbrace{l \cdot \log_2(l) \cdot f \cdot x}_{\text{FFT}} + \underbrace{f \cdot x}_{\text{multiplication}} + \underbrace{x}_{\text{summation}}] =$$

$$= y \cdot x \cdot f[l + x + \frac{1}{f}], \tag{8}$$

where l is the number of data points inside a synthetic aperture dependent on the range and the aperture angle. As explained above there is a huge gap between the two ways of motion compensation regarding processing time increasing with the aperture angle. A possible solution for time critical systems where only one target per synthetic aperture is expected is to process the image at first with the fast algorithm and postprocess the parts of the image where an aim is detected with the slow algorithm. But note, that this method fails if more than one target appears inside one synthetic aperture. If this is not guaranteed, one has to decide between image quality and processing time.

7 Conclusions

In this paper we dicussed the motion compensation introduced to an SAR system working in short-range with wide-beam antennas. We explained the setup and application of our radar system and show the difference between the algorithms charged by a motion compensation term and a new one, which is a derivation of the backprojection algorithm combined with the range stacking algorithm. The gap between high resolution and fast processing is analyzed and a proposal for processing special target scenes made. Simulation results are depicted and processing time is compared to the original algorithm.

References

Gerbl, F.: Evaluation of wide-beam, short-range synthetic aperture radar imaging, Ph.D. thesis, TU Muenchen, 2007.

Klausing, H. and Holpp, W.: Radar mit realer und sysnthetischer Apertur, 2000.

Soumekh, M.: Synthetic Aperture Radar Signal Processing with MATLAB Algorithms, 1std edn., 1999.

Reichthalhammer, T. and Biebl, E. M.: Specification and Design of a Fast-Tuning Stepped-Frequency Synthesizer for a Short-Range-SAR-System in the K-Band, 2010.

4

Leaky wave antenna with amplitude controlled beam steering based on composite right/left-handed transmission lines

M. A. Eberspächer and T. F. Eibert

Lehrstuhl für Hochfrequenztechnik, Technische Universität München, Germany

Abstract. An antenna comprising two different composite right/left-handed transmission line structures is proposed which enables easy beam steering at an operation frequency of 10 GHz. The composite right/left-handed transmission lines are based on planar, periodically arranged via free unit cells, implemented in microstrip technology. Both transmission lines exhibit the infinite wavelength phenomenon which occurs at 9.72 GHz and 9.89 GHz, respectively. Thus, operating the different leaky wave structures at 10 GHz, radiation with azimuth angles of $\pm 8°$ and $\pm 17°$ can be achieved depending on the selected input port. In order to obtain a tunable main beam direction, the radiation patterns of both structures are superimposed by feeding them simultaneously. The influence of each guiding structure, and hence the direction of the main beam, can be controlled via the feeding amplitude. As a result of this, the beam can be steered between $\pm 17°$ with a gain of up to 10 dBi. The guiding structures are arranged in parallel with a clearance of a=12.2 mm which is less than half of the wavelength in free space. This allows in a further step the attachment of additional guiding structures in order to increase the tunable angle range or creating an antenna array with a small beamwidth in the elevation plane without the occurrence of grating lobes. An antenna prototype was fabricated and validated by measurements.

1 Introduction

Since the advent of automotive radar systems, highly directive antennas supporting easy beam steering have received considerable attention. As shown in Balanis (2005), the beamwidth of an antenna becomes smaller when its aperture is increased. Thus, realising a small beamwidth requires a sufficiently large aperture. Travelling wave antennas are attractive candidates for this type of application due to the fact, that their aperture sizes do not adhere with the operation frequency, and hence the aperture can be chosen almost arbitrarily. The mapping between frequency and radiation direction of the antenna can be controlled during the design process by dispersion engineering (Walter, 1965). However, the main beam direction of a leaky wave antenna is tightly coupled to the operation frequency and cannot be changed dynamically. Thus, many research efforts have been done and versatile concepts have been presented to allow beam steering at a constant frequency, for example with a rotating dielectric grating structure as in Schneider and Wenger (2003), with a movable dielectric slab as in Matsuzawa et al. (2006), or by using lens antennas as in Binzer et al. (2007). All these presented structures are proven to work well, however, their realisation and production is very costly.

Correspondence to: M. Eberspacher
(mark.eberspaecher@tum.de)

2 Theory

2.1 Principle of leakage radiation

A leaky wave is a travelling wave which progressively leaks out power as it propagates along a waveguiding structure (Caloz and Itoh, 2006). Consequently, a structure supporting one or more leaky waves is referred to as leaky wave structure. In the following, a planar waveguide located in the xy-pane is considered. The waveguide supports an electromagnetic wave travelling along the structure in the x-direction. By suppressing the time dependency, this wave can be described as $\psi(x) = \psi_0 e^{-\gamma x}$, where $\gamma = \alpha + j\beta$ represents its complex propagation constant. The field above the guiding structure which is excited by the travelling wave can be written as

$$\psi(x,z) = \psi_0 e^{-\gamma x} e^{-jk_z z}. \tag{1}$$

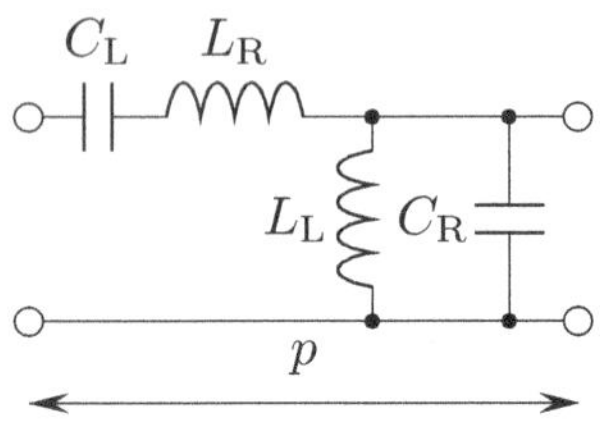

Fig. 1. Equivalent circuit of a CRLH TL unit cell with series impedance $Z = j\omega L_R + \frac{1}{j\omega C_L}$ and the parallel admittance $Y = j\omega C_R + \frac{1}{j\omega L_L}$.

In order to fulfill Maxwell's equations, the wavenumber in z-direction is given as

$$k_z = \sqrt{k_0^2 - (\beta - j\alpha)^2}, \tag{2}$$

where $k_0 = \omega/c$ is the free space wavenumber. If we consider a lossless antenna with relatively small leakage, α can be neglected and the wavenumber simplifies to $k_z = \sqrt{k_0^2 - \beta^2}$. Depending on β, this leads either to a purely imaginary k_z or a purely real k_z. Accordingly, the first case represents an evanescent wave exponentially decaying whereas the latter stands for leakage radiation.

The main beam direction of this leakage radiation can be calculated as shown in Walter (1965) according to

$$\Theta_{MB} = \arcsin \frac{\beta(\omega)}{k_0}. \tag{3}$$

Analysing the argument of the angular function yields that a real angle can only be found for $|\beta(\omega)| \leq k_0$. This accords exactly to the prerequisites in Eq. (2) for a purely real propagation constant k_z.

The relationship between the beamwidth Θ_{BW} and the aperture length L of the leaky wave antenna is given in Johnson (1993) by

$$\Theta_{BW} \approx \frac{1}{(L/\lambda_0)\cos\Theta_{MB}}. \tag{4}$$

For a given value of α, the antenna length L is usually chosen so that approximately 90% of the power is radiated. As expected, the beamwidth becomes smaller for a longer aperture.

2.2 Artificial transmission line theory

In order to design leaky wave antennas with predefined behaviour, guiding structures with certain dispersion characteristics are required. A very powerful approach to accomplish this are periodic structures. This type of structures are composed of periodically arranged small sections, which can be considered as unit cells. It is shown that the analysis of the whole structure can be restricted to the analysis of one unit cell (Pozar, 2005). Hence, in the ideal case also the design process of a guiding structure is reduced to the design effort of a single cell. The dispersion behaviour of such a guiding structure is given by

$$(\alpha + j\beta)p = \operatorname{arccosh} \frac{A+D}{2}, \tag{5}$$

where p stands for the physical length of the unit cell and the parameters A and D are its elements of the ABCD-Matrix representation. Since the result of the arccosh-function is ambiguous, the phase constant has multiple solutions $\beta_n p = \beta_0 p + \frac{2\pi n}{p}$ with $n = 0, \pm1, \pm2, \pm3, \ldots$. Besides the fundamental mode β_0, an infinite number of so-called space harmonics are obtained. In general, a structure can be designed in such a way that either radiation occurs at the fundamental mode or at any arbitrarily chosen space harmonic as long as Eq. (3) leads to a real angle Θ_{MB}. However, as it is mentioned in Paulotto et al. (2008) radiation at broadside using space harmonics can be disturbed due to coupling effects of a pair of these harmonics resulting in an open stop band. This problem can be overcome by radiating only with the fundamental mode. Consequently, β_0 is forced to be zero to achieve broadside radiation at a certain frequency $\omega \neq 0$, which can be realised by using composite right/left-handed (CRLH) transmission lines.

As shown in Fig. 1, the CRLH-unit cell is composed of a series and a parallel resonant circuit with the resonant frequencies ω_s and ω_p, respectively. In both resonant circuits the elements denoted with index L are accountable for the left-handed behaviour of the arrangement. On the other hand, components with index R make the unit cell to behave as right-handed. According to the equivalent circuit of a conventional transmission line, the right-handed contributions can be implemented either as lumped elements or as sections of transmission lines. Additionally, the occurring parasitics can be included in these elements.

Calculating the corresponding ABCD-Matrix elements of the circuit in Fig. 1, inserting them in Eq. (5), and performing a taylor series expansion yields $j\beta = \frac{\sqrt{YZ}}{p}$. It is shown in Caloz and Itoh (2006), that $\beta = \Re\{\beta\}$, if $\omega_s = \omega_p$, and hence no stop band occurs. This case is referred to as the balanced case in which the phase constant is written as $j\beta = \frac{j}{p}\left(\omega\sqrt{L_R C_R} - \frac{1}{\omega\sqrt{C_L L_L}}\right)$. Obviously, a given value of the phase constant at a certain frequency can be realised by choosing the appropriate network elements L_R, C_R, L_L and L_L. Thus, it is possible to create a leaky wave structure possessing a certain main beam direction at a given frequency.

The counterpart of the characteristic impedance of a homogeneous transmission line is the Bloch impedance of a periodic artificial structure. In contrast to the characteristic impedance, the latter is usually given at the terminals of the unit cell only. According to Pozar (2005), the Bloch impedance can be determined based on the ABCD-Matrix elements as

$$Z_L = \sqrt{\frac{Z}{Y}}. \tag{6}$$

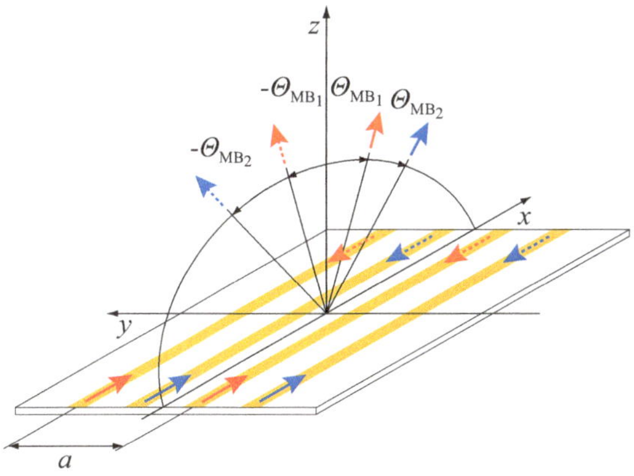

Fig. 2. Arrangement of two times two identical leaky wave structures. Excitation according to the red solid and dashed arrow leads to main beam direction Θ_{MB1} and $-\Theta_{\mathrm{MB1}}$, respectively. Excitation according to the blue solid and dashed arrow leads to main beam direction Θ_{MB2} and $-\Theta_{\mathrm{MB2}}$, respectively.

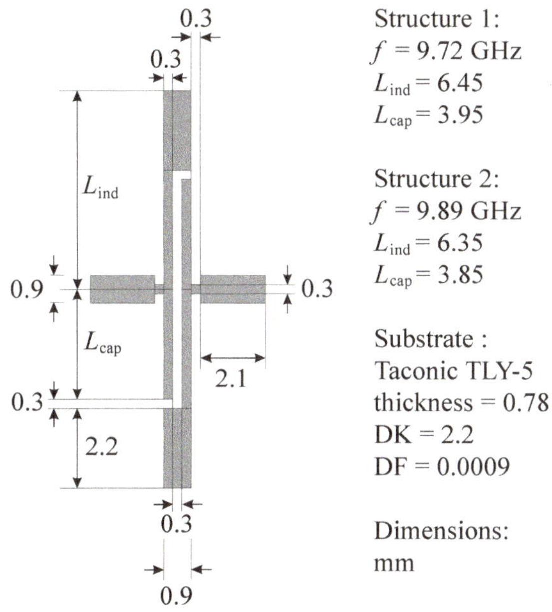

Fig. 3. Layout of the two realised unit cells.

Applying Eq. (6) to the circuit in Fig. 1 results in the frequency independent Bloch impedance $Z_{\mathrm{L}} = \sqrt{\frac{L_{\mathrm{L}}}{C_{\mathrm{L}}}} = \sqrt{\frac{L_{\mathrm{R}}}{C_{\mathrm{R}}}}$.

2.3 Concept of beam steering

Naturally, the main beam direction of a leaky wave antenna can be steered by changing its operation frequency. However, a concept allowing different main beam directions at a constant frequency is investigated in this work.

To clarify the basic idea, let us consider a leaky wave structure with two ports located at their opposite ends. The ports are referred to as port 1 and port 2, respectively. Operating the structure at the frequency $f_0 = \omega_0/2\pi \neq 0$ at port 1, a leakage mode is excited which propagates along the structure to port 2 at which it is supposed to be absorbed. Thus, radiation occurs with an azimuth main beam Θ_{MB1} according to Eq. (3) and the propagation constant of the excited mode. The structure can be designed in such a way that this angle $\Theta_{\mathrm{MB1}} \neq 0$ at the frequency f_0.

If the driven port and the terminated port are interchanged, the phase constant will be negative with respect to port 1 and hence, radiation with the main beam direction $-\Theta_{\mathrm{MB1}}$ will occur. As it is seen in Eq. (4), the beamwidth Θ_{BW1} of the antenna in the azimuth plane is determined by the aperture length L. Thus, the beamwidth can be manipulated by the amount of cascaded unit cells along the x-axis.

Further main beam angles are achievable by adding other leaky wave structures in parallel which possess the angles $\pm\Theta_{\mathrm{MB}n}$. This can be done any number of times if the beamwidth in the elevation (yz-) plane can be broad. However, the requirement of a small beamwidth in this plane limits this approach, since leaky wave structures exhibiting the same azimuth angle must be repeated periodically. Figure 2 shows such an arrangement of two times two leaky wave structures. Structures possessing the same main beam direction are marked with arrows of the same colour. The distance between identical antennas is a. In order to avoid grating lobes in the yz-plane, a has to be smaller than λ_0 for radiation in the upper hemisphere only. This limitation makes the design process troublesome, since reducing the clearance leads to increased mutual coupling. The increase of the coupling effects can grow up to such an extent that the structure becomes inoperative.

According to the antenna depicted in Fig. 2, there are four different main beam directions possible, which can be chosen via the corresponding feeding. An excitation of the antenna according to the red solid and dashed arrows leads to radiation into the direction of Θ_{MB1} and $-\Theta_{\mathrm{MB1}}$, respectively. The angles Θ_{MB2} and $-\Theta_{\mathrm{MB2}}$ can be adjusted by feeding according to the blue solid and dashed arrows, respectively.

Furthermore, the radiation patterns of two structures can be superimposed by feeding them simultaneously in order to gain additional main beam directions. The influence of each guiding structure, and hence the direction of the main beam, can be controlled via the feeding amplitude. To ensure that each combination leads to only one main beam, the particular patterns have to fulfill

$$\Theta_{\mathrm{MB1}} + \Theta_{\mathrm{BW1}}/2 \geq \Theta_{\mathrm{MB2}} - \Theta_{\mathrm{BW2}} \qquad (7)$$

with $\Theta_{\mathrm{MB1}} \leq \Theta_{\mathrm{MB2}}$.

3 Realisation

The aim is to create an antenna comprising two different leaky wave structures, which possess at the operation

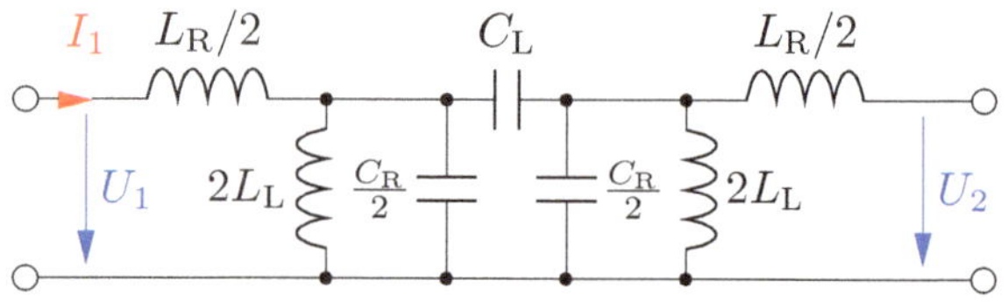

Fig. 4. Simplified equivalent circuit model of the realised unit cell with neglected losses.

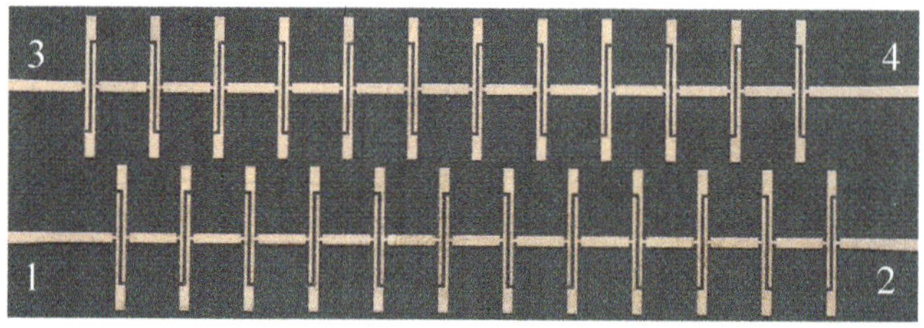

Fig. 6. Fabricated leaky wave structures. The numbers are indicating the ports.

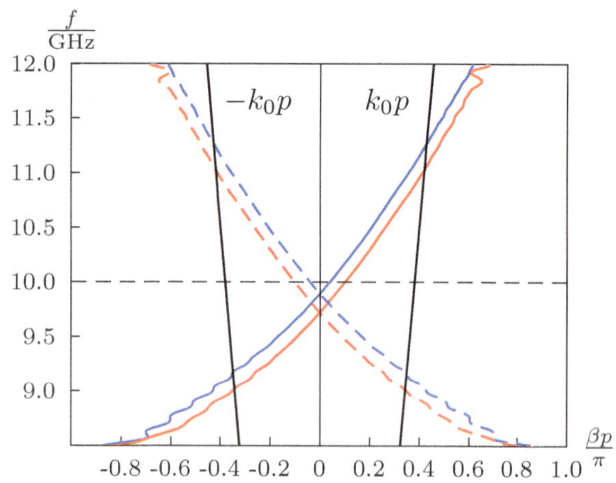

Fig. 5. Dispersion diagram of the unit cells with the transition frequencies $f_1 = 9.72\,\text{GHz}$ (red line) and $f_2 = 9.89\,\text{GHz}$ (blue line). At 10 GHz the unit cells exhibit a phase constant of $\beta_1 = 0.098\frac{\pi}{p}$ and $\beta_2 = 0.038\frac{\pi}{p}$, respectively.

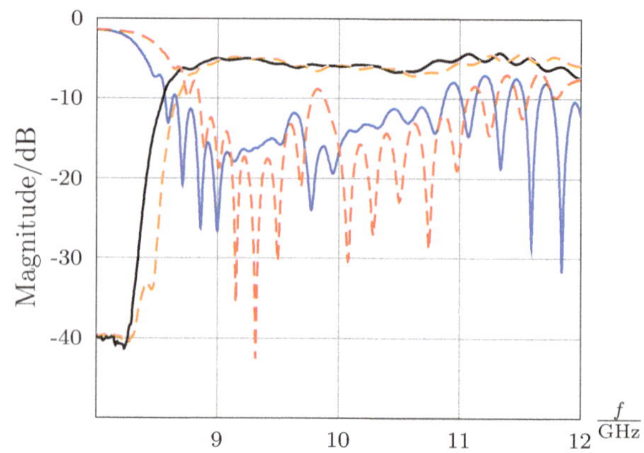

Fig. 7. Measured S-parameters of the leaky wave structures: S_{11} blue line, S_{21} black line, S_{33} red dashed line and S_{43} orange dashed line.

frequency of 10 GHz the main beam directions Θ_{MB1} and Θ_{MB2}, respectively. In order to allow the extension of the aperture in y-direction subsequently, the distance between adjacent structures should be less than $\lambda_0/2$. Furthermore, an easy to fabricate design is wanted which can be produced using a standard photolithography process.

The unit cell presented in Eberspächer et al. (2009) is easily realisable and needs neither an interlayer connection nor a multilayer fabrication process. Hence, it can be produced in an entirely printed circuit technology. Consequently, the concept of this unit cell is taken as a basis for this antenna design. However, it cannot be implemented directly, since strong mutual coupling between adjacent leaky wave structures occurs. Therefore, the unit cell must be narrowed in y-direction without changing the electromagnetic behaviour. This means that the open ended inductive stubs must be shortened while the electrical length must be unchanged. This can be achieved by introducing impedance steps along the stub. Beginning at the open end, a stepwise change of the characteristic impedance from a low to a high level is required. In microstrip technology, this is easily achieved by a step in the width of the line. The resulting design is shown in Fig. 3.

In the next step, the unit cells were balanced upon the transition frequencies 9.72 GHz and 9.89 GHz. According to the dispersion diagram seen in Fig. 5, one obtains the phase constants of $\beta_1(2\pi\,10\,\text{GHz}) = 0.098\frac{\pi}{p}$ and $\beta_2(2\pi\,10\,\text{GHz}) = 0.038\frac{\pi}{p}$, respectively. Employing Eq. (3), this leads theoretically to the main beam directions $\Theta_{MB1} = 15°$ and $\Theta_{MB2} = 5.8°$. Balancing a unit cell consisting of discrete elements can be done easily by choosing the appropriate values. However, the proposed cell requires a more sophisticated balancing procedure. Therefore, information are required which describe whether the unit cell is balanced or not. Furthermore, it is important to know, which resonant circuit should be tuned. These information are provided by the **Z** and **Y** matrix elements obtained by the numerical computation software CST Microwave Studio (CST).

During the balancing process of the structure losses are not considered. According to the resulting simplified equivalent circuit model, seen in Fig. 4, the element Z_{21} can be calculated analytically. Per definition $Z_{21} = \frac{U_2}{I_1}\Big|_{I_2=0}$. From this follows that neither of the series inductances must be taken into account. Thus, this impedance reads

$$Z_{21} = \frac{\mathrm{j}\omega^3 4L_L^2 C_L}{(1-\omega^2 L_L C_R)(1-\omega^2 L_L C_R - \omega^2 4 C_L L_L)}. \tag{8}$$

Searching for roots of the denominator gives

$$\omega_{fp} = \sqrt{\frac{1}{(C_R+4C_L)L_L}}$$

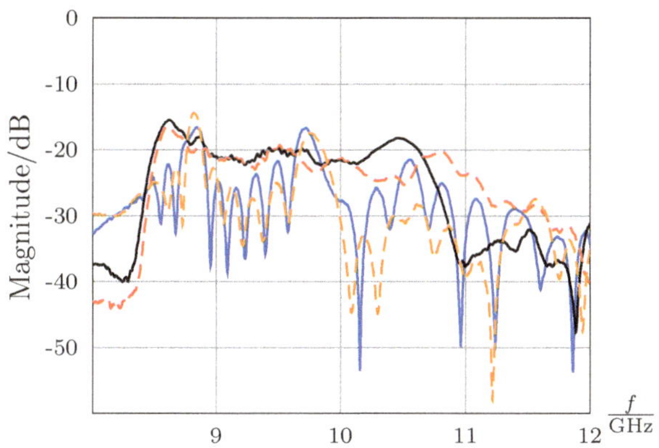

Fig. 8. Measured S-parameters of the leaky wave structures: S_{31} blue line, S_{41} black line, S_{32} red dashed line and S_{42} orange dashed line.

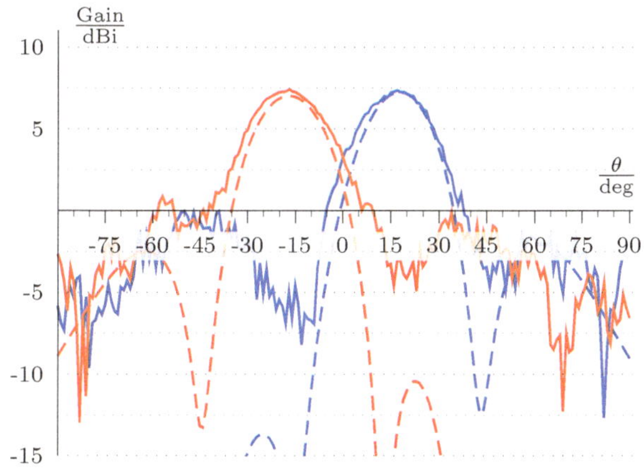

Fig. 9. Radiation pattern in the xz-plane (x-direction polarised). Measurement and simulation results are drawn in solid lines and dashed lines, respectively. Excitation at port 1 (blue line) and at port 2 (red line).

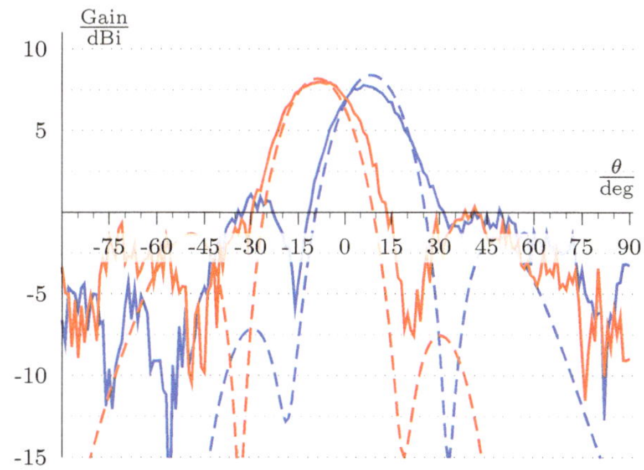

Fig. 10. Radiation pattern in the xz-plane (x-direction polarised). Measurement and simulation results are drawn in solid lines and dashed lines, respectively. Excitation at port 2 (blue line) and at port 4 (red line).

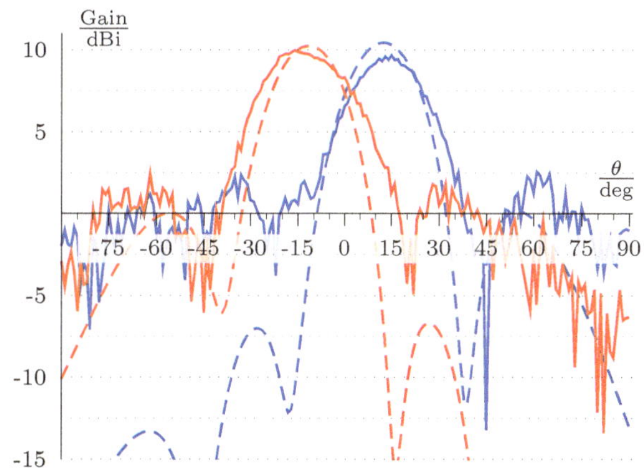

Fig. 11. Radiation pattern in the xz-plane (x-direction polarised). Measurement and simulation results are drawn in solid lines and dashed lines, respectively. Excitation at port 1+3 (blue line) and at port 2+4 (red line).

and

$$\omega_{fp_{res}} = \sqrt{\frac{1}{L_L C_R}}.$$

Since $C_L > 0$, it is found that $\omega_{fp} < \omega_{fp_{res}}$. The numerator is unequal to zero at both frequencies which leads to two maxima of $|Z_{21}|$. The position of the peak occurring at the higher frequency identifies the parallel resonant frequency.

Determining the series resonant frequency can be done based on Y_{11}. With the prerequisites that the parallel resonant frequency is tuned correctly, its conductance is insignificant around this frequency. Thus, the admittance matrix element simplifies to

$$Y_{11} = \left.\frac{I_1}{U_1}\right|_{U_2=0} \approx \frac{j\omega L_R}{1-\omega^2 L_R C_L}. \quad (9)$$

Only one maxima can be found for $|Y_{11}|$ which identifies exactly the series resonant frequency $\omega_{fs_{res}} = \sqrt{\frac{1}{L_R C_L}}$. Following these steps, both unit cells were balanced and tuned correctly to the desired frequencies. Their exact dimensions are shown in Fig. 3.

It should be mentioned, that due to mutual coupling the behaviour of a cascaded unit cell changes and detuning occurs. Thus, it is advisable to determine the elements Y_{11} and Z_{21} out of a cascaded arrangement by performing a deembedding process.

4 Results

The designed structure was fabricated, as seen in Fig. 6, and successfully tested. The measured reflection and transmission coefficients are shown in Fig. 7. Additional measurements were done determining the mutual coupling. As

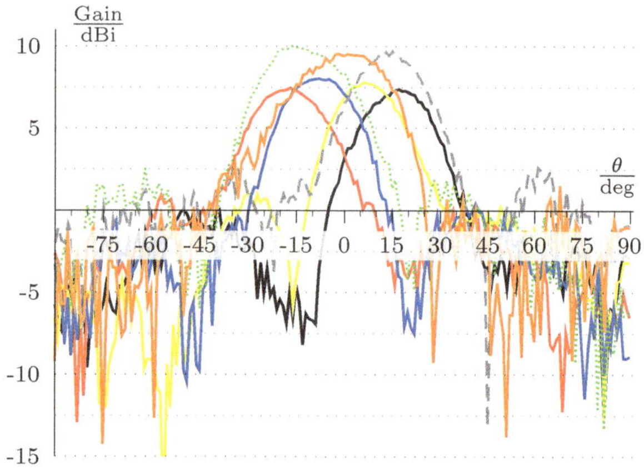

Fig. 12. Measured radiation pattern in the xz-plane (x-direction polarised). Excitation and colors are related as follows: Port 1 black, port 2 red, port 3 yellow, port 4 blue, port 1+3 gray dashed, port 2+4 green dotted, port 3+4 orange. Detailed pattern information are given in Table 1.

Table 1. Measured radiation pattern characteristics of the realised antenna.

Excitation	Gain	Θ_{MB}	Θ_{BW}
Port 1	7.4 dBi	17°	26°
Port 2	7.5 dBi	−17°	28°
Port 3	7.8 dBi	7°	25°
Port 4	8.0 dBi	−8°	27°
Port 1+3	9.7 dBi	14°	30°
Port 2+4	10.1 dBi	−15°	28°
Port 3+4	9.5 dBi	−1°	35°

shown in Fig. 8, this analysis yields an attenuation greater than 15 dB between neighbouring structures over a broad frequency range. The dispersion diagram obtained by simulation and the measured radiation patterns are depicted in Figs. 5, 9–12. Details of the achieved results are provided in Table 1. All parameters possess a good agreement with the analytical results from Eqs. (3) and (4). As expected, the gain increases when both structures are driven simultaneously, since the aperture in y-direction is enlarged. Consequently, the beamwidth in the yz-plane becomes smaller and gain enhancement takes place.

5 Conclusions

A beam steering concept was presented, which is based on two leaky wave structures with different main beam directions. The structures are realised as composite right/left-handed transmission lines with different transition frequencies. Thus, the main beam can be switched according to the chosen feeding port or steered by superimposing the different radiation patterns. The presented antenna possesses a scan angle range of $-17° \leq \Theta \leq 17°$ with a gain of up to 10 dBi at the operation frequency of 10 GHz.

References

Balanis, C. A.: Antenna theory – analysis and design, John Wiley & Sons, 3rd edn., 2005.

Binzer, T., Klar, M., and Groß, V.: Development of 77 GHz Radar Lens Antennas for Automotive Applications Based on Given Requirements, 2nd International ITG Conference on Antennas, 2007.

Caloz, C. and Itoh, T.: Electromagnetic Metamaterials, John Wiley & Sons, 2006.

CST Computer Simulation Technology, http://www.cst.com/, access: 31 March 2010.

Eberspächer, M. A., Eccleston, K., and Eibert, T. F.: Realization of Via-free Microstrip Composite Right/Left-Handed Transmission Lines, Proceedings of the German Microwave Conference GeMiC 2009, 2009.

Johnson, R.: Antenna Engineering Handbook, chap. Leaky-Wave Antennas, McGraw Hill, 3rd edn., 1993.

Matsuzawa, S., Sato, K., Inoue, Y., and Nomura, T.: W-Band Steerable Composite Right/Left-Handed Leaky Wave Antenna for Automotive Applications, IEICE TRANSACTIONS on Electronics, 2006.

Paulotto, S., Baccarelli, P., Frezza, F., and Jackson, D. R.: Full-Wave Modal Dispersion Analysis and Broadside Optimization for a Class of Microstrip CRLH Leaky-Wave Antennas, IEEE Transaction on Microwave theory and techniques, 2008.

Pozar, D. M.: Microwave Engineering, John Wiley & Sons, 3rd edn., 2005.

Schneider, R. and Wenger, J.: High resolution radar for automobile applications, Adv. Radio Sci., 1, 105–111, 2003, http://www.adv-radio-sci.net/1/105/2003/.

Walter, C. H.: Traveling wave antennas, Dover, 1st edn., 1965.

5

A cooperative transponder system for improved traffic safety, localizing road users in the 5 GHz band

B. Schaffer, G. Kalverkamp, M. Chaabane, and E. M. Biebl

Fachgebiet Höchstfrequenztechnik, Technische Universität München, Munich, Germany

Correspondence to: B. Schaffer (schaffer@tum.de)

Abstract. We present a multi-user cooperative mobile transponder system which enables cars to localize pedestrians, bicyclists and other road users in order to improve traffic safety. The system operates at a center frequency of 5.768 GHz, offering the ability to test precision localization technology at frequencies close to the newly designated automotive safety related bands around 5.9 GHz. By carrying out a roundtrip time of flight measurement, the sensor can determine the distance from the onboard localization unit of a car to a road user who is equipped with an active transponder, employing the idea of a secondary radar and pulse compression. The onboard unit sends out a pseudo noise coded interrogation pulse, which is answered by one or more transponders after a short waiting time. Each transponder uses a different waiting time in order to allow for time division multiple access. We present the system setup as well as range measurement results, achieving an accuracy up to centimeters for the distance measurement and a range in the order of hundred meters. We also discuss the effect of clock drift and offset on distance accuracy for different waiting times and show how the system can be improved to further increase precision in a multiuser environment.

1 Introduction

Each year, thousands of pedestrians are killed in traffic accidents: in German statistics alone, 3648 fatalities are reported for the year 2010 (Statistisches Bundesamt Deutschland, 2010). Therefore, one goal of the research initiative "Ko-FAS" is to find ways to lower this number by assisting drivers in recognizing and avoiding critical traffic situations. In the project "Ko-TAG", which is a part of the research initiative Ko-FAS, a new cooperative transponder system is developed. The system is able to detect and localize pedestrians, bicyclists and other vulnerable road users even if they are hidden behind obstacles so that the driver cannot see them. Based on the localization data of the road users around the ego-vehicle and its own driving status, the onboard computer can set up risk models for every user and can prevent dangerous situations by either warning the driver or applying autonomous maneuvers (Schwarz et al., 2011). A first approach to a cooperative sensor system for pedestrian detection and localization was developed in the AMULETT project (Morhart and Biebl, 2009). The AMULETT system can determine the range to a transponder with a standard deviation of less than 10 cm, using an RF-bandwidth of 60 MHz and a center frequency of 2.44 GHz. Signal processing is done by an FPGA in real-time, thus an update rate of more than 50 Hz can be accomplished. Since the 2.4 GHz ISM Band is already heavily used by wireless LAN, Bluetooth and other unlicensed applications, and due to the latest designations of separate frequency bands in the 5.8 to 5.9 GHz range for automotive safety and intelligent transport systems (Electronic Communications Committee, 2008), current localization technology must be tested at these new frequencies. Therefore, we designed an improved system, based on the AMULETT technology. This new system, called "SafeTAG 1.0" is considerably smaller, consumes less power and for the first time can demonstrate multi-user detection and localization with 20 units available. Originally, the SafeTAG 1.0 system operates at 2.44 GHz, but the frequency of operation can be moved to 5.768 GHz by connecting a frequency converter module.

2 Theory of operation

Many different approaches to detect and localize pedestrians in traffic scenarios have been proposed in the past, including radar, lidar and image processing via cameras. An overview over the use of these technologies is given in Gandhi and Trivedi (2007). The use of active transponders allows for simultaneous detection, classification and localization of pedestrians, other cars and transponder-equipped landmarks in real-time and with high accuracy, but at the cost of a more complex cooperative system. In the field of cooperative distance measurement systems, multiple realizations have been shown in the past. A ranging system based on frequency modulated continous wave chirps is presented in Roehr et al. (2008), for example. The system proposed in this work is based on the well known principle of pulse compression by pseudo noise codes. Since the basics of the distance measurement are already discussed in detail in Morhart and Biebl (2009), only a short overview of the principle of operation shall be given here.

A measurement cycle begins with the transmission of an interrogation pulse from the onboard-unit. The pulse consists of a short continuous wave preamble and a 256 bit binary maximum-length sequence, which modulates the carrier signal by on-off-keying. The bandwidth of the pseudo noise code is approximately 60 MHz. Once a transponder receives the interrogation pulse and detects its presence, a wait timer is started with a preset delay time. After the waiting time, the signal is sent back to the onboard unit, using the amplified original received samples to avoid jitter due to clock phase offset between the onboard unit and the transponder. The sign of the received signal is thereby inverted to make interrogation pulses and transponder responses distinguishable. This ensures that a second transponder does not erroneously respond to the answer of the first transponder. Each transponder has a different tag ID and therefore uses a different waiting time, defined as

$$t_{\text{wait}} = 5\,\mu\text{s} + (\text{tagID} - 1) \cdot 100\,\mu\text{s}, \tag{1}$$

so that every answer to the interrogation pulse can be sent in a different time slot, enabling the system to support many users by a time division multiple access scheme. For every time slot, the onboard unit calculates the signal roundtrip time of flight by correlating the transmitted sequence with the received signal, taking advantage of the special correlation characteristics of the used m-sequences (Lüke, 1992). From the interpolated maximum of the correlation result, the signal roundtrip time of flight t_{RTOF} and therefore the distance R to a transponder can be calculated using its assigned delay time t_{wait} and the speed of light c_0:

$$R = \frac{c_0}{2} \cdot (t_{\text{RTOF}} - t_{\text{wait}}) \tag{2}$$

At the same time, a trigger signal is issued to a direction-of-arrival unit, which then estimates the angle to each transponder by using the buffered receive signals of a 6-element antenna array and a super-resolution algorithm. To ease the direction-of-arrival estimation, the transponders append a 65 μs long continuous wave burst to the end of the answered interrogation pulse. The distance measurement module and the direction-of-arrival estimation module send their localization data to a central onboard computer, which sets up the environment model and reacts to critical traffic scenarios by either warning the driver or applying autonomous avoidance or mitigation maneuvers (Geissler et al., 2011).

3 Hardware setup

The SafeTAG 1.0 system is based on the AMULETT technology, but the transceiver has been redesigned and optimized. The operation frequency of the base system is still 2.44 GHz, but it can be converted to 5.768 GHz by using a frequency converter unit, offering the ability to test the localization technology at both bands. The SafeTAG 1.0 base system provides separate transmit- and receive connectors, whereas the commercial converter used for changing the center frequency only offers a single 5 GHz and a single 2.4 GHz connector, switching the direction of the conversion internally after autosensing the input power at the 2.4 GHz port. Therefore, it is necessary to implement a TX/RX switch between the SafeTAG and the converter unit. This switch is controlled by the internal TX/RX enable signals of the FPGA which also switch the TX and RX components on and off. Since the converter unit is attached to the power amplifier output of the base system, a 6 dB attenuation pad has to be inserted into the 2.44 GHz transmit path to avoid overloading the converter input and to set the output power of the 5 GHz system to 22 dBm. Figures 1 and 2 show a simplified block diagram and a photograph of the transceiver and converter unit.

4 Measurements

In the following section, laboratory and real-world measurements will be shown in order to analyze the performance of the transponder system at the new operating frequency of 5.768 GHz. The main reasons for distance measurement errors will be identified by example measurements and ideas to improve the system performance will be given.

4.1 Calibration

The distance between an onboard unit and a transponder causes a time delay in the roundtrip time of flight, which is measured by the onboard unit. But components in the transceivers like filters, amplifiers and transmission lines add additional delay to the signal, making calibration necessary before useful measurements can be carried out. Calibration is achieved by measuring distance at an arbitrary position, and then determining the offset between the mean displayed and the true distance. This offset is then subtracted from all following measurements, leaving possibilities for errors only due to clock noise and drift, signal noise, and effects of multipath propagation. The severity of these effects will be discussed in the following section, and example values from measurements will be given.

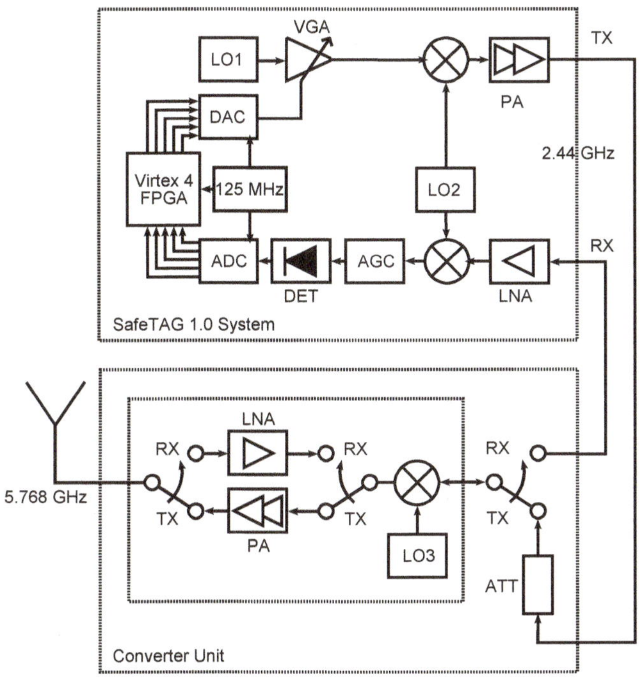

Fig. 1. Simplified block diagram of the transponder system.

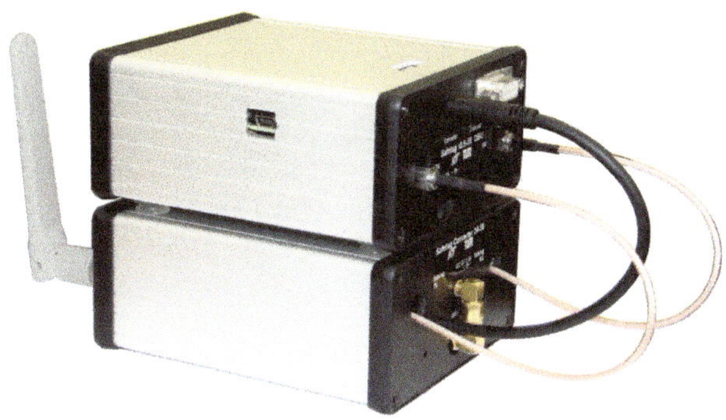

Fig. 2. Photograph of a transponder (top) with the converter unit (bottom) and a 5 GHz antenna attached.

4.2 Impact of clock phase and signal noise

Phase noise and short term changes in reference oscillator frequency as well as signal noise can affect the quality of the distance measurement, even directly after calibration. The standard deviation over many acquired samples at a static distance relative to the calibration distance can be used as an indicator for the error. For comparison, the results of four measurements are shown in Fig. 4, two with tag ID 1 ($t_{\mathrm{wait}} = 5\,\mu$s) and two with tag ID 11 ($t_{\mathrm{wait}} = 1005\,\mu$s). In all four cases, the transponder is connected to the onboard unit via a 5.5 m long coaxial cable to exclude effects of multipath propagation, as shown in Fig. 3. Additional attenuators between the onboard unit and the transponder are used to set

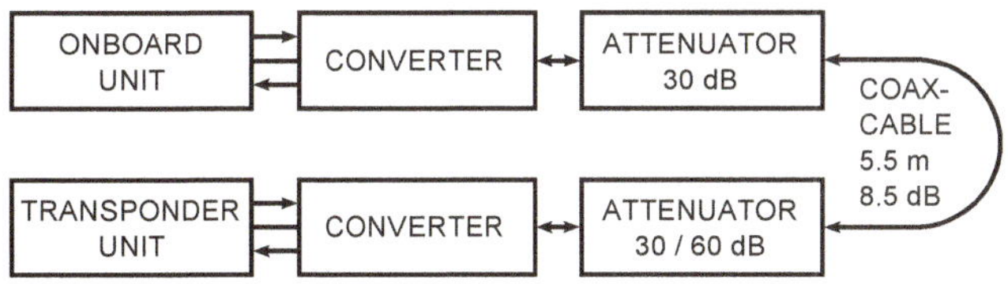

Fig. 3. Measurement setup.

a total attenuation of 68.5 dB in the first two and 98.5 dB in the second two measurements. Note that these attenuators will affect signal power in both signal directions, from the onboard unit to the transponder and back. Both units are warmed up and calibration is carried out before measurements are taken. At 68.5 dB attenuation, the standard deviation over 10 000 samples is 3.8 mm for tag ID 1 and 4.9 mm for tag ID 11. These values show that using higher tag IDs and therefore longer waiting times in a multi-user environment will not considerably harm relative distance accuracy caused by clock phase noise. While the measurements with 68.5 dB one-way path loss represent real scenarios with a distance of approximately 10 m to the road user in a line of sight condition, measurements with 30 dB additional path loss are taken to model longer distances and hidden positions, where additional attenuation due to diffraction occurs. At 98.5 dB attenuation, the standard deviation is considerably larger due to a lower signal-to-noise ratio, but the determined standard deviations of 83.6 mm for tag ID 1 and 87.0 mm for Tag ID 11 still allow for a localization with decimeter accuracy. Again, the values for the standard deviation do not differ much between different waiting times.

4.3 Impact of frequency offsets

In the last section, we discussed short term variations around the mean value at static distances, but – by calibration – neglected distance offsets in our measurements. In fact, a clock frequency offset between the transponder and the onboard unit will lead to slightly different waiting times in both units, affecting the absolute measured distance. This can for example be caused by reference oscillator manufacturing tolerances, aging, or temperature drift. All these change the frequency of the oscillator slowly in time, so the frequency difference can be calculated after determining the mean distance error over a row of samples. Since the absolute timing error will be larger for longer waiting times, higher tag IDs will show greater distance offsets. To determine an example value for the offset error, the first two measurements in Fig. 4 are calibrated with the same offset value – the calibration value determined with tag ID 1 – so the mean distance error of tag ID 1 is zero. Tag ID 11 then shows a mean distance error of $\delta R = -1.457$ m, which is due to the combined clock frequency offset between the onboard unit and the transponder, accumulated over the additional waiting time of $1000\,\mu$s

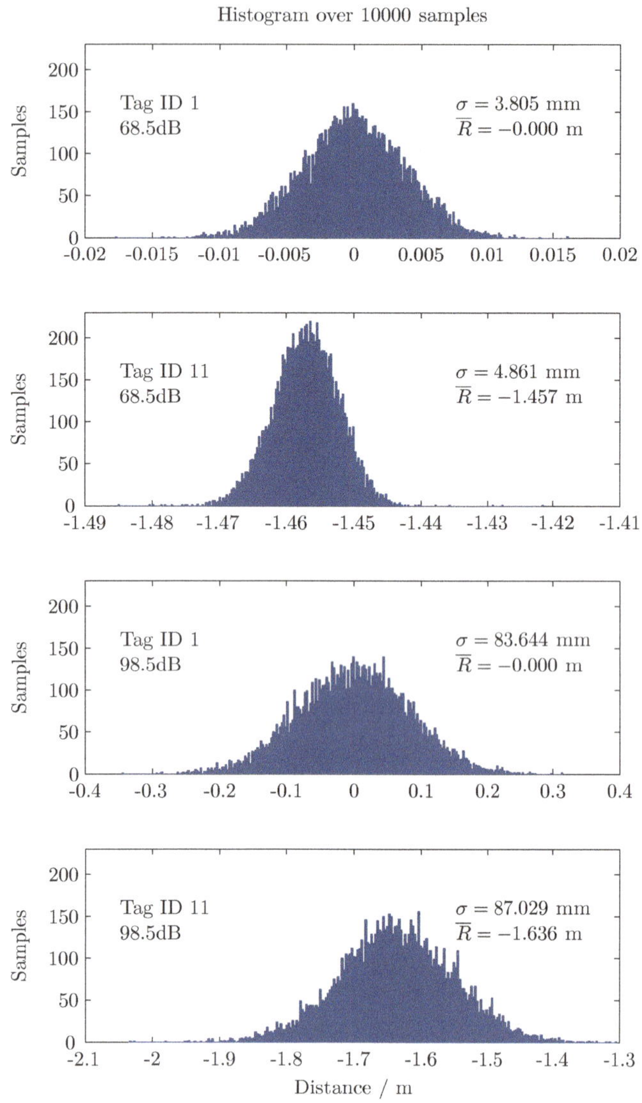

Fig. 4. Histograms for measurements with 68.5 dB one-way attenuation in the first and second graph, and 98.5 dB in the third and fourth graph. Tag ID = 1 for the first and third graph, and Tag ID = 11 for the second and fourth graph.

between timeslots 1 and 11. For this specific case, the relative clock error can be calculated to be

$$\frac{\delta R \cdot 2/c_0}{1000\,\mu s} = -9.7\,\text{ppm}, \tag{3}$$

which is well within the specified range of ±50 ppm of the used crystal oscillator. To remove or minimize these errors, different approaches are possible in future systems: of course, a transponder could be recalibrated whenever the tag ID is changed. While this is possible in the current prototype stage of development, this method will not be practicable in a future system, when tag IDs are allocated dynamically in a traffic ad-hoc network via data communication. However, calibration against a "golden standard" could be carried out for every single tag ID at factory. The calibration values could be saved in a lookup table and read whenever a new tag ID was assigned. This technique would include deterministic offsets, but not temperature and aging effects. In general, the effects of clock offset and drift should therefore be reduced by using high quality reference oscillators with good temperature and aging stability, and by keeping time slots as short as possible. Using a temperature compensated crystal oscillator as a reference clock with a guaranteed accuracy of 0.5 to 1 ppm instead of the currently used quartz oscillator could already improve the situation by a factor of 5 to 10 compared to the results presented above. Currently, Tag IDs differ in waiting time by $100\,\mu s$ each, mostly due to the $65\,\mu s$ long direction-of-arrival burst. In a future system, roundtrip time-of-flight and direction-of-arrival measurements will be separate and independent from each other, so the time-of-flight time slots could be reduced to a length of $10\,\mu s$. By combining a better oscillator with shorter timeslots, up to 100 transponders instead of 11 could be addressed with a precision ten times better than presented in the measurements above. In addition, an algorithm in the onboard unit could dynamically assign low tag IDs and therefore short waiting times to those transponders which are already close to the ego vehicle, keeping the less accurate timeslots for those which are still far away and not critical. Another option would be an on-the-fly measurement of the relative clock offset between the onboard unit and a transponder by carrying out two distance measurements in one cycle, one with a very short waiting time and one with a long one. The clock offset could then be calculated in the same way as it is calculated in the analysis above, and the determined value could be considered in the computed distance values.

4.4 Multipath propagation

Another effect on distance precision can be multipath propagation. Due to the limited bandwidth of the system, only multipath responses that differ by more than ΔR in path length can be distinguished by the system, while signal paths with a smaller separation cannot be resolved and will lead to a distance error due to their combination into one correlation maximum. In theory, the minimum resolvable path length separation of two signals, consisting of rectangular pulses with a duration of $\tau_p = 1/B$ is (Klausing and Holpp, 2000)

$$\Delta R = \frac{c}{B}, \tag{4}$$

which evaluates to 5 m for a pulse bandwidth of 60 MHz. Figure 5 shows the correlation results for 5 measurements with two signal paths, introduced by a power splitter and combiner with different cable lengths between the onboard unit and the transponder. The length of the second path is changed in 5 steps, while the system is calibrated so that the first path shows a distance of zero. It must be noted that the displayed distance for the second path is half the true value of

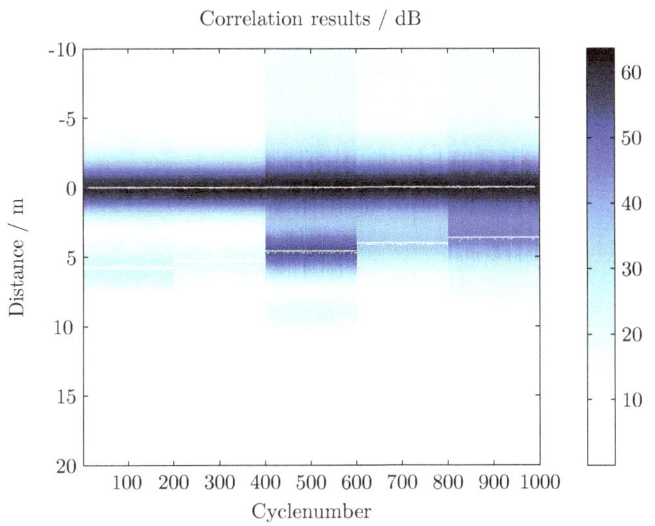

Fig. 5. Correlation results for 5 different multipath lengths, varying from 11 to 6.68 m. The white lines show the distance output of the system for the first and second maximum. The different intensities for the multipath maxima are a result of interference between the two paths.

the multipath length, since the system calculates the distance to the transponder, which is one half of the travelled distance of the signal. The shortest path difference at which the system can still resolve the two signals is found to be at 6.68 m. To improve the multipath resolution, superresolution algorithms could be used in future systems. According to previous research (Morhart and Biebl, 2010), these could improve the resolution of the system by a factor of approximately two. In practice, the multipath situation changes rapidly due to the movement of the ego vehicle and the transponders in traffic, while the line of sight path, if available, changes more slowly. By applying tracking algorithms in the onboard computer, some multipath situations can thus be resolved after a short time and measurement errors can be reduced.

4.5 Outdoor measurements

Finally, after the preceding laboratory measurements, a real world measurement shall be presented in order to show the functionality of the 5.768 GHz system in an outdoor traffic scenario. Figure 6 depicts the correlation results as output by the onboard unit, in raw format, without any filtering or tracking applied. For the shown setup, a pedestrian starts at a distance of 10 m from the car which is equipped with the onboard unit. The pedestrian is then slowly moving away to a distance of 100 m, before he comes back. At all times, the person is blocking the line of sight between the car and the transponder, including body attenuation and diffraction effects to the measurement. Even with the transponder screened by the human body, detection ranges of more than 100 m can easily be reached.

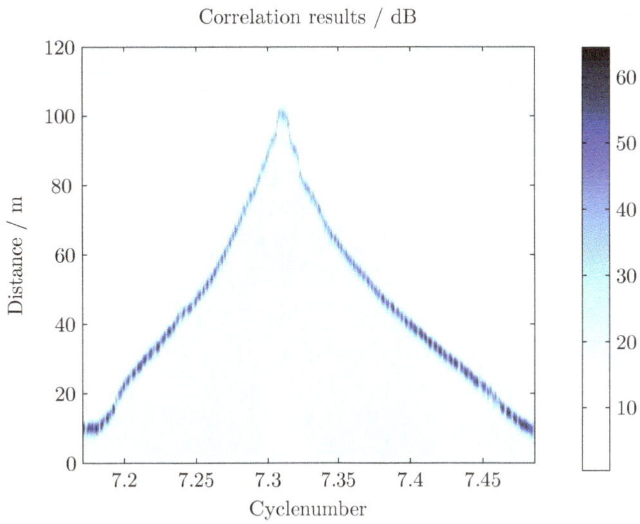

Fig. 6. Correlation results for an outdoor measurement, where the transponder is screened by the human body.

5 Conclusions and outlook

In the preceding sections, we presented a cooperative sensor system which is able to localize vulnerable road users in order to improve traffic safety. The system is usable at 2.44 and 5.768 GHz to offer the ability to evaluate localization technology at automotive safety related frequency bands. In outdoor measurements, a range of more than 100 m can be reached, while laboratory measurements show a precision in the range of one centimeter at high signal to noise ratios and with low tag IDs. Naturally, lower signal to noise ratios will degrade accuracy, but even at a path loss of 98.5 dB one-way, standard deviations below ten centimeters can be observed. Using higher tag IDs, which means using longer waiting times before retransmission of the localization signal, will introduce bigger measurement errors due to clock offset and drift, but a list of countermeasures is given to minimize their impact.

The current system only supports a number of fixed timeslots, and every transponder needs to have a fixed tag ID assigned. In future systems, tag IDs will need to be allocated dynamically in order to accept upcoming users down the road into the ad-hoc network. Dynamic management of the tag IDs will then not only provide precise distance measurements to those transponders that need accurate localization because they are in a close or critical situation, but will also allow for anonymization of the transponders. This dynamic allocation mechanism requires fast and robust data communication capabilities. With the current system, these would be limited to simple modulation schemes, due to the use of hardware amplitude modulators and demodulators. In order to support modern modulation schemes for data transmission, a new system is currently being developed. With the so-called SafeTAG 2.0 system, it will be possible to use three channels – one for distance measurement, one for data communication and one for network management – to improve the performance of the sensor system. In this new system, a TCXO-based reference clock with a guaranteed frequency tolerance of ±0.5 ppm over temperature will be used in order to reduce offset errors in the distance measurement. Additionally, the direction-of-arrival estimation will be accomplished in the data channel, allowing shorter timeslots in the roundtrip time of flight transmission scheme, thereby reducing distance measurement errors even further.

Acknowledgement. This work results from the joint project Ko-TAG, which is part of the project initiative Ko-FAS, and has been funded by the German Bundesministerium für Wirtschaft und Technologie (Federal Department of Commerce and Technology) under grant number 19S9011C.

References

Electronic Communications Committee: ECC Decision of March 14 2008 on the harmonised use of the 5875–5925 MHz frequency band for Intelligent Transport Systems (ITS), available at: http://www.cept.org/ecc/deliverables, 2008.

Gandhi, T. and Trivedi, M.: Pedestrian Protection Systems: Issues, Survey, and Challenges, Intelligent Transportation Systems, IEEE Transactions on, 8, 413–430, doi:10.1109/TITS.2007.903444, 2007.

Geissler, U., Trebing, A., Westhofen, D., and Zecha, S.: Intelligent collision risk prediction using cooperative sensor systems, Proceedings of the International Workshop on Intelligent Transportation (WIT), 8, 41–44, 2011.

Klausing, H. and Holpp, W.: Radar mir realer und synthetischer Apertur, Oldenbourg Verlag, 1st Edn., 2000.

Lüke, H. D.: Korrelationssignale, Springer Verlag, 1992.

Morhart, C. and Biebl, E. M.: Cooperative multi-user detection and ranging based on pseudo-random codes, Adv. Radio Sci., 7, 55–59, doi:10.5194/ars-7-55-2009, 2009.

Morhart, C. and Biebl, E. M.: High resolution time of arrival estimation for a cooperative sensor system, Adv. Radio Sci., 8, 61–66, doi:10.5194/ars-8-61-2010, 2010.

Roehr, S., Gulden, P., and Vossiek, M.: Precise Distance and Velocity Measurement for Real Time Locating in Multipath Environments Using a Frequency-Modulated Continuous-Wave Secondary Radar Approach, Microwave Theory and Techniques, IEEE Transactions on, 56, 2329–2339, doi:10.1109/TMTT.2008.2003137, 2008.

Schwarz, D., Kloeden, H., and Rasshofer, R.: Ko-TAG – Cooperative Sensor Technology for Traffic Safety Applications, Proceedings of the International Workshop on Intelligent Transportation (WIT), 8, 31–34, 2011.

Statistisches Bundesamt Deutschland: Verkehr – Verkehrsunfälle 2010, Fachserie 8 Reihe 7, Statistisches Bundesamt Deutschland, 2010.

6

Pedestrian recognition using automotive radar sensors

A. Bartsch, F. Fitzek, and R. H. Rasshofer

BMW Group Research and Technology, Munich, Germany

Correspondence to: A. Bartsch (armin.bartsch@bmw.de)

Abstract. The application of modern series production automotive radar sensors to pedestrian recognition is an important topic in research on future driver assistance systems. The aim of this paper is to understand the potential and limits of such sensors in pedestrian recognition. This knowledge could be used to develop next generation radar sensors with improved pedestrian recognition capabilities. A new raw radar data signal processing algorithm is proposed that allows deep insights into the object classification process. The impact of raw radar data properties can be directly observed in every layer of the classification system by avoiding machine learning and tracking. This gives information on the limiting factors of raw radar data in terms of classification decision making. To accomplish the very challenging distinction between pedestrians and static objects, five significant and stable object features from the spatial distribution and Doppler information are found. Experimental results with data from a 77 GHz automotive radar sensor show that over 95 % of pedestrians can be classified correctly under optimal conditions, which is compareable to modern machine learning systems. The impact of the pedestrian's direction of movement, occlusion, antenna beam elevation angle, linear vehicle movement, and other factors are investigated and discussed. The results show that under real life conditions, radar only based pedestrian recognition is limited due to insufficient Doppler frequency and spatial resolution as well as antenna side lobe effects.

1 Introduction

The use of radar sensors in automotive pedestrian recognition systems is of special interest since radar sensors are less influenced by environmental conditions (e.g. fog, rain, etc.) as other systems like video cameras (Wenger, 2007). Moreover, high resolution radar sensors are available in many modern vehicles as a part of Adaptive Cruise Control (ACC) systems.

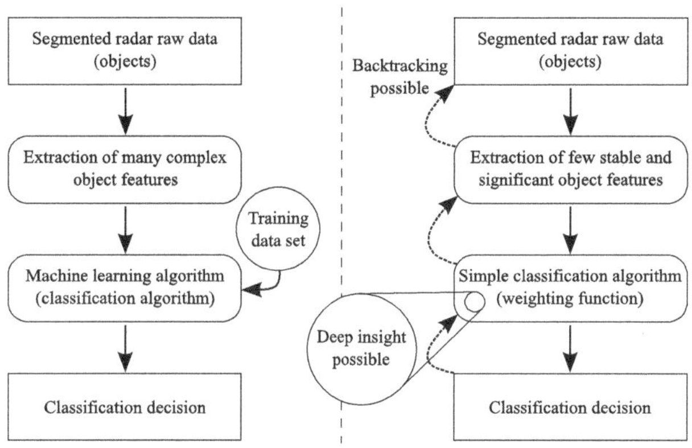

Fig. 1. A common recognition system with no backtracking possibility is shown on the left. On the right, the new recognition method with possible backtracking of classification errors is shown.

If ACC systems and pedestrian recognition systems could use the same radar sensor, hardware costs for producing these systems are kept to a minimum. In this paper, the potential of a modern series production automotive radar sensor, designed for ACC systems, for pedestrian recognition is explored. In particular, the limits of the radar sensor regarding decision making in pedestrian classification are investigated to see what future developments of automotive radar sensors are necessary to improve pedestrian recognition systems.

A radar based pedestrian recognition system consists of two main components, a radar sensor and a signal processing unit, i.e. radar raw data preprocessing combined with a classification algorithm. In this paper, a radar signal processing unit is developed that allows the investigation of the potential and limits of the radar sensor in an online system. Thus, in case of a classification error, it can be examined if the sensor data was not sufficient to be able to choose the correct object class or if the classification algorithm could not squeeze the necessary information out of the sensor's data because of a non-optimal classification algorithm. Misleading sensor raw data of an object is statistically not distinguishable from data of an object belonging to a different class. This could be data from a slowly moving pedestrian for example, that is almost identical to raw data from a small static object like a traffic sign because of quantization errors.

Previous approaches to pedestrian recognition with radar sensors (e.g. Benitez, 2011; Rohling, 2007; Freund, 2007; Büchele, 2008; Kouemou, 2008; He, 2010) mostly used complex signal features, machine learning for classification and often human gait models for interpreting micro Doppler signatures.

But classification decisions made by machine learning algorithms are non-transparent. Thus, it is not possible to examine why exactly the algorithm decided for the classification result in every case. The significance of the different object features, extracted from the sensor's data, was learned while processing a training data set. This curcial information is buried deep in the algorithm and cannot be easily obtained. Hence, machine learning does not allow for consistent backtracking from classification decisions to the critical information in the radar raw data that caused the decision (see Fig. 1). Human gait models (e.g., Nanzer, 2009, Kim, 2008, Ritter, 2007, Hornsteiner, 2008), which explain micro Doppler signatures, are limited to sensors with very high Doppler frequency resolution and need radar data over a longer, continous period of time instead of single frames.

That is why a different approach in pedestrian recognition without using machine learning and complex object features is needed. In this paper, a knowledge based pedestrian recognition system, consisting of simplified components, is shown that allows transparent classification decisions. The selection, significance testing, weighting of object features and classification algorithm details were found empirically.

A focus will be on boundary and surrounding conditions and their impact on radar raw data. The pedestrian's direction of movement, occlusion, radar elevation angle and other factors will be investigated to see how real life scenarios affect the recognition results in comparison to recognition under optimal conditions.

2 The radar sensor's raw data

In this paper, a 77 GHz band automotive scanning radar sensor is used. The sensor is designed for series production ACC systems. Table 1 shows some more detailed technical characteristics. Every 66 ms the sensor measures the received signal strength values and Doppler frequency shifts for every spatial resolution cell. A resolution cell is 1° by 0.25 m.

The Doppler frequency shift is proportional to the radial component of the relative velocity between the sensor and the reflecting object and, therefore, the sensor can easily convert the Doppler shift information to radial velocity information. It is important to keep in mind that an object moving equidistant to the sensor does not cause any Doppler shift in the reflected signal. The sensor output for the Doppler information is not limited by frequency measurement resolution because it is internally interpolated to minimize quantization errors.

Table 1. Technical details of the radar sensor used in this paper.

Frequency range	76–77	GHz
Resolution of Doppler frequency/ radial relative velocity	390/0.77	Hz/m s^{-1}
Spatial discretization in azimuth	1	°(degree)
Radial spatial discretization	0.25	m
Max. distance of objects	50	m
Beam steering range (azimuth)	17	°(degree)
Beam width (3 dB, azimuth)	2.5	°(degree)
Cycle time	66	ms
Modulation technique	chirp sequence	–

In every measurement cycle, one data frame consisting of two matrices is created. The so called "intensity image" contains the received signal strength values for all spatial resolution cells and the "frequency image" contains the radial relative velocities of reflection centers in every resolution cell. Figure 2 shows some example radar raw data of a pedestrian. These data matrices constitute the low-level radar raw data used for pedestrian recognition.

3 The pedestrian recognition algorithm

In this chapter, a novel method for pedestrian recognition avoiding machine learning and tracking is developed. The main concern in this step is to keep the signal processing from radar raw data to the classification result as simple and transparent as possible. This guarantees easy backtracking from classification decisions to the radar raw data. An even more powerful advantage of simple signal processing is that object features can be tested for their significance in classification by hand with low effort.

Tracking objects in radar raw data for better classification was discarded to benefit from single frame processing. For driver assistance systems, it is most essential to achieve minimum latency between appearance of a pedestrian in the raw radar data and the output of the correct classification decision. Tracking would introduce such latency and since radar based pedestrian recognition will be used together with other sensor systems in practice, tracking is more benefitial on a higher application level.

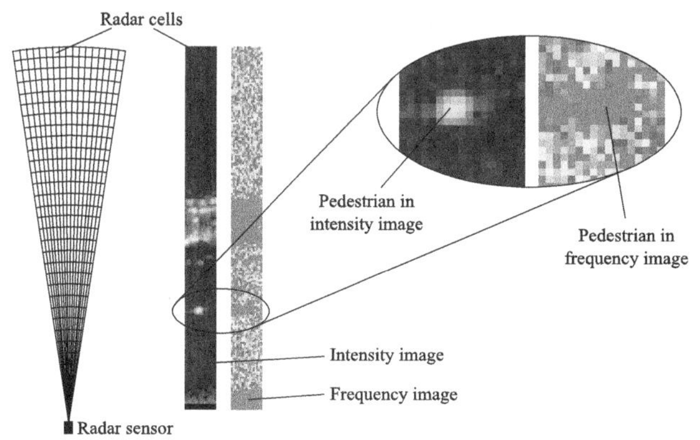

Fig. 2. Example of a pedestrian's radar raw data. The principle, how the sensor divides its spatial measurement range into resolution cells, is depicted on the left. The sensor measures the reflected power and Doppler shift in the reflected signal for every resolution cell, giving two data matrices, the intensity image and the frequency image (middle). The grey value of pixels in the intensity image refers to the reflected power while the grey value of pixels in the frequency image correspondes to the measured Doppler shift. A grey value of 128 indicates a Doppler shift of 0 Hz.

Furthermore, this paper highlights the differentiation between static objects and pedestrians. This is challenging because walking pedestrians have very low speed compared to the sensor's velocity measurement resolution of $0.77\,\mathrm{m\,s^{-1}}$. Other objects in typical traffic scenarios, e.g. cars, exhibit significant higher velocities and can be distinguished from pedestrians with less effort.

3.1 Raw data preprocessing

The signal processing chain starts with noise reduction and segmentation in the intensity image. Because simple low-pass filtering would blur the critical border between objects and surrounding, a combined bit-depth reduction and thresholding approach was used. The segmentation algorithm was developed in (Freund, 2007) and creates objects, i.e. sets of spatial neighboured sensor data that is likely from one physical object. This segmentation step has big influence on the classification result because it decides where the border is drawn between objects and the noisy surrounding. In the border area of objects, the frequency image contains artefacts that can extensively bias the classification algorithm. If the border between objects and the surrounding is drawn too close to the reflection center of objects, not all related resolution cells get attached to the object and important information about the object gets lost. The optimum in this trade-off is highly dependent on background noise in the intensity image.

In the frequency image, the ego-speed of the measurement vehicle v_{ego} has to be compensated. If the vehicle moves with v_{ego}, a static object in front of the vehicle has a relative velocity of $-v_{\mathrm{ego}}$, which is measured by the sensor. Experiments showed that the vehicle's ego-speed signal, generated by the vehicle's dynamic stability control system, is too noisy for satisfactory compensation. To eliminate this error, an object which is known to be static is needed in the radar raw data, e.g. a tree or a road barrier that was identified by another sensor system like a camera. Using this information, the frequency image can be recompensated with sufficient accuracy for the following recognition process. In this paper, the information about which objects in the radar data are static was provided by hand to allow experiments with $v_{\mathrm{ego}} \neq 0$.

3.2 Feature extraction

In this section we propose a set of five object features that change reliably and significantly when the object, represented by the segmented raw radar data, is a pedestrian or a static object. Since pedestrians are radar point targets (Yamada, 2005), their shape in the intensity image is the same as the one of any small reflective object. Yamada also shows that the strength of a radar signal, reflected by a pedestrian, and thus the intensity in the intensity image, is highly fluctuative (about 20 dB). Hence, with information from the intensity image only, it is impossible to distinguish between pedestrians and small arbitrary objects. Nevertheless, in this paper two object features from the intensity image, a size and a shape factor, are calculated to exclude big or elongate objects from potential pedestrians.

Figure 3 reveals that in a resolution cell the reflected power value from the intensity image and the radial relative velocity value from the frequency image are statistically uncorrelated. The graphs depict the frequency distributions of resolution cells with certain reflection intensity and radial relative velocity for different types of objects. The statistical uncorrelation means that the mean and variance of a resolution cell's radial velocity value are not correlated to its intensity value. Therefore, principal component analysis (PCA) or similar methods to transform possibly correlated data sets into uncorrelated data sets are obsolete. Another interesting point is that the relative velocity information is not correlated with the distance from the object center since resolution cells with high intensity are located at the center of an object.

While the variance of the intensity values is almost equal for moving pedestrians and static objects, the variance of the radial velocity values varies significantly. The classification algorithm distinguishes pedestrians from static objects mainly using the higher variance of the relative radial velocity components in the frequency image data of pedestrians (see Sect. 3.2.3). Three object features from the frequency image data are calculated for every object, giving in combination with two object features from the intensity image an only five dimensional feature space which is sufficient for proper classification.

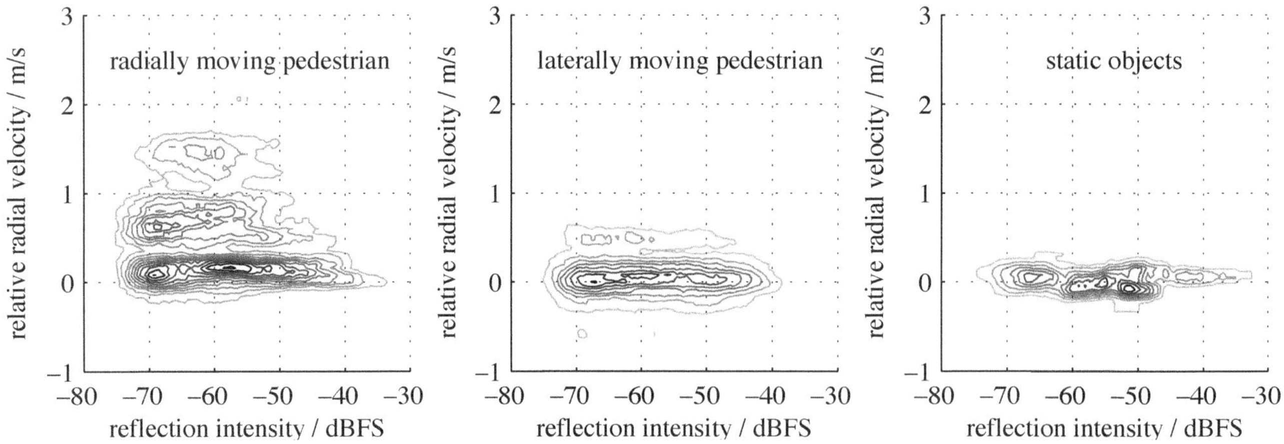

Fig. 3. Relative frequency distributions of resolution cells with certain reflection intensity and relative radial velocity for different object types. The frequency distributions were retrieved from approximately 150 data frames giving about 2000 data points per graph. Since mean and variance of the relative radial velocity values are independent of the reflection intensity, information from the intensity image and from the frequency image is statistically uncorrelated.

3.2.1 The size of an object

The size of an object is a measure that is proportional to an object's mean radius and simply the square root of the distance-compensated number of resolution cells assigned to the object. Distance compensation is necessary because with increasing distance r from the sensor, the received power P_D decreases with $1/r^4$ and the width of the spatial resolution cells get larger since they are always 1°in azimuth. With n' as the distance-compensated number of resolution cells of the object, $m_1 = \sqrt{n'}$ is the first object feature.

3.2.2 The shape of an object

The shape of a point target in the radar raw data is determined by the sensor's antenna pattern. In the intensity image, a small object has an elliptical shape with the longer axis in azimuthal direction and an axis ratio of about $D/d = 1.8$. To obtain an object's shape feature, the difference of the object's shape compared to an ideal ellipse is calculated. An ideal ellipse of equal area and half-axes a, b is placed over the object's center (r_c, ϕ_c) (see Fig. 4). In this coordinate system, the i-th resolution cell of an object with n cells has a local coordinate (x_i', y_i'). A distance measure for a cell to the ellipse's border is given by

$$\delta_i = \frac{y_i'^2}{a^2} + \frac{x_i'^2}{b^2} - 1 \,, \quad a = \frac{D}{d} \cdot b \,, \quad b = \sqrt{\frac{n \cdot d}{\pi \cdot D}} \,. \tag{1}$$

The size-normalized cumulative shape error gives the second object feature m_2:

$$m_2 = \frac{1}{\sqrt{n}} \sum_{i:\delta_i > 0} \delta_i^2 \tag{2}$$

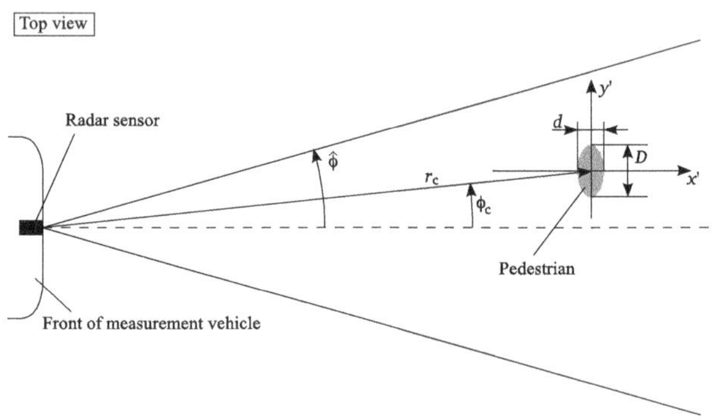

Fig. 4. An illustration of the sensor coordinate system, an object with center at (r_c, ϕ_c) and a local coordinate system for calculating the object's shape factor using (x', y') coordinates. The maximum beam steering angle $\hat{\phi}$ is the border of the azimuthal measurement range.

3.2.3 Doppler spectrum

The Doppler spectrum (DS) is the histogram of an object's Doppler shift information, extracted from all resolution cells assigned to the object. It is a density function that tells how prominent different relative velocities are in the object. The relative radial velocity axis v is divided into small intervals. Resolution cells with radial relative velocity v_{rr} increase the counter for the corresponding intervals $[v_i, v_j[$ with $v_i \le |v_{rr}| < v_j$. To get a densitiy function D_{Doppler}, normalization is necessary such that $\int D_{\text{Doppler}} \mathrm{d}v = 1$ holds.

Figure 5 depicts sample Doppler spectra, averaged over several hundred independent data frames. The wavy

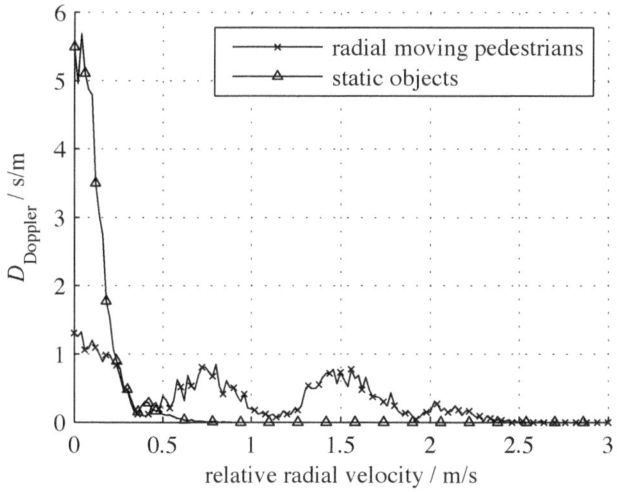

Fig. 5. Averaged Doppler spectra of radial moving pedestrians and static objects. The Doppler spectrum function D_{Doppler} is plotted over the relative radial velocity axis v. Ego-speed compensation is not necessary because the data was recorded with a non-moving radar sensor.

structure of the DS, especially of walking pedestrians, is not caused by movements of the object itself. The maxima of the wavy structure are always located at the theoretical frequency bins (if interpolation would not be used), and the local minima right inbetween, independent from the object's speed. Thus, the wavy structure is an artefact caused by the interpolation algorithm in the Doppler frequency measurement inside the radar sensor.

It can be clearly seen in Fig. 5 that a radial moving pedestrian has a broad DS from 0–2.5 $\mathrm{m\,s^{-1}}$. In contrast, DS of static objects are very narrow since they only cover velocities from 0–0.4 $\mathrm{m\,s^{-1}}$.

To boil down the shape of a DS to key figures, three intervals $[v_0, v_1], [v_1, v_2], [v_2, v_3]$ are considered (see Table 2 for actual values of v_0-v_2). The Doppler spectrum density function D_{Doppler} is integrated over these intervals, giving three object features m_3–m_5, representing the proportional distribution of radial relative velocities in the object over the three intervals.

$$m_3 = \int_{v_0}^{v_1} D_{\text{Doppler}} \mathrm{d}v \tag{3}$$

$$m_4 = \int_{v_1}^{v_2} D_{\text{Doppler}} \mathrm{d}v \tag{4}$$

$$m_5 = \int_{v_2}^{v_3} D_{\text{Doppler}} \mathrm{d}v \tag{5}$$

Radial velocities greater than 3 $\mathrm{m\,s^{-1}}$ are neglected since they are unlikely to be in pedestrians' DS and thus not the matter in this paper.

Table 2. Relative radial velocity interval borders for Doppler spectrum feature extraction.

v_0	v_1	v_2	v_3
$0\,\mathrm{m\,s^{-1}}$	$0.38\,\mathrm{m\,s^{-1}}$	$1.15\,\mathrm{m\,s^{-1}}$	$3\,\mathrm{m\,s^{-1}}$

3.3 Classification

The challenge in the classification step is to decide whether an object is a pedestrian or a static object by using the corresponding feature vector $\boldsymbol{m} = [m_1, \ldots, m_5]^{\mathrm{T}}$ only. The decision process is based on the knowledge that some values of these features are more probable if the object is of a certain class C. For example, a static object is very unlikely to have a significant fraction of Doppler information in the $[v_2, v_3]$ interval. Hence, a value of m_3 close to one is very unlikely for the class "Static Object".

Both valid classes, P for pedestrians and S for static objects, have a vector valued fuzzy membership function $\boldsymbol{f}_C : \mathbb{R}^5 \to [0,1]^5$, $C \in \{P; S\}$ that maps the feature vector $\boldsymbol{m}$ of an object to a membership value vector $\boldsymbol{u}_C = [\mu_{C,1}, \ldots, \mu_{C,5}]^{\mathrm{T}}$ for the corresponding class.

$$\boldsymbol{u}_C = \boldsymbol{f}_C(\boldsymbol{m}), \qquad \mu_{C,i} = f_{C,i}(m_i) \tag{6}$$

A value of $\mu_{C,i}$ close to one states a high probability of m_i for the corresponding class C while a small value of $\mu_{C,i}$ near zero indicates that the value of m_i is unlikely for that class. In this way, the membership values $\mu_{C,i}$ of the five object features of every object are calculated for both classes P and S. Graphs of the membership functions $f_{C,i}$ are plotted in Fig. 6. The functions $f_{S,1}$ and $f_{S,2}$ are not defined because static objects can be of any shape and size. Hence, all values of m_1 and m_2 have the same likelihood for the class S and thus their membership values for class S are suppressed in further calculations by giving them zero weight in the weighting step.

To decide whether class P or S is more appropriate for an object, given fuzzy membership vectors $\boldsymbol{u}_P$ and $\boldsymbol{u}_S$, the membership vectors are mapped to real membership values $\overline{\mu}_P, \overline{\mu}_S \in [0,1]$. This is done by a weighted average of the membership vectors' components

$$\overline{\mu}_C = \frac{\boldsymbol{w}_C^{\mathrm{T}} \cdot \boldsymbol{u}_C}{\sum_i w_{C,i}}, \tag{7}$$

where $\boldsymbol{w}_C = [w_{C,1}, \ldots, w_{C,5}]^{\mathrm{T}}$ is the weighting vector for a class $C \in \{P; S\}$.

Experiments have shown that constant weighting vectors $\boldsymbol{w}_C$ are not sufficient for proper classification. Instead, the combination of the membership values $\mu_{C,i}$ must be closer to a logical conjunction. This is achieved by allowing the weight vector $\boldsymbol{w}_C(\boldsymbol{u}_C)$ to be dependent on $\boldsymbol{u}_C$, meaning that

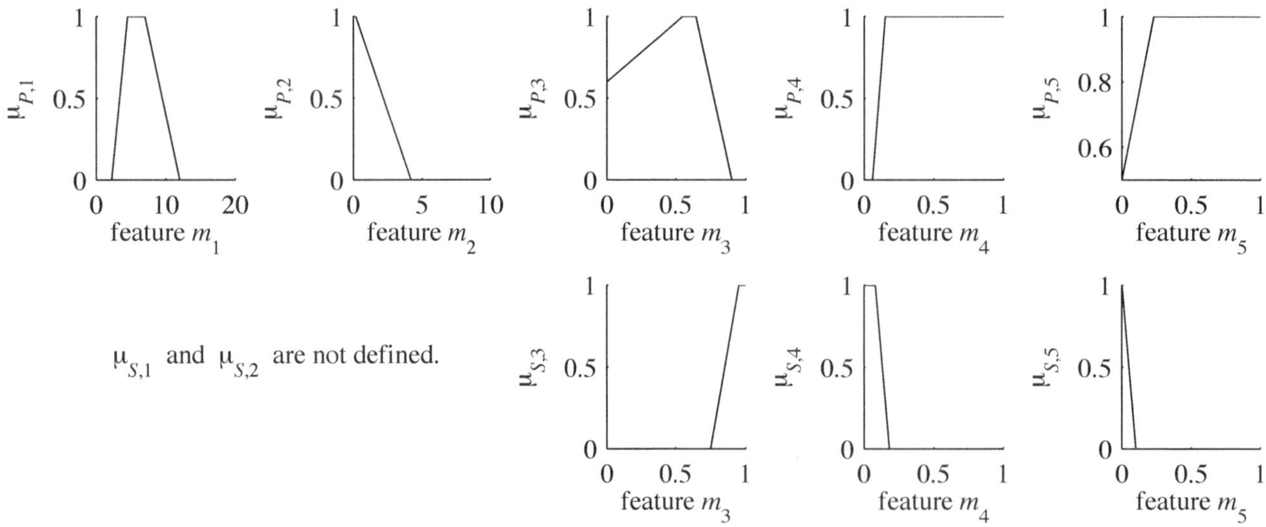

Fig. 6. Plots of membership functions $f_{P,i}$ and $f_{S,i}$ that map values of object features m_i to membership values $\mu_{P,i}$ and $\mu_{S,i}$ for the classes P for pedestrians and S for static objects.

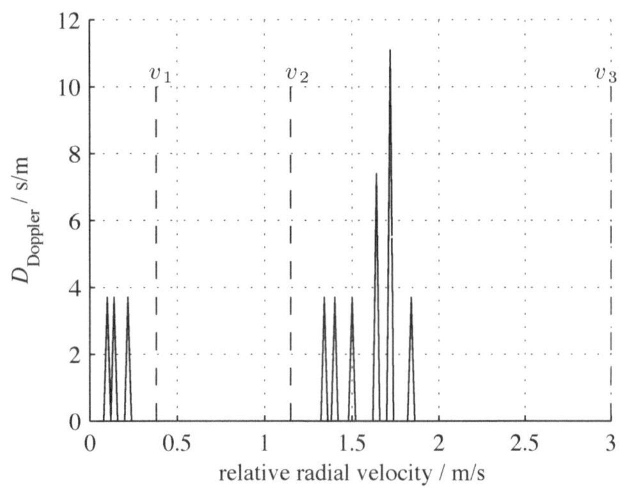

Fig. 7. Example of a single frame Doppler spectrum of a pedestrian.

low membership values $\mu_{C,i}$ are weighted stronger by increasing $w_{C,i}$ exponentionally:

$$w_{C,i} = \xi_{C,i} \cdot w_{C,i}^{0} \cdot \left(1 + \gamma \cdot \mathrm{e}^{-\frac{\mu_{C,i}}{\tau}}\right) . \tag{8}$$

The exponential function for increasing the base weight $w_{C,i}^{0}$ was chosen empirically. While γ and τ parametrize this value dependent weighting, $\xi_{C,i}$ is used for compensating a certain effect when a pedestrian has an unusual Doppler frequency distribution. In Fig. 7, a DS of a walking pedestrian is depicted, obtained from a single data frame. In this case the object does not contain a resolution cell with a relative radial velocity in the $[v_1, v_2]$ interval. Hence, the object feature m_4 will be zero and since there are some cells in the $[v_2, v_3]$ interval, $m_5 > 0$ will hold. Because m_4 equals zero, its weight $w_{P,4}$ will rise exponentially, and the overall probability $\overline{\mu}_P$ for this object to be a pedestrian will be low. This conclusion would be wrong because the characteristic Doppler information for a pedestrian is just concentrated in the $[v_2, v_3]$ interval rather than evenly distributed over the $[v_1, v_3]$ interval. In this case, where m_4 has a very low and m_5 a high value, $\xi_{P,4}$ is chosen small ($\xi_{P,4} \approx 1/\gamma$) to reverse the value dependent weighting. Analogously, $\xi_{P,5}$ is chosen if m_5 is very low in value while a significant share of Doppler information is in the $[v_1, v_2]$ interval. Since this cross-bonded value dependent weighting affects only $w_{P,4}$ and $w_{P,5}$, all $\xi_{C,i}$ except $\xi_{P,4}$ and $\xi_{P,5}$ have a constant value of one.

The decision whether an object is classified as a pedestrian or as a static object and, therefore, is assigned to the class P or class S, is made by comparing the united membership values $\overline{\mu}_P$ and $\overline{\mu}_S$. If $\overline{\mu}_P > \overline{\mu}_S$, the object is assigned to the class P and otherwise to class S. If both membership values $\overline{\mu}_P$ and $\overline{\mu}_S$ are small and hence the object is unlikely to be a pedestrian as well as a static object, it is declared as an unknown object. This happens almost exclusively in case of segmentation errors.

4 Experimental results

Because a binary classifier is used and the number of unknown declared objects is negligible, the quality of the pedestrian recognition system can be measured by only two key figures, r_{TP} and r_{FP}. The true positive rate r_{TP} tells what

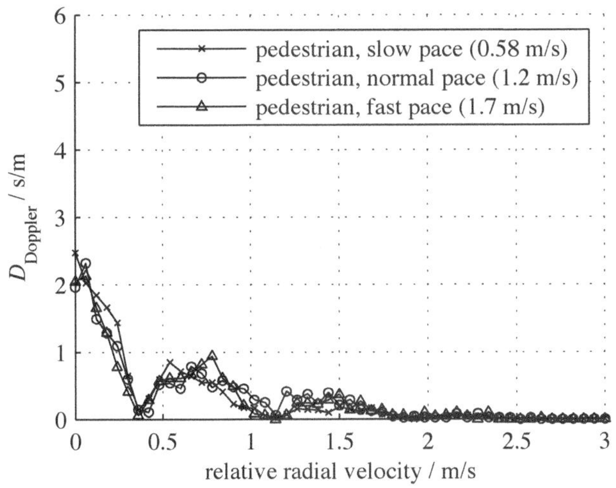

Fig. 8. Averaged Doppler spectra of a walking pedestrian for different walking paces. The angle of movement was $\varphi = 45°$ in all cases. In the investigated range of walking paces from $0.58\,\mathrm{m\,s^{-1}}$ to $1.7\,\mathrm{m\,s^{-1}}$, the DS is independent from the pedestrian's walking pace.

fraction of real pedestrians is correctly classified as such. On the other hand, the false positive rate r_{FP} or false alarm rate tells what fraction of static objects are classified incorrectly as pedestrians. The true negative rate $r_{TN} = 1 - r_{FP}$ and the false negative rate $r_{FN} = 1 - r_{TP}$ are implied under the above-mentioned assuptions.

The test sample set consisted in every experiment of about 150 to 600 samples if not mentioned otherwise.

4.1 Influences on the Doppler spectrum

A walking pedestrian's DS turned out to be very stable in terms of outside influences compared to other tested object features from the frequency image. It is not affected by the height in which the radar beam hits the person, due to different angles of elevation of the antenna beam as long as enough power is reflected for proper segmentation in the intensity image. Furthermore, the DS is fairly independent from the walking pedestrian's pace (see Fig. 8). Walking paces from $0.58\,\mathrm{m\,s^{-1}}$ to $1.7\,\mathrm{m\,s^{-1}}$ were tested and did not affect the pedestrian's DS because many different velocities within the segmented object occured due to arm and leg swing.

Nevertheless, the DS obviously is dependent on the pedestrian's angle of movement. A radar sensor can only measure the radial component v_{rr} of an object's relative velocity v_r. A good approximation is to assume the pedestrian centered in front of the sensor, i.e. $\phi = 0$ (see Fig. 9). By using this approximation, the radial velocity component can be simplified to

$$v_{rr} = v_r \cdot \cos\varphi \qquad (9)$$

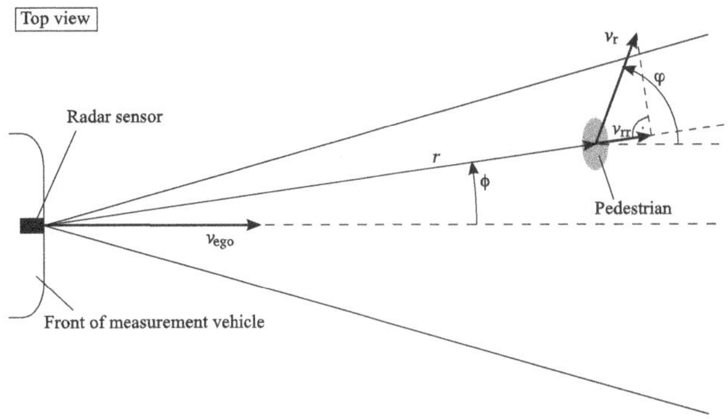

Fig. 9. A walking pedestrian with velocity v_r relative to the sensor with angle of movement φ has a radial relative velocity component v_{rr} that is measurable by the radar sensor.

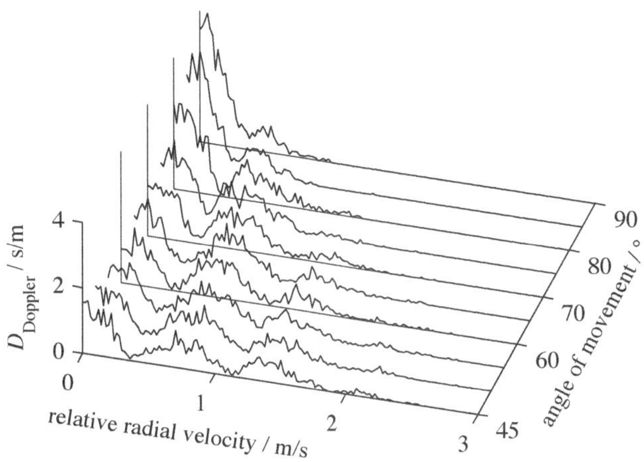

Fig. 10. Averaged Doppler spectra of a walking pedestrian over various angles of movement φ.

and it approaches zero for angles of movement near $\varphi = 90°$. Hence, lateral moving pedestrians have DS very similar to static objects. In Fig. 10, averaged DS of a walking pedestrian are plotted for different angles of movement φ. The closer φ approaches 90°, the DS shows increasing similarity to the one of static objects (compare to Fig. 5). Hence, the true positive rate r_{TP} after classification reaches 95.3 % for radially moving pedestrians ($\varphi = 0°$) and drops to 39.5 % for laterally moving pedestrians ($\varphi = 90°$). Figure 11 exhibits that the classification result is better than 88 % for $|\varphi| \leq 75°$.

Backtracking from classification decisions to radar raw data, to see whether the sensor data or the classification algorithm is responsible for misclassified data frames, was extensively used in empirically optimizing the classifier. As a result, almost all misclassifications of pedestrians as static

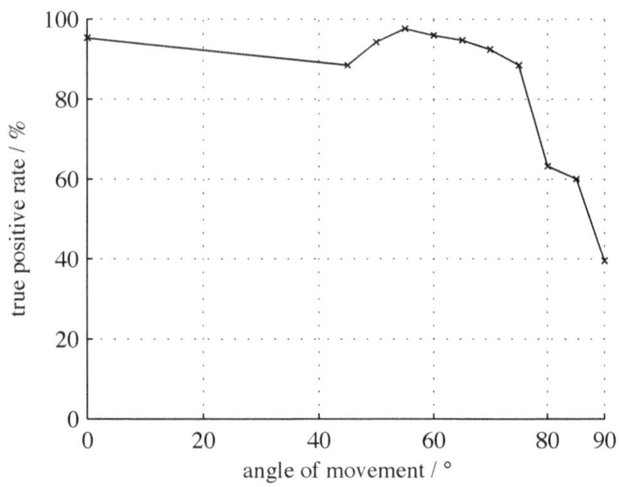

Fig. 11. Classification result r_{TP} for pedestrians walking with different angles of movement φ.

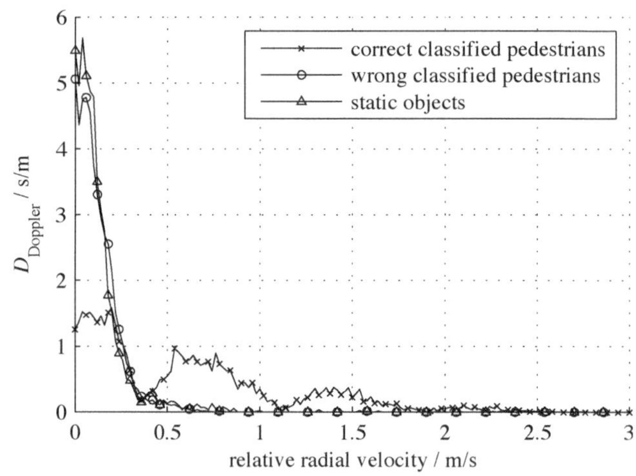

Fig. 12. Averaged Doppler spectra of correct classified and incorrect classified pedestrians walking with various angles of movement. The similarity of Doppler spectra of incorrect classified pedestrians and static objects is a limitation of the radar sensor in pedestrian recognition, since the sensor does not deliver statistical different in this case.

objects can now be tracked down to sensor data that is indistinguishable, even by hand, from a static object's raw data. In Fig. 12, the averaged DS of correct and misclassified pedestrians are plotted as well as the averaged DS of static objects for comparison. The DS of the misclassified pedestrians is almost the same as the one of static objects. Thus, in the case of misclassified pedestrians, the sensor does not deliver statistically different data from static objects and hence the developed classification algorithm is optimal within the restrictions of the chosen method for pedestrian recognition.

4.2 Occluded pedestrians

In this paper, two types of occlusions were investigated. In the first case, a pedestrian is hidden behind a single parked car and in the second case the pedestrian is in the gap between two parked cars (see Fig. 13). In both cases the pedestrian moves laterally while being occluded such that the pedestrian is not in the line of sight to the sensor. The lateral movement is typical for pedestrians who try to cross the street without looking.

In the case where the pedestrian is in the shadowed area hidden behind one parked vehicle at the side of the road, the radar waves cannot hit the pedestrian directly. Therefore, the signal reflected by the pedestrian and received by the sensor is very weak. As a result, the pedestrian blurs with the blocking car and often the segmentation algorithm is not able to seperate the two objects. In the classification step the true positive rate drops to $r_{\mathrm{TP}} = 6.6\%$, which is only slightly better than the false alarm rate of $r_{\mathrm{FP}} = 1.4\%$.

If the pedestrian is moving laterally in a gap between two parked cars at the side of the road (see Fig. 13), the radar wave can propagate in this gap because of multiple reflections. Thus, the pedestrian reflects more power back to the radar sensor and can be segmented properly in the radar raw data if the gap is not smaller than 50 cm. Also, the recognition result is much better, as the true poistive rate rises to values up to 29.4%. This result is remarkable because now the recognition rate is only ten percent less than $r_{\mathrm{TP}} = 39.5\%$ in the case of a lateral moving pedestrian in line of sight (see Fig. 11). The recognition results for different sizes l of the gap are shown in Table 3. Note that the false alarm rate r_{FP}

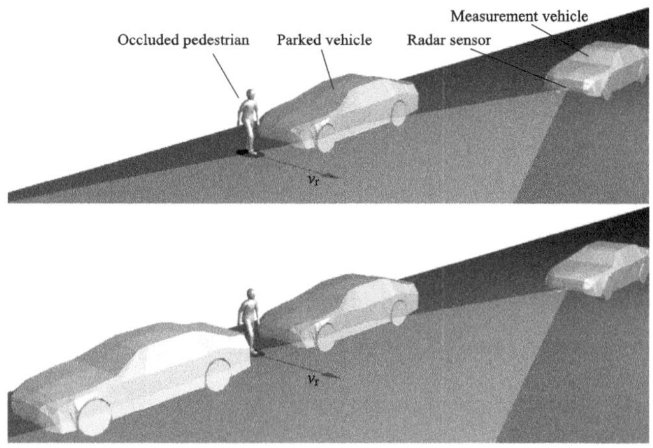

Fig. 13. Top: A lateral moving pedestrian is occluded behind a single parked car. Bottom: Now the lateral moving pedestrian is in a gap between two parked cars, where the radar wave can propagate due to multiple reflections.

Table 3. Classification result for a lateral moving pedestrian in a gap between two parked cars.

Size of gap l	30 cm	50 cm	70 cm	90 cm	110 cm
r_{TP}	20.9 %	29.4 %	28.5 %	20.1 %	25.2 %
r_{FP}	17.6 %	9.5 %	2.4 %	5.4 %	1.4 %

Fig. 14. A real urban traffic scene used for measuring the false alarm rate. On the left a photography of the scene is depicted and on the right the corresponding intensity image from the radar raw data.

is significantly higher in case of narrow gaps up to 50 cm due to segmentation problems in the blurred occluded areas.

4.3 False alarms

False alarms, meaning that static objects are incorrectly classified as pedestrians, are consequences of three root causes. Firstly, parasitic antenna side lobe effects can cause significant fractions in the DS of static objects that seem to have a higher velocity than static objects should have. Secondly, errors in the ego velocity compensation can offset the DS and finally, in case of errors in the segmentation step, resolution cells from the noisy surroundings of an object can get assigned to the object, resulting in DS fractions with higher velocity than possible for static objects. To see how good the specifity of the pedestrian recognition system is in practice, a real urban traffic scene was chosen (see Fig. 14). The vehicle drives along a road with lots of highly reflective static objects, such as cars, in the measurement range, which is the worst case scenario in terms of false alarms, since highly reflective objects cause high amplitudes of antenna side lobe effects. As a result, in the described traffic scenario, a false alarm rate of $r_{FP} = 4.9\%$ was obtained.

4.4 Driver assistance system scenario

A practical use of the introduced pedestrian recognition system would be collision avoidance by warning the driver of a pedestrian which moves towards the road, occluded in a gap in a row of roadside parked cars. This scenario consists of parked cars as highly reflective static objects, lateral movement of the pedestrian, occlusion and linear movement of the measurement vehicle with about $8.5\,\mathrm{m\,s^{-1}}$. With the combination of all those influences in one real scenario, a recognition rate of $r_{TP} = 19.3\%$ was achieved. The false alarm rate in this particular measurement was $r_{FP} = 3.2\%$.

5 Discussion

The gathered results reveal three main development areas for future radar sensors used for pedestrian recognition: Doppler frequency resolution, antenna side lobes and spatial resolution.

Doppler frequency resolution: An object at the position (r, ϕ) with relative velocity v_r to the radar sensor (see Fig. 9) has a relative radial velocity component given by

$$v_{rr} = v_r \cdot \cos(\varphi - \phi)\,. \qquad (10)$$

Equation 9 is the approximation of this equation assuming $\phi = 0$. If the object moves on a straight lateral path ($\varphi = 90°$), it still causes a small Doppler shift given by $v_{rr} = v_r \cdot \sin\phi$ with absolute maxima at the azimuthal borders of the measurement range $\pm\hat{\phi}$. The described experimental results for recognizing lateral walking pedestrians (see Sect. 4.1) are based on a walking pace of about $v_r = 1.2\,\mathrm{m\,s^{-1}}$. The theoretical maximum of the radial relative velocity component in this case is $v_{rr} = 0.17\,\mathrm{m\,s^{-1}}$ at $|\phi| = \hat{\phi}$, which is only about a fifth of the smallest measurable radial velocity of $0.77\,\mathrm{m\,s^{-1}}$, given by the radar sensor's Doppler frequency measurement resolution. Nevertheless, in the experiment the recognition rate of $r_{TP} = 39.5\%$ is much higher than the expected rate of about $r_{TP} \approx 4\%$, caused by false alarms, even if the pedestrian is central in front of the sensor ($\phi = 0$) where Eq. (10) gives $v_{rr} = 0$. This leads to the conclusion that micro Doppler effects (see (He, 2010)), e.g. swinging arms and legs, cause artefacts in the DS, which lead to these classification results.

For a particular driver assistance system, a use case could be to require direct measurement (without the help of micro Doppler effects) of a lateral walking pedestrian's radial velocity component v_{rr} in a range of $|\phi| \geq 3°$, given a walking pace of $v_r = 1.0\,\mathrm{m\,s^{-1}}$. This requirement gives a blind spot of 5.2 m width centered in front of the vehicle in 50 m distance. This blind spot could be observed by other sensor systems since it is usually directly visible. To fulfill these requirements, a radar sensor with v_{rr}-measurement resolution of at least $0.052\,\mathrm{m\,s^{-1}}$ or, respectively, Doppler frequency measurement resolution of 26.5 Hz is needed. The required measurement resolution is about fifteen times higher than the resolution of the radar sensor employed in this paper.

Antenna side lobe effects: Parasitic effects because of antenna side lobes affect the intensity image (ghost targets appear) as well as the frequency image (see Sect. 4.3), especially in the surrounding of highly reflective objects. The impact of antenna side lobes on the intensity image was modeled using a measured antenna pattern. Although in theory compensation of this effect would be possible, further improvements of todays radar sensors concerning side lobes are needed. The knowledge of these effects is the key to optimize the segmentation algorithm.

Spatial resolution: If a pedestrian is in line of sight and not more than about 35 m distant from the sensor, the radar sensor's spatial resolution is sufficient. However, if the pedestrian is occluded behind other objects, the corresponding radar data starts to blur with nearby objects. A finer spatial resolution might improve the separability of the objects in this case. If the pedestrian is further away from the sensor, the spatial resolution cells grow, since their width is a function of the azimuth angle. Hence, a distant pedestrian covers only few resolution cells, resulting in a DS of just few single peaks, which might not represent the DS well. A finer spatial resolution would improve the accuracy of the DS in this matter.

6 Summary and conclusions

Because pedestrians are point targets for radar sensors, object features from the intensity image cannot be used to distinguish between pedestrians and small static objects. Hence, information from the Doppler frequency image is needed. The DS, which is the histogram of the measured radial relative velocities of an object, turned out to be a stable object feature, holding the required information to reliably seperate pedestrians from static objects. A simple classification algorithm, based on five object features, was found with classification results compareable to modern machine learning algorithms. Radial moving pedestrians in line of sight can be recognized up to 95.3 % of the samples. The recognition rate drops to 39.5 % for lateral moving pedestrians and to 29.4 % for lateral moving pedestrians occluded in a gap between two parked cars. In case of misclassifications of pedestrians as static objects, the radar sensor does not deliver statistically different data as from static objects, making it impossible for classification systems (machine learning as well as the proposed solution) to decide for the correct class using only information from single radar raw data frames. Hence, pedestrian recognition using only single radar raw data frames shows poor classification results.

The radar sensor's main limitations in this use case turned out to be insufficient Doppler shift measurement resolution and antenna side lobe effects. Possibly, a finer spatial resolution would also improve the sensor's performance in recognizing occluded pedestrians. Our findings indicate that pedestrian recognition using this low-level approach is limited. Only radar data with higher Doppler frequency measurement resolution (independent of the used technology, e.g. 24 GHz, 77 GHz) would make major recognition improvements possible. This requirement of future automotive radar sensors is in line with claims in (Rasshofer, 2007). In state-of-the-art research on pedestrian recognition, today's radar sensors are mostly used to support camera-based systems (Gerónimo, 2010).

References

Benitez, D. and Zhaozhang, J.: Method for Human Only Activity Detection Based on Radar Signals, U.S. Patent Application US2011/0032139A1, 2011.

Büchele, M.: Optimierte Radarsignalverarbeitung für Fahrerassistenzsysteme, M. Sc. thesis, University of Ulm, Institute of Theoretical Computer Science, Ulm, Germany, 2008.

Freund, S.: Object Recognition from Radar Raw Data Using Image Processing Methods, Diploma thesis, Ludwigs-Maximilians-Universität Munich, Institute of Computer Science, Munich, Germany, 2007.

Gerónimo, D., López, A. M., Sappa, A. D., and Graf, T.: Survey of Pedestrian Detection for Advanced Driver Assistance Systems, IEEE Transactions on Pattern Analysis and Machine Intelligence, 32, pp.1239–1258, 2010.

He, F., Huang, X., Liu, C., Zhou, Z., and Fan, C.: Modeling and Simulation Study on Radar Doppler Signatures Of Pedestrian, in: Proc. of the 2010 IEEE Radar Conference, Washington DC, USA, 10–14 May 2010, 1322–1326, 2010.

Hornsteiner, C. and Detlefsen, J.: Characterisation of human gait using a continuous-wave radar at 24 GHz, Adv. Radio Sci., 6, 67–70, doi:10.5194/ars-6-67-2008, 2008.

Kim, Y. and Ling, H.: Human Activity Classification Based on Micro-Doppler Signatures Using an Artificial Neural Network, Proc. of the 2008 IEEE Symposium on Antennas and Propagation, San Diego, USA, 5–11 July 2008, 1–4, 2008.

Kouemou, G. and Opitz, F.: Impact of Wavelet Based Signal Processing Methods in Radar Classification Systems Using Hidden Markov Models, in: Proc. of the 2008 International Radar Symposium, Wroclaw, Poland, 21–23 May 2008, 1–4, 2008.

Nanzer, J. A. and Rogers, R. L.: Bayesian Classification of Humans and Vehicles Using Micro-Doppler Signals From a Scanning-Beam Radar, IEEE Microwave and Wireless Components Letters, 19, 338–340, 2009.

Rasshofer, R. H.: Functional Requirements of Future Automotive Radar Systems, in: Proc. of the European Radar conference, Munich, Germany, 10–12 October 2007, 259-262, 2007.

Ritter, H. and Rohling, H.: Pedestrian Detection Based on Automotive Radar, 2007 IET International conference on Radar Systems, Edinburgh, UK, 15–18 October 2007, 1–4, 2007.

Rohling, H., Heuel, S., and Ritter, H.: Pedestrian Detection Procedure Integrated into an 24 GHz Autmotive Radar, in: Proc. of the 2010 IEEE Radar Conference, Washington DC, USA, 10–14 May 2010, 1229–1232, 2010.

Wenger, J. and Hahn, S.: Long Range and Ultra-Wideband Short Range Automotive Radar, in: Proc. of the 2007 IEEE International Conference on Ultra-Wideband, Singapore, 24–26 September, 2007, 518–522, 2007.

Yamada, N., Tanaka, Y., and Nishikawa, K.: Radar Cross Section for Pedestrian in 76 GHz Band, in: Proc. of the 35th European Microwave Conference, Paris, France, 3–7 October 2005, 1–4, 2005.

7

Design and analysis of an isotropic two-dimensional planar Composite Right/Left-Handed waveguide structure

M. A. Eberspächer, M. Bauer, and T. F. Eibert

Lehrstuhl für Hochfrequenztechnik, Technische Universität München, Germany

Abstract. A two-dimensional isotropic Composite Right/Left-Handed (CRLH) waveguide structure is proposed which is designed for operation in X-band. The balanced structure possesses left-handed behaviour over a large bandwidth from 7.5 GHz up to its transition frequency at 10 GHz. Above this region, the unit cell behaves in a right-handed manner up to 13.5 GHz. Operating the structure within these bands yields a frequency dependent index of refraction ranging from $-2.5 \leq n \leq 0.8$. Isotropic characteristics are obtained between $8.5\text{GHz} \leq f \leq 12\text{GHz}$ resulting in $-1.5 \leq n \leq 0.8$. The planar CRLH structure is designed based on transmission line theory, implemented in microstrip technology and optimized using full-wave simulation software. An equivalent circuit model is determined describing the electromagnetic behaviour of the structure whose element values are obtained by even and odd mode analysis. The design of the unit cell requires an appropriate de-embedding process in order to enable an analysis in terms of dispersion characteristics and Bloch impedance, which are performed both.

1 Introduction

A structure possessing at least one frequency band in which the vectors $\boldsymbol{E}$, $\boldsymbol{H}$ and $\boldsymbol{k_g}$ form a left-handed triad while a right-handed one is found in the remaining frequency range is known as a Composite Right/Left-Handed structure. Operation of the guiding structure in the left-handed region results in wave propagation with negative phase velocity. In contrast, the group velocity associated with the flow of energy is positive in all regions. Since materials possessing such a behaviour have not been found in nature yet, they have to be produced artificially. One of the first structures exhibiting CRLH behaviour was realised by split-ring resonators located in the proximity of a resonant wire arrangement (Shelby et al., 2001). Due to the loose coupling of the resonators, this configuration suffers from high losses and only a small operative bandwidth of approximately 10% is achieved. It was shown that these problems can be overcome by synthesizing CRLH structures utilizing the transmission line approach (Lay et al., 2004) which is based on periodically arranged unit cells emulating a homogeneous structure. If mutual coupling between the unit cells is not considered, the waveguiding characteristics of the entire cascade is solely governed by the characteristics of the constitutive unit cells. As a consequence, the most challenging part in the artificial material synthesis is the design of appropriate unit cells which fulfil predefined requirements. Besides meeting the electrical characteristics, the structures must be feasible with an acceptable complexity. It has been shown that cascading planar CRLH unit cells in a specific manner enables alternative concepts for a multitude of applications like power dividers, beam forming concepts and antennas (Eccleston, 2010), (Caloz and Itoh, 2006).

Correspondence to: M. Eberspächer
(mark.eberspaecher@tum.de)

2 Theory

2.1 Synthesis

In order to realise a two-dimensional waveguiding structure which behaves within a certain bandwidth isotropically, the unit cell is supposed to be a fully symmetric and reciprocal four-port with a pair of ports for each direction of space. Using scattering matrix representation, this can be written as

$$\mathbf{S} = \frac{1}{2}\mathrm{e}^{-\mathrm{j}\phi}\begin{bmatrix} -1 & 1 & 1 & 1 \\ 1 & -1 & 1 & 1 \\ 1 & 1 & -1 & 1 \\ 1 & 1 & 1 & -1 \end{bmatrix} \tag{1}$$

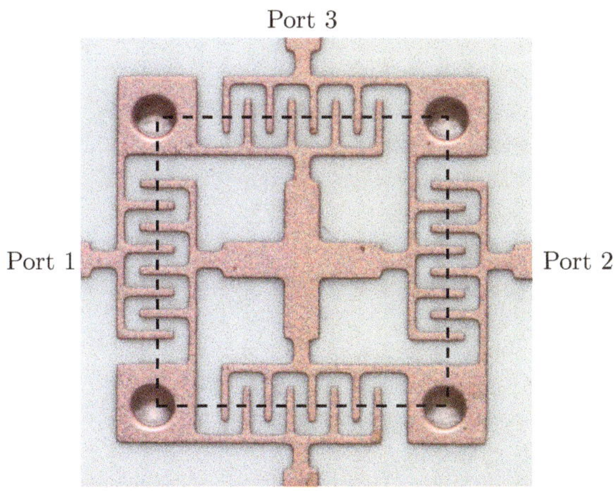

Fig. 1. Picture of the fabricated structure which is produced on Rogers RO4350B substrate (thickness $h = 0.762\text{mm}$ and $\epsilon_r = 3.66$). The 2d CRLH unit cell is within the square marked by the dashed line (side length $l = 4.54\text{mm} \approx \lambda/4$). Parts located outside the square belong to either adjacent cells or the feeding network.

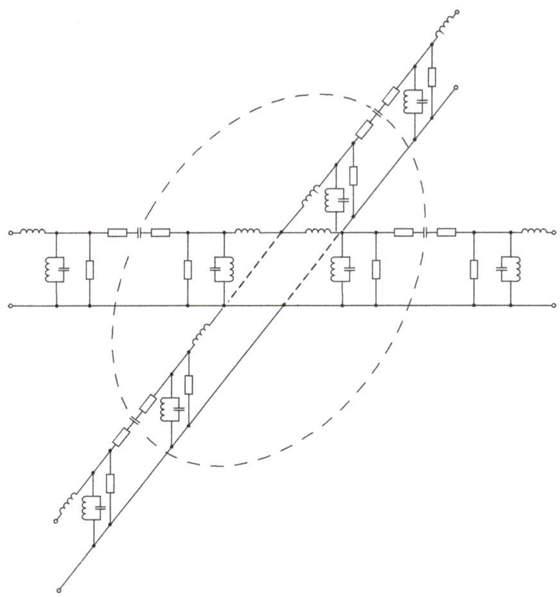

Fig. 2. Equivalent circuit model of the structure seen in figure 1 with the series elements C_L, $L_R/2$ and $R_S/2$ and the shunt elements $C_R/2$, $2L_L$ and $2R_P$. The equivalent circuit model of the 2d CRLH unit cell is within the region marked by the dashed line. Elements located outside the marked region belong to either adjacent cells or the feeding network.

whereas the phase variation experienced by a wave travelling through the cell is described by the exponential function $e^{-j\phi}$. For the synthesis process the impedance matrix representation of $\mathbf{S}$ is of further interest, which can be determined by

$$\mathbf{Z}_S = Z_0(\mathbf{I}+\mathbf{S})(\mathbf{I}-\mathbf{S})^{-1} \tag{2}$$

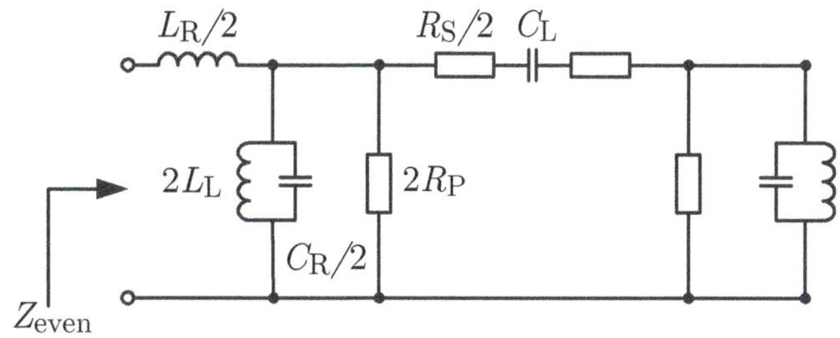

Fig. 3. Equivalent circuit model of the impedance Z_{even} obtained by even mode excitation of the unit cell with the feeding network.

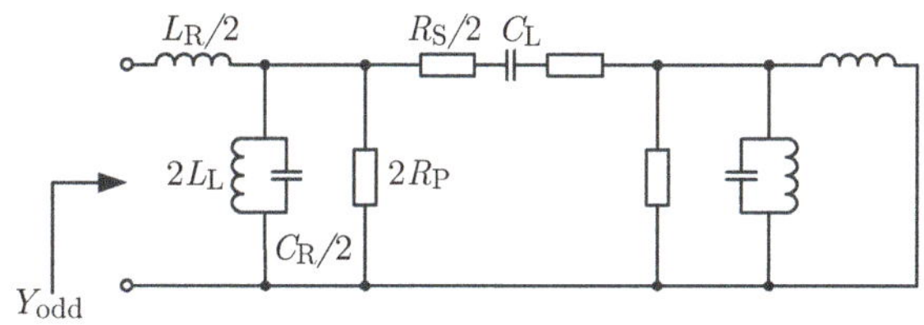

Fig. 4. Equivalent circuit model of the admittance Y_{odd} obtained by odd mode excitation of the unit cell with the feeding network.

with $\mathbf{I}$ being the identity matrix and Z_0 the reference impedance. Since the four-port network which is to be realised is assumed lossless its impedance matrix is purely imaginary. Thus, taking the imaginary part of $\mathbf{Z}_S$ results in

$$\Im\{\mathbf{Z}_S\} = -jZ_0 \begin{bmatrix} \frac{2\cos\phi-1}{2\sin\phi} & \frac{1}{2\sin\phi} & \frac{1}{2\sin\phi} & \frac{1}{2\sin\phi} \\ \frac{1}{2\sin\phi} & \frac{2\cos\phi-1}{2\sin\phi} & \frac{1}{2\sin\phi} & \frac{1}{2\sin\phi} \\ \frac{1}{2\sin\phi} & \frac{1}{2\sin\phi} & \frac{2\cos\phi-1}{2\sin\phi} & \frac{1}{2\sin\phi} \\ \frac{1}{2\sin\phi} & \frac{1}{2\sin\phi} & \frac{1}{2\sin\phi} & \frac{2\cos\phi-1}{2\sin\phi} \end{bmatrix}. \tag{3}$$

Now, the goal is given by finding a network comprising lumped elements which emulates $\mathbf{Z}_S$, at least within a certain bandwidth. In the simplest case this could be achieved by extending a T-shaped network with the series impedance Z and the shunt admittance Y to a four-port structure. The resulting network is given by

$$\mathbf{Z}_T = \frac{1}{2}Z\mathbf{I} + \frac{1}{Y}\mathbf{1} \tag{4}$$

with $\mathbf{1}$ being a matrix with all elements equal to 1. Consequently, the elements of $\mathbf{Z}_T$ may be determined and yield

$$Z = j2\tan\frac{\phi}{2} \tag{5}$$

and

$$Y = j\frac{2}{Z_0}\sin\phi. \tag{6}$$

Since we want the synthesized structure to appear as a homogenous one, only small phase angles ϕ are considered.

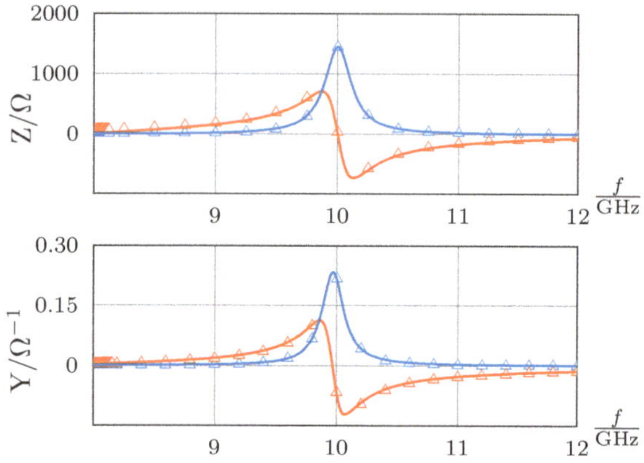

Fig. 5. Comparison between even impedance and odd admittance obtained from EM simulation (solid lines) and circuit simulation (triangles), respectively. Imaginary parts are displayed in red colour, real parts are drawn in blue colour.

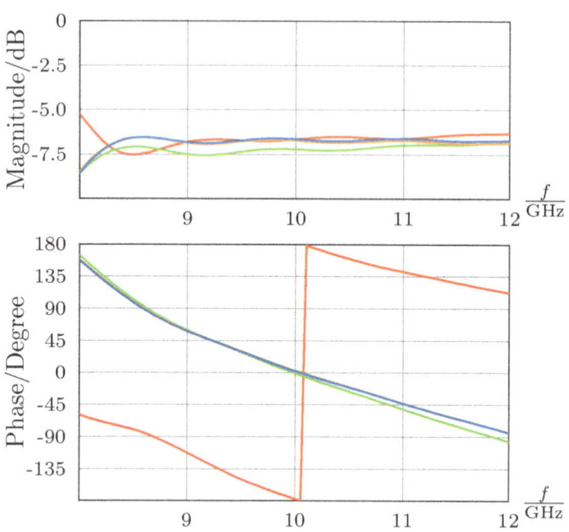

Fig. 6. Simulated scattering parameters of the structure seen in figure 1: S_{11} red line, S_{21} green line, S_{31} orange line, S_{41} blue line.

Hence, a first order Taylor series expansion in ϕ may be performed around $\phi = 0$ resulting in

$$Z \approx \mathrm{j} Z_0 \phi \tag{7}$$

and

$$Y \approx \mathrm{j}\frac{2}{Z_0}\phi. \tag{8}$$

Composite Right/Left-Handed behaviour is achieved if at least one frequency band with $\partial_\omega \phi(\omega) < 0$ and another with $\partial_\omega \phi(\omega) > 0$ exists. This can be accomplished by choosing appropriate network elements for Y and Z, respectively. For example, a parallel resonant circuit is inserted for Y and a series resonant circuit for Z.

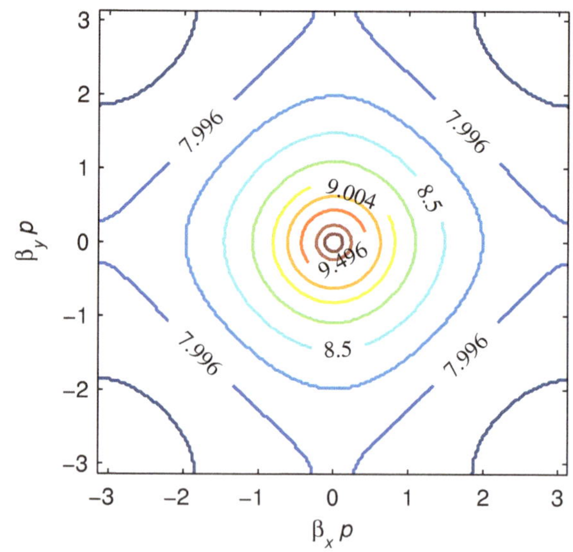

Fig. 7. Isofrequency contour plot (in GHz) of the left-handed mode with respect to two-dimensional wave propagation in the x and y directions.

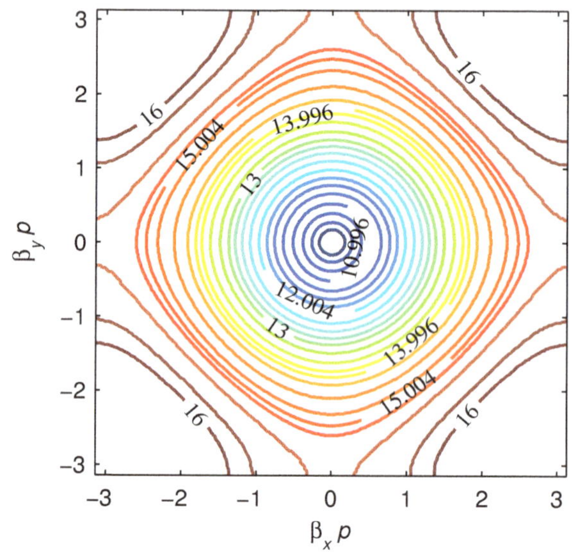

Fig. 8. Isofrequency contour plot (in GHz) of the right-handed mode with respect to two-dimensional wave propagation in the x and y directions.

2.2 Dispersion characteristics

As shown in Eqs. (3) and (4) the impedance matrix of the ideal unit cell comprises only two different types of elements. These are on the one hand the input impedances $Z_{\mathrm{I}} = Z_{nn}$ for $n \in 1,2,3,4$ and on the other hand all the remaining entries denoted by Z_{T}. However, we want to take possible anisotropic behaviour into account for the calculation of the dispersion characteristics. Thus, a third element is introduced which allows to distinguish between the impedance

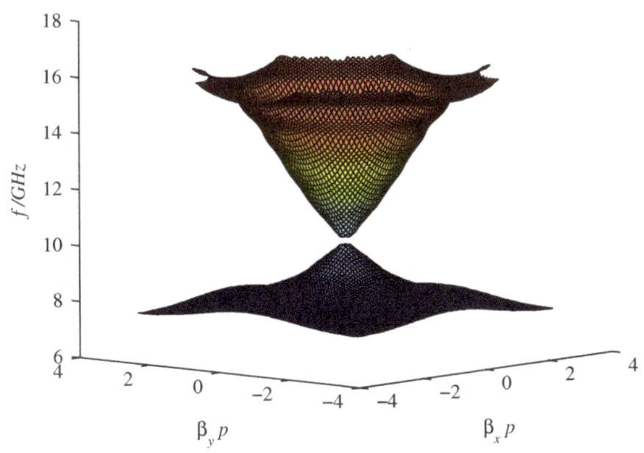

Fig. 9. Surface plot of the two-dimensional wave vector $\boldsymbol{\beta}$ exhibited by the CRLH structure.

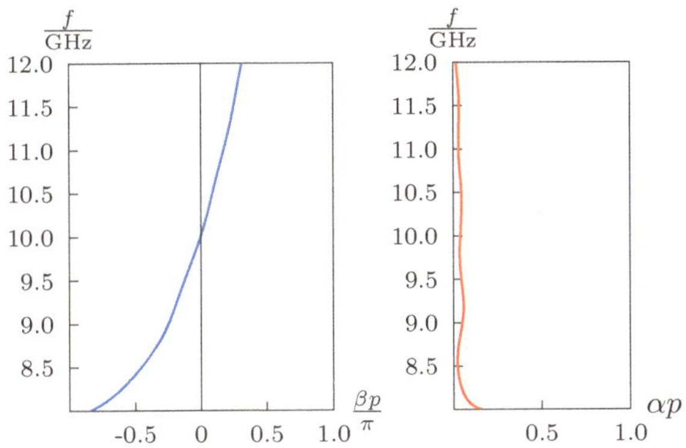

Fig. 10. Simulated wave propagation characteristics. Dispersion diagram for the region $\Gamma - X$ (left) and the corresponding attenuation constant (right).

of two ports which are arranged oppositely and perpendicularly, respectively. Consequently, the impedance matrix for such a four-port network reads

$$\mathbf{Z}_{\mathrm{r}} = \begin{bmatrix} Z_{\mathrm{I}} & Z_{\mathrm{T}} & Z_{\mathrm{X}} & Z_{\mathrm{X}} \\ Z_{\mathrm{T}} & Z_{\mathrm{I}} & Z_{\mathrm{X}} & Z_{\mathrm{X}} \\ Z_{\mathrm{X}} & Z_{\mathrm{X}} & Z_{\mathrm{I}} & Z_{\mathrm{T}} \\ Z_{\mathrm{X}} & Z_{\mathrm{X}} & Z_{\mathrm{T}} & Z_{\mathrm{I}} \end{bmatrix}. \tag{9}$$

It is assumed here that the two ports $1,2$ and $3,4$ are arranged along one Cartesian coordinate axis. Performing the Floquet ansatz based on $\mathbf{Z}_{\mathrm{r}}$ yields the dispersion characteristics

$$\begin{aligned}&[\cosh(\gamma_y p + \gamma_x p) + \cosh(\gamma_y p - \gamma_x p)](Z_{\mathrm{T}}^2 - Z_{\mathrm{X}}^2)\\ &\quad -[\cosh(\gamma_x p) + \cosh(\gamma_y p)](2Z_{\mathrm{I}}Z_{\mathrm{T}} - 2Z_{\mathrm{X}}^2)\\ &\quad\quad -2Z_{\mathrm{X}}^2 + 2Z_{\mathrm{I}}^2 = 0.\end{aligned} \tag{10}$$

For wave propagation along a principle axis, i.e. the region $\Gamma - X$ in the dispersion diagram, this expression simplifies to

$$\cosh(\gamma_x p) = \cosh(\gamma_y p) = Z_{\mathrm{I}}/Z_{\mathrm{T}}. \tag{11}$$

In this case all nodal points of the periodic arrangement being on a line perpendicular to the propagation direction are

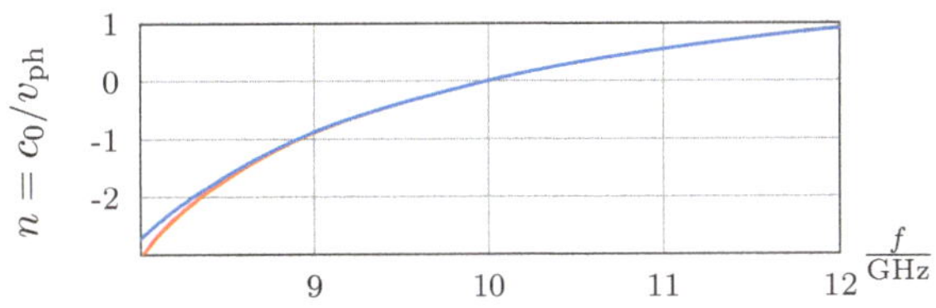

Fig. 11. Index of refraction obtained by EM simulations for the regions $\Gamma - X$ (red line) and $M - \Gamma$ (blue line).

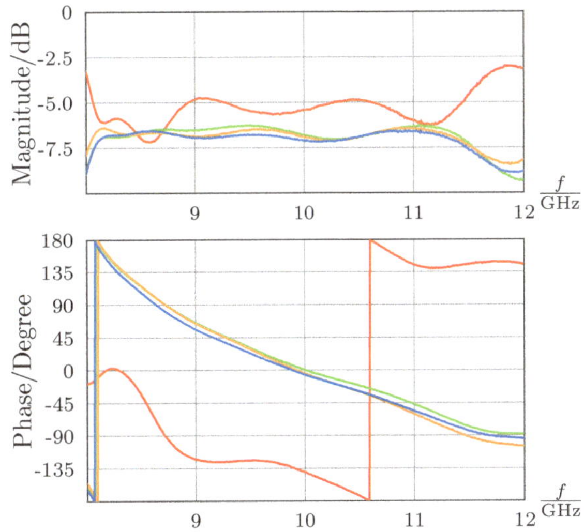

Fig. 12. Measured scattering parameters of the fabricated structure seen in figure 1: S_{11} red line, S_{21} green line, S_{31} orange line, S_{41} blue line.

equipotential points. Thus, no current is flowing across nodal points in transverse direction and the four-port matrix can be reduced to a two-port impedance matrix given by the upper left submatrix of $\mathbf{Z}_{\mathrm{r}}$ comprising the elements Z_{I} and Z_{T}. Consequently, the Bloch impedance reads

$$Z_{\mathrm{Bloch}_x} = \sqrt{{Z_{\mathrm{I}}}^2 - {Z_{\mathrm{T}}}^2}. \tag{12}$$

3 Design

The goal is to realise an isotropic two-dimensional CRLH unit cell which exhibits a left-handed mode for $f < 10\,\mathrm{GHz}$ and a right-handed mode for $f > 10\,\mathrm{GHz}$ fulfilling Eq. (3) around the transition frequency of $f_{\mathrm{T}} = 10\,\mathrm{GHz}$. Due to the given operation frequencies the use of discrete elements, for example surface mounted devices, may lead to difficulties. Hence, the design of the 2d unit cell is based on the 1d structure presented in (Eberspächer et al., 2009) which is fabricated in an entirely printed circuit technology based on transmission line theory. The original design can be considered as an interdigital capacitor whose arms supporting the individual fingers are extended to open-ended stubs in order to be-

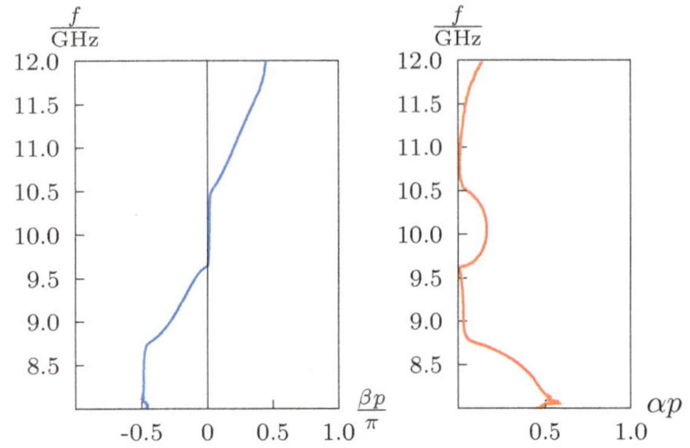

Fig. 13. Measured wave propagation characteristics. Dispersion diagram for the region $\Gamma - X$ (left) and the corresponding attenuation constant (right).

have inductively. Thus, this configuration realises a shunt inductor and a series capacitor simultaneously. Both elements are completed by the occurring parasitics to a series resonant circuit and a parallel resonant circuit, respectively. Adjusting the resonant frequencies as well as the Bloch impedance can be achieved easily by modifying either the stublength or the fingers of the arrangement allowing to meet given specifications and balancing the structure. Ensuring the artificial structure to behave homogeneously, the length of the unit cell must not exceed the quarter wave length limit (Lay et al., 2004). Dealing with 2d structures, this requirement must be fulfilled in both directions. Along with the desired isotropic characteristic this results in a quadratic footprint of the unit cell possessing a sidelength which is smaller or equal to a quarter of the wavelength. Although the absence of interlayer connections makes the initial design easy to fabricate and hence preferable, it cannot be extended to a two-dimensional unit cell in a straightforward manner, since the open-ended stubs cannot be reduced to fit into the footprint without losing its inductive behaviour. Therefore, the open-ended stubs are substituted by short-circuited ones which require to be reduced in length to maintain their electromagnetic properties. This modification influences strongly the field distribution within the interdigital capacitor. Whereas in the original case the maximum of capacitance was realised close to the open ends, in the modified version this appears close to the hostline. In order to reduce the amount of interlayer connections, these are placed at the corners of the unit cell so that they may be used by adjacent cells simultaneously, as seen in Fig. 1.

3.1 Analysis

As seen in Figs. 1 and 2, the boundary of the unit cell is located right between the interdigital capacitor.

In terms of an equivalent circuit model this can be considered as two capacitors connected in series representing together the capacitance of the interdigital capacitor separated by the cell boundary. However, excitation of the structure may only be performed with a quasi TEM mode bound to the microstrip lines. Consequently, the reference plane for determining the scattering parameters cannot be set correctly. As a result of this, the determinable scattering parameters always contain contributions of adjacent cells, in case of a periodically arranged cascade, or contributions of the feeding network, in case of analysing a single section. Thus, to obtain the properties of the unit cell itself the reference plane for the scattering matrix analysis must be placed between two inseparable capacitors which is not possible in practice. This difficulty can be overcome by modelling the unwanted contributions with an equivalent circuit model which is used to de-embed the unit cell parameters. For this, the values of the elements $L_R/2$, $2L_L$, $C_R/2$, $2R_P$, $R_S/2$ and $2C_L$ must be determined based on the simulated or measured network response which can be done by performing even and odd analysis.

Starting with the impedance matrix representation of the four-port network referred to as extended unit cell $\mathbf{Z}_{euc}$, which comprises the unit cell and the feeding network, allows to calculate

$$Z_{even} = \mathbf{Z}_{euc}(1,1) + \mathbf{Z}_{euc}(1,2) + 2\mathbf{Z}_{euc}(1,3) \quad (13)$$

and

$$Y_{odd} = 1/[\mathbf{Z}_{euc}(1,1) - \mathbf{Z}_{euc}(1,2)]. \quad (14)$$

It is assumed here that $\mathbf{Z}_{euc}$ can be described by a matrix possessing the same characteristics as the matrix shown in Eg. (9). Z_{even} is the impedance which appears at the terminals if the network is excited simultaneously at all ports with an identical signal. In this case no current is flowing across the nodal point and the equivalent circuit model simplifies to the one seen in Fig. 3. Likewise, exciting the structure at each two ports with signals possessing a phase shift of 180° leads to the odd admittance whose equivalent circuit model is seen in Fig. 4. In order to determine the values of the containing elements both equivalent networks are simulated with the circuit simulation software AWR Microwave Office (AWR). The results obtained from the EM structure are then compared with the results obtained by circuit analyses and in an iterative process the elements are modified until both responses match. This fitting sequence can either be performed with an automatic optimizer or by tuning the element values manually by performing the steps as follows: Starting with the observation of Z_{even} shows that at the resonant frequency of the shunt resonator the impedance of Z_{even} is expected to peak since the susceptances of the resonator cancel each other out.

Consequently, the product of L_L and C_R is immediately known. Due to the fact that the unit cell realises a band pass structure with only weak losses, the parallel resistance R_P must be highly resistive whereas its series counterpart R_S must be of low resistance. Hence, we can conclude that $R_P >> R_S$ and for this reason $\Re\{Z_{\text{even}}\} \approx R_P$ at the parallel resonant frequency. Now the quality factor of the shunt resonator can be determined which allows a first estimation of L_L and C_R. As seen in the equivalent circuit, depicted in Fig. 3, the series resonance frequency occurs at $\omega_{\text{rs}} = \sqrt{2}/\sqrt{L_R C_L}$ which is, if a balanced structure is assumed, far above the parallel resonant frequency. Thus, the series resonant circuit behaves capacitively, depending on the elements C_L and $L_R/2$, since it is operated below its resonance frequency. By modifying these values, the imaginary part of Z_{even} of the equivalent circuit model may be adjusted until it fits to the desired one.

After these first steps we can evaluate the odd admittance Y_{odd} at the shunt resonance frequency. Obviously, the parallel resonator appears as an open-circuit and the remaining network is almost the series resonator, as it can be seen in Fig. 4. Again, utilizing the quality factor and the resonance frequency allows to estimate the remaining element values. Since both adjustment steps are not decoupled the approximation of the element values can require some further iterations. Figure 5 shows the comparison of the even impedance and the odd admittance obtained by EM simulations and circuit simulations after the element values have been determined correctly.

The determined elements can be considered either as the feeding network, in case of analysing a single section, or as a part of adjacent cells in case of a periodically arranged cascade. Thus, by finding these element values the equivalent circuit model of the CRLH unit cell is fully determined.

4 Results

A well balanced unit cell is achieved after optimizing the structure supported by EM full-wave simulation using CST Microwave Studio (CST) combined with the procedure introduced in the previous section. The S-parameters of the resulting unit cell including the feeding network are shown in Fig. 6.

Although this is not the response of the unit cell itself, the isotropic behaviour and the power balance must be conserved anyway. As seen, the result is very close to the theoretical expectations of an isotropic structure which requires magnitudes of -6 dB, identical phase shifts of all transmission paths as well as a phase difference of 180° between transmitted and reflected waves. The waveguiding properties of the de-embedded unit cell, possessing a Bloch impedance $Z_{\text{Bloch}_x} \approx 60\,\Omega$, are shown in Figs. 7, 8 and 9. In regions where an isotropic behaviour is exhibited, the wave vector must have an absolute value which is independent of the direction. In terms of Fig. 9 this is represented by cone shaped characteristics or relating to the isofrequency plots this results in circles around the point $\beta_x p = \beta_y p = 0$ which can be observed for $8.5\,\text{GHz} \leq f \leq 12\,\text{GHz}$. It should be mentioned that the gap in Fig. 9 is an artefact of the visualization and does not represent a stop band as validated by the dispersion diagram depicted in Fig. 10. In the region $\Gamma - X$ and $M - X$ the index of refraction is determined and presented in Fig. 11 which also reveals the isotropy of the unit cell.

A prototype, seen in Fig. 1, was fabricated and measured. The results of the measurements are presented in Fig. 12 and the corresponding dispersion diagram in Fig. 13. As seen from the increasing attenuation constant between $9.5\,\text{GHz} \leq f \leq 10.5\,\text{GHz}$ the fabricated prototype is not balanced correctly. Further analysis reveals that the series resonant frequency is $f_{\text{rs}} = 10.57\,\text{GHz}$ and the parallel resonant frequency is $f_{\text{rp}} = 9.67\,\text{GHz}$. This can be explained by tolerances within the fabrication process.

5 Conclusions

A two-dimensional CLRH unit cell was presented which possesses an isotropic behaviour between $8.5\,\text{GHz} \leq f \leq 12\,\text{GHz}$. Within this frequency region a frequency dependent index of refraction ranging from $-1.5 \leq 0.8$ is achieved. A systematic procedure based on even and odd mode analysis was presented which allows to determine the element values of the equivalent circuit model of the proposed structure as well as its feeding network.

Acknowledgements. The authors gratefully acknowledge the support of the TUM Graduate School's Faculty Graduate Center EI at Technische Universität München, Germany.

References

AWR Microwave Office, http://www.awrcorp.com/, Last accessed: 15.02.2011.

CST Computer Simulation Technology, http://www.cst.com/, Last accessed: 15.02.2011.

Caloz, C. and Itoh, T.: Electromagnetic Metamaterials, Wiley, 2006.

Eberspächer, M. A., Eccleston, K., and Eibert, T. F.: Realization of Via-free Microstrip Composite Right/Left-Handed Transmission Lines, GeMiC, 2009.

Eccleston, K.: Beam forming transition based upon a zero-phase-shift metamaterial, IET Microw. Antennas Porpag., 4, 1639–1646, 2010.

Lay, A., Itoh, T., and Caloz, C.: Composite right/left-handed transmission line metamaterials, IEEE Microwave Magazine, 5, 34–50, 2004.

Shelby, R. A., Smith, D. R., and Schultz, S.: Experimental verification of a negative index of refraction, Science, 292, 77–79, 2001.

8

Suppression of grating lobes for MMW sparse array setups

S. Bertl, A. Kirschner, and J. Detlefsen

Fachgebiet Hochfrequente Felder und Schaltungen, Technische Universitt München, Arcisstr. 21, 80333 München, Germany

Abstract. For arrays the placement of the single elements determines the angular resolution and the unambiguity interval. The width of the total array determines the resolution capabilities. The wider the elements are placed from each other, the more space in Fourier domain is covered by the measurement and the resolution in time domain will improve. On the other hand the density of the elements has an effect on the angular interval in which objects can be detected unambiguously. For objects within the unambiguous interval grating lobes will appear outside this area while objects outside result in grating lobes in the interval of interest. In this paper the properties of arrays regarding resolution and unambiguity interval will be discussed and methods for the suppression of ambiguous grating lobes are suggested. One approach to suppress the influence of the grating lobes lies in the evaluation of different frequency bands.

1 Introduction

To realise a sensor with range and azimuth resolution capabilities, a single antenna with high gain that is moved mechanically can be used. To avoid mechanical movements arrays can be considered instead. Their performance depends on the number of elements and their placement. However the hardware effort is considerable higher such that the amount of elements is usually limited. A concept that reduces the effort to a certain part exploits the concept of virtual apertures. This means every combination of transmitter and receiver forms a virtual element of an array that contains overall more elements than the single arrays for transmit and receive. The properties of these arrays and possible applications in MMW-domain are discussed.

Correspondence to: S. Bertl
(sebastian.bertl@tum.de)

2 Properties of arrays

2.1 Concept of virtual arrays

Virtual arrays can be formed from an arbitrary transmitting array with N antenna elements and a receiving array with M elements. The position of the single antenna elements is known and denoted with $\boldsymbol{x}_{Tn}$ for the n-th transmitting antenna and $\boldsymbol{x}_{Rm}$ for the m-th receiving antenna.

For being able to separate the transmitted signals of the different Tx-antennas at the receivers it is required that the transmitted waveforms s_i are orthogonal such that

$$\int s_n(t)\cdot s_m^*(t)dt = \begin{cases} 1 & n=m \\ 0 & n\neq m \end{cases}. \tag{1}$$

In the receiving antennas the orthogonal waveforms are extracted by N matched filters. The total number of extracted signals is then $N\cdot M$. The received signal of a point scatterer with a reflection coefficient ρ can be written as

$$s_{m,n}(t) = \rho\cdot\exp(jk_0\boldsymbol{u}\cdot(\boldsymbol{x}_{Tn}+\boldsymbol{x}_{Rm})) \tag{2}$$

$$= \rho\cdot\exp\left(j2k_0\boldsymbol{u}\cdot\left(\frac{\boldsymbol{x}_{Tn}+\boldsymbol{x}_{Rm}}{2}\right)\right). \tag{3}$$

The vector $\boldsymbol{u}$ is the unit vector pointing from the center of the array towards the scattering centre. In this description it is assumed that the distance between the point scatterer and the two arrays is much larger than the dimensions of the array. Equation (3) shows that the response is the same as for an array with $M\cdot N$ elements where each element is transmitting and receiving just its own signal located in the middle of a pair of transmit and receive antenna at

$$\frac{\boldsymbol{x}_{Tn}+\boldsymbol{x}_{Rm}}{2} \quad \text{with} m=1\ldots,M, n=1,\ldots,N. \tag{4}$$

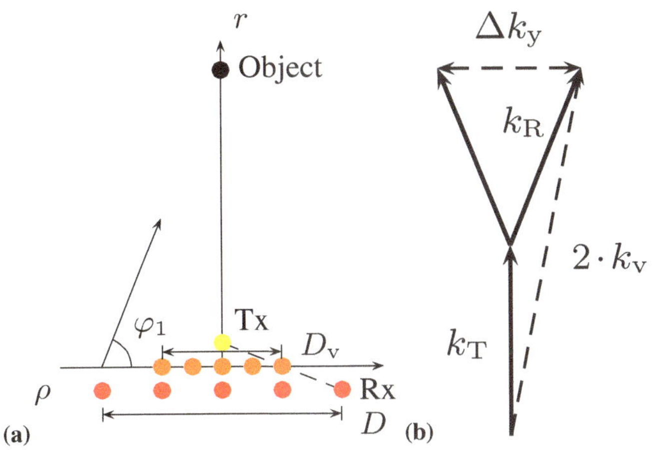

Fig. 1. Virtual array formed of one Tx and n Rx antennas. **(a)** Array geometry. **(b)** k-space representation.

2.2 Realisation of orthogonal signals

The separation of the transmitted signals at the receivers is essential for the realisation of virtual arrays. Different approaches can be considered for the orthogonalisation. In general the principles known from communication can be applied for the radar case too. The approach that does not require a high hardware effort is to separate the signals in time by transmitting the signals at a given sequence. The separation of the signals at the receiver is then done by applying rectangular windows to the specific time slots of the corresponding signals of the different transmitters.

When the used bandwidth B is sufficient B can also be split into N parts and each transmitter uses only a fraction B/N. The drawback of this solution is that the resolution of the system will be reduced and the hardware effort will be higher too, since the filtering of the signals at the receivers has to be done. In addition a separate down-conversion for each band has to be realised at the receiver.

Also a modulation of the phase of each transmitted signal is possible. In this case the separation can be done by correlation of the received signals with the signal of a specific transmitter in order to separate the corresponding pair.

2.3 Description of arrays in k-space

It can be shown, that a virtual array that is formed by a transmit and a receive array has the same resolution properties as the monostatic array with Tx and Rx placed in the centre of a pair of Tx and Rx antenna of the respective arrays. In the following an example with one transmit antenna and n receive antennas, shown in Fig. 1, is considered. The considerations can be applied to configurations with arbitrary amount of antennas in both arrays. A point scatterer placed at the perpendicular bisector of the Rx array at a distance r is considered.

For a fully monostatic array with Tx and Rx antennas at the same position the width of the occupied k-space becomes

$$\Delta k_y = \frac{D}{2r} \cdot |\boldsymbol{k}_{\mathrm{Rx}} + \boldsymbol{k}_{\mathrm{Tx}}| \tag{5}$$

$$= \frac{D}{r} |\boldsymbol{k}|. \tag{6}$$

The azimuth or y-resolution can then be determined approximately by

$$\Delta y = \frac{2\pi}{\Delta k_{\mathrm{tot}}} = \frac{2\pi}{2\Delta k_y} \tag{7}$$

$$= \frac{2\pi r}{2D|k|} = \frac{\lambda}{2D} \cdot r. \tag{8}$$

For the situation in case of the virtual array set up by a single transmit element and an object placed on the perpendicular bisector only the positions of the transmit array elements lead to an occupation of the k-space in y-direction. For the different elements the projection of the resulting k-vector $2 \cdot \boldsymbol{k}_{\mathrm{v}} = \boldsymbol{k}_{\mathrm{T}} + \boldsymbol{k}_{\mathrm{R}}$, shown in Fig. 1, on the k_y axis is

$$k_{y1} = \sin\varphi_1 \cdot |\boldsymbol{k}| = \frac{D}{2r} \cdot |\boldsymbol{k}| \tag{9}$$

$$k_{yn} = \sin\varphi_n \cdot |\boldsymbol{k}| = \frac{D}{2r} \cdot |\boldsymbol{k}|. \tag{10}$$

Taking the two outer elements into account since they result in the largest extension in k_y-direction, the size of the occupied k-space is

$$\Delta k_y = k_{y1} + k_{yN} = \frac{D}{r} \cdot |\boldsymbol{k}|. \tag{11}$$

and the resulting azimuth resolution results in

$$\Delta y = r \cdot \frac{\lambda}{D}. \tag{12}$$

Comparing this with the monostatic case, from in (8)

$$\Delta y = r \cdot \frac{\lambda}{D} = r \cdot \frac{\lambda}{2D_{\mathrm{v}}} \tag{13}$$

it turns out that the resolution of the virtual array is the same as for a monostatic array with half the array size: $D = 2 \cdot D_v$.

2.4 Realisation of virtual arrays

For the realisation of a virtual array different possibilities can be considered. In general an equal spacing of the virtual elements is desired. Possible goals could be minimum total space of the overall setup or a specific distribution of the virtual elements with respect to the positions of the real array elements. Setups for those two mentioned realisations are shown Fig. 2. For all realisation a virtual array with $M \cdot N$ virtual elements is generated by $M + N$ antennas. The geometric dimensions can be different for the same virtual extension.

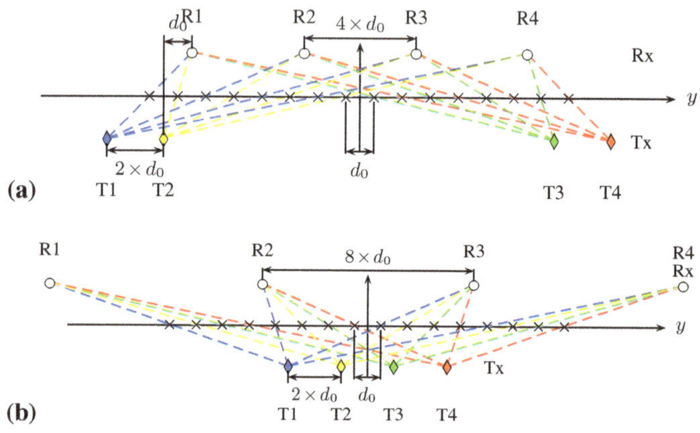

Fig. 2. Different options for setting up a virtual array. **(a)** Placement of Rx- und Tx-antennas according to Wei (2009), Setup 1. **(b)** Alternative placement of Rx- und Tx-antennas to separate left and right part of the virtual array with a maximum number of virtual elements, Setup 2.

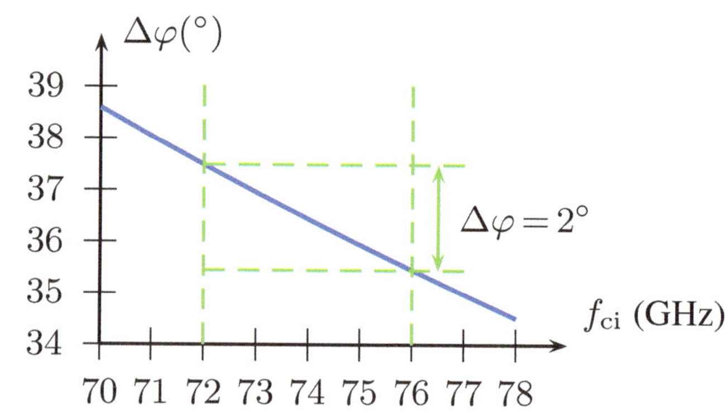

Fig. 3. Distance between object's position and grating lobe as a function of centre frequency f_{ci} (object at the perpendicular bisector of the array).

3 Methods for grating lobe suppression

3.1 Choice of the antenna pattern

A first approach to reduce the influence of objects that are outside the unambiguity interval on this interval, is to adapt the antenna pattern such that the main lobe of the single antenna elements mainly covers the unambiguity interval. Objects outside will then be received already with lower amplitudes. This approach will not be discussed in detail here.

3.2 Azimuth ambiguity and resolution

As shown in the previous section the azimuth resolution capabilities depend on the maximum dimensions of the array, see Eq. (13). The more space the elements occupy, the higher the resolution. On the other hand the number of virtual array elements determines the number of azimuth sample points and therefore together with the resolution the interval where scatterers can be localised uniquely. Let d_0 be the distance between the virtual array elements. The unambiguity interval then becomes

$$\varphi_{\text{unamb}} = \pm\sin^{-1}\left(\frac{c_0}{4 f_c d_0}\right). \tag{14}$$

When the distance between the virtual elements is e.g. $d_0 = 0.8\cdot\lambda$, then the unambiguity interval becomes $\varphi_{\text{unamb}} = \pm 18.2°$. At an angular offset of

$$\Delta\varphi_{\text{grating}} = 2\cdot\arcsin\left(\frac{c_0}{4 f_{c0} d_0}\right) \tag{15}$$

grating lobes will appear.

3.3 Evaluation of sub bands

As shown in Eq. (15) the position of the grating lobes is varying with a change of the centre frequency of the used frequency band. In contrast to this the results of different sub bands overlap at the correct angular position of the object. When the spacing between the virtual elements is expressed in terms of the wavelength $\lambda_{c,0}$ that corresponds with $f_{c,0}$ and $d_0 = \alpha\lambda_{c,0}$, then the distance between main and grating lobe in another frequency band with a centre frequency of $f_{c,i}$ becomes

$$\Delta\varphi_{\text{grating,i}} = 2\cdot\arcsin\left(\frac{f_{c0}}{4 f_{ci}\alpha}\right). \tag{16}$$

When the azimuth resolution cell of the array is smaller than the separation of the grating lobes, they will be placed in different resolution cells. The grating lobes can then be distinguished from the main lobe by comparing the responses obtained for two different frequency bands.

In Fig. 3 the distance between grating lobes and main lobe is shown as a function of the centre frequency at the frequency range of interest $f = 72\ldots78\,\text{GHz}$. When e.g. a distance between the virtual elements $d_0 = 0.8\lambda$ is used, the resolution capabilities in azimuth are around $\Delta\varphi = 2°$. The distance of the grating lobes for $d_0 = 0.8\lambda$ would be $\Delta\varphi = 2°$ when the centre frequencies are separated by 4 GHz, such that the grating lobes could be resolved at different positions, when the two sub-bands are evaluated.

4 Measurements

4.1 Evaluation of one band

For first measurements an existing MMW-sensor operating at a slightly higher frequency range $f = 90.5\ldots100.5\,\text{GHz}$ with 1 Tx and Rx channel is used. In order to simulate the data of an array, the sensor is mounted on a linear positioner and the

Fig. 4. Measurement scene in a corridor with several objects.

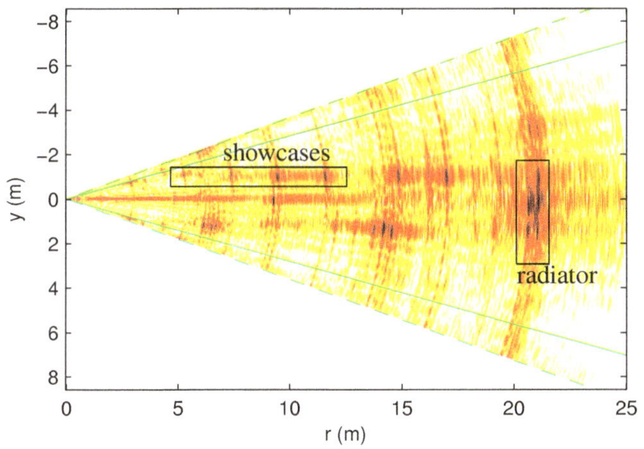

Fig. 5. Reconstruction result of the scene according to the photo shown in Fig. 4, logarithmic scale with 50 dB dynamic range, green line: unambiguous interval, $f = 90.5 \ldots 93.5\,\text{GHz}$.

measurement samples are taken at different positions. The measurements are taken in steps of $\Delta y = 3\,\text{mm}$, which corresponds to a spacing of $d_0 = 0.92 \times \lambda_c$ of the equivalent array. The measurements were done along a corridor of a length of approx. 20 m. A picture of the measured scene is shown in Fig. 4. The major objects that appear are: a trihedral, a copper roll on a mount, a fire extinguisher, showcases and a radiator. The reconstruction result is shown in Fig. 5. The results are represented in a logarithmic scale with 50 dB dynamic range. The used frequency range for the measurement is $f = 90.5 \ldots 93.5\,\text{GHz}$.

4.2 Processing of sub bands

For the evaluation of sub bands two parts of the sensor's available bandwidth are used. The lower band is $f_1 = 90.5 \ldots 92.8\,\text{GHz}$ and the upper one covers $f_2 = 98.2 \ldots 100.5\,\text{GHz}$. When the two sub bands are evaluated it shows that the result of them don't overlap exactly in range. In addition the grating lobes are not at the exact position in azimuth according to the used frequencies. A possible reason could be inaccuracies in the data acquisition (trigger), but also a mismatch in tuning voltage of the VCO and expected frequency, which leads to inaccurate output frequencies. As a consequence the evaluation of the measured data has to be adapted. The two images can not be compared point by point at the exact position of a peak. Instead of this a mask is generated, where main-lobes are identified. This is done by determining the object responses produced by grating lobes and main lobes in the two sub-bands. The main lobe is at the same position in the two pictures, grating lobes are at different positions. The mask contains only these points where main lobes have been identified, for other coordinates it is set to zero. A result of this processing scheme is shown in Fig. 6. It shows that outside the unambiguous region grating lobes have been removed for the most part.

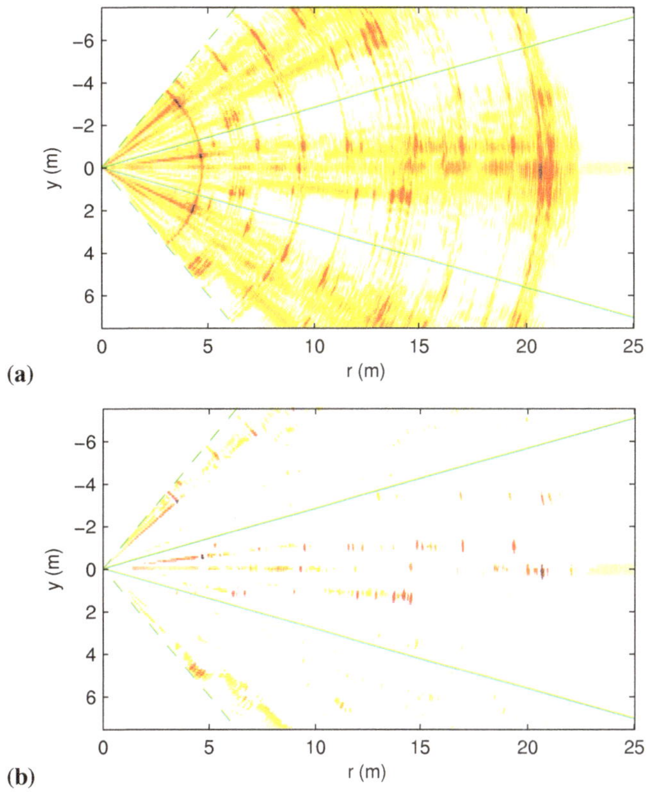

Fig. 6. Comparison of original processed array data and results after applying a mask with the position of the main lobes. **(a)** Unprocessed measurement result. **(b)** Result after evaluation of the sub bands.

5 Conclusions

The concept of virtual arrays has been presented. It is useful to implement arrays with several virtual elements while keeping the hardware effort at a lower level compared to the realisation of an array with Tx and Rx at each position of the array. To demonstrate the expected imaging properties an existing MMW-sensor has been used to setup an array with 16 elements. In order to reduce the presence of azimuth ambiguities, two different sub bands have been evaluated. A suppression of grating lobes is possible with this processing step. For the considered bandwidth of 72...78 GHz in the realisation of the system, two sub-bands with 2 GHz bandwidth and a separation of the centre frequencies f_{ci} of 4 GHz is considered. Evaluation of sub-bands allows to distinguish between main and grating lobes. Using this approach arrays can be tailored to the desired field of view and a good azimuth resolution.

References

Wei, M. Digital Antennas, in: Multistatic Surveillance and Reconnaissance: Sensor, Signals and Data Fusion, pages 5-1–5-29, RTO-EN-SET-133, 2009. Available from http://www.rto.nato.int.abstract.aps

Li, J., Stoica, P. MIMO radar signal processing, John Wiley & Sons Inc., Hoboken, New Kersey, 2009.

Ender, Joachim H. G. MIMO-SAR, in: Proceedings of IRS-07, International Radar Symposium 2007, Kln, Germany, September 2005–2007.

9

Ray tracing with multi-radiation transmitters

R. Brem and T. F. Eibert

Lehrstuhl für Hochfrequenztechnik, Technische Universität München, 80290 Munich, Germany

Correspondence to: R. Brem (rbrem@tum.de)

Abstract. A restriction in using electromagnetic ray tracing for field prediction is given by the far-field condition: the results are only valid in the far-field region of the radiator. In this paper, it will be shown how ray tracing for accurate field computation can also be applied in the near-field regions of transmitters. The reduction of required large distances between transmitter and receiver is achieved by subdividing the transmitter in smaller subtransmitters. Even for complex transmitters, e.g. antennas with objects in close proximity such as metallic carrier platforms, subtransmitter models can be very efficiently generated by using the Multilevel Fast Multipole Method (MLFMM). This well-known integral equation solving technique makes very large problems in computational electromagnetics manageable. The subtransmitters can be directly generated based on this algorithm. A simulation example will show the improved modeling accuracy and options for simplification and refinement will also be discussed.

1 Introduction

Electromagnetic ray tracing is a widespread technique for field prediction in large scenarios. It is based on Geometrical Optics (GO), where the assumption is that electromagnetic energy propagates along straight lines. The underlying physical properties of these rays are energy conservation and Fermat's principle. With the extensions Geometrical Theory of Diffraction (GTD) and Uniform Theory of Diffraction (UTD) the diffraction behavior of rays can be explained (Balanis, 1989). Ray tracing finds its use for electrically large problems where exact solvers, e.g. integral equation solvers based on the Method of Moments (MoM) or finite element solvers, would be far too inefficient due to the complexity. A typical example is the use of ray tracing in the design phase of mobile networks. Besides the feasibility of field prediction for such scenarios, further benefits of ray tracing compared to exact field computation methods are the nonexistence of memory and simulation time increase with increasing frequency and an easy parallelization (Cavalcante et al., 2006; Epstein and Rhodes, 2010).

Two approaches exist: in deterministic ray tracing, all rays between transmitter and receiver are exactly computed by image theory. For this, the necessary preprocessing can be quite time-consuming. In ray launching (also called shooting and bouncing rays (SBR) (Ling et al., 1989)), rays are sent out in arbitrary directions where no preprocessing is required. In order to avoid endless running rays which never reach the receiver in ray launching, rays will be neglected after a certain number of interactions or attenuation. Due to the fact that a ray will normally not hit a receive point exactly in ray launching, a tolerance is introduced by using a receiver sphere. Both methods will give the same results if enough rays in the ray launching method are used to cover all relevant paths. Many concepts were introduced to improve the accuracy of ray-tracers (Mohtashami and Shishegar, 2008; Mohtashami et al., 2010) and to speed up the computation (Mohtashami and Shishegar, 2009; Park et al., 2011).

The simplest transmitter model is a localized source with the corresponding far-field radiation pattern. This is an appropriate model when the distance between transmitter and receiver is large enough, i.e. the far-field condition has to be fulfilled. A reduction of this distance increases the error in field computation which can lead to very inaccurate simulation results. For areas of application such as indoor radio channel measurements or car-to-car communication where distances between transmitter and ray-intersecting objects are small or the radiators are large, respectively, this can deteriorate simulation performance. In order to overcome this problem, the transmitter is partitioned in subtransmitters which are also modeled by their far-field radiation patterns.

In the next section, the modeling procedure is presented. A simulation example verifies the better field computation results and an option for model improvement is also shown.

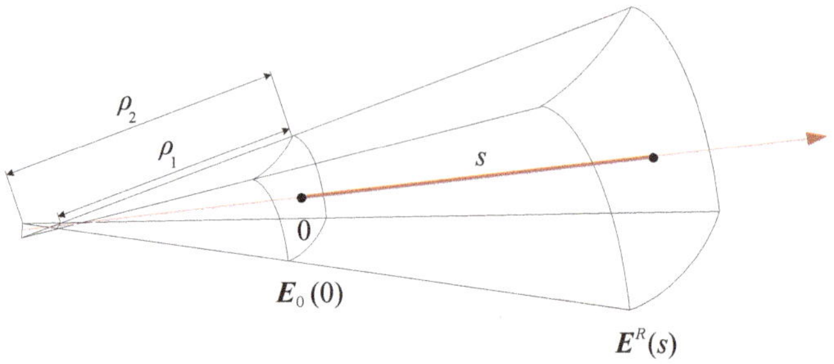

Fig. 1. Astigmatic ray tube.

2 Procedure

In general, the electric field $\boldsymbol{E}$ dependent on an electric source current density $\boldsymbol{J}$ can be written according to (Balanis, 1989)

$$\boldsymbol{E}(\boldsymbol{r}) = -jkZ \iiint_V \left(\bar{\boldsymbol{I}} + \frac{1}{k^2}\nabla\nabla\right) \frac{\mathrm{e}^{-jk|\boldsymbol{r}-\boldsymbol{r}'|}}{4\pi|\boldsymbol{r}-\boldsymbol{r}'|} \cdot \boldsymbol{J}(\boldsymbol{r}')\mathrm{d}v', \quad (1)$$

where $k = \omega\sqrt{\varepsilon\mu}$ is the wave number and $Z = \sqrt{\mu/\varepsilon}$ is the impedance of the considered homogeneous medium, respectively. Typically, free space is considered so that $Z = 377\,\Omega$. With far-field assumptions, i.e. for $r = |\boldsymbol{r}| \gg |\boldsymbol{r}'| = r'$, the field can be written as

$$\boldsymbol{E}^{FF}(\boldsymbol{r}) = -jkZ\frac{\mathrm{e}^{-jkr}}{4\pi r}\left(\hat{\vartheta}\hat{\vartheta} + \hat{\varphi}\hat{\varphi}\right) \cdot \iiint_V \mathrm{e}^{jk\hat{r}\cdot\boldsymbol{r}'}\boldsymbol{J}(\boldsymbol{r}')\mathrm{d}v', \quad (2)$$

where $\hat{r}$ is the unit vector in direction of $\boldsymbol{r}$. Such far-fields describe spherical waves, which can be seen by the term $\mathrm{e}^{-jkr}/4\pi r$. Ray tracing can be accomplished by taking these fields as sources, where they are a special form of the more general astigmatic ray representation according to (Balanis, 1989)

$$\boldsymbol{E}^R(s) = \mathrm{e}^{-jks}\sqrt{\frac{\rho_1\rho_2}{(\rho_1+s)(\rho_2+s)}}\boldsymbol{E}_0(0). \quad (3)$$

This equation is valid along a straight ray path parameterized by s. ρ_1, ρ_2 are the radii of curvature of the astigmatic ray tube visualized by Fig. 1.

Equation (1) does not have ray field character, so that ray tracing cannot be applied here. It can be shown that the far-field of a source distribution contains also the complete near-field information. However, it is not possible to access this near-field information by a conventional ray tracing approach. Ray tracing compatible far-field representations which describe the near-field of a radiator are necessary to solve this problem. Important is the fact that the transition from near- to far-field region of a transmitter depends on its dimension, where the typical condition for the far-field is $r > 2D^2/\lambda$ (with distance r, max. extension D of radiator and wavelength λ) (Balanis, 2005). With smaller radiator dimensions, the far-field starts at a smaller distance. Therefore, a subdivision of the radiating object into smaller subtransmitters is helpful. The far-field assumption for the subtransmitter fields is valid at a smaller distance compared to the complete transmitter, so that ray tracing can be applied. This approach enables a ray based far-field representation of the transmitter's near-field where the validness of this modeling approach for decreasing distances to the transmitter can be guaranteed by decreasing also the subtransmitter dimensions, i.e. using a finer model with more subtransmitters.

The determination of current distributions on electric conducting objects like antennas, is typically the first step in solving radiation problems. Fast integral equation solvers like the efficient MLFMM algorithm are used by default for this. Due to the fact that MLFMM (Chew et al., 2001) is based on a hierarchical subdivision of the object into levels and groups, it is an ideal method to generate the subtransmitter models. The subtransmitters are represented by the groups (cubic boxes) which hold a part of the object. The entire object is enclosed in a hierarchical oct-tree of boxes with different sizes on the different levels. The high efficiency of MLFMM relies on describing the interaction between well-separated surface currents by a hierarchical plane wave representation of the radiated and received fields instead of computing the interaction of each current with each other current, as done in the classical MoM (Method of Moments). For each group, such a plane wave representation is computed and the field at a receiver group is obtained by appropriate translation operators (Chew et al., 2001).

In order to compute the electromagnetic near- and far-fields from a known current distribution, the same efficient hierarchical scheme can be employed (Tzoulis and Eibert, 2005). For all lowest level MLFMM groups (smallest boxes), the computed currents are used to generate a plane wave representation of the corresponding radiated fields (Eibert, 2005). A new subtransmitter of the next higher level is obtained by combining eight subtransmitter boxes on the level below. On the higher level, the new subtransmitter has the double box side length and its source point is in the middle of the eight boxes. The field is computed by an aggregation and interpolation operation of the eight subtransmitter fields. The simple single source point transmitter model is obtained by repeating this procedure until only the box on the highest level remains.

All subtransmitters of one level, representing in sum the entire object, are chosen for a ray tracing simulation. The field samples from each transmitter serve as start values $\boldsymbol{E}_0$ in Eq. (3). Different subtransmitter models can be easily generated by choosing the lowest level box size and a certain level.

3 Simulation example

Consider a $\lambda/2$-dipole, oriented along the z-axis in a Cartesian coordinate system. The dipole is placed 5 cm over the center of a perfectly conducting, quadratic plate of 1 m^2 in the xy-plane, as shown in Fig. 2a. One could imagine a vehicular antenna modeled by such a configuration, where the plate represents a part of the vehicle which significantly influences the radiation. The dipole is fed in the middle by a 1 GHz voltage source. The electric field shall be computed at different distances d from the antenna. For a meaningful error investigation, receiver planes of 1 m^2 providing $N = 441$ probe points each, are used. The planes are placed along the line $y = x$ and $z = \text{const} = 1$ m, illustrated by Fig. 2b.

First, the currents on antenna and plate are computed by an MLFMM integral equation solver. Then, the complex electric field is exactly computed for the receiver planes at distances $d = \{10\,\text{m}, 30\,\text{m}, 50\,\text{m}, 100\,\text{m}, 150\,\text{m}, 200\,\text{m}\}$. The solution is compared to a field computation in the same planes by ray tracing with three different discretizations of the radiator: one, four and nine subtransmitters are used. Fig. 3 shows the average relative error, expressed as

$$\frac{1}{N}\sum_{i=1}^{N}\left|\frac{E^i_{\text{RT}}}{E^i_0}-1\right| \tag{4}$$

for the different values of d and the three numbers of transmitters. E^i_{RT} and E^i_0 are the i-th complex field value of the ray tracing and the exact (MLFMM) solution, respectively.

The figure shows that a distinctive lower error in field computation near the transmitter is achievable by four and nine subtransmitters compared to the traditional model. For this improvement in accuracy, an increase in simulation time must be accepted. However, the following remarks show that this increase does not emerge as critical problem. When the same number of rays for all transmitters and ray launching are used, the simulation time will only linearly increase with the number of subtransmitters. Due to the fact that the scope of this method is the near-field region of the radiator, the distances the rays have to travel are naturally short, which makes the use of more transmitters less problematic. Furthermore, not many subtransmitters are necessary. With only a small number, the accuracy can be considerably increased.

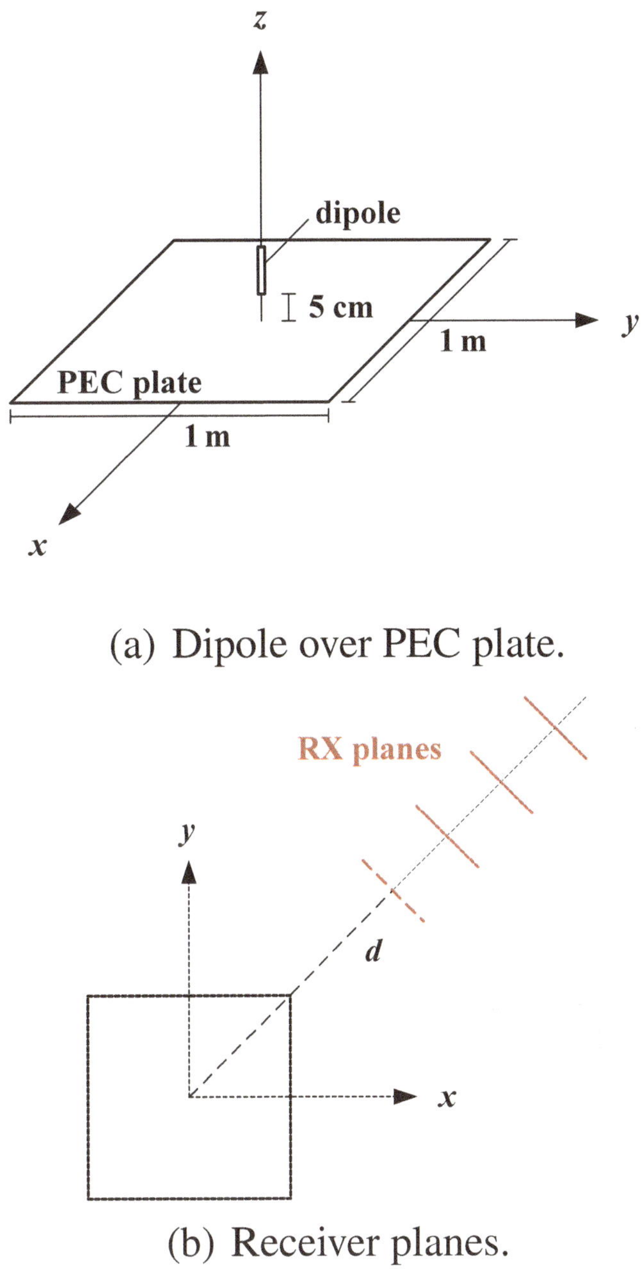

Fig. 2. Simulation test.

4 Model improvement

An easy way to speed up ray tracing simulations is to disregard subtransmitters whose radiated electromagnetic powers are much smaller than those of others. Large radiators may contain subtransmitters with weak currents and hence also fields so that their contribution to the total field is low. This could be extended to neglect subtransmitters which contribute little in a certain direction of interest.

Important for further improvement of ray tracing accuracy is the choice of the subtransmitter radiation centers. Instead of using the center of a subtransmitter box, the source point should be positioned at the location of the main current contribution. This position is denoted as current gravity center and computed by

$$\boldsymbol{r} = \frac{\sum_{i=1}^{N} w_i \boldsymbol{v}_i}{\sum_{i=1}^{N} w_i} \qquad w_i = \frac{I_i}{\max_i(I_i)}, \tag{5}$$

where N is the number of current basis functions in the subtransmitter, $\boldsymbol{v}_i$ is the edge midpoint of the current basis function i and the weights w_i are computed as current magnitude I_i over maximum current of the subtransmitter. Prior to a

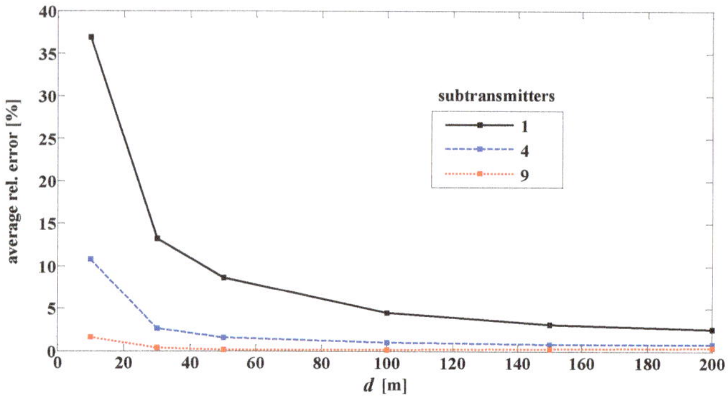

Fig. 3. Average relative error for different numbers of subtransmitters and distances.

ray tracing simulation, the subtransmitter fields are related to these new source points, i.e. the phases must be correctly adapted.

In the simple simulation example of Fig. 2a benefit of these new radiation sources can only be observed for very small distances compared to the values of d due to the small subtransmitters. However, for larger radiating objects a considerable better accuracy for relevant distances can be achieved by this method.

5 Conclusions

An approach to produce finer transmitter models for electromagnetic ray tracing was shown. So, accurate field prediction in the near-field region of radio transmitters is made possible. By utilization of the Multilevel Fast Multipole Method, the subtransmitters can be very efficiently generated. Reason for this is the inherent hierarchical subdivision of a radiating object as basic principle of the algorithm. By adjusting the subtransmitter box size on the lowest level or choosing a certain level, great freedom in transmitter model generation exists. The simulation result shows the quality of this method.

References

Balanis, C.: Advanced Engineering Electromagnetics, John Wiley & Sons, New York, 1989.

Balanis, C.: Antenna Theory, John Wiley & Sons, Hoboken, 2005.

Cavalcante, A. M., de Sousa, M. J., Costa, J. C. W., Frances, C. R. L., Protasio dos Santos Cavalcante, G., and de Souza Sales, C.: 3D ray-tracing parallel model for radio-propagation prediction, in: International Telecommunications Symposium, 269–274, doi:10.1109/ITS.2006.4433282, 2006.

Chew, W., Jin, J., Michielssen, E., and Song, J.: Fast and Efficient Algorithms in Computational Electromagnetics, Artech House, Boston, MA, 2001.

Eibert, T. F.: A diagonalized multilevel fast multipole method with spherical harmonics expansion of the k-space Integrals, IEEE Trans. Antennas Propag., 53, 814–817, doi:10.1109/TAP.2004.841310, 2005.

Epstein, B. R. and Rhodes, D. L.: GPU-accelerated ray tracing for electromagnetic propagation analysis, in: IEEE International Conference on Wireless Information Technology and Systems (ICWITS), 1–4, doi:10.1109/ICWITS.2010.5611922, 2010.

Ling, H., Chou, R.-C., and Lee, S.-W.: Shooting and bouncing rays: calculating the RCS of an arbitrarily shaped cavity, IEEE Trans. Antennas Propag., 37, 194–205, doi:10.1109/8.18706, 1989.

Mohtashami, V. and Shishegar, A. A.: A new double-counting cancellation technique for ray tracing using separation angle distribution, in: IEEE International RF and Microwave Conference (RFM), 306–310, doi:10.1109/RFM.2008.4897435, 2008.

Mohtashami, V. and Shishegar, A. A.: Application of wavefront decomposition to ray tube tracing, in: First Conference on Millimeter-Wave and Terahertz Technologies (MMWaTT), 5–9, doi:10.1109/MMWATT.2009.5450465, 2009.

Mohtashami, V., Parsa, S., and Shishegar, A. A.: Efficient enhancement of the accuracy of Ray tracing, in: IEEE Antenn. Prop., 1–4, doi:10.1109/APS.2010.5560920, 2010.

Park, H.-G., Kim, H.-T., and Kim, K.-T.: Beam tracing for fast RCS prediction of electrically large targets, PIER, 20, 29–42, doi:10.2528/PIERM11060702, 2011.

Tzoulis, A. and Eibert, T. F.: Efficient electromagnetic near-field computation by the multilevel fast multipole method employing mixed near-field/far-field translations, IEEE Antenn. Wirel. Pr., 4, 449–452, doi:10.1109/LAWP.2005.860195, 2005.

Reconfigurable ultra-wideband transmitter for generation of arbitrary impulse shapes and modulation schemes

A. T. Ott[1], **C. J. Eisner**[1], **M. D. Blech**[2,*], **and T. F. Eibert**[1]

[1]Lehrstuhl für Hochfrequenztechnik Technische Universität München 80290 Munich, Germany
[2]Institut für Hochfrequenztechnik Universität Stuttgart, 70550 Stuttgart, Germany
[*]now at: Sensing Systems Laboratory, Sony Deutschland GmbH, 70327 Stuttgart, Germany

Correspondence to: A. T. Ott (hft@ei.tum.de)

Abstract. In this paper a reconfigurable ultra-wideband (UWB) impulse radio (IR) transmitter is presented. The IR signal is synthesized at an intermediate frequency (IF) by employing a multi-Nyquist digital-to-analog converter (DAC) with 12 bit resolution and an update rate of 2.3 GHz. Digital generation of signals in a field programmable gate array (FPGA) guarantees very high flexibility of the reconfigurable design. For upconversion to the radio frequency (RF) band, a first order bandpass (BP) sampling concept and an alternative conventional concept with mixer stages, have been realized. The system enables to generate signals with arbitrary modulation schemes and techniques at an external host personal computer (PC) employing MATLAB. Different measurements using a digitizing oscilloscope have been conducted to demonstrate the performance of the transmitter.

1 Introduction

Since the approval of the UWB frequency band by the Federal Communications Commission (FCC) (FCC, 2002), UWB technology has gained enormous research interest (Reed, 2005). Today UWB multiband orthogonal frequency division multiplexing (OFDM) (Batra et al., 2004) and direct sequence spread spectrum (DSSS) systems (Fisher et al., 2004) have established itself in the market. Actually only a few companies like Time Domain (Time Domain Corporation, 2011) or Decawave (DecaWave, 2011) provide IR systems commercially. Their products are characterized by a very low power consumption and they offer the possibility of high resolution tracking combined with low data rate communication.

For planning and evaluation purposes, reconfigurable transceiver designs are an ideal choice as they offer a considerable flexibility with respect to different applications. Recently, software defined radio (SDR) concepts providing common IR modulation schemes and adjustable data rates have been published (Verhelst et al., 2008). For the transmission of IR signals in the RF band, direct conversion UWB concepts have been presented, which overcome the problems of non ideal mixer stages (Anderson, 2006; Chen and Brodersen, 2007; Keller et al., 2007). However these concepts are mainly focusing on the receiver implementation whereas adjustable analog pulse generators are used on the transmitter side.

The design and realization of an exemplary subsampling UWB direct conversion SDR IR transceiver has been presented in (Blech et al., 2009) and (Blech et al., 2010). This concept includes a SDR receiver and transmitter solution which increases the flexibility and reduces the impact of non ideal hardware significantly. However the focus of research was on the design of the hardware platform and only binary phase shift keying (BPSK) and quadrature phase shift keying (QPSK) modulation schemes have been analyzed by measurements.

An improved redesign of the transmitter with considerably increased capabilities of the digital signal processing (DSP) stage is presented in this paper. The novel design offers the possibility to generate arbitrary 2-ary and 4-ary IR modulation schemes on an external host PC with MATLAB as programming language. Main parts of the transmitter are an FPGA (Avnet, 2006) and an ultra high speed multi-Nyquist DAC (Maxim Integrated Products, Inc., 2006). Hence direct conversion of the output signal to RF is possible by applying a bandpass filter. Furthermore, an additional frontend with analog mixer stages can be used for upconversion to RF of the analog UWB signal, synthesized at an intermediate frequency (IF). Due to hardware limitations the prototype is designed for laboratory use. The prototype is currently designed to transmit impulses at a centre frequency of 2.5 GHz and an impulse bandwidth of up to 1 GHz. However, the results are used to draw conclusions for an implementation with RF output frequencies between 6 GHz and 10 GHz.

This paper presents the hardware design of the transmitter in Sect. 2. In Sect. 3, a signal model for characterizing possible IR modulation schemes supported by the transmitter is provided. Measurement results demonstrating the performance of the concept are depicted in Sect. 4 before the paper is concluded in Sect. 5.

2 Transmitter design

The DAC features 12 bit resolution at an update rate up to $r_U = 2.3\,\mathrm{GS\,s^{-1}}$ and thus generation of signals of up to 1.15 GHz bandwidth. For upconversion the transmitter either uses first order sampling plus filtering in the 2nd or 3rd Nyquist zone or a conventional architecture with mixer stages. To achieve comparable results, signals are synthesized for the RF frequency range between 2 GHz and 3 GHz. This is due to the hardware restrictions of the DAC, as signal distortion increases significantly for the 4th and higher Nyquist zones. The current implementation provides variable pulse repetition frequencies (PRF) reaching from 0.5 MHz to 62.5 MHz. In principle higher repetition rates are possible on the transmitter side, however a maximum PRF of 62.5 MHz is reasonable, as the PRF is limited by the correlation receiver due to multipath propagation, which causes overlapping impulses (Win and Scholtz, 1998), (Reed, 2005). Although the transmitter operates outside the approved UWB frequency band, the concept is adequate for testing the performance in a laboratory environment.

2.1 Digital signal processing architecture

The block diagram of the DSP stage of the transmitter is depicted in Fig. 1. The clock signal of the DAC is generated externally and has been chosen to 2 GHz. A Virtex 5 FPGA has been applied for digital generation of pulse trains. Up to four different impulse shapes are stored in a random access memory (RAM) offering the possibility of arbitrary modulation of the incoming bit stream. To relax timing constraints, the FPGA clock rate for transmitting the digitized waveforms from the FPGA to the DAC has been chosen to $f_{\mathrm{FPGA}} = f_s/8$. Configuration of the FPGA as well as streaming of data is implemented by connecting an external host PC with a high speed universal serial bus interface (USB) 2.0 controller to the FPGA. By accessing the USB interface of the PC in MATLAB, impulse shapes and modulation schemes can be designed and visualized in MATLAB. The prototype of the developed transmitter is depicted in Fig. 2.

2.2 Frontend concept

Block diagrams for signal generation with a conventional and a subsampling concept are illustrated in Fig. 3. The subsampling concept uses replicas of the signal in higher Nyquist zones (see Fig. 3 top) whereas in the conventional concept, mixer stages are applied for shifting signals to the RF band. Both concepts have been realized and investigated in this work.

Each discrete signal theoretically provides a periodic spectrum. However the shape of the replicas in higher Nyquist zones are strongly affected and attenuated by the non-ideal sampling impulse shape of the DAC. For optimization of the output power and gain flatness in different Nyquist zones the DAC exhibits different modes of operation. According to the shape of the impulse response in the time domain the DAC has a so called non-return-to-zero (NRZ), a return-to-zero (RZ) and a RF mode (Blech et al., 2009). With respect to the required output frequency the NRZ mode provides the highest power in the first Nyquist zone, the RZ mode offers the flattest gain in the third Nyquist zone whereas the RF mode has the highest output power in the third Nyquist zone. Adjustable power amplifiers (PAs) increase the output signal power according to the effective isotropic radiated power (EIRP) requirements of the regulation.

In general the RF mode is applied in the subsampling concept due to the highest signal-to-noise-ratio (SNR) in the third Nyquist zone. An anti-aliasing bandpass (BP) filter yields the desired output signal in the RF band. As subsampling concepts for UWB transmitters are currently restricted by hardware limitations to signals below 3 GHz, mixer stages enable the desired upconversion to frequencies beyond 6 GHz. For generation of signals on the IF the NRZ mode is applied. To compare the system performance with the subsampling concept and to keep the oscillator circuit as simple as possible the mixer frontend is designed for RF frequencies between 2 GHz–3 GHz likewise. Equivalent to subsampling concepts a pseudo–coherent design is achieved by coupling of the oscillator signals of the DAC and the mixer. For upconversion of signals a passive diode ring double-balanced mixer is utilized (Mini-Circuits, 2011). By using a low pass (LP) anti-aliasing filter and the NRZ mode of the DAC, an IF signal with high SNR and significant suppression of spectral components in higher Nyquist zones is achieved. Using a BP filter for image rejection yields the desired RF output signal.

3 Analytical rationale of impulse generation

Finding and testing of appropriate impulse shapes for IR systems under real conditions is one of the purposes of the transmitter design. Fulfilling the EIRP constraints of the FCC modulated Gaussian pulses, Gaussian monocycle pulses or numerically generated pulses (Parr et al., 2003) are practicable for sending data from the transmitter to the receiver. Therefore, a signal model adapted to the hardware capability in terms of telecommunication engineering and signal bandwidth is presented.

Many signal models have been published in recent years (Win and Scholtz, 1998; Romme and Piazzo, 2002; Benedetto and Giancola, 2004). They offer the possibility of predicting the EIRP or to decide whether modulation schemes appear to be feasible for certain applications. With respect to the performance of the presented

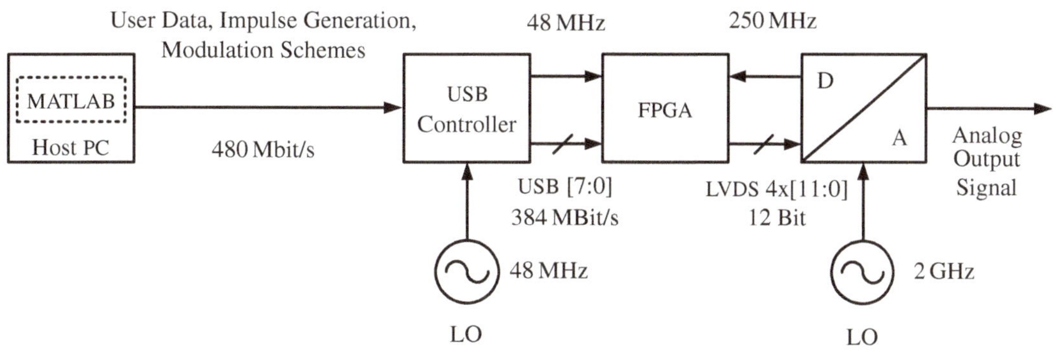

Fig. 1. Block diagram of the digital signal synthesis in the transmitter.

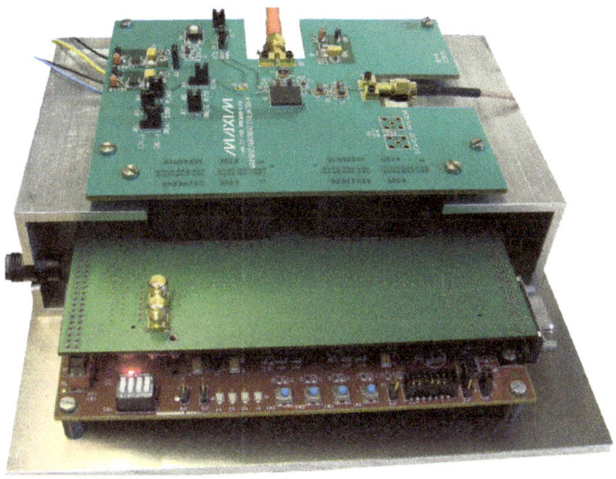

Fig. 2. Developed IR SDR UWB transmitter including FPGA and DAC development boards.

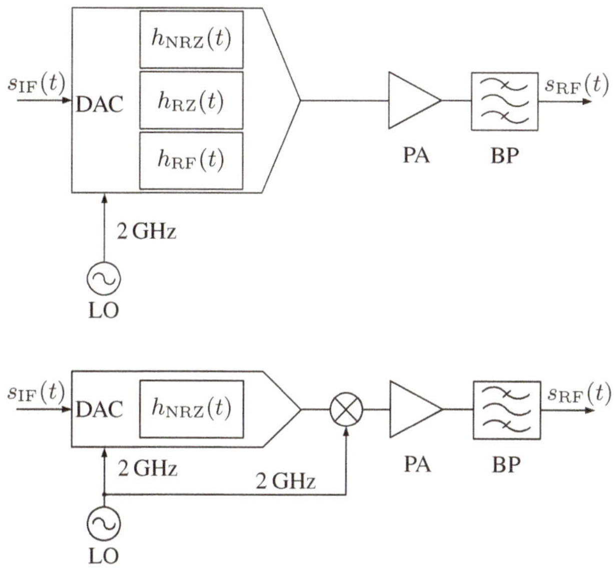

Fig. 3. Block diagram of the digital signal synthesis in the transmitter.

transmitter a signal model similar to (Win and Scholtz, 1998) is introduced. The transmitter supports classical 2-ary and 4-ary modulation schemes like on-off-keying (OOK), pulse-code-modulation (PCM), binary-pulse-amplitude-modulation (BPAM), quadrature-pulse-amplitude-modulation (QPAM), pulse-position-modulation (PPM) a combination of 2-ary modulation schemes or novel modulation schemes. Based on 4-ary modulation schemes a mapping of the binary input sequence to the output sequence $c(n) \in \{1,2,3,4\}$ is necessary. Hence the UWB signal

$$s(t) = \sum_{n=-\infty}^{\infty} a^{(c(n))} p^{(c(n))}(t - nT_S + \epsilon^{(c(n))}) \quad (1)$$

can be derived from an infinite sum of the amplitude $a^{(c(n))}$, an additional time shift $\epsilon^{(c(n))}$ and pulse shape $p^{(c(n))}(t)$ of each symbol. The pulse repetition time is denoted with T_S. Hence the impulse shape of a certain symbol $c(n)$ is described by

$$p_S^{(c(n))}(t) = a^{(c(n))} p^{(c(n))}(t + \epsilon^{(c(n))}) \quad (2)$$

and the resulting transmitted signal

$$s(t) = \sum_{n=-\infty}^{\infty} p_S^{(c(n))}(t - nT_S) \quad (3)$$

is obtained from an infinite series of different impulses.

As Gaussian impulses exhibit the smallest time to bandwidth product, they are perfectly suited for UWB applications. In (Blech et al., 2008) a method of defining Gaussian impulses according to the regulatory aspects of the FCC has been published. Following this proposal the shape of Gaussian impulses are defined in the frequency domain and adapted to the present signal model. Impulses with a Gaussian shape exhibit Gaussian characteristic in the time and frequency domain. By applying the Fourier transform, Gaussian impulses in the time and frequency domain

$$\mathrm{e}^{-\alpha t^2} \circ\!\!-\!\!\bullet \sqrt{\frac{\pi}{\alpha}}\mathrm{e}^{-\frac{1}{4\alpha}\omega^2} \tag{4}$$

are directly related to each other (Poularikas, 2000). In (4) the factor α scales the shape of the Gaussian impulse with respect to time. To fulfill the regulatory aspects it is evident to determine the amplitude of the time domain output signal. According to the linearity of the Fourier transform the correspondence

$$\hat{U}\mathrm{e}^{-\alpha t^2} \circ\!\!-\!\!\bullet \hat{U}\sqrt{\frac{\pi}{\alpha}}\mathrm{e}^{-\frac{\pi^2 f^2}{\alpha}} \tag{5}$$

is introduced, where $\hat{U}$ denotes the peak amplitude of the time domain signal. According to the FCC regulations UWB impulses have to exceed a bandwidth B of 500 MHz whereby the factor G_F at the cutoff frequency f_C scales the amplitude to -10 dB below the maximum amplitude of the impulse. Employing these initial conditions, the unknown variable α is derived from Eq. (5) according to

$$\alpha = -\frac{B^2\pi^2}{4\ln(G_\mathrm{F})}. \tag{6}$$

Replacing α in Eq. (5) the Gaussian impulse in the frequency domain

$$P_\mathrm{S}(f) = \hat{U}\sqrt{\frac{-\ln(G_\mathrm{F})}{\pi(B/2)^2}}\mathrm{e}^{\ln(G_\mathrm{F})(\frac{f^2}{(B/2)^2})} \tag{7}$$

can be directly adapted to the bandwidth definition of the FCC. In Fig. 4 a Gaussian impulse in the frequency domain with a -10 dB bandwidth of $B = 500\mathrm{MHz}$ is depicted. Thus the Gaussian baseband impulse in the time domain

$$p_\mathrm{S,BB}(t) = \hat{U}\mathrm{e}^{\frac{(B/2)^2\pi^2}{\ln(G_\mathrm{F})}t^2} \tag{8}$$

is likewise related to $\hat{U}$ and B. Based on Eq. (8) a Gaussian baseband impulse in the time domain is depicted in Fig. 5. As the Gaussian impulses have to be shifted from the baseband to the RF band, the baseband impulse is modulated with the carrier frequency f_RF according to

$$p_\mathrm{S,RF}(t) = \hat{U}\mathrm{e}^{\frac{(B/2)^2\pi^2}{\ln(G_\mathrm{F})}t^2}\cos(2\pi f_\mathrm{RF} t). \tag{9}$$

Gaussian waveforms with IF and RF carrier frequencies are also illustrated in Fig. 5. The IF impulse corresponds to the digitally synthesized impulse shape in the first Nyquist zone. The Gaussian impulses at the center frequency $f_\mathrm{RF} = 2.5\mathrm{GHz}$ is equivalent to the transmitted output signal.

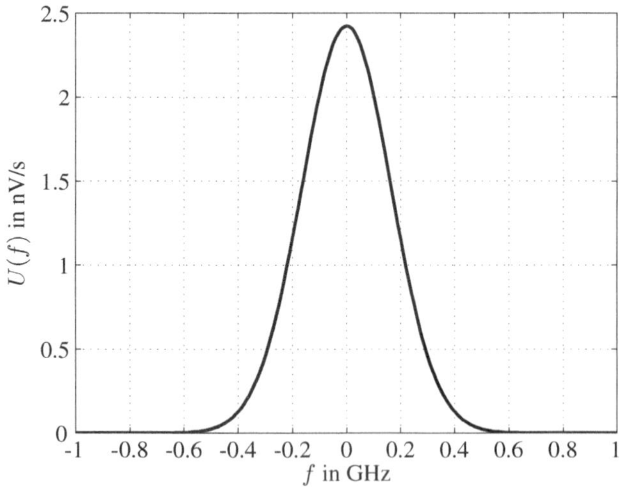

Fig. 4. Gaussian baseband impulse with a -10 dB bandwidth of 500 MHz in the frequency domain.

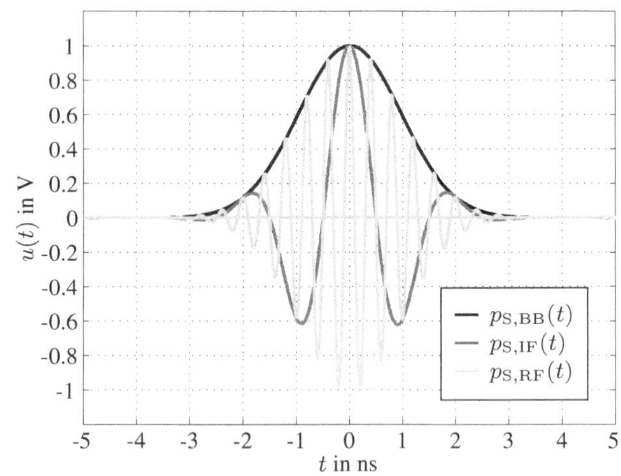

Fig. 5. Gaussian impulse modulated with different carrier frequencies and a −10 dB bandwidth of 500 MHz in the time domain.

Comparing the shapes of the baseband and RF Gaussian impulses, it is obvious that the envelope of the RF output signal maintains the shape of the baseband impulse response. The peak amplitude depends on the equivalent isotropically radiated power (EIRP) and is chosen individually for different modulation and coding schemes (Ott et al., 2012).

4 Results

Measurements of exemplary modulation schemes have been carried out to show the multifunctional performance of the transmitter. To achieve an obvious distinction between the performance of different modulation schemes, impulses are depicted in the first Nyquist zone and the symbol rate is

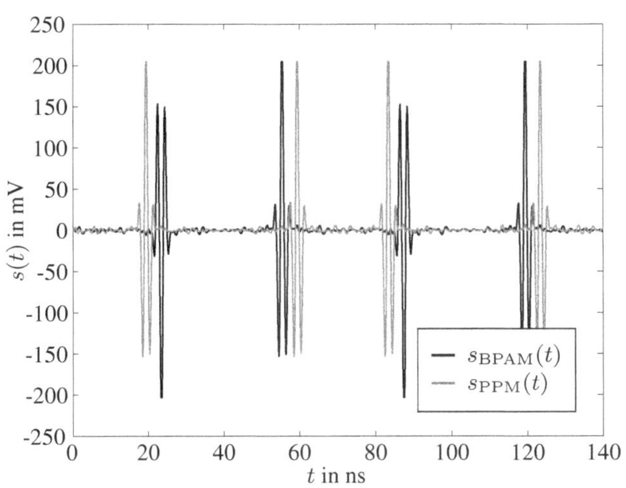

Fig. 6. Measurement of exemplary pulse trains with QPSK modulation scheme and TR-BPAM modulation technique provided by the transmitter.

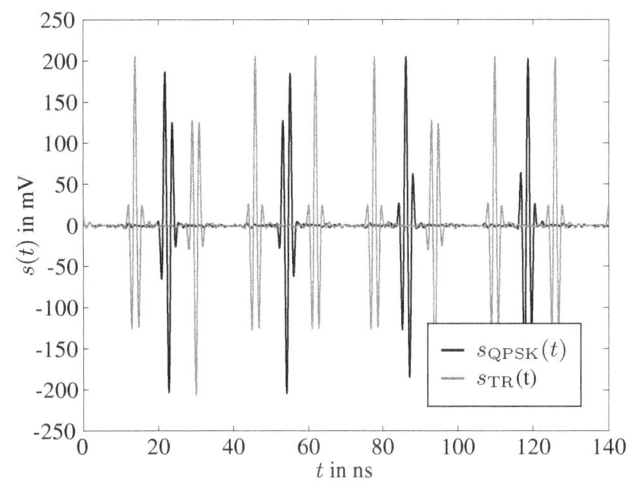

Fig. 7. Measurement of exemplary pulse trains with QPSK modulation scheme and TR-BPAM modulation technique provided by the transmitter.

chosen to 62.5 Mbit s^{-1}. A LeCroy WaveMaster 808 Zi-A oscilloscope has been employed for recording pulse trains with different modulation schemes. Representing 2-ary modulation schemes BPAM and PPM have been used and are depicted in Fig. 6. Four impulses of a binary sequence with changing sign have been extracted. As depicted in Fig. 6 sign or position of pulses changes when employing BPAM or PPM, respectively.

In equivalent way 4-ary modulation schemes are applicable by the transmitter. In Fig. 7 four individual waveforms are mapped on the input sequence and hence two bits are transmitted via one impulse. The orthogonal shape of the pulses is achieved by shifting the phase information of the carrier frequency by 90°. This IR modulation scheme has been introduced in (Blech et al., 2008) and can be seen as quadrature-phase-shift-keying (QPSK) pulse modulation or as PCM as every of the four impulses has an individual shape. Furthermore the transmitter enables to produce a combination of two 2-ary modulation schemes. Besides a various number of supported modulation schemes the transmitter offers transmitted reference (TR) modulation technique. Hence the afore mentioned modulation schemes can be additionally adapted to the TR modulation technique e.g. TR-PCM, TR-BPAM or TR-PPM. As an example TR-BPAM (TR-BPSK) is depicted in Fig. 7. In a TR system a reference as well as a data impulse is transmitted, whereas the first impulse is employed to demodulate the later impulse by cross correlation. As an advantage, no channel estimation is necessary at the receiver. However, the data rate reduces by a factor of two.

The performance analysis of the frontend is important for the future design of the transmitter. Gaussian impulses synthesized by the subsampling and the conventional concept are presented in Fig. 8. Using the RZ-mode of the DAC results in the best SNR performance of the IR SDR transmitter. Although the frequency response of the DAC is non-ideal, significant distortions of the impulse shape can not be recognized. However, a high fidelity of the Gaussian impulse is achieved by employing a mixer design likewise. As the NRZ mode has been applied for synthesizing the IF signal, a higher output signal amplitude compared to the subsampling concept is achieved. The correlation coefficient

$$\rho = \frac{\int_{-\infty}^{\infty} p_{\mathrm{S,RF}}(t)\, p_{\mathrm{S,RF,M}}(t)\mathrm{d}t}{\sqrt{\int_{-\infty}^{\infty} \left|p_{\mathrm{S,RF}}(t)\right|^2 \mathrm{d}t \int_{-\infty}^{\infty} \left|p_{\mathrm{S,RF,M}}(t)\right|^2 \mathrm{d}t}} \tag{10}$$

of the measured and the ideal Gaussian impulses is a criterion for the performance of the transmitter. In (10) $p_{\mathrm{S,RF,M}}(t)$ is either the measured impulse shape of the conventional or the subsampling concept. A very accurate impulse shape of the generated pulses with both frontend concepts is achieved as $\rho_{\mathrm{SDR}} = 0.991$ and $\rho_{\mathrm{Mixer}} = 0.986$ are the resulting correlation coefficients of the SDR and the conventional concept, respectively.

In order to utilize a transmitter operating in a frequency range between 6 GHz and 10 GHz, a pseudo coherent design including mixer stages will be applied in the future. The main reason is that there is currently no DAC available for shifting signals into the UWB frequency band. However, a frontend with high complexity is necessary to guarantee the coherent design, to minimize clock jitter and to avoid signal distortion.

5 Conclusions

For many applications IR technology is an appropriate solution. A reconfigurable development platform offers the opportunity for enhanced investigation of IR technology. In this

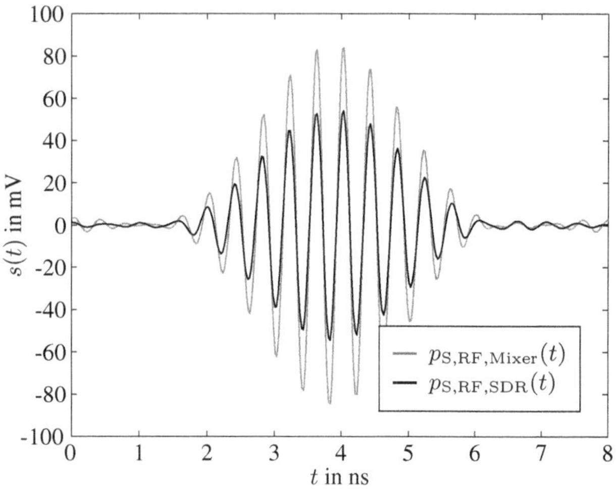

Fig. 8. Typical impulse synthesized by the SDR and the conventional concept in the 3rd Nyquist zone.

paper a prototype of an IR UWB transmitter has been presented. For RF transmissions either a conventional approach including mixer stages or a subsampling approach has been implemented. Both approaches have been compared and it turned out that a solution with a conventional architecture is preferred for the following design of the transmitter. The digital part of the transmitter offers a very high flexibility with respect to telecommunication engineering. Practical measurements of pulse trains with different modulation schemes and techniques have been performed and show the excellent performance of the transmitter.

Acknowledgement. The authors thank LeCroy for supporting the work.

The work described in this paper is related to research activities carried out during the affiliation of M. Blech with the Universität Stuttgart.

References

Batra, A., Balakrishnan, J., Dabak, A., et al.: IEEE Multi–band OFDM Physical Layer Proposal for IEEE 802.15 Task Group 3a, IEEE P802.15 Wireless Personal Area Networks, 2004.

Anderson, C. R.: A software-defined ultra-wideband transceiver testbed for communications, ranging, and imaging, Ph.D. thesis, Virginia Polytechnic Institute; State University, Blacksburg, VA, USA, 2006.

Avnet: Xilinx Virtex-5 LX50 Development Kit, http://www.avnet.com, 2006.

Benedetto, M. D. and Giancola, G.: Understanding Ultra Wide Band Radio Fundamentals, Prentice Hall, 1 edn., 2004.

Blech, M. D., Geier, D., and Eibert, T. F.: Concept of an UWB impulse radio B-/QPSK transmitter based on standard logic components, in: International Conference on Ultra-Wideband (ICUWB), 1, pp. 225–228, 2008.

Blech, M. D., Neumaier, P., Ott, A. T., Zan, A. A., and Eibert, T. F.: Performance analysis of a software defined subsampling ultra-wideband B-/QPSK impulse radio transceiver, in: European Wireless Technology Conference (EuWIT), 112–115, 2009.

Blech, M. D., Ott, A. T., Neumeier, P., Möller, M., and Eibert, T. F.: A reconfigurable software defined ultra-wideband impulse radio transceiver, Adv. Radio Sci., 8, 67–73, doi:10.5194/ars-8-67-2010, 2010.

Chen, S. W. and Brodersen, R. W.: A Subsampling Radio Architecture for Ultrawideband Communications, IEEE T. Signal Proces., 55, 5018–5031, 2007.

DecaWave: ScenSor, http://www.decawave.com, 2011.

FCC: Revision of Part 15 of the Commission's Rules Regarding Ultra-Wideband Transmission Systems,http://www.fcc.gov, 2002.

Keller, C. M., Burkhart, J. M., and Phuong, T. T.: Ultra-Wideband Direct Sampling Receiver, in: International Conference on Ultra-Wideband (ICUWB), 387–392, 2007.

Maxim Integrated Products, Inc.: 12-Bit, 2.3Gsps, Multi-Nyquist DAC with Selectable Frequency Response, http://www.maxim-ic.com, 2006.

Mini-Circuits: ADE-30+ Frequency Mixer, http://minicircuits.com, 2011.

Ott, A. T., Eisner, C. J., and Eibert, T. F.: Enhanced Investigations on Effective Isotropic Radiated Power Emissions of Impulse Radio Devices, in: 6th European Conference on Antennas and Propagation (EuCAP), 1–5, 2012.

Parr, B., Cho, B., Wallace, K., and Ding, Z.: A novel ultra-wideband pulse design algorithm, Communications Letters, 7, 219–221, 2003.

Poularikas, A. D.: The Transforms and Applications Handbook, CRC Press, 2 edn., 2000.

Fisher, R., Kohno, R., Ogawa, H. et al.: DS-UWB Physical Layer Submission to 802.15 Task Group 3a, IEEE P802.15 Wireless Personal Area Networks, 2004.

Reed, J. H.: An Introduction to Ultra Wideband Communication Systems, Prentice Hall, 2005.

Romme, J. and Piazzo, L.: On the power spectral density of time-hopping impulse radio, in: Conference on Ultra Wideband Systems and Technologies, 241–244, 2002.

Time Domain Corporation: PulsOn 400 RCM, http://www.timedomain.com, 2011.

Verhelst, M., Ryckaert, J., Vanderperren, Y., and Dehaene, W.: A Low Power, Reconfigurable IR-UWB System, in: International Conference on Communications (ICC), 3770–3774, 2008.

Win, M. Z. and Scholtz, R. A.: Impulse radio: how it works, Communications Letters, 2, 36–38, 1998.

11

Influences of weather phenomena on automotive laser radar systems

R. H. Rasshofer[1], **M. Spies**[2], **and H. Spies**[2]

[1]BMW Group Research and Technology, Hanauer Str. 46, 80992 Munich, Germany
[2]Ingenieurbüro Spies, Seibersdorf 18, 86558 Hohenwart, Germany

Abstract. Laser radar (lidar) sensors provide outstanding angular resolution along with highly accurate range measurements and thus they were proposed as a part of a high performance perception system for advanced driver assistant functions. Based on optical signal transmission and reception, laser radar systems are influenced by weather phenomena. This work provides an overview on the different physical principles responsible for laser radar signal disturbance and theoretical investigations for estimation of their influence. Finally, the transmission models are applied for signal generation in a newly developed laser radar target simulator providing – to our knowledge – worldwide first HIL test capability for automotive laser radar systems.

1 Introduction

Advanced driver assistance systems (ADAS) develop from comfort enhancement to safety applications. With a constantly growing need for higher perception sensor data quality, laser-based sensors tend to dominate many experimental intelligent vehicle systems with applications ranging from pedestrian protection (see Walchshäusl et al., 2006) to fully autonomous driving (see Darms et al., 2009). New, promising signal processing approaches like e.g. occupancy grid based methods adopted from robotics (see Thrun et al., 2005) heavily rely on laser radar sensors and pave the way for highly automated driver assistance. Compared to millimeter-wave radar, laser radar (lidar) systems provide higher angular resolution in the azimuth plane being able to separate targets spaced less than 1 deg apart. This is a key feature for many ADAS applications since high angular resolution is crucial for the determination of object width and shape. This information provides valuable input to object classification algorithms and tracking systems that precisely determine the object's orientation. Moreover, laser radar systems using more than one detection layer are providing angular resolution even in elevation. This feature is especially important for automatic alignment procedures, for robust pitch angle compensation, for lane marking detection and for estimation of object height. Although measurement capability in elevation is theoretically possible with millimeter-wave radar, too, no such products have shown up on the automotive radar market so far. This is mostly due to the very high complexity of radar front-ends exhibiting two-dimensional millimeter-wave beam steering. Although from a performance point of view, laser radar seems to be ideally suited for use in ADAS, its sensitivity to weather phenomena such as fog, rain or snow has been considered a major drawback for these perception sensors and limited their use on the mass market to low-cost systems. This work deals with the primary physical effects influencing laser radar systems under bad weather conditions. From laser radar theory, we derive a model for the raw signals to be expected and subsequently estimate the performance degradation. Moreover, the developed signal model is used for signature generation and verification in a novel laser radar HIL test system which provides for the first time the possibility to evaluate and optimize weather-based influences on laser radar performance in the laboratory in real time.

Correspondence to: M. Spies
(martin.spies@ib-spies.de)

2 Laser radar fundamentals

2.1 Target detection

Laser radar systems, sometimes called LADAR (Laser radiation detection and ranging), LIDAR (light detection and ranging) and LIVAR (laser-illuminated viewing and ranging) use a modulated, intense source of light to illuminate the sensor's field of view containing objects to be detected. From the objects in the sensor's field of view, a certain part of optical power is reflected back towards the sensor where the signal is detected by an optical receiver. In most automotive laser radar systems, optical signal transmitters and receivers are located close to each other or even use the same optics in a coaxial beam configuration. For successful detection, the received optical signal has to be larger than the sensitivity limit of the optical receiver. For automotive applications, only incoherent detection has been realized so far. As optical receivers, PIN-photo diodes (PIN doping profile: p^+, i (intrinsic), n) and avalanche photo diodes (short: APD) typically are used. Both types of diodes feature higher bandwidth than standard *pn*-photo diodes. Compared to PIN-diodes, APDs usually provide a superior signal-to-noise ratio (SNR) due to their intrinsic gain, however require complicated biasing (see Smith et al., 1978).

2.2 Transmitter design

Eye safe operation is a must for automotive laser radar systems. As these systems should be laser class 1 or class 1M products, their design has to comply with the rules given in IEC 60825-1:2007 (2007) which limit the peak power of a single pulse and the average power depending on field of view, pulse burst rate and lens aperture. Since most automotive laser radar systems use semiconductor laser diodes as optical transmitters, the system's wavelength is either $0.850\,\mu$m or $0.905\,\mu$m. For these wavelengths, low-cost pulse laser diodes and low-noise APDs are readily available.

2.3 Range measurement

Due to the modulation of the transmitted signal, the time difference t (sometimes called TOF time of flight) between illumination signal emission and detection of the signal reflected by the target can be measured. Since t is directly proportional to the distance the signal traveled, the range R of the reflection point can be deduced, assuming c to be the known propagation velocity of the optical signal in the given transmission medium.

$$R = ct/2. \tag{1}$$

From fundamental pulse radar physics (see e.g. Skolnik, 1990), it is known that the system's range resolution δR, – i.e. the capability of the laser radar to separate two detections which are located at a radial distance δR from each other within one beam – is limited by the bandwidth of the transmitted signal and the bandwidth of the optical detector.

2.4 Angular measurement

To measure the angle between sensor and reflection point, different techniques can be applied. Most laser radar sensors are using a sharply limited illumination beam which only illuminates an area limited by $\pm\vartheta_T$ in elevation and $\pm\varphi_T$ in azimuth. The receiver's optics, however uses a projection only observing a sharply limited detection area characterized by $\pm\vartheta_R$ in elevation and $\pm\varphi_R$ in azimuth. Object detection can only take place if a reflection point is located within the illumination beam of the transmitter and the detection area of the receiver. The projection of both the illumination beam and the reception area typically is not an ellipse, however. A rectangular assumption can be used in most cases in order to enable analytic calculation as shown by Spies and Spies (2006b). In most laser radar sensors, the field-of-view is divided into N detection sectors. In multi-beam systems, the sectors are illuminated by a number of N_T transmitters and observed by N_R receivers. For coaxial optics using the same optical path both for transmit and receive, $\vartheta_T = \vartheta_R$ and $\varphi_T = \varphi_R$ yields optimal performance. For bistatic configurations, offsets and differences between illuminated area and detection area have to be taken into account in the vicinity of the sensor as will be shown in Sect. 3.3.

To cover the complete field of view, two basic system types have been developed: systems using mechanical scanning of the detection area (so called "laser scanners") and systems using multiple transmitters and receivers (so called "multi-beam laser radar" or "staring laser radar"). Laser scanners tend to provide higher angular resolution than multi-beam laser radar, however due to their larger mechanical complexity often exhibit higher system costs and larger package size.

For identification of ground clutter, to compensate for horizontal misalignment and for target classification, some laser scanners use multiple layer techniques to provide some degree of resolution in elevation as stated by Spies and Spies (2006a).

2.5 Velocity measurement

With incoherent light detection typically used in automotive laser radar systems, a direct measurement of a target's velocity due to the DOPPLER effect is not possible. Coherent detection, often used in military DOPPLER laser radar (see Owens, 1969) is not feasible for automotive use since the coherence length of the semiconductor lasers employed by all sensors on the market today is not large enough to enable low-noise carrier frequency DOPPLER measurement. Today, velocity determination typically is performed by tracking of reflection points' movements over time. For optimized tracking, data update rates well above 15 Hz are necessary to cope with the dynamic requirements of future driver assistant systems. Attention has to be paid to the selection of the reflection points to be tracked. Sudden jumps of reflection points of periodic structures within the sensor's field of view tend to introduce velocity measurement errors. Moreover, for the detection of object velocity, also the shape of the object has to be taken into account. Especially in the vicinity of the sensor, laser radar often tends to see more than one reflection point on the surface of extended objects like e.g. cars. In this case, shape-based tracking techniques (see Kirchner and Ameling, 2000) like e.g. L-Shape tracking (see Fayad and Cherfaoui, 2007) have to be applied to obtain valid and precise results for lateral and longitudinal velocity of the tracked objects.

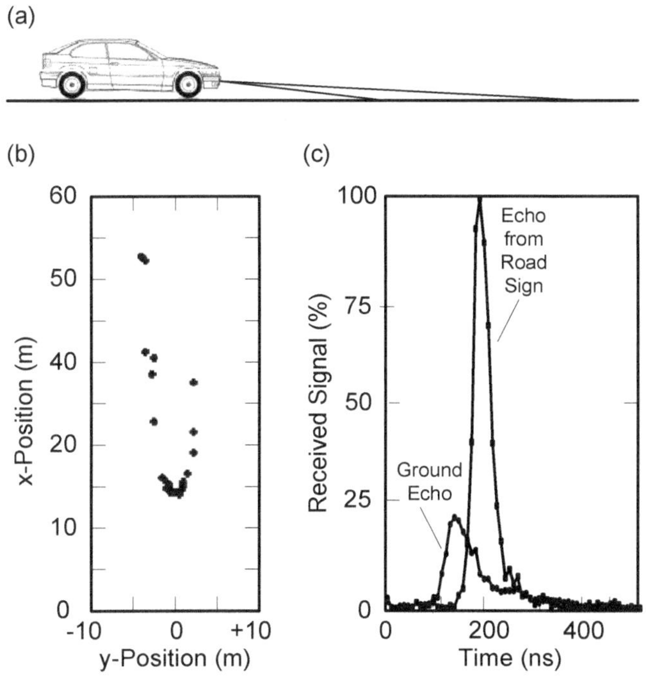

Fig. 1. **(a)** Beam configuration leading to detection of ground echo. **(b)** Typical spatial distribution ground echo detection, basis of shape-based ground echo classification. **(c)** Time domain classification of ground echo based on different time domain response of rigid targets (e.g. road signs, other cars etc.) and ground signal with different time domain response as shown by Issa (2007).

2.6 Raw data based target classification

Besides target detection and location, classification based on raw reflection data has become very important in many driver assistance and active safety systems. Generally speaking, the following classes of raw detections have to be distinguished:

Type 1. Raw detections from solid objects like cars, pedestrians or other infrastructure. In this case, the scattering plane has almost no depth-/range distribution.

Type 2. Raw detections from noise in the optical detector or detector amplifier chain. These detections can be dealt with by the application of constant false alarm rate (CFAR) techniques in the target detector, residual detections will be of statistical nature and thus will be eliminated in the tracker.

Type 3. Raw detections from electromagnetic interference (EMI) from inside or outside of the sensor e.g. power supply spikes, clock interference, laser diode modulator crosstalk etc. These detections must be carefully resolved on the hardware level by proper system design and sophisticated mechanical setups including highly effective shielding techniques. Especially periodic interference tends to be hard to deal with on a system level since these kinds of detections will not be removed in the tracker because of their periodic, deterministic nature.

Type 4. Raw detections from large signal effects in the receiver chain. These detections arise from clipping or ringing in the receive amplifier chain after application of very large optical receive signals which might be due to e.g. retro-reflecting objects. To minimize these detections, gain control in the receive amplifier chain and/or control of the transmit power must be utilized.

Type 5. Targets due to ground echo/ground clutter, often generated by the car's pitch or by bumpy roads. Since these detections do come from real world reflections, their classification must be based on additional information. If a multi layer laser radar sensor is used, a comparison between detections in different layers might be used to identify the ground echo. If the laser radar exhibits only a single layer, two methods of classification are known as described in the the work of Issa (2007): a shape-based approach, taking advantage of the special location of the ground echo detections which is based on the beam geometry and the density distribution of the transmitters (e.g. detections matching parabolic shape, see Fig. 1b and ground echo recognition based on the distributed nature of the reflection points like shown in Fig. 1.

Type 6. Detections generated by reflections in rain, fog, dust or snow. The following sections will show how the return signals for these kinds of detections look like.

Automotive laser radar sensors have to distinguish between the different types of the raw detections listed above and should only pass detections of type 1 to the subsequent clustering and tracking algorithms. All other detections should either be minimized by system design or should be clearly recognized as not relevant for the driver assistance function, providing valuable information for the subsequent signal processing (e.g. data association and tracking) Many laser radar systems show deficiencies within this classification so careful evaluation has to take place during the sensor test. Especially detections of type 6 are hard to test since environmental test conditions like rain, snow or fog are difficult to reproduce under lab conditions.

3 Modeling of laser radar performance

3.1 Review of the laser radar equation

The laser radar equation describes the received power $P_R(R)$ at a laser radar's detector as a function of a target's range. For non-elastic scattering, the received signal power can be calculated as a convolution between the transmit signal P_T and the spatial impulse response $H(R)$ of the optical channel and the target as stated e.g. by Zhao et al. (1988):

$$P_R(R) = C_A \int_{t'=0}^{2R/c} P_T(t')\ H(R - ct'/2)dt' \tag{2}$$

The system constant C_A is not depending on range or time and is given by:

$$C_A = c\eta A_R/2. \tag{3}$$

A_R represents the aperture area of the sensor's optical receiver and η characterizes losses in the receiver's optic.

In the subsequent sections, a detailed review of all factors appearing in Eq. (2) will be carried out, considering automotive application's constraints.

3.2 Transmit pulse modeling

In classic laser radar theory, a DIRAC-shaped transmit pulse with a total pulse energy E_P according to Eq. (4) is assumed:

$$P_{T\delta}(t) = E_P \delta(t) \tag{4}$$

Using the rather simple transmission pulse model given in Eq. (4), the classic laser radar equation is obtained:

$$\begin{aligned} P_{R\delta}(R) &= C_A \int_{t'=0}^{2R/c} E_P \delta(t')\, H(R - ct'/2) dt' \\ &= C_A E_P H(R). \end{aligned} \tag{5}$$

This model is valid for target ranges $R \gg c\tau_P$ with τ_P denoting the duration of the transmit pulse.

In most automotive sensor systems found on the market today, τ_P is in the range of 10 to 20 ns resulting in a geometrical pulse length $c\tau_P$ of 3 to 6 m. As the sensitivity of automotive laser radar systems to fog and rain is crucial, a sound understanding of particle scattering in the range from 5 to 15 m in front of the laser radar is of special importance. For this target range, the simple laser radar equation is not valid and a careful evaluation of Eq. (2) has to be carried out.

A simple but yet realistic model for the time signature $P_T(t)$ of automotive lidars' transmit pulses is a $\sin^2$-function according to Eq. (6):

$$P_T(t) = \begin{cases} P_0 \sin^2\left(\frac{\pi}{2\tau_H} t\right) & 0 \le t \le 2\tau_H \\ 0 & \text{else} \end{cases} \tag{6}$$

This model is valid for most semiconductor laser diodes powered by state-of-the-art driver circuits. P_0 denotes the peak power of the laser radar pulse and τ_H its half-power pulse width. The total energy E_P of a single $\sin^2$-impulse as defined by Eq. (6) is:

$$E_P = P_0 \tau_H \tag{7}$$

Typical values of today's pulse laser radar systems exhibit a peak power P_0 of up to 80 W resulting in pulse energy up to 1.6 μJ.

A generic model for the pulse shape of laser radar transmit signals is given in Steinvall et al. (2006) and analyzed for its impact on laser radar's performance.

3.3 Spatial impulse response function

The spatial impulse response function H can be split into two functions describing the spatial impulse responses of the optical channel H_C and the spatial impulse response of the target H_T: $H = H_C H_T$.

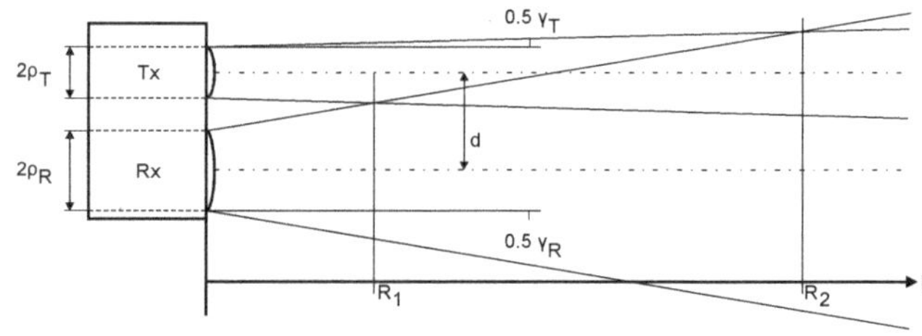

Fig. 2. Laser radar using a bistatic beam configuration.

The channel's spatial impulse response H_C is modeled as

$$H_C(R) = \frac{T^2(R)}{R^2} \xi(R). \tag{8}$$

In Eq. (8) $T(R)$ represents the total one-way transmission loss introduced by losses in the transmission medium:

$$T(R) = \exp\left(-\int_{r=0}^{R} \alpha(r) dr\right) \tag{9}$$

$\alpha(R)$ represents the local extinction coefficient characterizing the local, differential power loss in a distance R. Details on the calculation of $\alpha(R)$ in fog, rain or snow will be given in Sects. 3.5 and 3.6.

In Eq. (8), $\xi(R)$ describes the crossover function defined by Halldorsson and Langerhoic (1978) or Sassen and Dodd (1982) as ratio between the area A_T illuminated by the transmitter and the area A_R observed by the optical receiver:

$$\xi(R) = \begin{cases} \frac{A_R(R) \cap A_T(R)}{A_T(R)} & \text{if } A_R(R) \cap A_T(R) < A_T(R) \\ 1 & \text{else.} \end{cases} \tag{10}$$

In case of a coaxial transmit/receive optics, $\xi(R)$ is a constant factor not depending on range.

In the important case of bistatic beam configuration using circular illumination and detection areas with parallel optical axis exhibiting a displacement d, the radius R_1 of first contact between transmit and reception area is given by

$$R_1 = \frac{d - \rho_T - \rho_R}{\tan(\gamma_T/2) + \tan(\gamma_R/2)} \tag{11}$$

Here, ρ_T and ρ_R denote the radius of the transmission and reception aperture and γ_T and γ_T the total opening angle of the transmitter's and detector's field of view, respectively. For a distance larger than R_2, the transmit- and reception area overlap completely which yields $\xi(R) = 1$ for $R > R_2$.

$$R_2 = \frac{d - \rho_R + \rho_T}{\tan(\gamma_R/2) - \tan(\gamma_T/2)} \tag{12}$$

It is clear that in this kind of beam configuration, $\xi(R)$ varies strongly with the distance. Defining

$$r_T = R\tan(\gamma_T/2) + \rho_T \quad \text{and} \quad r_R = R\tan(\gamma_R/2) + \rho_R \tag{13}$$

$$\phi_{\mathrm{T}} = 2\arccos\left(\frac{r_{\mathrm{T}}^2 - r_{\mathrm{R}}^2 + d^2}{2 d r_{\mathrm{T}}}\right) \tag{14}$$

$$\phi_{\mathrm{R}} = 2\arccos\left(\frac{r_{\mathrm{R}}^2 - r_{\mathrm{T}}^2 + d^2}{2 d r_{\mathrm{R}}}\right) \tag{15}$$

the crossover function $\xi(R)$ is given as

$$\xi(R) = \begin{cases} 0 & \text{if } R \leq R_1 \\ \frac{r_{\mathrm{T}}^2(\phi_{\mathrm{T}} - \sin\phi_{\mathrm{T}}) + r_{\mathrm{R}}^2(\phi_{\mathrm{R}} - \sin\phi_{\mathrm{R}})}{2\pi r_{\mathrm{T}}^2} & \text{if } R_1 < R < R_2 \\ 1 & \text{if } R \geq R_2 \end{cases} \tag{16}$$

Please note that this simplified calculation is based on a homogeneous intensity distribution over transmit beam. If laser radar systems with more complex beam configurations have to be analyzed, the same calculation steps might be used, however $\xi(R)$ might not always be available in closed form.

3.4 Modeling hard targets

Properties of laser radar targets are described by the spatial impulse response function $H_{\mathrm{T}}(R)$. Laser radar pulses reflected at plain surfaces of solid objects are said to hit a hard target if the time signature of the laser pulse is not changed upon reflection. In this case, a simple model for the hard target spatial impulse response $H_{\mathrm{T}\delta}(R)$ can be derived. A hard target located at $R = R_0$ featuring a target area of A_{TA} exhibits a spatial impulse response $H_{\mathrm{T}\delta}$ given as:

$$H_{\mathrm{T}\delta}(R) = \begin{cases} \beta_0 \delta(R - R_0) & \text{if } A_{\mathrm{TA}} \geq A_{\mathrm{T}}(R_0) \\ \beta_0 \delta(R - R_0)\frac{A_{\mathrm{TA}}}{A_{\mathrm{T}}} & \text{if } A_{\mathrm{TA}} < A_{\mathrm{T}}(R_0). \end{cases} \tag{17}$$

$A_{\mathrm{T}}(R_0)$ denotes the cross-sectional area of the transmit beam at R_0 and β_0 describes the differential reflectivity of the target. In the important case of LAMBERT reflection characteristic with a reflectivity of $0 < \Gamma \leq 1$, it is given by:

$$\beta_0 = \Gamma/\pi. \tag{18}$$

In laser radar systems featuring very high range resolution, solid objects might not appear as hard laser radar targets if the laser beam – due to its lateral expansion – is hitting points of different range on the solid object. Errors introduced to range measurements by this effects have been analyzed by Grönwall et al. (2007) but might be considered of lower importance for automotive laser radar systems on the market today.

Depending on the geometry of the target, the angle of incident and the target's surface properties, the energy reflected by the target's surface has different angular distributions. Concerning the automotive environment, three main characteristics of angular power distribution can be found:

Type 1. Speckling reflections on flat, clean surfaces like glass or polished acrylic lacquer. Here, most of the energy is not reflected back to the laser radar receiver. This effect tends to largely increase laser radar target reflectivity variation and can be held responsible for target loss even for small target distances along with false detections due to unintended reflections from the environment.

Type 2. Diffuse reflections from e.g. dirty sheet metal or the road surface. These reflections can be modeled by a LAMBERT reflection characteristic.

Type 3. Retro reflection taking place at tail lights, license plates or road signs. In this case, almost all incident energy is directly reflected back to the location of the laser radar receiver. This kind of reflection usually yields best laser radar long-range performance but might overload or saturate the laser radar receiver chain for short range targets. In case of retro reflection, the target's reflectivity Γ might become significantly larger than 10.

Real world traffic objects usually represent a mixture of these 3 types of reflection characteristics. A generalized model for the angular energy distribution of reflected laser radar pulses is given by the bidirectional reflection distribution function (BRDF).

3.5 Influences of rain and fog

Soft laser radar targets provide distributed scattering of the transmit pulse. Their spatial impulse response function H_{T} is described by the backscattering coefficient $\beta(R)$: $H_{\mathrm{T}}(R) = \beta(R)$.

Laser pulse transmission through rain and fog has been subject to intense research in recent years. As shown by many authors, both rain and fog consist of small water droplets. Scattering at a single droplet of water is modeled by scattering of an electromagnetic wave at a dielectric sphere exhibiting a diameter D and a complex refraction index $\underline{n}_W$ which is depending on the wavelength. A typical value is $\underline{n}_W = 1.323520 + \mathrm{j}\,5.150 \times 10^{-7}$ for $\lambda = 905$ nm.

For a single, spherical water droplet, the extinction efficiency $Q_{\mathrm{EXT}}(D)$ and the backscattering efficiency $Q_{\mathrm{B}}(D)$ can be calculated using MIE'S theory described in Bohren and Huffman (1983). Figure 3.5 shows the corresponding values for Q_{EXT} and Q_{B} as a function of droplet diameter D.

To model rain or fog, a statistical distribution of water droplets featuring different diameters D can be used. The probability to hit a droplet of diameter D is said to be $N(D)$. Assuming only single, non-elastic scattering, i.e. the light scattered by a water droplet is not scattered by another droplet and no energy is converted to other wavelength, the extinction coefficient α and the backscattering coefficient β can be calculated to be:

$$\alpha = \frac{\pi}{8} \int_{D=0}^{\infty} D^2 Q_{\mathrm{EXT}}(D) N(D) dD \tag{19}$$

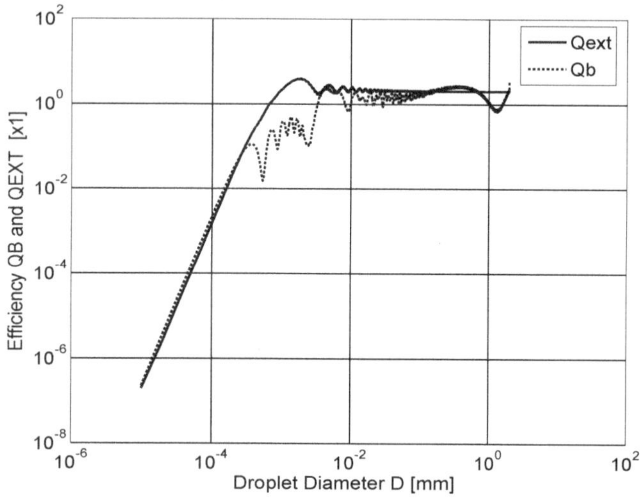

Fig. 3. Extinction efficiency $Q_{\mathrm{EXT}}(D)$ and backscattering efficiency $Q_{\mathrm{B}}(D)$ for a water droplet of diameter D at $\lambda = 905$ nm.

$$\beta = \frac{\pi}{8} \int_{D=0}^{\infty} D^2 Q_{\mathrm{B}}(D) N(D) dD \tag{20}$$

In the past, many authors studied the distribution of droplet diameters for rain and fog resulting in different models for $N(D)$. Marshall and Palmer (1948) modeled the diameter of rain droplets as an exponential function of rain rate R_{R} which is expressed in mm h^{-1}:

$$N(D) = 8000 \frac{1}{\mathrm{m\,mm}} \exp\left(-4.1\, R_{\mathrm{R}}^{-0.21} D\right) \tag{21}$$

As turned out during our investigations (see Adelsperger, 2008), this simple model does not hold true for laser radar pulse transmission due to the short measurement time in which large deviation from the exponential distribution (21) are likely. Feingold and Levin (1985) used a Log-normal distribution to improve the model of Marshall and Palmer (1948), however we found that using a Gamma distribution according to Deirmendjian (1969) provides most flexibility to model both rain and fog (see Isaac et al., 2001). In their model, four parameters α, γ, ρ and r_{C} describe the distribution $N(D)$:

$$N(D) = \frac{\gamma \rho b^{\frac{\alpha+1}{\gamma}}}{\Gamma\left(\frac{\alpha+1}{\gamma}\right)} \left(\frac{D}{2}\right)^{\alpha} e^{-b\left(\frac{D}{2}\right)^{\gamma}} \tag{22}$$

with

$$b = \frac{\alpha}{\gamma\,(D_{\mathrm{C}}/2)^{\gamma}} \tag{23}$$

In Eq. (22), $\Gamma(\cdot)$ denotes the Gamma function and D_C denotes the droplet radius having maximum probability.

Table 1. Parameters of droplet size distribution in rain using a gamma-function model.

Weather condition	ρ [m^{-3}]	α	γ	r_{C} [mm]
Rain (coast)	1000	1	0.5	0.05
Rain (continental)	1000	2	0.5	0.07

Table 2. Parameters of droplet size distribution in fog using a gamma-function model.

Weather condition	ρ [cm^{-3}]	α	γ	r_{C} [μm]
Haze (coast)	100	1	0.5	0.05
Haze (continental)	100	2	0.5	0.07
Strong advection fog	20	3	1.0	10.00
Moderate advection fog	20	3	1.0	8.00
Strong spray	100	6	1.0	4.00
Moderate spray	100	6	1.0	2.00
Fog of type "Chu/Hogg"	20	2	0.5	1.00

Tables 1 and 2 show the parameters of the droplet size distribution for various types of rain and fog. In the automotive environment, however, the given distributions are strongly influenced by spray from other cars or trucks and thus tend to vary strongly from the given model. Although the given models are not accurate in these cases, they provide a good starting point for further investigations and a good first guess on the backscattered power levels and the transmission power loss factors to be expected.

3.6 Snow

Snowflakes may consist of single ice crystals or aggregates of ice crystals featuring varying mass densities ranging from 0.005 g cm^{-3} to 0.5 g cm^{-3} depending on temperature and various other conditions see Oguchi (1983); Rasmussen et al. (1999). The size distribution $N(D)$ of the snowflake aggregates can be modeled to be

$$N(D) = N_0 \exp(-\Lambda D). \tag{24}$$

depending on the precipitation rate R_{P} measured in millimeters of equivalent liquid water per hour, see Nebuloni and Capsoni (2008). Table 3 shows parameters of this distribution as determined by different authors.

Electromagnetic modeling of single snowflakes turns out to be very difficult due to their complex geometric shape and their strongly varying water content. Investigations were carried out by different authors modeling ice crystals as aggregates of dielectric spheres as investiged by Neshyba (2003), hexagonal slabs considered by Yang and Liou (1996) or using an extension to MIE'S theory as done by Zhou et al. (2003). A rough estimation of the extinction coefficient α_{Snow} in case of dry and wet snow is given in Eqs. (25) and (26), respectively.

Table 3. Parameters of size distribution of snowflakes.

Author	Reference	N_0	Λ
Gunn et al.	Gunn and Marshall (1957)	$7.6 \times 10^3 \times R^{-0.87}$	$2.55 \times R^{-0.48}$
Sekhon et al.	Sekhon and Srivastava (1970)	$5.0 \times 10^3 \times R^{-0.94}$	$2.29 \times R^{-0.45}$

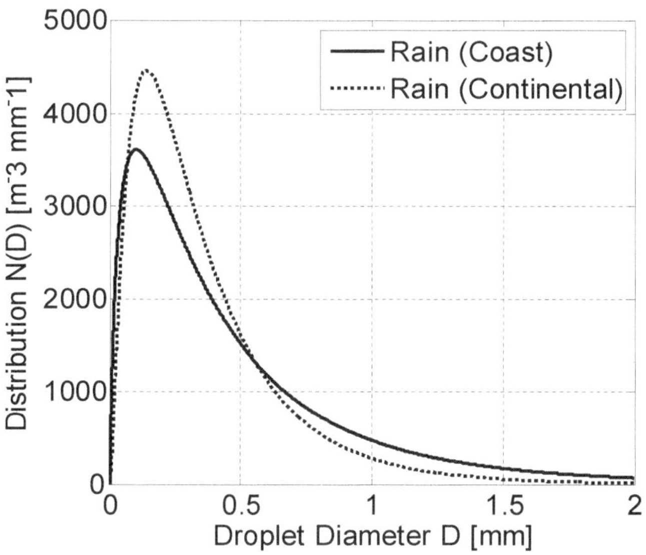

Fig. 4. Distribution of rain droplets according to model Eq. (22) with parameters of Table 1.

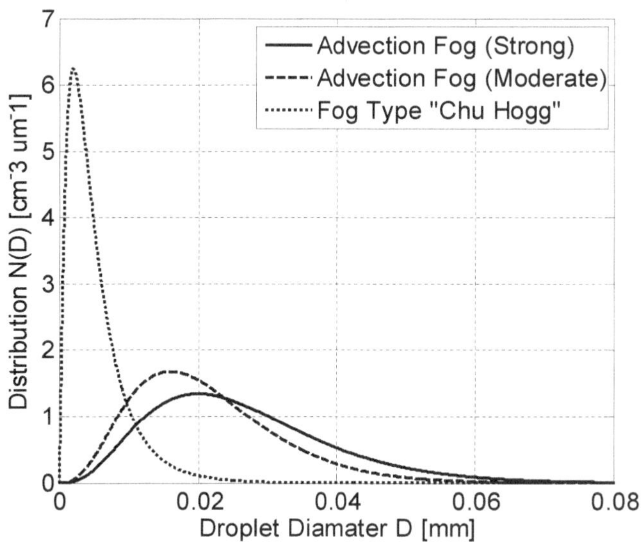

Fig. 5. Distribution of fog droplets according to model Eq. (22) with parameters of Table 2.

$$\alpha_{\text{Snow,Dry}} \approx 15\,\text{h}\,\text{mm}^{-1}\,\text{dB}\,\text{km}^{-1} R_\text{P} + 1\,\text{dB}\,\text{km}^{-1} \tag{25}$$

$$\alpha_{\text{Snow,Wet}} \approx 2\,\text{h}\,\text{mm}^{-1}\,\text{dB}\,\text{km}^{-1} R_\text{P} - 0.1\,\text{dB}\,\text{km}^{-1} \tag{26}$$

3.7 Unwanted raw detection due to weather phenomena

Backscattering from snowflakes or water droplets turns out to be a source of unwanted detections in automotive laser radar systems as explained in Sect. 2.6. In a critical distance range $0 < R < R_2$ ($R_2 < 10$ m in most sensor systems) where the crossover function $\xi(R)$ (see Eq. 10) is rising from 0 to 1, the transmit beam power density is very high so snowflakes or water droplets acting as small but efficient reflectors are likely to generate false alarms. Although – at first glance – these detections might be easily removed by means of tracking filters, the necessity for filtering introduces time delays or rises detection threshold levels while still not guaranteeing total removal of unwanted detections which are not tolerable in safety-critical vehicle functions like e.g. autonomous emergency braking. The situation becomes worse in case of coaxial beam configurations where $R_2 = 0$. The receive power levels generated by snowflakes or water droplets in these cases are surprisingly high and tend to easily overload the optical receiver chain. Adaption of transmit power between pulses (inter pulse transmit power control) is a common cure, however leads to further reduction of the maximum detection range which is already low in case of heavy rain and snow anyway. Another approach to improve the situation is using a receiver architecture featuring enhanced dynamic range and using fast gain control mechanisms in every amplifier stage which, however, adds cost and complexity to the sensor.

During a winter-time measurement campaign, we determined the false alarm rate of a scanning laser radar featuring a coaxial beam configuration. As can be seen in Fig. 6, the large transmit power density of the laser radar and the high reflectivity of snowflakes generates a false alarm rate of up to $2.5\,\text{s}^{-1}\,\text{m}^{-2}$ in a range of 5 m in front of the sensor. Using a bistatic beam configuration with properly designed opening angles $\gamma_{\text{T/R}}$ and TX/RX spacing d, the false-alarm rate can be reduced, however still will require careful attention. This is especially true if the sensor user does not have access to the raw detections of the system but has to rely on the analysis of tracked data during sensor evaluation.

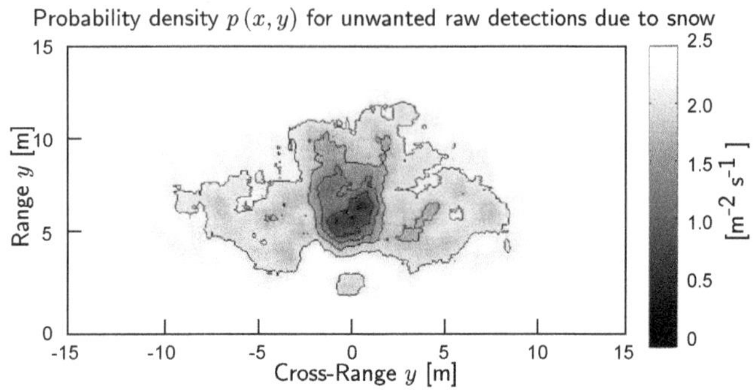

Fig. 6. Probability density $p(x,y)$ of unwanted raw target detections due to heavy snowfall, measured with an automotive laser radar system.

3.8 Verification and conclusions on signal modeling

The predicted laser radar return signals have been compared with real-world measurements in fog, rain and snow along with reference measurements with hard targets of known reflectivity. For hard laser radar reference targets (see Sect. 4.1), about 10–20% absolute accuracy was reached. Long-term tests in fog showed very good agreement concerning the laser radar return signal (see Fig. 7). Residual deviations between measured data and predicted values are due to simplifications in modeling of the overlap function, the power homogeneity across the beam and the assumption of uniform fog distribution.

In practice, it turns out to be very hard to predict the local distribution functions of rain or fog which are strongly varying within the the field of view and over time. Moreover, experiments showed that the droplet distribution in the vicinity of the road surface is far from being uniform due to spray from other cars, large trucks or vans. In these cases, it might make sense to rely on pre-recorded data from previous test sessions and rather use the models to scale or extrapolate the signatures for different environmental conditions than to predict the laser radar returns from scratch.

4 Conventional laser radar target simulation

4.1 Standard targets

Road testing of a laser radar provides a realistic insight on the typical performance of the sensor, however comparison of the obtained test results between different sensor types or the same sensor with different software- and/or hardware configuration can only be done on a statistical basis and therefore needs a large amount of test data. To provide some traffic-independent figures of merit, the use of standard targets of known size and light reflectivity have turned out to be helpful. During our tests, we used aluminum plates featuring different types of coatings to realize standard targets of known properties. Our reference targets all had a size of of $0.8\,\mathrm{m}^2$. To model low reflectivity targets, we used a coating featuring $\Gamma = 0.07$, as normal reflection a coting featuring $\Gamma = 0.2$ and as high reflecting target a coting providing $\Gamma = 4$, see Table 4.

Table 4. Standard targets for laser radar evaluation.

Target	Size	Reflectivity Γ	Comment
1	$0.8\,\mathrm{m}^2$	>4	High reflectivity target
2	$0.8\,\mathrm{m}^2$	0.2	Normal reflectivity target
3	$0.8\,\mathrm{m}^2$	0.07	Low reflectivity target

4.2 Measurement linearity

Laser radars tend to introduce distance measurement errors depending on target reflectivity. This is due to optical overloading of the receiver chain or due to the detection process itself. This effect must be watched carefully during sensor evaluation. A simple test to check for the measurement linearity of a laser radar is using a plain, large concrete wall painted in white as a target. Under ideal conditions, the wall will be measured as a straight line of reflection points, one in each beam.

4.3 Road testing under adverse weather conditions

A classic approach to test the performance of laser radar systems under adverse weather conditions is doing road testing in heavy rain, dense fog or during snowfall. Although this approach enables accurate and realistic judgement of the sensor system's adverse weather performance, the results tend to depend heavily on the exact conditions present during test time and cannot be easily reproduced. Moreover, some weather conditions like heavy snowfall or dense fog can only be tested regionally or seasonally and thus can't be tested as required. Because of this, the improvement of the adverse weather performance of today's laser radar sensors is limited to the availability of reproducible test conditions. Overcoming this problem was the main goal of the optical laser radar target simulator which will be presented in Sect. 5.

4.4 Artificial simulation of rain

Experiments were carried out to simulate rain under lab conditions using an artificial indoor rain simulator. The systems consists of 32 nozzles placed in a height of about 6 m distributing water over an area of $19 \times 12\,\mathrm{m}$. A high-performance pump system is enabling a maximum flow of water up to $6\,\mathrm{m}^3\,\mathrm{h}^{-1}$ which equals a maximum rain rate of $27\,\mathrm{mm}\,\mathrm{h}^{-1}$. The system proved to be quite helpful during sensor evaluation and comparison, however the spatial distribution of the artificially produced rain turned out to be unbalanced featuring pronounced peaks of rain density under the

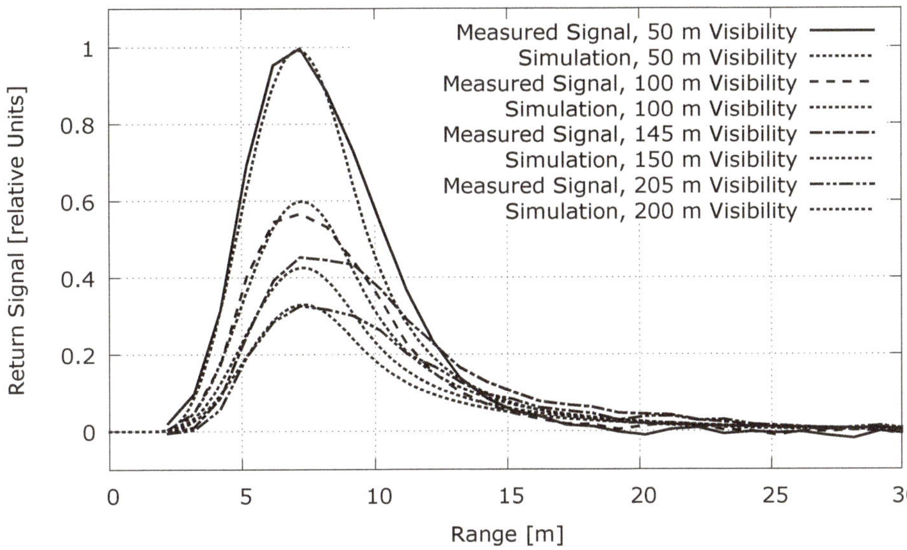

Fig. 7. Measured laser radar return signal in case of different visibility of 50 m, 100 m, 150 m and 200 m compared to numerically simulated return signals based on the theory presented in this paper. We assumed $\alpha = \frac{3}{\text{Visibiliy}}$ and $\beta = \frac{0.046}{\text{Visibiliy}} \cdot \text{sr}^{-1}$ as parameters for this simulation.

Fig. 8. Test session using indoor rain simulator without rain (left) and with a rain rate of 10 mm h^{-1} (right).

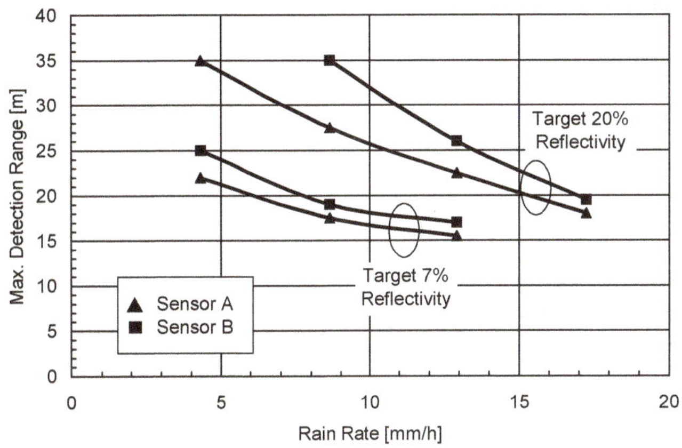

Fig. 9. Results of sensor tests using indoor rain simulator. The graphs show the maximum target detection range R_{MAX} as a function of rain rate for standard targets of 7% and 20% reflectivity and for sensors from two different manufacturers A and B.

nozzles. Figure 8 shows a test session using the described indoor rain simulator, Fig. 9 shows results obtained using the rain simulator.

5 Electro-optical laser radar target simulation

5.1 Introduction

Frequently, sensor developers and users are faced with the task to evaluate software and hardware improvements of laser radar sensors. In this case, tests under adverse weather conditions usually cannot be done due to a lack of corresponding weather conditions (e.g. lack of snow or fog) during the evaluation phase. Moreover, if sensors from different manufacturers have to be evaluated, it turns out to be very hard to exactly reproduce the same environmental conditions for both sensors.

To overcome these difficulties, a novel electro-optical laser radar target simulator system (short: OSS) has been developed and used as an automotive laser radar target simulator. The OSS is capable to exactly reproduce the optical return signals measured by reference laser radars under adverse weather conditions by highly accurate replication of pulse shape, wavelength and power levels. It can handle multiple reflections in one sensor beam and might be extended to be used with scanning laser radar systems, too. With the OSS system, it became possible to measure weather performance enhancements due to hardware and/or software changes during the laser radar design process thus shortening the design cycle and improving the sensor quality. The test signals can

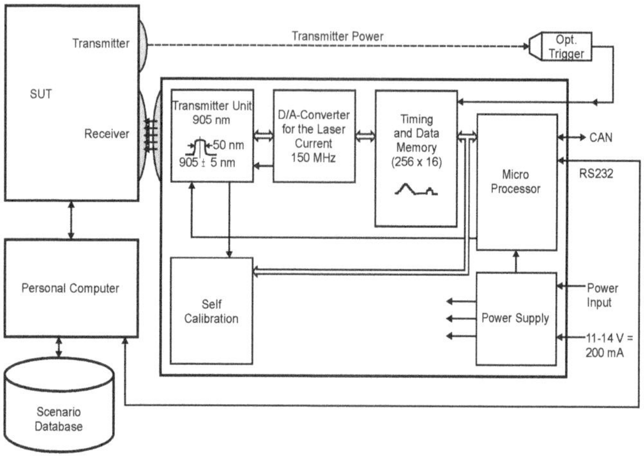

Fig. 10. Block diagram of an optical laser radar target simulator.

either be pre-recorded by the DUT itself or a by reference laser radar. Moreover, the test signals might be generated synthetically using the theory presented in Sect. 3. The system is not directly comparable to known millimeter-wave radar target generators (see e.g. Anritsu, 2009) since those only take signal delay and signal attenuation into account and cannot change the pulse's time-domain signature.

5.2 Technical description

Figure 10 shows a block diagram of the OSS. The DUT's transmit pulse is received by the optical trigger unit of the OSS. Upon reception, it triggers data transfer of a stored backscatter signal from the data memory of the OSS towards a DAC at a clock rate of 150 MHz. The output of the DAC drives a laser diode by means of a fast precision power amplifier. The backscatter data is either stored in a test computer or in the microprocessor unit of the OSS and can be repeated statically or dynamically with a maximum repetition rate of about 560 kHz which is equal to a maximum target range of 250 m for the OSS simulation, taking an offset of 6 m due to trigger delay into account. The output power of the OSS transmitter unit is radiated through its output lens towards the DUT, constant illumination can be added to emulate disturbing background light. With an evaluation software, the backscatter signal output of the DUT can be recorded on the test computer and compared with the expected response of the DUT.

The OSS is capable of generating a calibrated power density of $0.5\,\mathrm{nW\,mm^{-2}}$ to $2\,\mu\mathrm{W\,mm^{-2}}$ at the optical receiver of the DUT, featuring an dynamic range of 36 dB.

The power density S_D on the DUT's photo detector generated by the OSS transmitter power density S_O is depending on the focal length f_D of the DUT's receiver, the diameter D_D of the DUT's lens and the focal length f_O of the OSS transmitter and its lens diameter D_O, respectively. Moreover, as the output power of the OSS can be set to discrete values

Table 5. Key technical data of OSS.

Criteria	Data
Wavelength	905 ± 5 nm
Bandwidth	40 nm
Deviation of optical axis to unit case	$< \pm 5$ deg
Focal length f_OSS of output lens	80 mm
Diameters of output lens D_OSS	38 mm
Diffuser's aperture	4 mm diameter
Spatial deviation of power density	$\pm 5\%$ in an area of 2 mm diameter around the center of diffuser's aperture
Minimum output power density	$0.5\,\mathrm{nW\,mm^{-2}}$
Maximum output power density	$2\,\mu\mathrm{W\,mm^{-2}}$
Output power density quantization	$5.2 \times 10^{-10}\,\mathrm{W\,mm^{-2}\,Bit^{-1}}$
Output power steps	4096 (12 Bit quantization)
Time discretization	6.67 ns
Trigger accuracy	3.33 ns
Optical trigger delay	≈ 50 ns
Range discretization	1 m
Accuracy of calibration	$\pm 10\%$ (18 to 24 °C)
Data interface	RS232
Power supply	11–14 V/typ. 200 mA

in steps of S_q, it is controlled by a digital value N taken from the signature memory ranging from 0 to 4095. Equation (27) describes the corresponding relations:

$$S_D = \begin{cases} N S_q \left(\frac{f_\mathrm{O}}{f_\mathrm{D}}\right)^2 \left(\frac{D_\mathrm{D}}{D_\mathrm{O}}\right)^2 & \text{for } D_\mathrm{D} < D_\mathrm{O} \\ N S_q \left(\frac{f_\mathrm{O}}{f_\mathrm{D}}\right)^2 & \text{for } D_\mathrm{D} \geq D_\mathrm{O} \end{cases} \qquad (27)$$

Figure 11 shows a picture of the first OSS prototype testing an automotive laser radar. To provide accurate calibration, a thermal secondary standard and a calibrated photo PIN diode receiver was used at an output power density of $2 \times 10^{-6}\,\mathrm{W\,m^{-1}}$. The accuracy of both secondary standards was $\pm 2\%$.

5.3 Measured results

To prove the basic functionality of the OSS, a 16-channel automotive laser radar was stimulated with both synthetic and pre-recorded real-world sensor signals.

With a reference laser radar sensor, the backscatter signal of a hard laser radar target located in fog was recorded (Fig. 12). This signal was replicated by the OSS and detected by a another laser radar in the lab. As can be seen (Fig. 13),

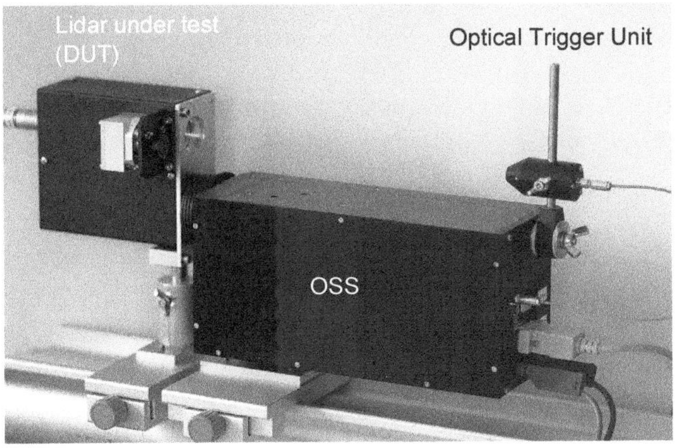

Fig. 11. OSS prototype hardware testing an automotive 16-beam laser radar.

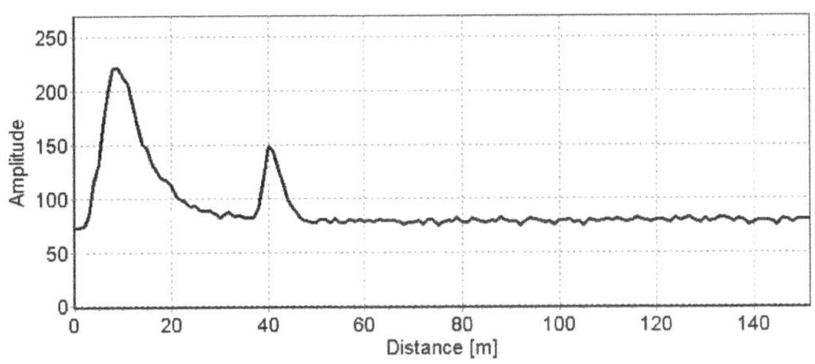

Fig. 12. Recorded signal of a target in fog used for sensor test.

very accurate simulation of the target located in foggy environment could be reached.

The examples show a limitation of the OSS target simulation: Since the OSS is triggered by the DUT's transmit laser pulse, a time delay is introduced for the simulated target signal. The delay is about 53 ns resulting in a virtual target shift of about 8 m. Although this delay will be of minor importance for long-range laser radar target simulation, it limits possibilities for short range target simulation which are of special interest when considering influences of rain and fog. To overcome this limitation, different approaches might be taken:

- Changing the distance offset adjustment in the sensor.
- Electrical triggering of the OSS using a special electrical trigger output from the DUT or using a trigger signal derived from internal DUT trigger signals which typically are 45 to 55 ns (6 to 8 m) prior to the actual transmit pulse output of the DUT's laser.
- Improvement of OSS' trigger circuitry minimizing delays within the OSS.
- Use of larger OSS trigger clock frequency.

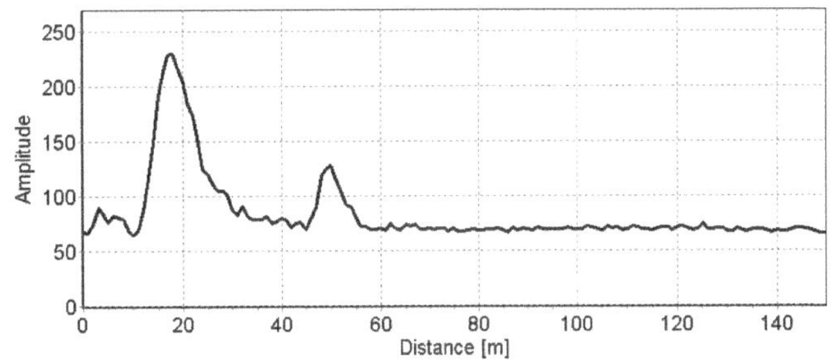

Fig. 13. Laser radar's response to stimulation according to Fig. 12.

The measured results showed that the OSS concept is a powerful approach for real-time HIL testing of automotive laser radar sensors. With the OSS concept, it is possible to test a laser radar's response to either real-world data obtained from test drives, or simulated data from hard or soft target models. Furthermore, a combination of measured data with numeric target modeling might be another option in many situations. Data once recored with a reference laser radar might be converted to signals stimulating other models of laser radar using Eq. (27).

5.4 Next steps

With a less-featured version of the OSS, an end-of-line testing and calibration of laser radars at the OEM's production line along with testing, adjustment and calibration is greatly simplified. To be compatible with laser radar sensors from different manufacturers featuring different wavelength, the transmit front-ends of the OSS might be easily changed. Complete sequences of recorded or simulated objects, including reduced visibility and rain or snow might be used to evaluate or compare laser radar based driver assistant systems' behavior using the OSS.

6 Conclusions

From laser radar theory and a model-based description of hard and soft laser radar targets, basic factors influencing the weather performance of modern automotive laser radar systems can be deduced. Based on this knowledge, improvement of laser radars' hardware and software is possible. To test the sensor performance improvements, three basic approaches are known: real world test drives under adverse weather conditions, tests with artificially induced weather phenomena (e.g. artificial rain) and electro-optical laser radar target simulation. Due to the drawbacks of first two approaches, a novel electro-optical laser radar target simulator system (short "OSS") has been developed, built and tested. The OSS prototype showed high accuracy in simulating laser radar performance in fog and proved to be very helpful during sensor optimization in the lab. Future improvements of the OSS concept should include a high-speed data interface enabling real-time, multi-channel playback of pre-recorded road scenarios. Generation of useful and accurate test data for rain, fog and snow should be based on prerecorded signatures scaled to the desired values using signal models or laser radar returns from snow, fog or rain. As experiments showed, signal modeling without calibration and scaling based on real world measurements might exhibit limited accuracy in the case of soft targets (e.g. fog) and should rather be used to model hard targets. Although of limited accuracy, soft target models turn out to be very useful for estimation of laser radar return signals during sensor system design.

References

Adelsperger, A.: Entwicklung innovativer Testverfahren für automobile LIDAR Sensoren, Diploma thesis, University of Applied Science, Munich, Germany, 2008.

Anritsu: ME7220A: Radar Test System (RTS) 76 to 77 GHz, http://www.eltm.ru/store/Anritsu/ME7220A.pdf, 2009.

Bohren, C. F. and Huffman, D. R.: Absorption and Scattering of Light by Small Particles, John Wiley and Sons, 1983.

Darms, M. S., Rybski, P. E., Baker, C., and Urmson, C.: Obstacle Detection and Tracking for the Urban Challenge, IEEE Trans. Intell. Transp. Syst., 10, 475–484, 2009.

Deirmendjian, D.: Electromagentic Scattering on Spherical Polydispersions, Elsevier, 1969.

Fayad, F. and Cherfaoui, V.: Tracking objects using a laser scanner in driving situation based on modeling target shape, in: Proc. of the 2003 IEEE Intelligent Vehicles Symposium, Istanbul, Turkey, 2007.

Feingold, G. and Levin, Z.: The Lognormal Fit to Raindrop Spectra from Frontal Convective Clouds in Israel, J. Clim. Appl. Meteorol., 25, 1346–1363, 1985.

Grönwall, C., Steinvall, O., Gustafsson, F., and Chevalier, T.: Influence of laser radar sensor parameters on range-measurement and shape-fitting uncertainties, Opt. Eng., 46, 106201-1 – 106201-11, 2007.

Gunn, K. L. S. and Marshall, J. S.: The Distribution with Size of Aggregate Snowflakes, J. Meteorol., 15, 453–455, 1957.

Halldorsson, T. and Langerhoic, J.: Geometrical form factors for the lidar function, Appl. Optics, 17, 240–244, 1978.

IEC 60825-1:2007: Safety of laser products – Part 1: Equipment classification and requirements, in: IEC Standards, 2007.

Issa, B.: Optimierte Signalverarbeitung im Bereich der Fahrerassistenzsysteme, Diploma thesis, Technische Universität München, Munich, Germany, 2007.

Isaac, K. I., McArthur, B., and Korevaar, E.: Comparison of laser beam propagation at 785 nm and 1550 nm in fog and haze for optical wireless communications, Proc. SPIE Vol. 4214, Optical Wireless Communications III, pp. 26–37, 2001.

Kirchner, A. and Ameling, C.: Integrated Obstacle and Road Tracking using a Laser Scanner, in: Proc. of the 2000 IEEE Intelligent Vehicles Symposium, Dearborn (MI), USA, 2000.

Marshall, J. S. and Palmer, W.: The Distribution of Raindrops with Size, J. Appl. Meteorol., 5, 165–166, 1948.

Nebuloni, R. and Capsoni, C.: Laser Attenuation by Falling Snow, in: Proc. of the 2008 Symp. on Communication Systems, Networks and Digital Signal Processing, pp. 265–269, Graz, Austria, 2008.

Neshyba, S. P.: Representation of a nonspherical ice particle by a collection of independent spheres for scattering and absorption of radiation: 2. Hexagonal columns and plates, J. Geophys. Res., 108, 6.1–6.18, 2003.

Oguchi, T.: Electromagnetic Wave Propagation and Scattering in Rain and Other Hydrometeors, Proc. of the IEEE, 71(9), 1029–1078, 1983.

Owens, J. C.: Optical Doppler Measurement of Microscale Wind Velocity, Proc. IEEE, 57(4), 530–536, 1969.

Rasmussen, R. M., Vivekanandan, J., Cole, J., Myers, B., and Masters, C.: The estimation of snowfall rate using visibility, J. Appl. Meteorol., 38, 1542–1563, 1999.

Sassen, K. and Dodd, G.: Lidar crossover function and misalignment effects, Appl. Optics, 21, 3162–3165, 1982.

Sekhon, R. S. and Srivastava, R. C.: Snow size spectra and radar reflectivity, J. Meteorol., 27(2), 299–307, 1970.

Skolnik, M.: Radar Handbook, McGraw-Hill Publishing Company, 1990.

Smith, D., Hooper, R., and Garrett, I.: Receivers of optical communications: A comparison of avalanche photodiodes with PIN-FET hybrids, Optical and Quantum Electronics, 10, 293–300, 1978.

Spies, H. and Spies, M.: Advanced sensor technology boosts fast passenger safety to new system design, in: 8th International Symposium and Exhibition on Car Occupant Safety Systems, Karlsruhe, Germany, 2006a.

Spies, M. and Spies, H.: Automobile Lidar Sensorik: Stand, Trends und zukünftige Herausforderungen, Adv. Radio Sci., 4, 99–104, doi:10.5194/ars-4-99-2006, 2006b.

Steinvall, O., Chevalier, T., and Larsson, H.: Data collection and simulation of high range resolution laser radar for surface mine detection, Proc. SPIE 6214, pp. 85–102, 2006.

Thrun, S., Burgard, W., and Fox, D.: Probabilistic Robotics, MIT Press, 2005.

Walchshäusl, L., Lindl, R., Vogel, K., and Tatschke, T.: Detection of road Users in Fused Sensor Data Streams for Collision Mitigation, in: Proc. Advanced Microsystems for Automotive Applications AMAA 2006, pp. 53–65, Berlin, Germany, 2006.

Yang, P. and Liou, K. N.: Geometric-optics – integral-equation method for light scattering by nonspherical ice crystals, Appl. Optics, 35, 6568–6584, 1996.

Zhao, Y., Lea, T. K., and Schotland, R. M.: Correction function for the lidar equation and some techniques for incoherent CO_2 lidar data reduction, Appl. Optics, 27, 2731–2740, 1988.

Zhou, X., Shusun, L., and Stamnes, K.: Geometrical-optics code for computing the optical properties of large dielectric spheres, Appl. Optics, 42, 4295–4396, 2003.

12

A fast three-dimensional deterministic ray tracing coverage simulator for a 24 GHz anti-collision radar

H. Azodi, U. Siart, and T. F. Eibert

Lehrstuhl für Hochfrequenztechnik, Technische Universität München, Arcisstr. 21, 80333 Munich, Germany

Correspondence to: H. Azodi (hossein.azodi@tum.de)

Abstract. A matrix-based approach for implementation of deterministic ray tracing is suggested and presented in this paper. The frequency-independent feature of the ray tracing in addition to the matrix-based implementation results in a reliable fast simulator for understanding the behavior of the electromagnetic fields in the vicinity of a collision avoidance radar. Results of this technique are in a good agreement with the results from method of moments integral equation solutions while the computations are about 100 times faster.

1 Introduction

Computer numerical control (CNC) machines are a group of manually programmed modern machines for drilling, milling, laser or plasma cutting, etc. (Fig. 1). They are ranged within the expensive workshop machinery. Due to human errors in coding their controls, disastrous collisions between the tool of the machine and surrounding metallic objects can easily cause very expensive damage. It is, therefore, wise to detect possible collisions in advance and stop the machine. For this reason, we suggest to encircle the tool of CNC machines with a number of radar modules to detect the possibility of harmful collisions and to stop the machine before collisions take place.

Figure 2 shows a mock-up of CNC tool and the radar modules surrounding it. We propose to use the minimum of 3 modules with 120° circular distance so that full coverage of the surrounding space can be achieved. Thanks to the simple and symmetrical geometry any calculation or simulation for one module can be easily expanded for other modules in different positions. The radiator of this module is a linear patch array antenna with 8 elements which operate in the ISM band around 24.125 GHz.

Due to frequent replacements of tool, object, etc., a CNC machine has a varying electromagnetic environment. Consequently, the radar system requires to calibrate itself throughout the operational mode of the machine. On the other hand, the design should remain consistent and practically compatible with any CNC machine. These highly-demanding features require an accurate, fast, and reliable radar system. False alarms, not desirable though, are accepted but collision detection failure is not permitted at all.

From an electromagnetic point of view, this problem can be categorized in near-field antenna propagation modeling and channel estimation. Understanding the behavior of electromagnetic (EM) waves in the vicinity of the tool plays the most important role in designing such a radar system. In the next section, we explain why ray tracing is a good candidate for simulating the fields and obtaining the channel model. In Sect. 3, the main idea of our fast deterministic ray tracing approach is explained and the results of this approach are presented in Sect. 4. Section 5 concludes this paper.

2 Electromagnetic modeling and solution technique

The mock-up, which contains the most important characteristics of the tool of the CNC machine, is modeled with two cylinders (Fig. 3). The antenna is located on the holder. As shown in Fig. 4, a single patch array antenna in an open PEC cavity is used. The electromagnetic modeling of the antenna itself and its local environment is decoupled from the propagation problem. The current electromagnetic model includes the most important aspects of the mock-up. A detailed model is simply created by adding more cylinders to the model. Therefore, we sticked to this simple model at this stage and for later stages we will add more complexities to the model. It is, indeed, possible due to flexibility of our simulation technique which becomes evident to the readers in the rest.

Finite Difference Time Domain (FDTD), Integral Equation (IE) and Ray Tracing are 3 candidates to simulate the fields of the electromagnetic wave propagation problem.

Fig. 1. Tool of a CNC and its machinery

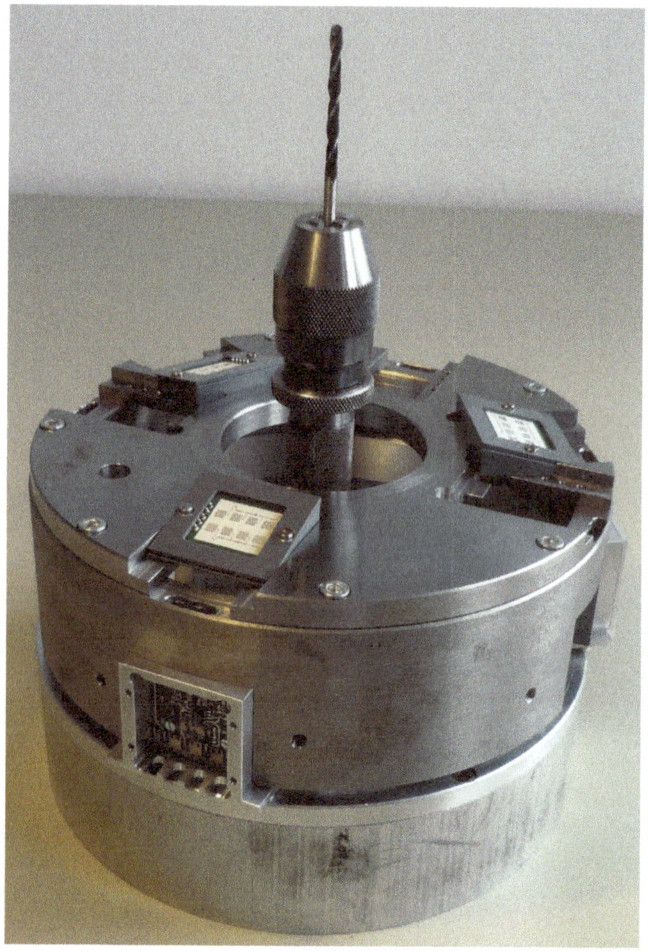

Fig. 2. Initiative measurement mock-up designed and manufactured in our workshop and the Radar module on the surrounding area

A typical FDTD code would need more than 10 lines/λ at $F_{CEN} = 24$ GHz leads to 38 million mesh-cells for the structure which takes several days of CPU time for every scenario on a quad-core computer with 12 GB of RAM. Moreover, 10 lines per wavelength is the minimum and for more accurate results an increase to 16 lines/λ is recommended. Consequently, simulations at higher ISM bands such as $F_{CEN} = 60$ GHz appear impractical.

The Method of Moments (MoM) – integral equation (IE) method is another potential solution candidate, particularly suitable due to the fully metallic surface of the geometry. The problem can therefore be discretized with about 5 million unknowns at 24 GHz. This can be handled with the aforementioned computer, although it is a big problem. Hence, the IE method is used as a standard method to compare and validate the results of the ray tracer as it is still a big problem for every-day optimization and design processes. For this purpose, an in-house IE-MoM code is used Eibert (2004).

The third candidate to solve this problem is ray tracing Ling et al. (1989). As mentioned in Sect. 1, this is a near-field problem. In contrary, ray tracing is a method where fields are assumed to propagate in the far-field region. This is the estimation which should be taken into account in order to achieve the benefits of ray tracing method. This estimation can be improved by considering more geometrical details of the objects such as curvature, multiple reflections, etc. In our approach the problem of antenna simulation and field calculations are decoupled. The antenna far-field pattern is obtained using FDTD technique and it is inserted into the wave propagation ray tracing code. Once the ray paths to an observation point is found, field calculation is a summation of the initial fields of antenna pattern multiplied by reflection or diffraction coefficients depends on path, free space loss and propagation phase shift.

In Shooting and Bouncing Rays (SBR) Ling et al. (1989), millions of rays are launched from the transmitter. The path of the ray is traced until reaching the observation point or leaving the simulation bounds. Even in an indoor scenario a very small portion of the rays will reach observation points. So, calculating paths of such a huge number of rays wastes resources in case one could anticipate the potential paths.

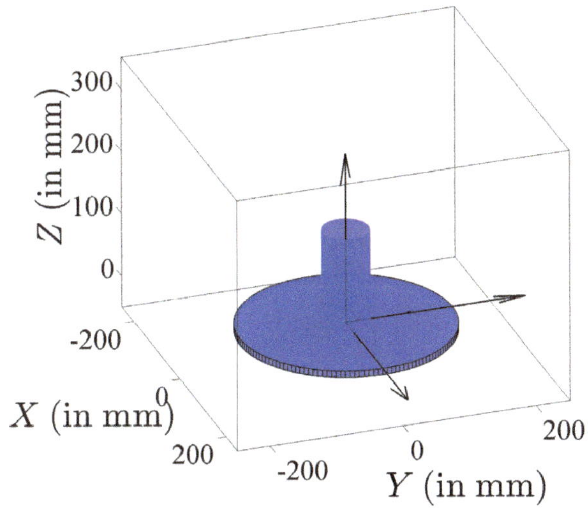

Fig. 3. Electromagnetic modeling of the initiative mock-up.The problem is decoupled in modeling the geometrical scenario and antenna, separately.

This will bring the idea of deterministic ray tracing while we have a priori information by knowing the geometry of the problem.

Another problem arises in SBR when a ray closely passes the observation point. As an observation point is theoretically dimensionless, these rays are usually not included. Nevertheless, it is seen that Flores et al. (1998); Noori et al. (2006) these missing rays might have an effect on the field of that point. Therefore, the idea of reception sphere is developed as a solution to this problem. In our deterministic ray tracing method, the observation point and all the other points remain dimensionless, i.e. ideal, and we are still able to calculate reaching paths to them accurately.

3 Fast matrix-based deterministic ray tracing

Deterministic ray tracing is suggested and developed in order to obtain the EM fields in the vicinity of the radar. This method enables us to have an overview of the coverage of the radiation strength and to estimate the channel response much faster than the aforementioned discretizing methods. As calculations are done using *same-size* matrices, a loopless MATLAB® code will provides with the results in a few minutes on the used PC.

From a general perspective, the idea of matrix-based deterministic ray tracing consists of two main parts. In the first part, the code calculates the required (mostly) geometrical matrices and the second part of the code is responsible for calculating the fields due the various rays and adding them together. These two parts are explained in Sects. 3.1 and 3.2, respectively.

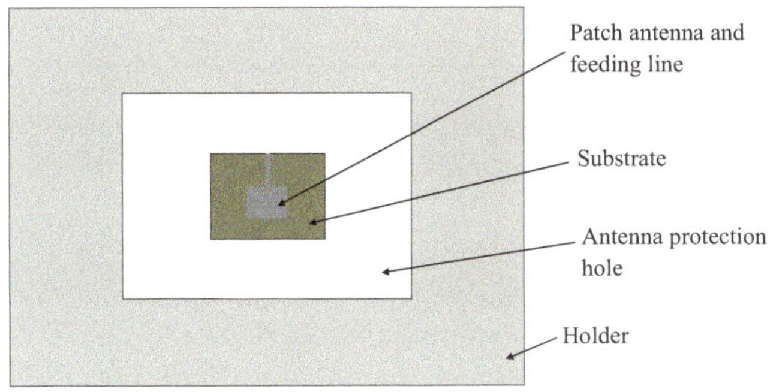

Fig. 4. Modeling of antenna as a single patch in a box of PEC. The box is empty and the antenna is a single patch fed with microstripline.

3.1 Pre-calculation and matrix formation

The entire observation space is discretized with equi-distant grids. For each grid in the observation space, there exists a unique element in every matrix. For example, $\mathbf{X}= [x_{mnp}]$ is the X-coordinate of the grid in the position of (m, n, p) of the space. The same applies to $\mathbf{Y}$ and $\mathbf{Z}$ matrices. Size of the matrices depends on the discretization fineness. The complexity of the problem increases with resolution as well but not with frequency.

Having these matrices, the rest of the pre-calculation phase is determining the geometrical matrices. For instance, the Euclidean distance to the origin:

$$\mathbf{R} = \left[\sqrt{x_{mnp}^2 + y_{mnp}^2 + z_{mnp}^2}\right] \tag{1}$$

where all operations are element-wise. Equation (1) can be implemented in MATLAB® without using a loop. Any other required matrix can be calculated with the same principle.

The most important matrices in the pre-calculation phase are *shadow (or illumination)* matrices. Zero, 0, in a shadow (or one, 1, in an illumination) matrix means the corresponding observation point, P_{mnp}, is illuminated, counterpart means it is shadowed. Obviously, shadow matrices are ray dependent. These matrices contain the information of ray paths. So, in computation of them the Fermat's principle is used. This means, for example, for the direct ray the direct path from the antenna to the observation point is considered.

To calculate these matrices numerically, a geometry data matrix, $\mathbf{B}_{\mathrm{GEO}}$, is created first. One, 1, in $\mathbf{B}_{\mathrm{GEO}}$ indicates that the corresponding point in the space is occupied with an object (or a part of an object). For empty space the matrix returns 0. Thus, the size of this matrix is the same as other matrices and if higher accuracy is required, finer discretization can be applied. The shadow matrix element for the observation point (m, n, p) is, therefore,

$$s_{mnp} = \left(\int_{T_{mnp}}^{P_{mnp}} \mathbf{B}_{\mathrm{GEO}} \, \mathrm{d}l \geq 1 \right) \tag{2}$$

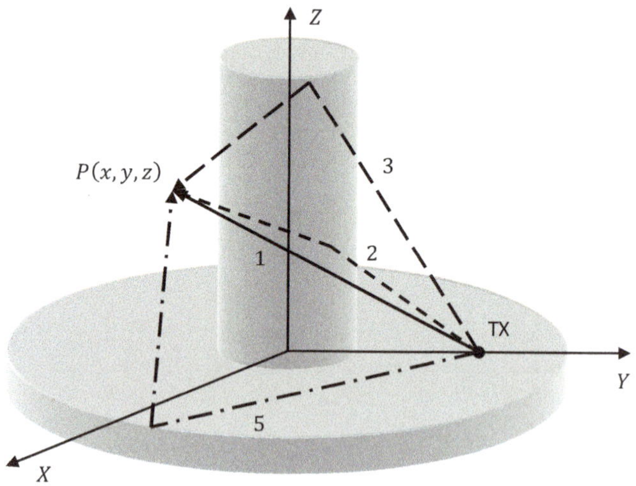

Fig. 5. 5 rays contribute to find out the field in one point. They are identified with numbers on each path. As creeping ray and direct ray cannot reach an observation point together, in this schematic the chosen point is not illuminated by creeping ray.

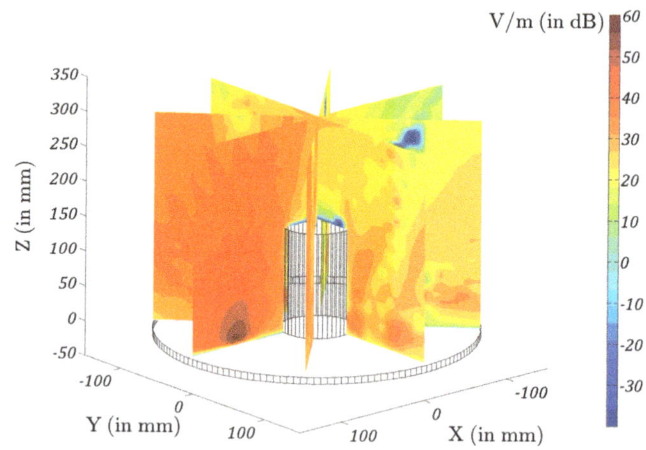

Fig. 6. Magnitude of E-field in surrounding space of the machine sliced every 45 degrees.

where T_{mnp} is the position of the transmitter corresponding to the observation point at P_{mnp} and the ray type. It should be noted again that depending on the ray type the transmitter position can be different from the antenna position. For instance in reflection from the cylinder, the transmitter is first the antenna but, later, when we need the path from the reflection point to the observation point, it becomes a point on the cylinder. The reflection point is computed based on Fermat's principle and is saved in the corresponding element of the "reflection point from the cylinder" matrix. The integral in Eq. (2) is over the connecting straight line between the points T_{mnp} and P_{mnp} and the integrand is the geometry data function of the space. By discretization,

$$s_{mnp} = \left(\sum_i [b_{m_i n_i p_i}] \geq 1 \right) \tag{3}$$

where $[b_{m_i n_i p_i}]$ runs over the elements of geometry data matrix which fall in the connecting line of transmitter and observation points. Calculating all s_{mnp}, the shadow matrix, $\mathbf{S} = [s_{mnp}]$, is created.

The pre-calculation process is finished by calculating the geometrical matrices, shadow (or illumination) matrices and incident field matrices. Incident field matrices correspond to amplitude and phase of the fields which are just propagated in the direction of the ray path from the antenna. These fields are generated by interpolating the spherical coordinates of the direction and the far-field pattern.

Contributions of 5 rays are summed up in our simulations. These rays are direct ray, reflected ray by the cylinder, diffracted ray by the upper edge of cylinder, creeping ray around the cylinder, and diffracted ray by the edge of the holder (Fig. 5).

3.2 Field calculation and summation

Except the direct path from the transmitted to the observation point, the other paths are going through several points on the structure. The coordinates of these points in each ray type are calculated and stored in separate matrices. For example for the reflection from the cylinder, there is a point on the cylinder where the reflection occurs, obviously. The coordinates of this point are stored in $\mathbf{X}_{\mathrm{N,REF}}$, $\mathbf{Y}_{\mathrm{N,REF}}$, and $\mathbf{Z}_{\mathrm{N,REF}}$, respectively. It should be noted again that the elements of matrices are in accordance with the observation point. This means that $\mathbf{X}_{\mathrm{N,REF}} = [x_{mnp}]$ has the first coordinate of the point N_{REF} in accordance with the observation point P_{mnp}. N_{REF} can be computed using the geometrical optics rule of reflection. Taking into account that N_{REF} is a point on the cylinder with, i.e. $0 \leq z_{mnp} \leq H_{\mathrm{CYL}}$ and $x_{mnp}^2 + y_{mnp}^2 = R_{\mathrm{CYL}}^2$, where H_{CYL} and R_{CYL} are height and radius of the cylinder, respectively, a quadratic equation is obtained which in principle has 4 roots. From them, only one is the desired root for P_{mnp}. This root is real and belongs to the illuminated side of the cylinder. Thus, a direct solution Tignol (2004) to the equation will provide us with the desirable result.

For the diffraction from the holder cylinder (the second cylinder), where the grazing rays of the antenna approaches the circular edge of the cylinder, the story is quite different. The geometrical equation is a non-linear trigonometric equation from degree of 6. Consequently, there are more than one point from which a path to the observation point, P_{mnp}, exists. In order to consider all the diffraction points on the

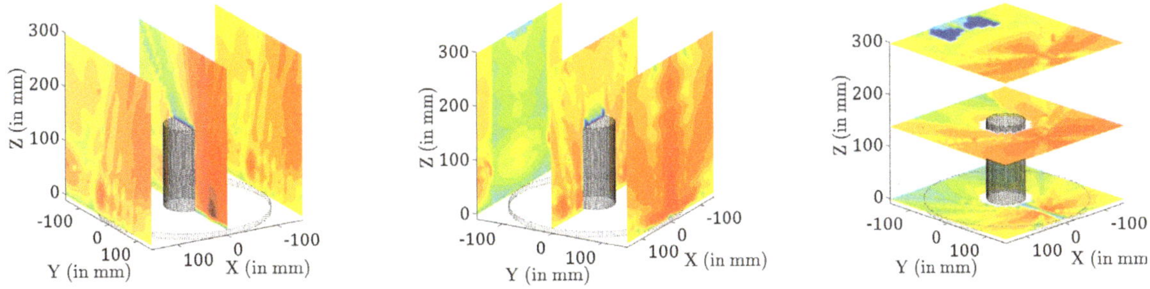

Fig. 7. Magnitude of E-field in surrounding space of the machine sliced w.r.t. X-, Y-, and Z-axes.

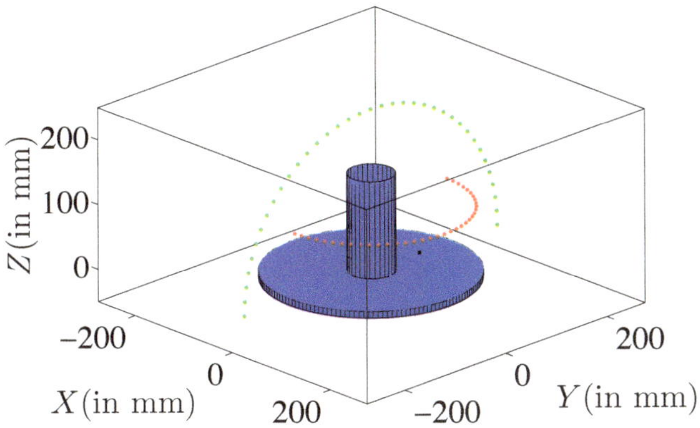

Fig. 8. Point series positions to compare the result of deterministic ray tracing and MoM-IE technique.

holder edge, the code finds 5 roots of about 30 000 equations simultaneously. Number of equations comes directly from the number of observation points for which the field is computed. It depends on the desired fineness of computation domain. The maximum number of roots (5 in this case) is only the maximum real roots of any equation. In order to have an efficient method regarding the computation time, it is necessary to solve all these equations and to retrieve their zeros simultaneously. Otherwise, our technique takes the same time as stochastic ray tracing techniques. To the best understanding of the authors, the iterative algorithm in Ying and Katz (1989) is the most reliable and general algorithm to find out the multiple roots of an equation. This algorithm is expanded in our code to solve the equations in matrix form, i.e. 30 000 equations altogether.

Creeping rays (James, 1986, Pages 195–203) are also taken into account to uniform the transition from illuminated to deep shadow region. On the cylinder, a creeping ray follows a geodesic path which is simply a helical curve. The characteristics of the helical path is subjected to geometrical ray optics with the condition that on the starting and ending parts the ray is tangent to the cylinder.

Finding the paths, the field in every observation point can be obtained. The antenna is simulated separately where it is buried in body of PEC and covered by a dielectric material with relative permittivity close to 1 (Fig. 4). The far-field pattern of the antenna is stored in a file so that for every direction of Θ and Φ in the space the amplitude and phase of the incident field is known. As our observation points do not fall exactly on the saved directions of the far-field pattern, we interpolate this file with the point directions to calculate the incident fields. Later, they are multiplied by the coefficients of that path, e.g. reflection or diffraction coefficients (James, 1986, Page 233) for reflection from the cylinder or diffraction from the edge of the cylinder, respectively, and so on. These coefficients have the effect of the astigmatic loss included. Each path, additionally, has its own propagation phase term which is also multiplied. Finally, the field matrix due to each ray is filtered by the corresponding illumination matrix. As the result, fields become null in shadowed points and when they are summed up, the result has the contributions of the physically reachable rays to that point.

4 Results

The deterministic ray tracer code is used to simulate the surrounding space of the radar system in $31 \times 31 \times 31 = 29\,791$ equi-distant grids in the single center frequency of $F_{\mathrm{CEN}} = 24.145$ GHz. This simulation covers the space within the range of -150 mm to 150 mm for X or Y and 0 mm to 300 mm for Z. The entire simulation process takes 42 min while this is almost the same time it takes for a C/C++ shooting and bouncing ray tracer to deliver the result of a single transmitter and single receiver scenario. The computer has a 2.67 GHz quad-core i7 Intel® CPU and is running Win 7 64 bit with 12 GB of RAM.

Figure 6 presents the magnitude of E-field in logarithmic scale. In this figure, fields are sliced every $\Delta\Phi = \pi/4$. The same scaling is used for the field magnitudes in Fig. 7. The magnitude of the fields in this figure are presented in different layers along the principal coordinate axes. The antenna is located at $(X, Y, Z) = (0, 95, 0)$ mm.

In another experiment, the result of deterministic ray tracer is compared with the result of MoM-IE technique. For this purpose, we used the calculation method of Eibert (2004) to simulate the fields in 2 series of points in Fig. 8. The

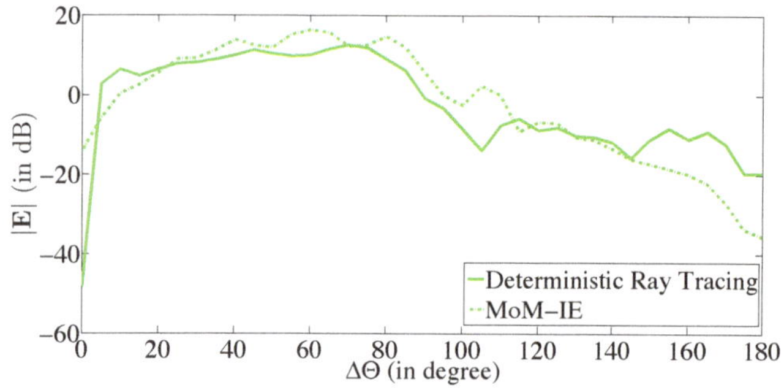

Fig. 9. Comparing the Ray Tracing and MoM-IE at the position of the green points in Fig. 8.

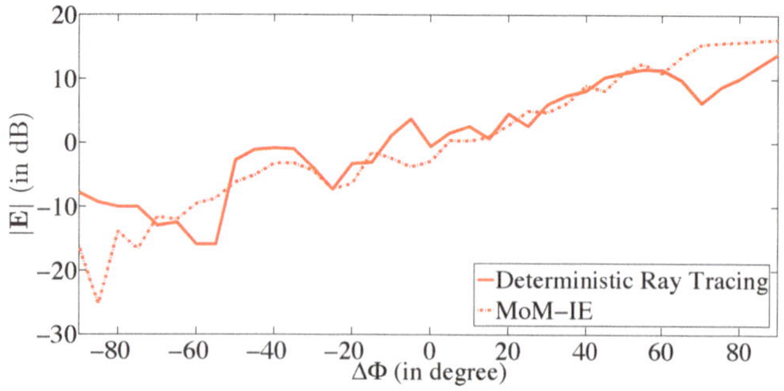

Fig. 10. Comparing the Ray Tracing and MoM-IE at the position of the red points in Fig. 8.

green points are located in the upper half-plane of $Y = 0$ in every 5 degrees on a circle with the radius of 250 mm and center of $(X, Y, Z) = (0, 0, 0)$. The red points are located in $Z = 100$ mm plane on a half circle with radius of 150 mm and center of $(X, Y, Z) = (0, 0, 100)$. The result of this comparison is presented in Figs. 9 and 10.

Comparing in both cases shows that the result of ray tracing stays in a good agreement with MoM-IE technique in the illumination and transition regions. In deep shadow region, ray tracer does not follow the same trend which results up to 10 dB difference. Higher accuracies are achievable by adding more details in the cost of higher runtime. In our application, we need a fast qualitative field distribution in an electrically large area. In this respect, the code performs well by computation of the electrical field in about 30 000 points which cover the surrounding space. This runtime will increase linearly with number of rays. However, the runtime does not depend on fineness of the geometry matrix. The latter feature is very helpful to increase the accuracy of computation.

5 Conclusions

In this work, a fast deterministic ray tracer technique with matrix-based approach was presented. It was coded for a simple geometry scenario, however, the main functions such as the simultaneous multiple root finder and shadow region detector are in their most general forms. This means with a priori knowledge of the number of needed rays, one may simulate the EM fields for any structure as the structure can be imported with a simple geometry boolean data matrix. Based on the simulation, one not only may investigate coverage but also calculate the channel impulse response complete Doppler or FMCW radar scenarios. The most important advantage of the code is the frequency-independent short time of computation which is needed for complex scenarios and scenarios including moving targets where many simulations should be carried out.

References

Eibert, T.: Iterative-Solver Convergence For Loop-Star And Loop-Tree Decompositions in Method-of-Moments Solutions of the Electric-Field Integral Equation, IEEE Antenn. Propag. M., 46, 80–85, doi:10.1109/MAP.2004.1374101, 2004.

Flores, S., Mayorgas, L., and Jimenez, F.: Reception Algorithms For Ray Launching Modeling of Indoor Propagation, in: Proceedings of IEEE Radio and Wireless Conference, 261–264, 1998.

James, G. L.: Geometrical Theory of Diffraction For Electromagnetic Waves, 3rd Edn., IET, 1986.

Ling, H., Chou, R.-C., and Lee, S.-W.: Shooting and Bouncing Rays: Calculating the RCS of an Arbitrarily Shaped Cavity, IEEE T. Antenn. Propag., 37, 194–205, 1989.

Noori, N., Shishegar, A., and Jedari, E.: A New Double Counting Cancellation Technique For Three-Dimensional Ray Launching Method, in: IEEE Antennas and Propagation Society International Symposium, 2185–2188, 2006.

Tignol, J.-P.: Galois' Theory of Algebraic Equations, 2nd Edn., World Scientific Publisher Co., 2004.

Ying, X. and Katz, I.: A Simple Reliable Solver For All the Roots of a Nonlinear Function in a Given Domain, Computing, 41, 317–333, 1989.

13

Near-field radiation pattern distortion of antenna attached to wall in through-the-wall radar imaging

B. Cetin, H.-R. Benedickter, and P. Leuchtmann

Institute for Electromagnetic Fields and Microwave Electronics, ETH Zurich, Switzerland

Correspondence to: B. Cetin (bora.cetin@ifh.ee.ethz.ch)

1 Introduction

Through-the-wall radar imaging (TWRI) has become subject of growing interest in the recent decade. It is supposed to be used for law-enforcement purposes during surveillance, reconnaissance and rescue missions (Ferris et al., 1999; Baranoski, 2008; Amin, 2011). Radar systems in general are based on the detection of reflected waves caused by scattering objects. According to shape and amplitude of the reflection, the measurement data are processed into a tomographic image. When the signal is propagating through-the-wall (TTW), the reflected signal's original shape and amplitude are altered. For the reconstitution of the reflection of interest (Thajudeen et al., 2011), the distortion causing signal path needs to be known regarding to its structure and electromagnetic material properties. Otherwise the image will be degraded and defocused (Li et al., 2010; Chang et al., 2010).

In many studies the properties of different kinds of construction material have been investigated (Muqaibel, 2003; Stavrou et al., 2003). Due to the natural origin of some materials (e.g. wood and brick), natural conditions (e.g. moisture) and not standardized material mixtures (e.g. concrete), published material values can differ significantly. To overcome this uncertainty in material parameters, on-site material extraction methods have been proposed (Ahmad et al., 2007; Jia et al., 2011). These methods are based on the single layer homogeneous wall model and have a limitation in accuracy. In a realistic inhomogeneous multilayer wall, however, multipath propagation and anisotropic wall characteristics influence the detected value of material properties.

Reduction of clutter caused by ceiling, floor and the back wall can be achieved by placement of the antenna close or even directly on the walls themselves (Jia et al., 2011). At the same time the wall penetrating electromagnetic wave's power density will be higher, resulting in a more sensitive sensing of reflections (Farwell et al., 2008). In addition, another uncertainty in the system gets omitted: the distance between wall and antenna (Wang et al., 2006). By the placement of the antenna directly on the wall and the decrease of the distance between antenna and target into the near field distance jointly with the use of large imaging apertures, the resolution of the imaging system can be optimized (Sheen et al., 2007).

This work analyzed the distortion of the radiation pattern when the antenna is placed on wall. Sensitivity analysis were made to investigate the impact of imprecise knowledge of material parameters on the radiation pattern. The principle setup is explained in Sect. 2. The used components including antenna and the examined walls are presented in Sect. 3. In Sect. 4 the simulation techniques are explained, whereas Sect. 5 shows the measurement setup. The results are presented and discussed in Sect. 6, followed by the conclusions in Sect. 7.

2 Principle setup

The radiation pattern is a characteristic parameter of an antenna. It describes the radiated electromagnetic wave properties (e.g. signal strength, phase, polarization) in dependence of the radiation direction. For obtaining the radiation pattern, a relative rotation between the antenna under test (AUT) and the monitoring system is required. In a simulational environment field monitors can spot the field strength on a sphere around the AUT. In measurements, the AUT and the probe antenna need to be rotated relatively against each other as drafted in Fig. 1. There are two technical implementations of this principle (Balanis, 2005). The first one is based on the rotation of the AUT around its own axis in azimuth plane. In the second method, the probe antenna is moved circularly around the AUT. Both methods are suitable for determining the radiation pattern of the antenna when put directly on the

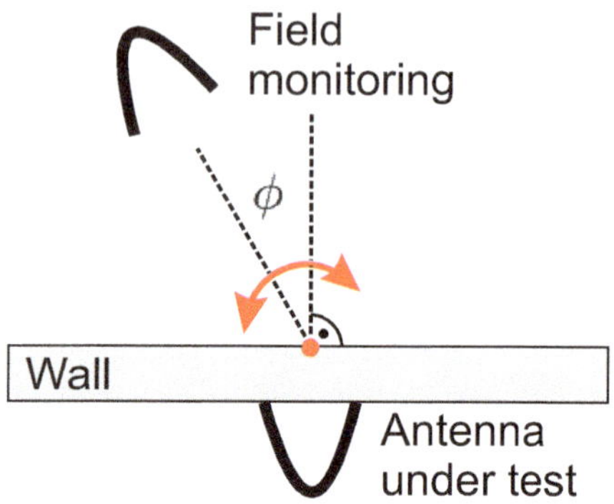

Fig. 1. Principle of obtaining the radiation pattern in simulations and measurements. It is based on the relative rotation ϕ between the antenna under test (AUT) and the monitoring system (field monitor in simulations, probe antenna in measurements). Here, the setup shows the antenna put directly on a wall.

wall. The method was chosen in dependence of the circumstances as described in Sect. 5.

Two datasets for each the simulations and the measurements are required for determining the influence of a wall on an antenna's radiation pattern: the first dataset is for determining the non-distorted radiation pattern of the AUT in free-space. The second one is for the distorted radiation pattern that results from the placement of the antenna directly on the wall. By comparison of both datasets, the distortion of the radiation pattern and the damping through-the-wall (TTW) Δs_{21} caused by the wall can be quantified

$$\Delta s_{21} = s_{21,\text{withwall}} - s_{21,\text{freespace}} \ [\text{dB}]. \tag{1}$$

3 Components

The distortion of an antenna's radiation pattern is being analyzed when the antenna is placed directly on a wall. The investigations include simulations and measurements in the frequency range 0.7–10.0 GHz as proposed for TWRI.

3.1 Antenna

The used antenna is the ETS-Lindgren 3164-08, a non-resonant ultra-wideband (UWB) antenna based on the quad-ridged horn design. It is dual-polarization capable and operational in the frequency range of interest. This antenna's reactive part within the near-field (NF) region is situated in the coupling region between the coaxial input and the ridges. As it vanishes within one wavelength, no reactive NF is present at the antenna's mouth and therefore only very limited mutual coupling between antenna and wall is possible (Seybold, 2005).

Table 1. Material parameters at $f = 1.75$ GHz derived from rectangular waveguide measurements and used for the simulations. The sensitivity analysis was carried out at 2.0 GHz.

Material	ε_r	$\tan\delta$
Wood	1.81	0.077
Sensitivity analysis	1.6 1.7 1.8	0.06 0.07 0.08
Brick	3.49	0.005
Sensitivity analysis	Spacing between bricks g: 1 cm 2 cm	
Sand lime brick	5.02	0.035
Insulating foam	≈ 1	≈ 0

3.2 Wall materials

The first class of material investigated is the homogeneous dielectric slab. Representatively for this class of material, glued laminated timber made of wood with the size of Thickness (T) 4.0 cm × Length (L) 200 cm × Width (W) 60 cm is chosen. Putting three specimens together yield in a 180 cm wide wooden slab. The second class of material is inhomogeneous brick. The size of the used bricks is L 29.0 cm × W 17.5 cm × Height (H) 19.0 cm (Swiss Modul 030419, Fig. 2). The third wall is a typical Swiss multilayer inhomogeneous exterior wall with a total thickness of 60 cm. It consists of three layers, namely sand lime brick (Creabeton 510604, L 25.0 cm × W 14.5 cm × H 19.0 cm), insulating foam (T 15.0 cm) and brick (Swissbrick, L 29 cm × W 17.5 cm × H 19 cm). A comparison of Fig. 2 and Fig. 3 shows that the internal structure of the used bricks are almost identical. Before measuring the material parameters at $f = 1.75$ GHz as listed in Table 1, the stones and bricks were dehumidified in an oven. In dependence of the permittivity ε_r, the wavelength of the electromagnetic wave reduces to

$$\lambda = \frac{\lambda_0}{\sqrt{\varepsilon_{r,\text{eff}}}} = \frac{\lambda_0}{\sqrt{\varepsilon_r}} \sqrt{\frac{V_{\text{total}}}{V_{\text{material}}}} \tag{2}$$

when propagating through material. The effective permittivity of an inhomogeneous material distribution can be calculated by the material's permittivity and the ratio of the material volume to total volume (see Fig. 2), provided that the cavities are much smaller than the wavelength.

4 Simulation technique

Simulations were carried out to cross-check with the measurements. As simulations are virtual, geometries and material parameters can be changed in the setup for sensitivity analysis purposes. Scenarios that cannot be measured experimentally still can be simulated, enabling investigations of extreme scenarios. The simulations were carried out using the Finite-Element-Method (FEM) field simulator HFSS (High

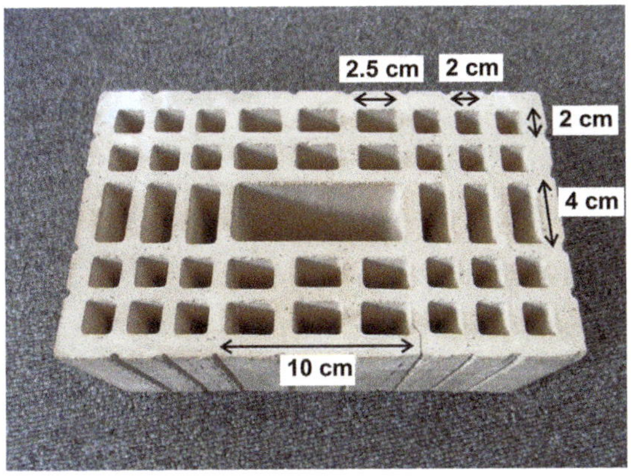

Fig. 2. Cross-section of the brick used in these investigations. Geometry is almost identical to brick used in the multilayer wall (see Fig. 3). The cavities cause the decrease of the effective permittivity.

Frequency Structure Simulator). For this, the antenna's geometry and the different kinds of wall were implemented in the programme. The use of the FEBI (Finite Element Boundary Integration) boundary condition helped reducing the computational volume (Edgar, 2011) for the simulation of the single antenna. For overcoming memory shortcomings for the simulation of the whole structure including antenna and wall, the FEBI and Datalink (a technique for separating the computational space into parts) functions were used jointly, facilitating the computational separation of antenna and wall and retaining the distance between them close to zero at the same time.

Sensitivity analysis

To evaluate the effects of differing material parameters in the case of wooden slab, a sensitivity analysis was carried out at $f = 2.0\,\mathrm{GHz}$ with three different values for the permittivity ε_r and three different values for the loss tangent $\tan\delta$, respectively (see Table 1). For this, the damping Δs_{21} TTW defined in Eq. (1) was determined for all nine material properties combinations and compared with the reference ($\varepsilon_r = 1.7$, $\tan\delta = 0.07$). In the case of brick wall, a sensitivity analysis was performed on the horizontal and vertical spacings g between the bricks. Two different values (see Table 1) have been used to evaluate the impact of brick spacing on the radiation pattern.

5 Measurement setup

The radiation pattern of the AUT was measured in different antenna configurations. These are described in this section and the specific setup parameters are summarized in Table 2. The probe antenna for recording the electromagnetic field strength on a circle around the AUT in azimuth plane is of the same model as the AUT itself (Sect. 3.1). All radiation patterns were recorded in the four possible polarization directions (co-polarizations HH and VV, cross-polarizations HV and VH; the letters denote the polarization of the AUT and probe antenna, respectively).

Fig. 3. Cross-section of the multilayer wall on the open-air test site Witaumatte (Thun, CH). The used materials are sand lime brick (left), brick (right) and insulating foam in between (not visible in this picture).

The measurements were carried out in the frequency domain, using an Agilent 8510C Vector Network Analyzer (VNA). The calibration was applied to the connection interface between coaxial cable and antenna. Therefore, the measurements include all effects due to frequency dependency of the antennas. However, they are cancelled by the subtraction of both datasets as described in Sect. 2.

5.1 Free-standing antenna in anechoic chamber

The radiation pattern of the AUT in free-space configuration was recorded in the anechoic chamber. It serves as a reference for evaluating the distorted radiation patterns of the AUT when put directly on different kinds of wall. The relative rotation between the AUT and the probe antenna was achieved by rotating the AUT around its own axis in azimuth plane. Due to the use of a robotic device, the adjusted azimuth angle ϕ is very precise.

5.2 Antenna on homogeneous wooden slab and inhomogeneous brick wall in anechoic chamber

The distorted radiation patterns of the AUT when put on homogeneous wooden slab and inhomogeneous brick wall (see Fig. 4) was recorded in the anechoic chamber. For this, the

Table 2. Geometrical setup parameters for both the simulations and the measurements. The simulations provide data for all required distances and for $\phi = [-180°, 180°]$ with $\Delta\phi = 1°$.

Antenna configuration	Distance AUT ↔ probe antenna	Recorded azimuth angle interval ϕ	Increment in azimuth angle $\Delta\phi$	Recorded frequency spectrum f
Free-standing (indoor)	$d_1 = 1.50$m	$[-178°, 180°]$	2°	0.7 – 10.0GHz
	$d_2 = 2.70$m	$[-178°, 180°]$	2°	0.7 – 10.0GHz
	$d_3 = 4.14$m	0°	–	0.7 – 10.0GHz
On wooden slab (indoor)	$d_1 = 1.50$m	$[0°, 70°]$	5°	0.7 – 10.0GHz
	$d_2 = 2.70$m	$[-35°, 70°]$	5°	0.7 – 10.0GHz
On brick wall (indoor, Fig.4)	$d_1 = 1.50$m	$[-25°, 60°]$	5°	0.7 – 10.0GHz
On multilayer wall (outdoor, Fig.6)	$d_1 = 1.50$m	$[-60°, 60°]$	10°	0.7 – 4.3GHz
	$d_4 = 3.00$m	0°	–	0.7 – 4.3GHz

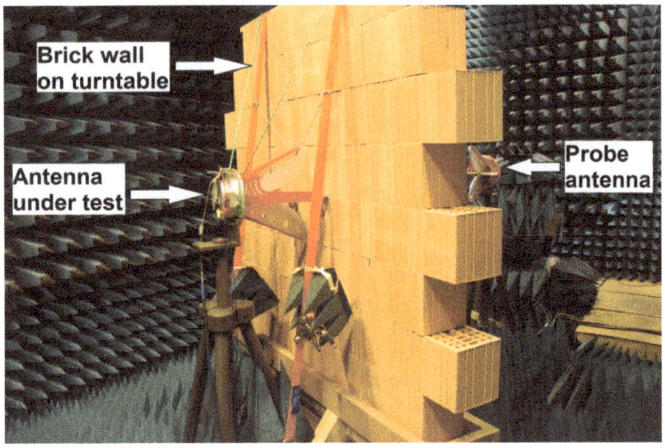

Fig. 4. Measurement setup of antenna placed on brick wall in position A (see Fig. 5). The structure comprising of wall and AUT is being rotated around its axis and the radiation pattern is recorded with the probe antenna. Measurements were carried out inside an anechoic chamber. The setup for the wooden slab is accordant.

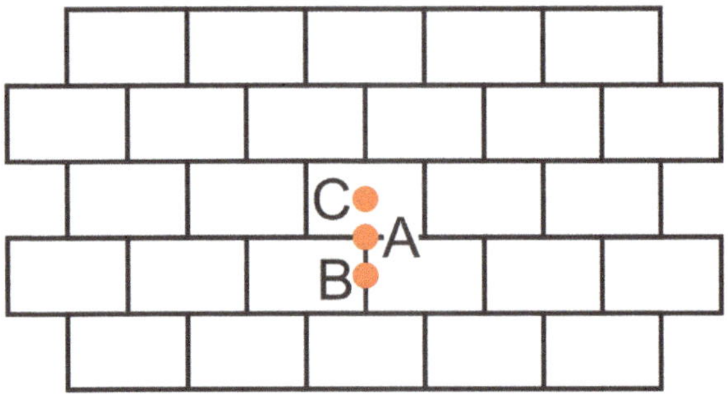

Fig. 5. Measurement positions for AUT on the brick wall. Two rows are omitted at the bottom and one row at the top.

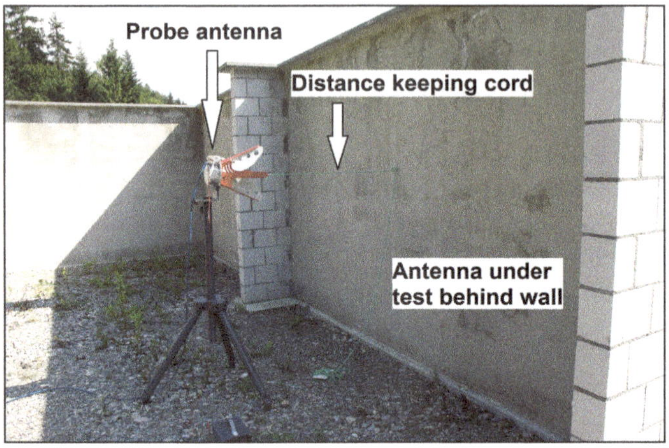

Fig. 6. Measurement setup of antenna on multilayer wall on the open-air test site Witaumatte (Thun, CH). The AUT is being placed directly on the wall. The radiation pattern is recorded with a probe antenna that is moved on a circle around the wall (distance keeping cord).

walls were erected on a table standing on a turntable inside the anechoic chamber. The AUT was fixed on the walls. By rotating this structure around its own axis, the distorted radiation pattern was recorded with the probe antenna being on a fixed spot. The use of an angle metre in connection with the special construction of the table helped keeping the step sizes $\Delta\phi$ very precise. Due to a mistake in the reference ϕ_{offset}, the data are recorded with an offset in angle that is constant for all ϕ. This mistake was corrected in Sect. 6.1 in case of the homogeneous wooden slab. Due to the brick's inhomogeneous profile (Fig. 2), transmissions TTW are sensitive to the location of measurement which makes corrections senseless as other effects are very much dominating the measurements. For quantification of the sensitivity, radiation pattern measurements have been carried out at the three differing positions on the brick wall as indicated in Fig. 5.

5.3 Antenna on inhomogeneous multilayer wall at open-air test site

In the case of the inhomogeneous multilayer wall at the open-air test site in Thun (CH), the probe antenna was moved circularly around the AUT as shown in Fig. 6. Even though the same angle metre was used here as well, a less precise adjustment of the azimuth angle ϕ is being expected. A deviation of a few degrees for each position is realistic. Due to being in a realistic environment, parasitic effects and disturbance sources can interfere with the measurements. An attenuator was used for suppressing these interferences (and parasitically the signal). A 10-Watt amplifier was used to raise the signal. This limited the frequency range to 0.7 – 4.3 GHz.

6 Results and discussion

For the calculation of the net damping Δs_{21} caused by the multilayer wall, free-space measurements and measurements on the wall at corresponding distances and the same VNA setting were executed. In this case the only change in the configuration is the presence of the wall and Δs_{21} can be calculated by subtracting the results of both measurements from each other as shown in Eq. (1).

6.1 Homogeneous wooden slab

In contrary to the perfectly symmetrical simulational setup, inaccuracies lead to an offset azimuth angle ϕ_{offset} in the measurement setup. This offset was quantified by the identification of the symmetry centre of the radiation pattern recorded at the distance $d_2 = 2.70$ m. Its value is $\phi_{\text{offset}} \approx 2°$, and the results were corrected accordingly. Simulated and measured damping Δs_{21} in the azimuth plane is plotted in Fig. 7 for $f = 1.0$ GHz and the distance $d_2 = 2.70$ m between AUT and observation point. Due to the perfect symmetry in the field simulator, the simulation results for both polarizations are symmetric. In the case of an infinite slab, perpendicular (VV) and parallel (HH) polarization directions within the plane of incidence fall together at normal incidence and show the same transmission characteristics. As the simulation results intercept at $\phi = 0°$, the effects of the slab edges are negligible at this angle. In contrary, for the measurements there is a difference of ≈ 0.15 dB between both polarizations at normal incidence, resulting from positioning imprecisenesses like inclination of the slab, adjustment and torsion of the antenna. For these reasons the recorded radiation pattern is affected irreversibly. With increasing azimuth angle both simulation and measurement data for the VV polarization show corrugation behaviour, in contrast to the HH polarization data that stays smooth. This results from the antenna's radiation pattern that is broader in the cut perpendicular to the excitation (ETS-Lindgren, 2010): as the data was recorded in azimuth plane, this complies with the VV data. Due to broader main lobe, the propagating wave interacts with the edges of the slab, causing wave interference in the observation point. The damping is almost constant for $\phi < 40°$. The wave propagating TTW gets attenuated only in the normal direction of the slab. Therefore, the longer propagation path of the electromagnetic wave described by Snell's law does not lead to bigger damping values Δs_{21}. Simulations and measurements are in good agreement for $\phi < 45°$. If the angle increases further, disturbing effects caused by the slab's edges dominate the results. Minimizing these effects are only possible by using a bigger wooden slab in simulations and measurements.

The results of the accomplished sensitivity analysis are plotted in Fig. 8. The deviation in damping Δs_{21} is shown in dependence of ε_r (x axis) and $\tan\delta$ (coloured bars). The values $\varepsilon_r = 1.7$ and $\tan\delta = 0.07$ are considered as reference. The maximum deviation for the values under study is approximately 0.25 dB, this means the relative uncertainty is about 3 %. If the assumed ε_r is correct, the maximum deviation caused by a wrong $\tan\delta$ is about 0.2 dB (relative uncertainty 2 %), in case the $\tan\delta$ is known and ε_r is unknown, the maximum deviation is about 0.1 dB (relative uncertainty 1 %). The deviation due to an overestimated ε_r gets partially compensated with an underestimated $\tan\delta$, and vice versa accordingly. In this case the deviation is limited to 0.1 dB (relative uncertainty 1 %).

As a recapitulation it can be said that the distortion of the antenna's radiation pattern is limited to damping Δs_{21} for measurement angles $\phi < 45°$. Due to the symmetry and homogeneous character of the material, this does not only apply to the azimuth plane but also to the inclination plane: HH simulations and measurements in the azimuth plane can also be interpreted as VV in the inclination plane. The good agreement of simulations and measurement show that transmission through a homogeneous slab is predictable. The sensitivity analysis shows that the quantified deviations in the result are smaller than the deviation in the material parameter. While the relative uncertainty of the permittivity ε_r is about 6 % and that of the loss tangent $\tan\delta$ is about 14 %, the combined maximum consequence in Δs_{21} is just 3 %. If one of the two material parameters (ε_r and $\tan\delta$) is overestimated and the other one is underestimated, this leads to a partial compensation. At the same time the deviations are quite small and do not affect the radiation pattern dramatically. If the importance of the signal amplitude is of minor interest, the deviation due to wrong material parameters within the range of this investigation can be neglected.

6.2 Inhomogeneous brick wall

In the first step, the measured data at all three positions on the wall as shown in Fig. 5 were compared regarding to the main radiation direction $\phi = 0°$. The damping Δs_{21} over the whole frequency range 0.7 – 10.0 GHz in VV polarization is shown in Fig. 9. All curves show very similar transmission characteristics at frequencies below 1 GHz. Beginning at this

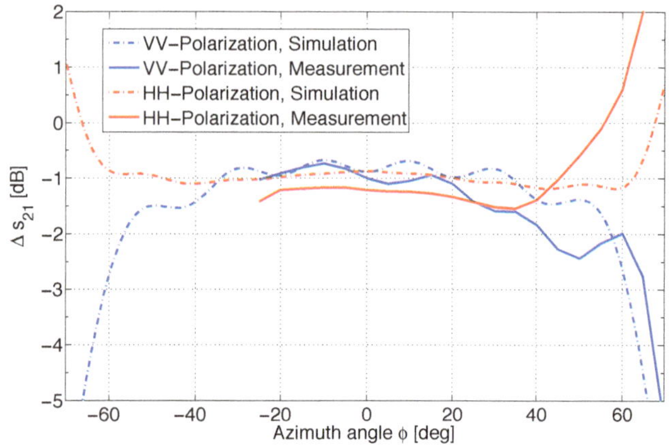

Fig. 7. Damping Δs_{21} caused by the wooden slab at $f = 1\,\text{GHz}$ and the distance $d_2 = 2.70\,\text{m}$ between slab and observing point. Comparison of simulations and measurements in VV and HH polarization.

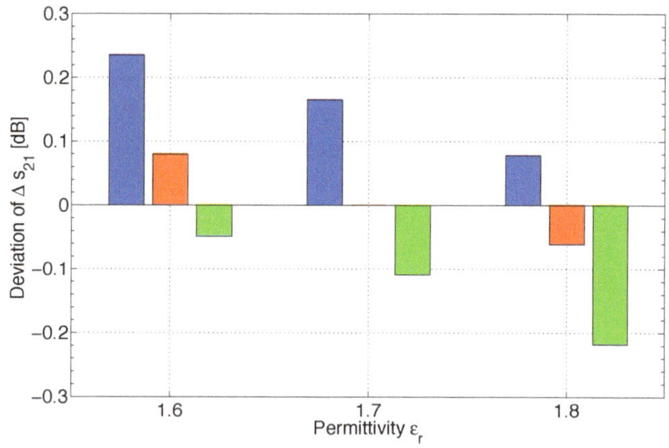

Fig. 8. Simulation of the transmission through the wooden slab with different material parameters (Table 1). Comparison of the deviation in damping Δs_{21} at $\phi = 0°$, $\theta = 0°$ with the reference deviation $\varepsilon_r = 1.7$, $\tan\delta = 0.07$. The bars represent $\tan\delta = 0.06,\ 0.07,\ 0.08$ (blue, red and green respectively).

frequency the curves start deviating from each other. At this frequency, the free-space wavelength λ_0 is in the range of 30 cm. During its propagation through the material it reduces to approximately 23 cm due to Eq. (2). At the same time the brick's biggest cavity has a diagonal length of $\approx 11\,\text{cm}$: this length is in the same range as the wavelength in the material. Thereby the cavity is now visible to the wave, causing electromagnetic interaction and causing scattering. The differences between the curves show the sensitivity of the transmission TTW regarding the positioning of the antenna on the wall. The results for the three positions differ among each other. At $f = 2.5\,\text{GHz}$, transmission TTW at antenna position B increases to $s_{21,B} \approx +4\,\text{dB}$, whereas the transmission at position C decreases drastically $s_{21,C} \approx -30\,\text{dB}$. These deviations are due to the interference of wave fractions reflected on the cavities: at the observed point the total electric field is measured and therefore the in-phase sum of all waves might sum up to almost zero. By changing the position of the antenna, the distances to the cavities change accordingly and with them the phases of the partial waves. As the consequence, the transmission shows different results for different antenna positions. In the case of antenna position B and C, Δs_{21} can reach values greater than zero. So even focusing effects at certain positions and certain frequencies are possible. It can also be observed that one of the curves always has an amplitude $\Delta s_{21} > -10\,\text{dB}$. This is a hint that many antenna positions can complement each other and the probability of higher Δs_{21} increases with the growing number of used antennas. As the transmission is very much dominated by other effects, ϕ_{offset} is negligible.

In the second step, measured data are compared at one single frequency $f = 2\,\text{GHz}$ in dependence of the azimuth angle ϕ. Both simulation and measurement data are shown in Fig. 10. As stated in Table 1, simulations were carried out for two different spacing values between the bricks ($g = 1\,\text{cm}$ and 2 cm). The curves are normalized in such a way that straight radiation loss is coded in the $\phi = 0°$ value of each curve. Big differences between the simulation and measurement curves can be observed. In the simulations, against prediction, transmission for the bigger gap between the bricks $g = 2\,\text{cm}$ lead to bigger damping Δs_{21}, even though the antenna is directly in front of the gap. Assumptions based on Fig. 9 and the highly sensitive transmission depending on geometry and frequency would require the radiation pattern to be similarly turbulent. As the simulations were carried out for a dense angular step $\Delta\phi = 1°$, the smoothness of the curves are truthful. This is a result of the perfect periodicity and geometry of the simulation setup. In contrast, the measurements were recorded in sparse angular steps $\Delta\phi = 10°$. The smoothness of the curves here can be a result of numerical interpolation, hiding local drops in field strength. In spite, the main radiation beam of the antenna is preserved.

As a result, making conclusions from measured reflections to the original objects are not possible as the damping caused by the wall is unpredictable as a high grade of determination is demanded. This includes the exact positions of the bricks and their detailed characteristics, which is practically impossible. Floquet mode analysis and measurement in (Honcharenko et al., 1994) on concrete cinder blocks that have a simpler cross-section and periodicity than brick showed the difficulties of accurate prediction of TTW transmission, revealing the strong dependence of diffraction on frequency and periodicity. Here, the manifold periodicity of the brick wall further increases the difficulty of analysis. The additional dependence on the positioning of the transmitting antenna makes the uniform illumination of the room impossible. This can cause problems in detection and classification

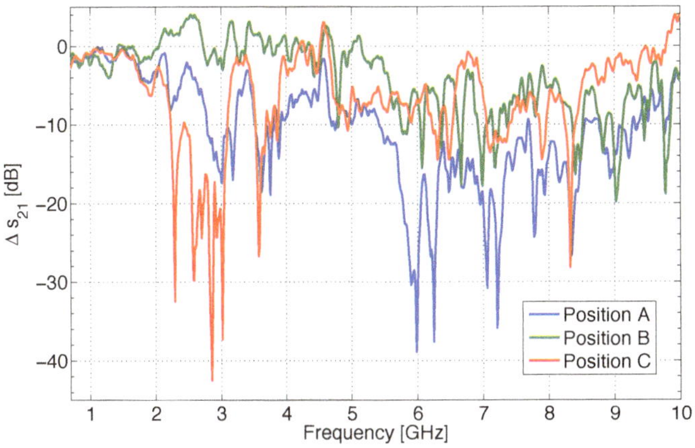

Fig. 9. Damping through brick wall at three different measurement points A, B, C (Fig. 5) with rotation angle $\phi = 0°$ for the frequency spectrum $0.7 - 10.0$ GHz in VV polarization.

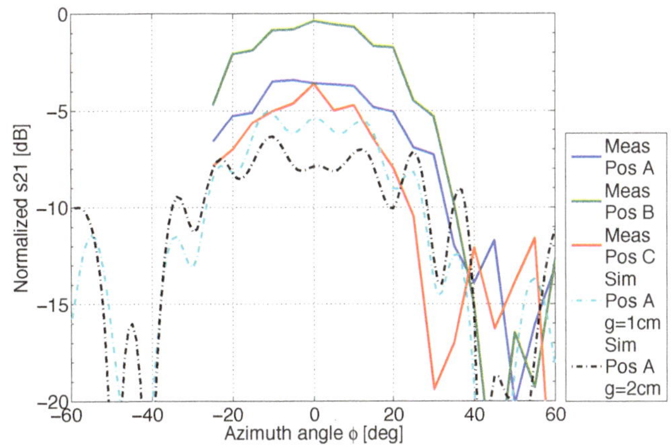

Fig. 10. Radiation pattern of brick wall for $f = 2$ GHz in VV polarization: comparison of measurement results in positions A, B, C and simulation results for different H and V brick spacings g.

of scatterers. In the worst case the radar system will be blind to several spots and will not be able detecting anything located there. Together with ultra-wideband (UWB) systems, multi-view approaches are most suitable for solving the problem of incalculable transmission: as the antennas are placed or shifted at different locations, diversities in signal strength get balanced through statistical variation. In the later image fusion process, superposition of the multiple view data will lead to a balance for all scatterers inside the room. The measurements have shown that the unreliable transmission at the different measurement positions is compensated by the increasing probability that one of the measurement positions will cause sufficient transmission. Here with three different measurement positions, there is always one $\Delta s_{21} > -10$ dB. With an increasing number of used antennas it is being expected that the probability of higher signal strength TTW increases as well.

6.3 Inhomogeneous multilayer wall

The result of the damping measurement through the inhomogeneous multilayer wall over the frequency $0.7 - 4.3$ GHz in VV polarization is shown in Fig. 11. For the calculation of the net damping Δs_{21}, free-space measurements in the anechoic chamber were used as the reference rather than those in the open test site. As free-space measurements in the anechoic chamber at the required distance $d_4 = 3.00$ m were not available, a quadratic interpolation between the available data for the distances 1.47 m, 2.67 m and 4.14 m was executed. The mistake committed with this procedure is below 1 dB, which is insignificant to the result. The blue graph represents the measured net loss, the red graph is a linear fit. It indicates that the loss increases approximately with 16 dB/GHz. For frequencies $f > 3.35$ GHz the signal falls below the noise level and data are not usable anymore.

The radiation patterns in VV polarization for different frequencies up to a maximum of $f = 3.3$ GHz are shown in Fig. 12. For comparison, measurement data of the free-standing antenna are shown as well. To demonstrate the relative amplitude difference in dependence of the frequency, the curves are normalized equally as in Figure 10: the central radiation direction $\phi = 0°$ indicates the net loss Δs_{21} due to the wall's presence. In the case of the free-standing antenna, net loss is 0 dB. In theory, these values for the damping should agree with the data presented in Fig. 11. Practically, the discrepancies show that the recorded transmission at two different distances ($d_4 = 3.00$ m and $d_1 = 1.50$ m) can be different, an indication that wave interference plays an important role. The radiation patterns of the antenna in free-space configuration and the antenna on the wall at 0.7 GHz show symmetry. The curves for the antenna on wall at higher frequencies do not show symmetry anymore. The transmission values seem to be randomly distributed within specific tolerances that increase with higher frequency, forming an approximately 12 dB band in the case of $f = 3.3$ GHz. Preference in radiation direction as shown for the antenna on brick wall vanishes. This and higher sensitivity with increasing frequency show that wave interference of the diffracted waves on the cavities dominates transmission TTW. In addition, the high damping Δs_{21} of the walls codetermine the transmission as visible from the amplitude distances between the individual curves.

Very similar to the sole brick wall, the field distribution behind the wall cannot be analyzed analytically nor by simulations as a high degree of determination for the wall is necessary. Sensitivity increases together with the frequency, depending on too many parameters at the relevant frequencies. Wave interference as well as spatial dispersion is even more dominant for the multilayer wall than for the brick wall as the main radiation direction of the wave vanishes. The high

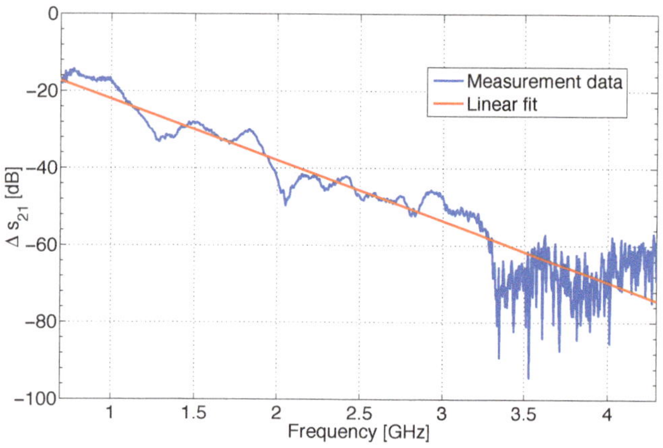

Fig. 11. Damping through multilayer wall in VV polarization for frequency spectrum 0.7–4.3 GHz. The linear fit indicates that Δs_{21} increases with 16 dB/GHz. Measurement in anechoic chamber used as reference. Distance between antennas $d_4 = 3.00$ m.

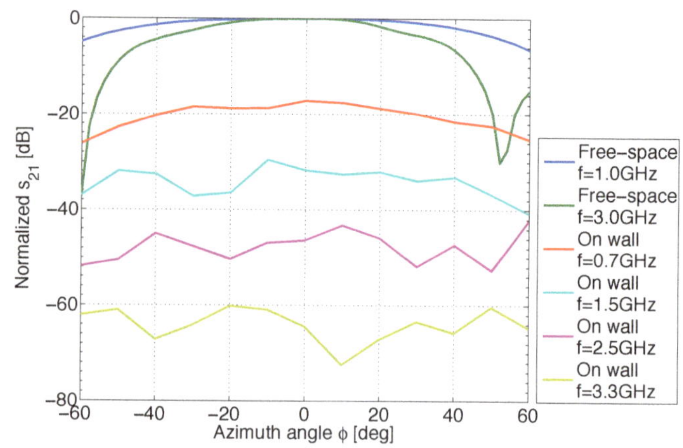

Fig. 12. Distorted radiation pattern of antenna on inhomogeneous multilayer wall in comparison to radiation pattern of free-standing antenna at different frequencies in VV polarization. Distance between antennas $d_1 = 1.50$ m.

damping value of the wall also plays a very important role. As the signal needs to travel back and forth, high dynamic range of the imaging system is required.

7 Conclusions

In this work, simulations and measurements for the determination of near-field radiation patterns and their distortion by three different kinds of wall have been presented.

It has been shown that simulations and measurements are in very good agreement in case of the homogeneous wooden slab, making transmission through a homogeneous slab predictable. Sensitivity analysis for different material properties have shown that the error range due to imprecise material knowledge is much smaller than the error range of the material properties taken as a basis in this work. Additionally they are within measurement inaccuracies and therefore could be neglected. This shows that simulations are reliable and can be used in the prediction of transmission TTW. In the case of the inhomogeneous brick wall, three different measurement positions lead to different near-field patterns. The results for the TTW transmission in normal incidence show the sensitivity of the brick wall both on the positioning of the AUT and on frequency: depending on these variables it is possible that the potential TWRI system is blind to certain spots. This is caused by interference of waves scattered at the cavities. Nevertheless, the main lobe of the radiation pattern is conserved qualitatively. This does not apply for the inhomogeneous multilayer wall: wave interference caused by the cavities garbles the radiation pattern in such a way that field distribution seems to be randomly distributed within certain aberrations, even destructing the symmetry of the radiation pattern. In addition, transmission is dominated by the high damping values. The field strength measured behind walls made of inhomogeneous material depends on many variables like frequency, position of the AUT, direction of radiation and the position of investigation. Simulations are not reliable because the determination of the whole wall is essential. As similar characteristics were reported for reinforced concrete walls (Dalke et al., 2000; Antonini et al., 2003), this problem spans the vast majority of wall structures.

The appearances of unpredictable signal nulls in the frequency spectrum make the use of narrowband single-view systems inadequate. Multi-view approaches (e.g. multi-static radar, MIMO, SAR) based on UWB architecture can solve the problem of unreliable transmission by statistically balancing the field distribution behind the wall.

Acknowledgements. This work was supported under grant number Aramis 041-19 by *Armasuisse W+T*, administrational agency of the Swiss Federal Department of Defense, Civil Protection and Sports.

References

Ahmad, F., Amin, M. G., and Mandapati, G.: Autofocusing of Through-the-Wall Radar Imagery Under Unknown Wall Characteristics, IEEE T. Image. Process., 16, 1785–1795, doi:10.1109/TIP.2007.899030, 2007.

Amin, M. G.: Through-the-Wall Radar Imaging, CRC Press, Boca Raton (FL), USA, 2011.

Antonini, G., Orlandi, A., and D'elia, S.: Shielding Effects of Reinforced Concrete Structures to Electromagnetic Fields due to GSM and UMTS Systems, IEEE T. Magn. , 39, 1582–1585, doi:10.1109/TMAG.2003.810327, 2003.

Balanis, C. A.: Antenna theory: analysis and design, 3 Edn, John Wiley & Sons, Hoboken (NJ), USA, 2005.

Baranoski, Edward J.: Through-wall imaging: Historical perspective and future directions, J. Frankl. Inst., 345, 556–569, doi:10.1016/j.jfranklin.2008.01.005, 2008.

Chang, P. C., Burkholder, R. J., and Volakis, J. L.: Adaptive CLEAN With Target Refocusing for Through-Wall Image Improvement, IEEE T. Antenn. Propag., 58, 155–162, doi:10.1109/TAP.2009.2036131, 2010.

Dalke, R. A., Holloway, C. L., McKenna, P., Johansson, M., and Ali, A. S.: Effects of Reinforced Concrete Structures on RF Communications, IEEE T. Electromagn. C., 42, 486–496, doi:10.1109/15.902318, 2000.

Edgar, D.: HFSS 13: Hybrid FE-BI for Efficient Simulation of Radiation and Scattering, Ansys. Inc., 2011.

ETS-Lindgren: Model 3164 Series – Open Boundary Quad-Ridged Horns – User Manual, 2010.

Farwell, M., Ross, J., Luttrell, R., Cohen, D., Chin, W., and Dogaru, T.: Sense through the wall system development and design considerations, J. Frankl. Inst., 345, 570–591, ISSN 0016-0032, doi10.1016/j.jfranklin.2008.01.004, 2008.

Ferris Jr, D. D. and Currie, N. C.: Survey of current technologies for through-the-wall surveillance (TWS), P. SPIE., 3577, 62–72, 1999.

Honcharenko, W. and Bertoni, H.: Transmission and Reflection Characteristics at Concrete Block Walls in the UHF Band Proposed for Future PCS, IEEE T. Antenn. Propag., 42, 232–239, 1994.

Jia, Y., Kong, L., and Yang, X.: A novel approach to target localization through unknown walls for through-the-wall radar imaging, Progr. Electromagn. Res., 119, 107–132, 2011.

Li, L., Zhang, W., and Li, F.: A Novel Autofocusing Approach for Real-Time Through-Wall Imaging Under Unknown Wall Characteristics, IEEE T. Geosci. Remote, 48, 423–431, doi:10.1109/TGRS.2009.2024686, 2010.

Muqaibel, A. H.: Characterization of Ultra Wideband Communication Channels, Ph.D. thesis, Virginia Polytechnic Institute and State University, Blacksburg (VA), USA, 2003.

Seybold, J. S.: Introduction to RF Propagation, John Wiley & Sons, Hoboken (NJ), USA, 2005.

Sheen, D. M., McMakin, D. L., and Hall, T. E.: Near Field Imaging at Microwave and Millimeter Wave Frequencies, IEEE/MTT-S International Microwave Symposium, Honolulu (HI), USA, 3-8 June 2007, 693–1696, doi:10.1109/MWSYM.2007.380033, 2007.

Stavrou, S. and Saunders, S. R.: Review of constitutive parameters of building materials, in: Proceedings of the 12th International Conference on Antennas and Propagation ICAP, Exeter, UK, 31 March – 3 April 2003, 1, 211–215, doi:10.1049/cp:20030052, 2003.

Thajudeen, C., Hoorfar, A., Ahmad, F., and Dogaru, T.: Measured complex permittivity of walls with different hydration levels and the effect on power estimation of twri target returns, Prog. Electromagn. Res. B, 30, 177–199, doi:10.2528/PIERB10091004, 2011.

Wang, G. and Amin, M. G.: Imaging Through Unknown Walls Using Different Standoff Distances, IEEE T. Signal Proces., 54, 4015–4025, 2006.

14

Electromagnetic diffraction and scattering of a complex-source beam by a semi-infinite circular cone

H. Brüns and L. Klinkenbusch

Institut für Elektrotechnik und Informationstechnik, Christian-Albrechts-Universität zu Kiel, Germany

Correspondence to: L. Klinkenbusch (lbk@tf.uni-kiel.de)

Abstract. A complex-source beam (CSB) is used to investigate the electromagnetic scattering and diffraction by the tip of a perfectly conducting semi-infinite circular cone. The boundary value problem is defined by assigning a complex-valued source coordinate in the spherical-multipole expansion of the field due to a Hertzian dipole in the presence of the PEC circular cone. Since the incident CSB field can be interpreted as a localized plane wave illuminating the tip, the classical exact tip scattering problem can be analysed by an eigenfunction expansion without having the convergence problems in case of a full plane wave incident field. The numerical evaluation includes corresponding near- and far-fields.

1 Introduction

Scattering and diffraction of a plane electromagnetic wave by a semi-infinite perfectly electrically conducting (PEC) circular cone has often been treated in the literature. A survey on this subject can be found in the classical monograph by (Bowman et al., 1987). More recently, the scattered field has been obtained by a multipole-based integration of the exact surface-currents for the case of an incident plane wave. However, that solution suffers from a missing convergence of the finally obtained multipole series, and the application of suitable series-transformation techniques has been necessary to asymptotically derive the corresponding limiting values (Klinkenbusch, 2007; Kijowski and Klinkenbusch, 2011). Particularly, we are interested in the fields related the cone's apex, as the corresponding scattered field can act as a further element to complete asymptotic methods like the Geometrical Theory of Diffraction (GTD) and the Uniform Theory of Diffraction (UTD).

For the analysis of the influence of the tip on the scattered field, we have used a localized plane wave, that is, a complex-source beam (CSB) as the incident field. This can be achieved by simply assigning a complex-valued location to a dipole source. The width, focus length, and direction of incidence of the CSB can be controlled nearly arbitrarily (Orlov and Peschel, 2010). Combining this powerful tool with the spherical-multipole expansion provides a convergent eigenfunction solution with focusses on the evaluation of the scattered related to the tip of the cone.

Section 2 summarizes the solution of Maxwell's equations in the presence of a PEC circular cone, and in Sect. 3 the complex-source technique is introduced. Finally, we present some numerical results in Sect. 4.

2 Solution of Maxwell's equations in the presence of a PEC semi-infinite circular cone

Consider a perfectly electrically conducting (PEC) circular cone as illustrated in Fig. 1 with a half outer opening angle ϑ_0 embedded in a homogeneous medium with permittivity ε and permeability μ. We are looking for a solution of the time-harmonic Maxwell's equations at a time factor $e^{+j\omega t}$.

2.1 Helmholtz equation in spherical coordinates

The homogenous scalar Helmholtz equation

$$\Delta\Phi(r,\vartheta,\varphi)+\kappa^2\Phi(r,\vartheta,\varphi)=0 \tag{1}$$

with the wave number $\kappa=\omega\sqrt{\varepsilon\mu}$ can be solved in spherical coordinates r,ϑ,φ (ϑ represents the polar coordinate.) by a classical separation ansatz. We finally obtain the following elementary solution

$$\Phi_\nu^m(r,\vartheta,\varphi)=z_\nu(\kappa r)Y_\nu^m(\vartheta,\varphi) \tag{2}$$

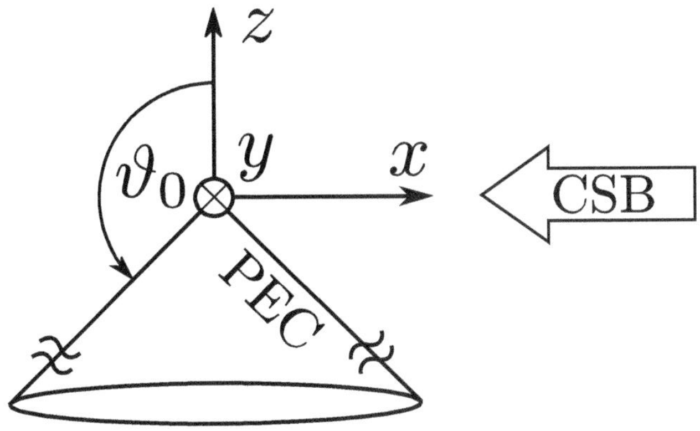

Fig. 1. A PEC, semiinfinite, circular cone located on the negative z-axis. The half outer opening angle is ϑ_0. A CSB incides from x-direction.

with z_ν being spherical Bessel functions related to (ordinary) Bessel functions by

$$z_\nu(\kappa r) = \sqrt{\frac{\pi}{2\kappa r}} Z_{\nu+1/2}(\kappa r) \tag{3}$$

Particularly we will need (everywhere regular) spherical Bessel functions of the first kind, $j_\nu(\kappa r)$, and spherical Hankel functions of the second kind, $h_\nu^{(2)}(\kappa r)$, satisfying the radiation condition at the chosen time factor $e^{+j\omega t}$.

The spherical harmonics $Y_\nu^m(\vartheta,\varphi)$ are defined as

$$Y_\nu^m(\vartheta,\varphi) = N_\nu^m P_\nu^m(\cos\vartheta) e^{jm\varphi}, \tag{4}$$

that is, as products of associated Legendre functions of the first kind $P_\nu^m(\cos\vartheta)$ in ϑ, and harmonic functions in φ. The normalization constant N_ν^m is chosen such that

$$\int_0^{\vartheta_0}\int_0^{2\pi} \left|Y_\nu^m(\vartheta,\varphi)\right|^2 \sin\vartheta \, d\vartheta d\varphi = 1. \tag{5}$$

2.2 Determination of eigenvalues

For the given geometry the solution has to be 2π-periodic in φ, hence the eigenvalues m have to be integral. In order to get solutions of the Helmholtz equation for a PEC circular cone as depicted in Fig. 1, these solutions have to fulfill the boundary conditions on the cone's surface. Therefore we need spherical harmonics that satisfy the Dirichlet condition as well as spherical harmonics that satisfy the Neumann condition at $\vartheta = \vartheta_0$. This can be achieved by finding eigenvalues ν_d and ν_n (Blume and Krebs, 1994) according to the transcendental equations

$$P_{\nu_d}^m(cos\ \vartheta)|_{\vartheta_0} = 0 \tag{6}$$

$$\left.\frac{\partial P_{\nu_n}^m(cos\ \vartheta)}{\partial\vartheta}\right|_{\vartheta_0} = 0 \tag{7}$$

which have been solved by means of a suitable numerical procedure, e.g. a bisection method.

Moreover, it has been shown by (Katsav et al., 2012) that for all valid eigenvalue pairs (ν, m) it holds that

$$m^2 \leq \nu(\nu+1). \tag{8}$$

Figure 2 shows the eigenvalue pairs for $\vartheta_0 = 135°$ and $\nu_{\max} = 40$ in the ν, m-plane.

2.3 Spherical-multipole expansion

Within a homogeneous domain, any solution of Maxwell's equations outside the sources can be constructed by means of a complete set of spherical-multipole functions defined by

$$\boldsymbol{M}(\boldsymbol{r}) = (\boldsymbol{r}\times\nabla)\,\Phi(\boldsymbol{r}) \tag{9}$$

$$\boldsymbol{N}(\boldsymbol{r}) = \left(\frac{1}{\kappa}\nabla\times(\boldsymbol{r}\times\nabla)\right)\Phi(\boldsymbol{r}) \tag{10}$$

provided that Φ is a solution of the scalar homogeneous Helmholtz Eq. (1). They are determined according to

$$\boldsymbol{M}_{\nu,m}(r,\vartheta,\varphi) = z_\nu(\kappa r)\ \boldsymbol{m}_{\nu,m}(\vartheta,\varphi) \tag{11}$$

$$\begin{aligned}\boldsymbol{N}_{\nu,m}(r,\vartheta,\varphi) = &-\frac{z_\nu(\kappa r)}{\kappa r}\ \nu(\nu+1)Y_{\nu,m}(\vartheta,\varphi)\boldsymbol{e}_r \\ &-\frac{1}{\kappa r}\frac{d}{dr}[r z_\nu(\kappa r)]\boldsymbol{n}_{\nu,m}(\vartheta,\varphi)\end{aligned} \tag{12}$$

where $\boldsymbol{m}$ and $\boldsymbol{n}$ denote the transvere multipole functions:

$$\begin{aligned}\boldsymbol{m}_{\nu,m}(\vartheta,\varphi) = &-\frac{1}{\sin\vartheta}\frac{\partial Y_{\nu,m}(\vartheta,\varphi)}{\partial\varphi}\boldsymbol{e}_\vartheta \\ &+\frac{\partial Y_{\nu,m}(\vartheta,\varphi)}{\partial\vartheta}\boldsymbol{e}_\varphi\end{aligned} \tag{13}$$

$$\begin{aligned}\boldsymbol{n}_{\nu,m}(\vartheta,\varphi) = &\frac{\partial Y_{\nu,m}(\vartheta,\varphi)}{\partial\vartheta}\boldsymbol{e}_\vartheta \\ &+\frac{1}{\sin\vartheta}\frac{\partial Y_{\nu,m}(\vartheta,\varphi)}{\partial\varphi}\boldsymbol{e}_\varphi.\end{aligned} \tag{14}$$

To accomplish that the tangential part of the electric field vanishes on the PEC cone's surface, the spherical-multipole expansion of the total field reads

$$\boldsymbol{E}(\boldsymbol{r}) = \sum_{\nu_d,m} a_{\nu_d,m}^{II(I)}\boldsymbol{N}_{\nu_d,m}^{II(I)}(\boldsymbol{r}) + \frac{Z}{j}\sum_{\nu_n,m} b_{\nu_n,m}^{II(I)}\boldsymbol{M}_{\nu_n,m}^{II(I)}(\boldsymbol{r}) \tag{15}$$

and for the magnetic field

$$\boldsymbol{H}(\boldsymbol{r}) = \frac{j}{Z}\sum_{\nu_d,m} a_{\nu_d,m}^{II(I)}\boldsymbol{M}_{\nu_d,m}^{II(I)}(\boldsymbol{r}) + \sum_{\nu_n,m} b_{\nu_n,m}^{II(I)}\boldsymbol{N}_{\nu_n,m}^{II(I)}(\boldsymbol{r}). \tag{16}$$

$$r > r',\ (r < r') \tag{17}$$

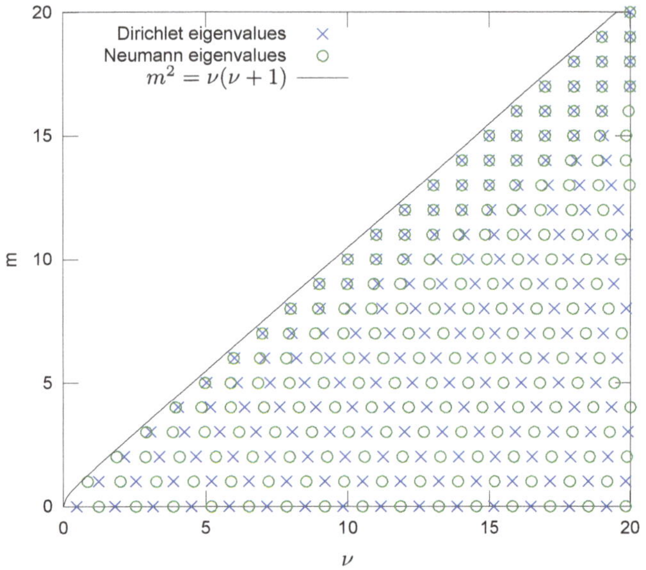

Fig. 2. Pairs of eigenvalues for $\vartheta_0 = 135°$, $\nu_{\max} = 20$.

where ν_d and ν_n are eigenvalues of the Dirichlet and Neumann type, respectively. $Z = \sqrt{\mu/\varepsilon}$ represents the intrinsic wave impedance of the homogeneous domain. The indices (I) and (II) stand for the use of spherical Bessel functions j_ν or spherical Hankel functions of the second kind $h_\nu^{(2)}$ are used according to Eq. (17) (see also Sect. 4). The multipole amplitudes $a_{\nu_d,m}$ and $b_{\nu_d,m}$ can be determined for a Hertzian dipole source at $\boldsymbol{r}'$ electrically polarized in $\boldsymbol{c}_e$ according to

$$a^{II(I)}_{\nu_d,m} = -\kappa_0^2 Z_0 \frac{(-1)^m}{\nu_d(\nu_d+1)} \boldsymbol{N}^{I(II)}_{\nu_d,-m}(\boldsymbol{r}') \cdot \boldsymbol{c}_e\,, \tag{18}$$

$$b^{II(I)}_{\nu_n,m} = -j\kappa_0^2 \frac{(-1)^m}{\nu_n(\nu_n+1)} \boldsymbol{M}^{I(II)}_{\nu_n,-m}(\boldsymbol{r}') \cdot \boldsymbol{c}_e\,. \tag{19}$$

The electric far-field is found from the asymptotic form of the spherical Hankel function of the second kind for large arguments as

$$\boldsymbol{E}_\infty(\boldsymbol{r}) = \left(\frac{e^{-j\kappa r}}{\kappa r}\right)\left[\sum_{\nu_d,m} -j^{\nu_d} a^{II}_{\nu_d,m} \boldsymbol{n}_{\nu_d,m} + \frac{Z}{j}\sum_{\nu_n,m} j^{(\nu_n+1)} b^{II}_{\nu_n,m} \boldsymbol{m}_{\nu_n,m}\right]. \tag{20}$$

For the evaluation of the scattered field, we simply need to subtract the incident field from the total field where the incident field is computed by a similar approach using a free space spherical-multipole expansion.

3 Complex-source beam

The complex-source beam (CSB) offers the possibility to describe a focussed beam analytically. If a complex valued location is assigned to a dipole source, a Gaussian beam is generated. The beam's focus, width, and direction can be controlled arbitrarily by the chosen complex location. The complex location vector of the source has the form

$$\boldsymbol{r}'_c = \boldsymbol{r}'_r - j\boldsymbol{r}'_i\,, \tag{21}$$

where $\boldsymbol{r}'_r$ gives the actual position of the beam's waist and $\boldsymbol{r}'_i$ the direction of incidence. The absolute value of $\boldsymbol{r}'_i$ correspond to the rayleigh length of the generated Gaussian beam. To clarify this, we insert a complex $\boldsymbol{r}'$ into the free space Green's function. We choose a purely imaginary location $\boldsymbol{r}'$:

$$\boldsymbol{r}' = -jz_0\boldsymbol{e}_z\,. \tag{22}$$

That means, the beam has a rayleigh length z_0, has it's waist in the origin, and propagates in z-direction. Assuming a small distance $\rho = \sqrt{x^2+y^2}$ from the z-axis, a second order taylor series expansion of $|\boldsymbol{r}-\boldsymbol{r}'|$ delivers a paraxial approximation:

$$|\boldsymbol{r}-\boldsymbol{r}'| = \sqrt{x^2+y^2+(z+jz_0)^2} \approx \left((z+jz_0) + \frac{1}{2}\frac{\rho^2}{(z+jz_0)}\right), \quad \rho \ll z, \quad z > 0 \tag{23}$$

Inserting Eq. (23) into the Green's function we get:

$$\begin{aligned} G(\boldsymbol{r},\boldsymbol{r}') &= \frac{1}{4\pi}\frac{e^{-j\kappa|\boldsymbol{r}-\boldsymbol{r}'|}}{|\boldsymbol{r}-\boldsymbol{r}'|} \\ &\approx \frac{1}{4\pi(z+jz_0)}\exp\left(-j\kappa(z+jz_0) - \frac{j\kappa\rho^2}{2(z+jz_0)}\right) \\ &= \frac{\exp(\kappa z_0)}{4\pi(z+jz_0)}\exp\left(-j\kappa z - \frac{j\kappa\rho^2}{2\left(z+\frac{z_0^2}{z}\right)} - \frac{\rho^2}{\frac{2z_0}{\kappa}\left(1+\frac{z^2}{z_0^2}\right)}\right). \end{aligned} \tag{24}$$

Ignoring the factor $\exp(\kappa z_0)/4\pi(z+jz_0)$, the result of Eq. (24) directly correspond to the form of the Gaussian beam:

$$\exp(-j\kappa z)\exp\left(-\frac{j\kappa\rho^2}{2R(z)}\right)\exp\left(-\frac{\rho^2}{w^2(z)}\right). \tag{25}$$

By comparing Eqs. (24) and (25) we can identify following beam parameters: The radius of curvature R is

$$R(z) = z + \frac{z_0^2}{z}\,. \tag{26}$$

The radius w of the beam is

$$w(z) = w_0\sqrt{1+\left(\frac{z}{z_0}\right)^2}\,, \tag{27}$$

where w_0 is the radius at the waist:

$$w_0 = \sqrt{\frac{2z_0}{\kappa}}\,. \tag{28}$$

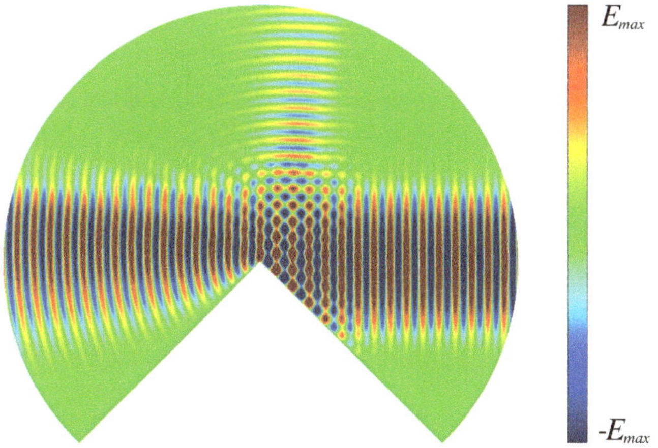

Fig. 3. E_y in the x, z-plane.

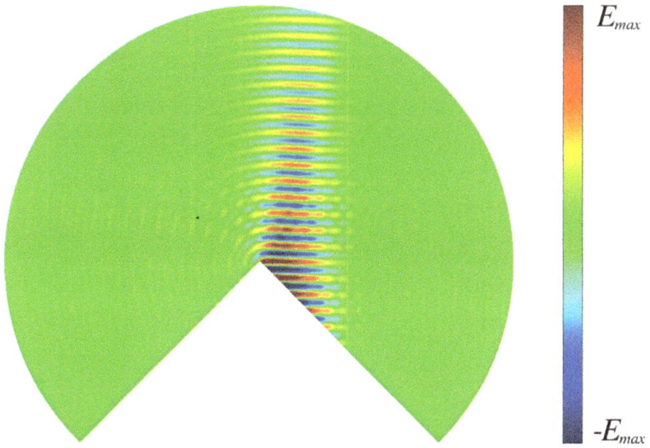

Fig. 4. H_x in the x, z-plane within a range of $r \leq 20$m.

As we can see in Eq. (26), the radius of curvature gets large for $z \ll z_0$. That means that there are nearly plane wave fronts close to the beam's waist which are – due to the transverse Gaussian profile – also localized.

4 Treatment of a complex-valued coordinate in the spherical-multipole expansion

The following equations are special cases of the Gegenbauer addition theorem applied to spherical Bessel functions of zeroth order

$$\frac{\sin(\kappa R)}{\kappa R} = \sum_{n=0}^{\infty}(2n+1)j_n(\kappa r)j_n(\kappa r')P_n(\cos\gamma), \tag{29}$$

$$-\frac{\cos(\kappa R)}{\kappa R} = \sum_{n=0}^{\infty}(2n+1)j_n(\kappa r)y_n(\kappa r')P_n(\cos\gamma). \tag{30}$$

They are valid for arbitrary complex r, r', γ, κ and provided that

$$|re^{\pm j\gamma}| < |r'| \tag{31}$$

and

$$R = \sqrt{r^2 + r'^2 - 2rr'\cos\gamma}. \tag{32}$$

Having two position vectors $\boldsymbol{r}(r, \vartheta, \varphi)$ and $\boldsymbol{r}'(r'\vartheta', \varphi')$, the distance $|\boldsymbol{r} - \boldsymbol{r}'|$ between this positions can be expressed by R if we choose

$$\cos\gamma = \cos\vartheta\cos\vartheta' + \sin\vartheta\sin\vartheta'\cos(\varphi - \varphi'). \tag{33}$$

By combining Eqs. (29), (30) and the relation

$$P_n(\cos\gamma) = \frac{4\pi}{2n+1}\sum_{m=-n}^{+n} Y_n^m(\vartheta, \varphi)Y_n^{m*}(\vartheta', \varphi') \tag{34}$$

we can construct a spherical multipole expansion of the scalar free space Green's function:

$$g(\boldsymbol{r}, \boldsymbol{r}') = \frac{1}{4\pi}\frac{e^{-j\kappa|\boldsymbol{r}-\boldsymbol{r}'|}}{|\boldsymbol{r}-\boldsymbol{r}'|} \tag{35}$$
$$= \sum_{n=0}^{\infty}\sum_{m=-n}^{+n} j_n(\kappa r)h_n^{(2)}(\kappa r')Y_n^m(\vartheta, \varphi)Y_n^{m*}(\vartheta', \varphi').$$

This expansion is valid for any complex location $\boldsymbol{r}'$, if the condition Eq. (31) is fulfilled. If it is not, $\boldsymbol{r}$ and $\boldsymbol{r}'$ can be permuted due to the Green's function symmetry. This paper only deals with CSBs that point directly to the tip of the cone, so there is only a complex r'-component (Katsav et al., 2012). In this case γ is real and the condition that has to be fulfilled is

$$|r| < |r'|. \tag{36}$$

This must be considered for the multipole expansions in section 2.3, particulary for Eq. (17), if a complex-source beam is used.

5 Numerical results

The following results are for a circular cone with a half opening angle of 45° ($\vartheta_0 = 135°$) and are calculated with multipole expansions according to the preceding sections using a maximum order $\nu_{max} = 40$. The coordinates of the CSB are $r' = (12 + j48)$ m, $\vartheta' = 90°$, $\varphi' = 0°$; this means that the beam has a rayleigh length of 48 m and propagates from x-direction to the tip of the cone. The electric field of the beam is polarized in y-direction. The wave number is $\kappa = 5$/m. Figure 3 shows the y-component of the electric field in a x, z-crosscut through the cone. The beam is partly reflected and and there is interference between the incident and the reflected field. Above the beam's part that has passed the tip we can observe less strong interference with scattered parts

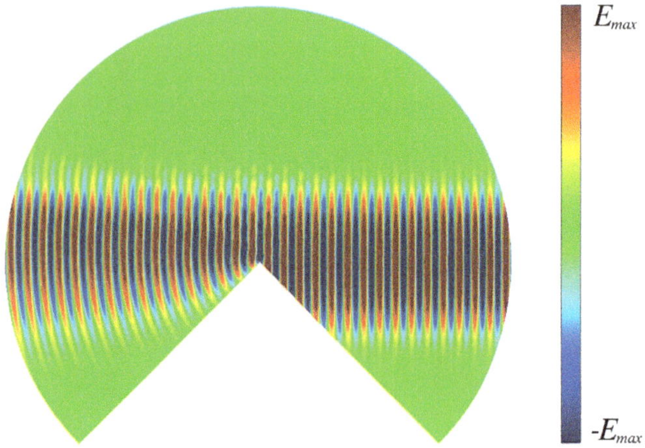

Fig. 5. H_z in the x, z-plane within $r \leq 20$m.

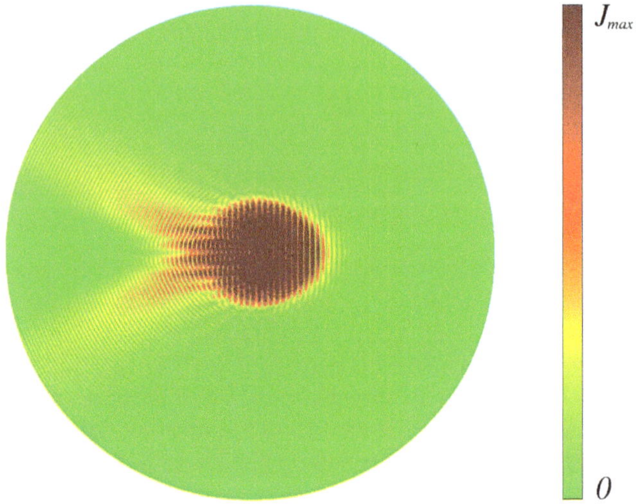

Fig. 6. Current density distribution on the cone's surface in the region $r \leq 40$m. View from z-direction. The beam incides from right-hand side.

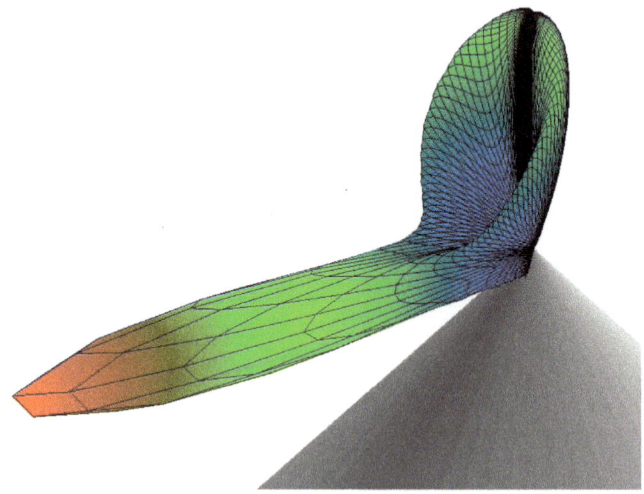

Fig. 7. Far-field pattern with the magnitude of the electric field.

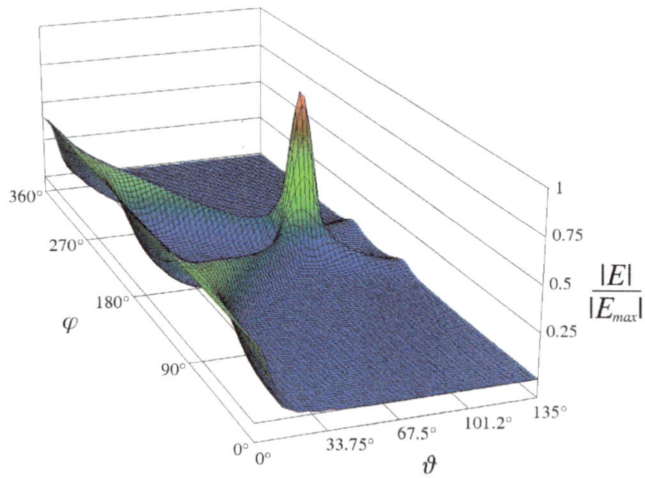

Fig. 8. Magnitude of the electric far-field plotted against ϑ, φ.

of the beam. In the shadow region of the cone we can clearly see diffracted spherically shaped wavefronts. Figure 4 and 5 show the x- and the z-component of the magnetic field. While the incident part of the magnetic field is polarized in z-direction, the reflected part is polarized in x. In both figures diffracted parts can be found. Figure 6 shows the current density distribution on the cone's surface, which can be calculated with

$$\boldsymbol{J}_s(\boldsymbol{r}) = -\boldsymbol{e}_\vartheta \times \boldsymbol{H}(\boldsymbol{r})|_{\vartheta=\vartheta_0}. \tag{37}$$

There are two notable creeping waves in the shadow region of the cone. Finally, the magnitude of the scattered far-field is examined; in Fig. 7 it is shown in the form of a far-field pattern, in Fig. 8 it is plotted against ϑ, φ. The beam is reflected in every direction of φ; the highest values are observed in the direction of propagation. For $90° < \vartheta < 135°$ we observe two relatively small maxima, particulary in Fig. 8. This might be a consequence of the creeping waves, which can be observed in Fig. 6 of the current density.

6 Conclusions

The scattering of an electromagnetic CSB by a semi-infinite circular cone has been analysed using spherical multipole expansions. Convergent results of the total or scattered field can be obtained for both near- and far-field. So far, only beams pointing directly to the tip of the cone have been considered. Future work will expand this approach to other geometries, e.g. elliptic cones. Furthermore the requirements for CSBs pointing in arbitrary directions will be investigated, so that other parts of the cone can be illuminated.

Acknowledgements. This work was supported by the Deutsche Forschungsgemeinschaft (KL815/10-1&2).

References

Bowman, J. J., Senior, T. B. A., and Uslenghi, P. L. E., Electromagnetic and acoustic scattering by simple shapes (Revised Printing), Hemisphere Pub. Corp., New York, 1987.

Blume, S. and Krebs, V.: Der Rückstreuquerschnitt des semiinfiniten Kreiskegels, Arch. Elektrotech., 77, 239–244, 1994

Katsav, M., Heyman, E., and Klinkenbusch, L.: Complex-source beam diffraction by an acoustically soft or hard circular cone, Electromagnetics in Advanced Applications (ICEAA), 2012 International Conference on, proceedings, Cape Town, South Africa, 2–7 September 2012, 135–138, 2012

Kijowski, M. and Klinkenbusch, L.: Eigenmode analysis of the electromagnetic field scattered by an elliptic cone, Adv. Radio Sci., 9, 31–71, 2011

Klinkenbusch, L.: Electromagnetic scattering by semi-infinite circular and elliptic cones, Radio Sci., 42, RS6S10, doi:10.1029/2007RS003649, 2007

Orlov, L. and Peschel, U.: Complex source beam: A tool to describe highly focused vector beams analytically, Phys. Rev. A, 82, 063820, doi:10.1103/PhysRevA.82.063820, 2010

15

A fully probe corrected near-field far-field transformation technique employing plane-wave synthesis

R. A. M. Mauermayer and T. F. Eibert

Lehrstuhl für Hochfrequenztechnik, Technische Universität München, 80290 Munich, Germany

Correspondence to: R. A. M. Mauermayer (raimund.mauermayer@tum.de)

Abstract. The far-field behavior of an antenna under test (AUT) can be obtained by exciting the AUT with a plane wave. In a measurement, it is sufficient if the plane wave is artificially generated in the vicinity of the AUT. This can be achieved by using a virtual antenna array formed by a probe antenna which is sequentially sampling the radiating near-field of the AUT at different positions. For this purpose, an optimal filter for the virtual antenna array is computed in a preprocessing step. Applying this filter to the near-field measurements, the far-field of the AUT is obtained according to the propagation direction and polarization of the synthesized plane wave. This means that the near-field far-field transformation (NFFFT) is achieved simply by filtering the near-field measurement data. Taking the radiation characteristic of the probe antenna into account during the synthesis process, its influence on the NFFFT is compensated.

The principle of the plane-wave synthesis and its application to the NFFFT is presented in detail in this paper. Furthermore, the method is verified by performing transformations of simulated near-field measurement data and of near-field data measured in an anechoic chamber.

1 Introduction

Today's wireless communication, radar or direction finding systems make use of electrically large antennas like parabolic reflectors or antenna arrays to generate far-field radiation characteristics suitable for the particular application. After fabrication the far-field of the antenna under test has to be examined by measurements to verify that it meets the requirements on phase and magnitude. Due to the large size of the antenna relative to wavelength, an accurate measurement under far-field conditions inside an anechoic chamber can hardly be accomplished, simply because of the enormous measurement distance required. Thus, performing measurements in a reduced distance in the radiating near-field region of the AUT and subsequently applying a near-field far-field (NFFF) transformation is an attractive alternative.

The radiated fields of the AUT and the probe are usually represented by a truncated series of orthogonal spherical, cylindrical or planar field modes to formulate a transmission equation describing the coupling between both antennas (Hansen, 1988). Solving this equation for the modal coefficients the far-field can be computed by evaluating the field modes for the radial distance going to infinity.

For the plane-wave synthesis approach (Hansen, 1988; Yamaguchi et al., 2009; Bennett and Schoessow, 1978), there is no explicit transmission equation needed. The probe which is sampling the near-field successively at different locations is considered to form a virtual array of probe antennas on a measurement surface. This array is then used to synthesize a plane wave in the vicinity of the AUT through a weighted superposition of the fields radiated by the elements of the virtual probe antenna array. The appropriate weights form a filter which is gained from the solution of an inverse problem.

The proposed method allows shifting most of the computational expense of the NFFF transformation and probe correction to a preprocessing step whose result is a set of filters containing the weighting factors for the probe signals. These filters are used to compute the far-field directly from the acquired measurement data in a near-field far-field transformation step.

The paper is structured as follows: in the following part the principle and the results of the plane-wave synthesis is shown. How the synthesis is employed for near-field far-field transformation is given in the third part. In the forth part, the method is applied to simulated and real near-field measurement data of a Yagi-Uda and a patch antenna. The results of the transformation together with some accuracy considerations are presented.

2 The probe corrected plane-wave synthesis

2.1 Principle

The principle of the synthesis of plane waves by a virtual array of probes which are positioned at different locations in the near-field of the AUT is visualized in Fig. 1. The probes are excited according to the filter vector $\boldsymbol{w}\left(\hat{k}, \boldsymbol{E}_0\right)$, which consists of N complex coefficients, one for each probe position and polarization orientation. The objective is to generate the field of the plane wave

$$\boldsymbol{E}_{\mathrm{pl}}(\boldsymbol{r}) = \boldsymbol{E}_0 e^{-jk\hat{k}\cdot\boldsymbol{r}} \tag{1}$$

$$\boldsymbol{H}_{\mathrm{pl}}(\boldsymbol{r}) = \frac{1}{Z_0}\hat{k} \times \boldsymbol{E}_{\mathrm{pl}}(\boldsymbol{r}) \tag{2}$$

in the vicinity of the AUT by superimposing the probe fields. The propagation direction of the plane wave is described by the unit vector $\hat{k}$ and its polarization is given by the electric field vector $\boldsymbol{E}_0$. Z_0 describes the free space wave impedance.

The virtual array is equivalent to a probe antenna with ideal polarization purity positioned in the far-field of the AUT in direction $-\hat{k}$ and exciting the AUT with a quasi plane-wave field with the same polarization as the synthesized plane wave. The power wave $b_1^{\mathrm{FF}}\left(\hat{k}, \boldsymbol{E}_0\right)$ at the feed point of the AUT in Fig. 1 is proportional to this far-field measurement.

For the wave field synthesis process inside a volume, the AUT is enclosed by a Huygens' surface. A filter must be found that enables the synthesis of the tangential field components of the desired plane wave on the surface. The uniqueness theorem (Harrington, 2001) guarantees that if the superimposed tangential field components of the virtual probe array correspond to those of the plane wave, the plane-wave field is also present inside the source free test volume.

The probe antenna is modeled by its equivalent electric $\boldsymbol{J}_n\left(\boldsymbol{r}'\right)$ and magnetic $\boldsymbol{M}_n\left(\boldsymbol{r}'\right)$ surface currents located at the n-th probe sampling position. Its characteristic electromagnetic field is computed from the surface integrals

$$\boldsymbol{E}_n(\boldsymbol{r}) = \iint\limits_A \left[\bar{\mathbf{G}}_J^E\left(\boldsymbol{r},\boldsymbol{r}'\right)\cdot\boldsymbol{J}_n\left(\boldsymbol{r}'\right) + \bar{\mathbf{G}}_M^E\left(\boldsymbol{r},\boldsymbol{r}'\right)\cdot\boldsymbol{M}_n\left(\boldsymbol{r}'\right)\right] dA' \tag{3}$$

$$\boldsymbol{H}_n(\boldsymbol{r}) = \iint\limits_A \left[\bar{\mathbf{G}}_J^H\left(\boldsymbol{r},\boldsymbol{r}'\right)\cdot\boldsymbol{J}_n\left(\boldsymbol{r}'\right) + \bar{\mathbf{G}}_M^H\left(\boldsymbol{r},\boldsymbol{r}'\right)\cdot\boldsymbol{M}_n\left(\boldsymbol{r}'\right)\right] dA', \tag{4}$$

where $\bar{\mathbf{G}}_{J/M}^{E/H}\left(\boldsymbol{r},\boldsymbol{r}'\right)$ are the dyadic Greens' functions for free space. The tangential components of the probe field are sampled at discrete locations $\boldsymbol{r}_i$ $(i = 1, \ldots, I)$ on the Huygens' surface. In order to avoid the excitation of resonance modes inside the test volume, the combined field vector (Hansen, 1988)

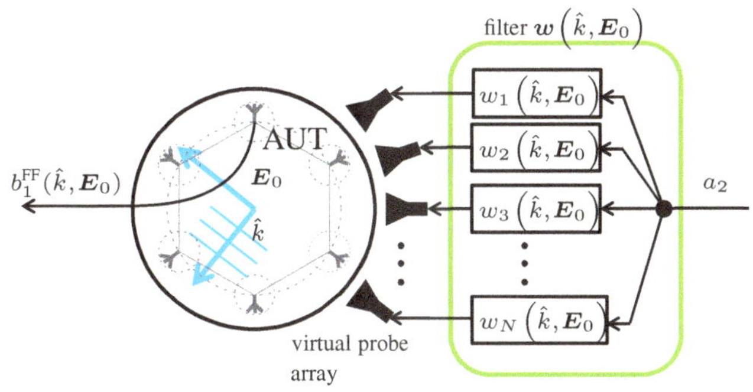

Fig. 1. The principle of plane-wave synthesis by a virtual array of probes positioned in the near-field region of the AUT. The AUT consists of six antenna elements. The power wave $b_1^{\mathrm{FF}}\left(\hat{k}, \boldsymbol{E}_0\right)$ is measured at the feed point of one antenna element.

$$\boldsymbol{F}_n(\boldsymbol{r}_i) = \boldsymbol{E}_n(\boldsymbol{r}_i)|_{\tan} - Z_0\boldsymbol{n}\times\boldsymbol{H}_n(\boldsymbol{r}_i) \tag{5}$$

is used, with $\boldsymbol{n}$ being the normal vector of the Huygens' surface. The constraint that the tangential components of the superimposed probe fields have to be equal to that of the plane wave leads to the equation system

$$\underbrace{\begin{pmatrix} \boldsymbol{F}_1(\boldsymbol{r}_1) & \ldots & \boldsymbol{F}_N(\boldsymbol{r}_1) \\ \boldsymbol{F}_1(\boldsymbol{r}_2) & \ldots & \boldsymbol{F}_N(\boldsymbol{r}_2) \\ \vdots & \ddots & \vdots \\ \boldsymbol{F}_1(\boldsymbol{r}_I) & \ldots & \boldsymbol{F}_N(\boldsymbol{r}_I) \end{pmatrix}}_{\mathbf{F}} \underbrace{\begin{pmatrix} w_1\left(\hat{k}, \boldsymbol{E}_0\right) \\ \vdots \\ w_N\left(\hat{k}, \boldsymbol{E}_0\right) \end{pmatrix}}_{\boldsymbol{w}\left(\hat{k}, \boldsymbol{E}_0\right)} = \underbrace{\begin{pmatrix} \boldsymbol{F}_{\mathrm{pl}}(\boldsymbol{r}_1) \\ \vdots \\ \boldsymbol{F}_{\mathrm{pl}}(\boldsymbol{r}_I) \end{pmatrix}}_{\boldsymbol{F}_{\mathrm{pl}}} \tag{6}$$

of the inverse problem with $\boldsymbol{F}_{\mathrm{pl}}(\boldsymbol{r}_i)$ being the combined field vector evaluated for the plane wave. As there are usually more sampling points I on the Huygens' surface than probe positions N, the equation system is over-determined. Therefore, it is solved for the filter coefficients in a least mean square error sense in form of

$$\boldsymbol{w}\left(\hat{k}, \boldsymbol{E}_0\right) = \mathbf{F}^{+}\boldsymbol{F}_{\mathrm{pl}} \tag{7}$$

using the pseudoinverse $\mathbf{F}^{+}$ of matrix $\mathbf{F}$ in Eq. (6). This solution also minimizes the norm of the filter vector and the power radiated by the virtual probe antenna array, respectively.

In order to avoid 3-D measurements, it is also investigated whether it is possible to synthesize a plane-wave field in a flat region around the AUT using a virtual array of probe antennas positioned on a closed curve around the test region. The motivation for this idea is to improve the accuracy of far-field measurement results for an antenna measurement facility that only allows to record near-field data for some great circles but not on a complete measurement surface.

For this synthesis problem, the field components of the probe are directly sampled in the flat area instead of sampling the tangential field components on a Huygens' surface. However the kind of inverse problem remains the same and it is solved in the same way as described above.

2.2 Measurement setups and sampling conditions

The measurement setup has to be defined for the synthesis process. There are two setups for the investigation of the plane-wave synthesis inside a volume and on a planar area.

For the synthesis in a volume the spherical measurement setup depicted in Fig. 2a is considered. The probe is positioned on a spherical measurement surface with radius 1 m. At each position of the regular sampling grid in ϑ and φ direction the probe is oriented with vertical and horizontal polarization. The probe sampling rate, which depends on wavelength λ and the radius ρ of the minimum sphere, is derived from the sampling condition (Yaghjian, 1986)

$$\Delta\vartheta = \Delta\varphi \leq \frac{\lambda}{2(\rho+\lambda)} \tag{8}$$

for conventional NFFFT using spherical multipoles. The Huygens' surface is formed by the minimum sphere which encloses the AUT and has a radius of $\rho = 0.5\,\mathrm{m}$.

For the synthesis of a plane wave on a planar area the setup shown in Fig. 2b is used. The intention here is to generate a field distribution on the disc with radius 0.5 m that is more similar to a plane wave than the field of the probe antenna itself. The probe is positioned on a measurement circle in the xy plane with radius 1.98 m. Again the probe sampling interval in φ direction is chosen following Eq. (8).

The sampling point distance of the probe fields on the Huygens' surface and on the disc is chosen to be smaller than $\lambda/2$ according to the Nyquist-Shannon sampling theorem.

2.3 Filter coefficients

For the spherical measurement, an open ended rectangular waveguide probe operating at 2 GHz is used. It is positioned on the measurement surface in 5° steps in ϑ and φ direction. The tangential field components on the Huygens' surface are also sampled on a regular grid in ϑ and φ direction in 6° intervals. For all probe positions the probe fields are computed on the Huygens' surface by Eq. (3) and Eq. (4). The sampled tangential components result in the columns of matrix $\mathbf{F}$ in Eq. (6). The plane wave that is to be synthesized is propagating in positive x direction and is vertically polarized as it is given in Fig. 2a. The necessary filter coefficients for the optimal synthesis of the plane wave are gained from the solution of the equation system according to Eq. (7). In Fig. 3 the resulting normalized magnitude of the filter coefficients is shown. As the probe is vertically and horizontally oriented there is one filter coefficient for each of the probe polarization states. It can be clearly seen that most of the power is radiated by the vertically polarized probes which are radiating in propagation direction of the plane wave whereas the horizontally oriented probes are excited with less power because their radiated fields have a dominant horizontal electric field component and fit less to the vertically polarized plane wave.

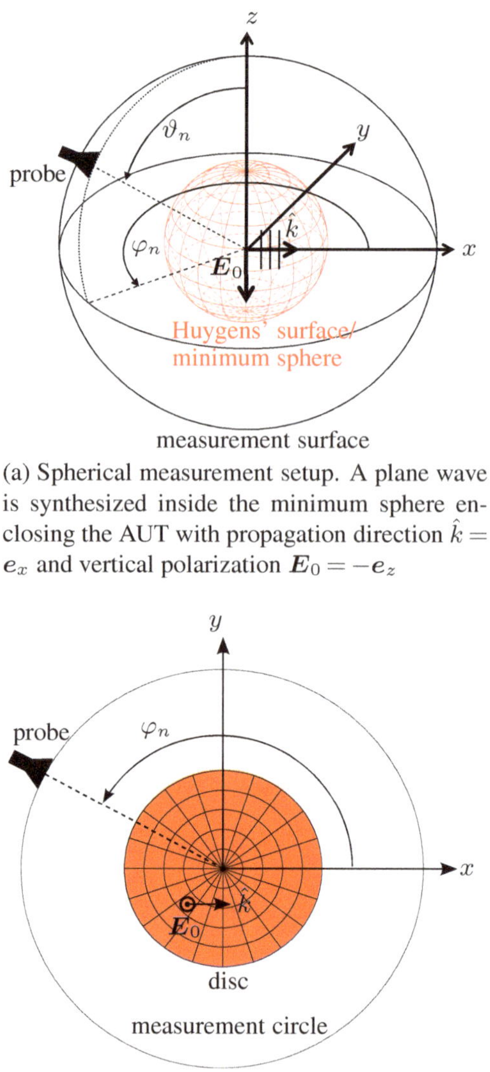

(a) Spherical measurement setup. A plane wave is synthesized inside the minimum sphere enclosing the AUT with propagation direction $\hat{k} = \boldsymbol{e}_x$ and vertical polarization $\boldsymbol{E}_0 = -\boldsymbol{e}_z$

(b) Measurement setup for plane-wave synthesis on a disc. The probe is rotated on the measurement circle.

Fig. 2. Measurement setups for plane-wave synthesis inside a volume **(a)** and on a disc **(b)**.

A horn antenna probe, also operating at 2 GHz, is used for the synthesis on a disc. It is rotated on the measurement circle in 5° steps. Due to the high polarization purity of the probe field on the disc, it is only oriented in vertical polarization to synthesize a vertically polarized plane wave propagating in positive x direction. Figure 4 shows the normalized magnitude of the resulting filter coefficients. As for the spherical

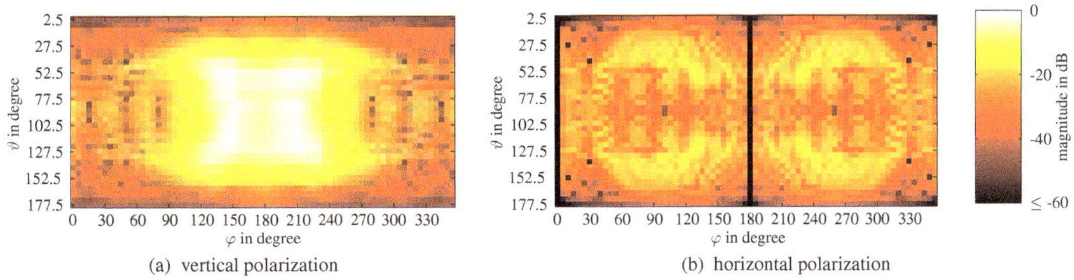

Fig. 3. Normalized magnitude of the filter coefficients for a vertically **(a)** and a horizontally **(b)** polarized open ended waveguide probe synthesizing a vertically polarized plane wave propagating in positive x direction.

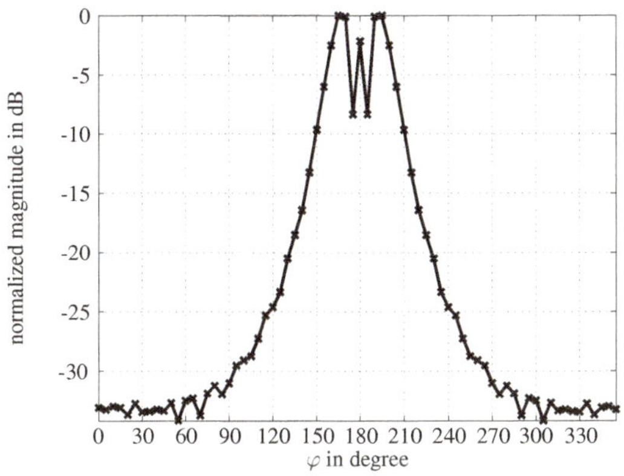

Fig. 4. Normalized magnitude of the filter coefficients for the synthesis on a disc.

setup, most of the power is radiated by the elements of the virtual probe antenna array which are already pointing in the propagation direction of the plane wave, i.e. probes on the measurement circle with φ_n around 180°.

2.4 Wave field synthesis results

The elements of the virtual probe array are excited according to the computed filter coefficients. The most important accuracy measure of the synthesis method is the quality of the wave field which is generated. Since the quality of the plane wave determines the accuracy of the following NFFF transformation process. Therefore, the deviation of the magnitude and phase of the synthesized wave field from the ideal plane wave is regarded.

In the case of the spherical setup, the errors are computed on lines which are crossing the minimum sphere in different directions to get an impression of the electrical field distribution inside the volume. The deviations are shown in Fig. 5. The magnitude error of the generated wave field is lower than -80 dB. The phase error is smaller than 0.006°. As only the z component of the synthesized electric field is considered, the polarization purity was checked to be above 80 dB inside the minimum sphere. It is possible to further increase the quality of the synthesized plane wave by increasing the sampling rate of the probe on the measurement surface.

The errors of the plane wave synthesized on the disc are shown in Fig. 6. The magnitude shows a maximum deviation of about -17 dB and the phase of about 2°. It should be mentioned that the errors increase below and above the considered planar region in z direction. Nevertheless, as it was intended, the superimposed field on the disc of the virtual array is more similar to a plane wave than the field of only one probe on the measurement circle, which would result in a maximum deviation of -11 dB in magnitude and 140° in phase.

In general, a closed boundary surface with the appropriate impressed equivalent surface currents is mandatory for the accurate generation of a desired wave field. For the synthesis of the plane wave on a disc, this assumption is violated since the virtual probe array only forms kind of a line current around the disc. Therefore an accurate synthesis of the desired wave field cannot be achieved.

3 NFFFT based on plane-wave synthesis

To find a relation between the near-field measurements and the AUT far-field, the signal flow in Fig. 1 is inverted so that the transposed system shown in Fig. 7 is obtained. This can be done due to the assumption of the reciprocity property of the whole system. The far-field of the AUT in direction $-\hat{k}$, which is now proportional to the power wave signal $b_2^{FF}\left(\hat{k}, \boldsymbol{E}_0\right)$, is computed by

$$b_2^{FF}\left(\hat{k}, \boldsymbol{E}_0\right) = \sum_{n=1}^{N} w_n\left(\hat{k}, \boldsymbol{E}_0\right) b_{2,n}^{NF} \tag{9}$$

applying the filter coefficients $w_n\left(\hat{k}, \boldsymbol{E}_0\right)$ to the power wave signals $b_{2,n}^{\mathrm{NF}}$ received by the probe at position n in the AUT near-field. From Eq. (9) it is obvious that the NFFF transformation employing plane-wave synthesis is just a filtering of the near-field data.

Although the filter is computed for only one specific direction $\hat{k}$, the propagation direction of the plane wave can be

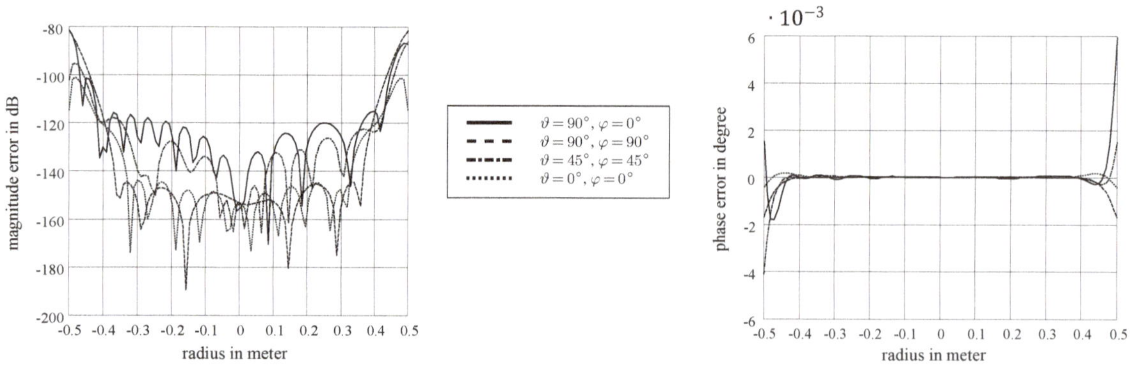

(a) magnitude error (b) phase error

Fig. 5. Magnitude and phase error of the synthesized plane wave evaluated on several lines crossing the minimum sphere in different directions.

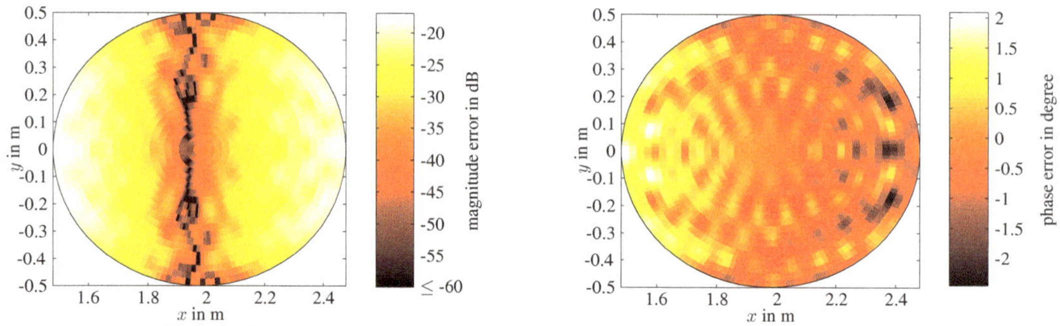

(a) magnitude error (b) phase error

Fig. 6. Magnitude and phase error of the synthesized plane wave on a disc.

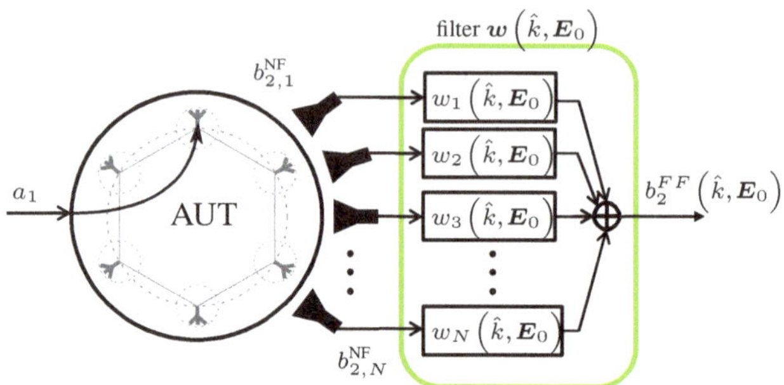

Fig. 7. Considered system for the NFFFT employing plane-wave synthesis.

rotated in φ direction by a cyclic rotation of the filter coefficients taking advantage of the rotational symmetry of the probe sampling points on the measurement and of the Huygens' surface. This allows to determine a horizontal cut of the AUT far-field pattern (E-plane) without having to solve the inverse problem of Eq. (6) for each φ angle of plane wave incidence again.

4 Application of the NFFFT based on plane-wave synthesis

4.1 Simulated and measured near-field data

A spherical near-field measurement is simulated with FEKO (EM Software and Systems, 2011) for the verification of the proposed method of synthesizing a plane wave inside a volume. A three element Yagi-Uda antenna composed of wire elements is used as AUT. It is vertically polarized and its main beam direction is pointing in positive x direction. The simple structure of the antenna makes a low computation time possible needed for the simulation of about 5000 near-field measurement points. Figure 8 shows the normalized magnitude of the received probe signal for copolar orientation of the waveguide probe.

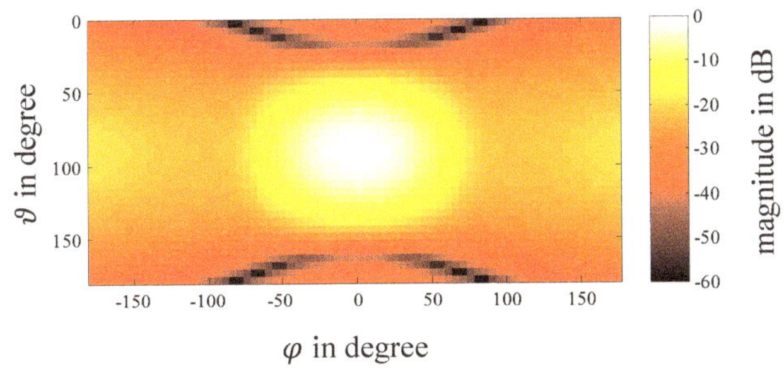

Fig. 8. Normalized magnitude of the probe signal for copolar orientation of the waveguide probe.

The NFFFT based on the wave field synthesis on a disc is verified applying it to simulated near-field data acquired by the horn antenna probe in the near-field of a patch antenna. This simulation scenario is also built up in an anechoic chamber, shown in Fig. 10, and real measurements are recorded. Since the AUT represents one element of a circular antenna array it is mounted to a construction made from Rohacell so that the patch antenna pattern is not disturbed by the mounting. The normalized, simulated and measured probe signals are plotted in Fig. 9.

Both AUT antennas are positioned in 0.3 m distance to the rotational axis as they represent the antenna element of a circular antenna array which is sketched in Fig. 1.

The simulated and real measurement setups, including the probe and sampling rates, are the same as for the synthesis processes, so the computed filters can be directly applied to the simulated and measured near-field data for NFFF transformation.

4.2 NFFF transformation results

For the spherical NFFF transformation the filter of Fig. 3 is directly applied to the acquired near-field data of Fig. 8. Evaluating Eq. (9) for the cyclically rotated versions of the filter coefficients leads to the horizontal cut (H-plane) of the far-field pattern. The transformation result compared to the reference solution computed by FEKO is given in Fig. 11. As it is to be expected from the high quality of the synthesized plane wave in the vicinity of the AUT there is no difference recognizable between reference and NFFFT results. Regarding the errors of the transformation in Fig. 12, it can be seen that the error of the normalized far-field magnitude is below −80 dB and the phase error is in the range of some thousandths of one degree. So the transformation errors are in the range of the magnitude and phase errors of the synthesized plane wave.

Already from the quality of the synthesized plane-wave field on the disc it can be derived that the transformation results of the near-field data acquired on the measurement circle cannot reach the accuracy as the one of the complete spherical measurement. Figure 13 shows the results of the

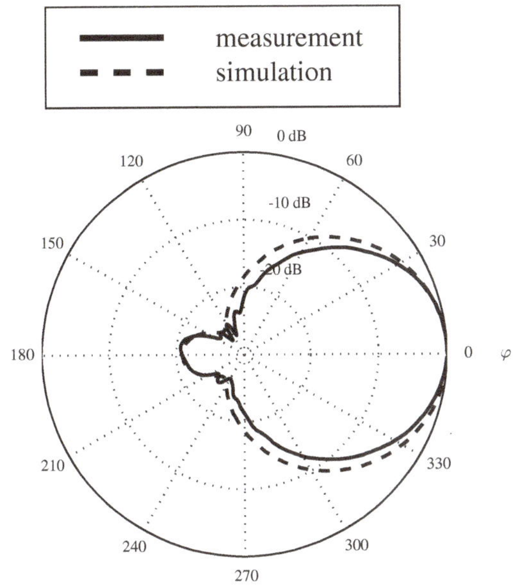

Fig. 9. Simulated and measured near-field data acquired on the measurement circle.

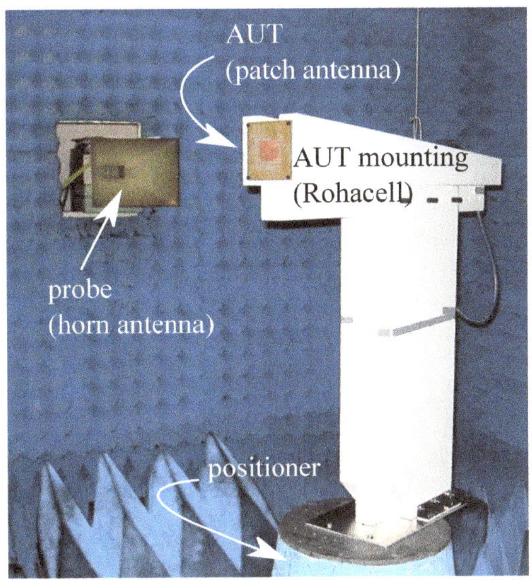

Fig. 10. Measurement setup in an anechoic chamber for the NFFFT based on the plane-wave synthesis on a disc.

NFFF transformation for the simulated and measured near-field data. For the simulated data, the transformation result matches very well to the reference of the patch antenna far-field pattern.

The positive effect of the transformation is also visible for the measured near-field data. However, there is still a major deviation from the reference compared to the transformation results of the simulated near-field data. The main reason for

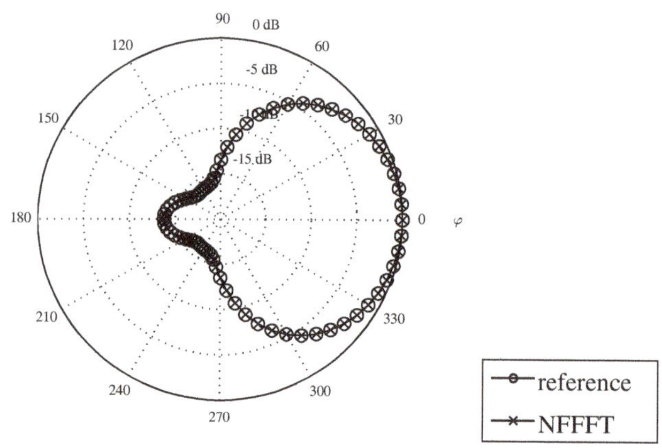

Fig. 11. Magnitude of reference and transformed far-field.

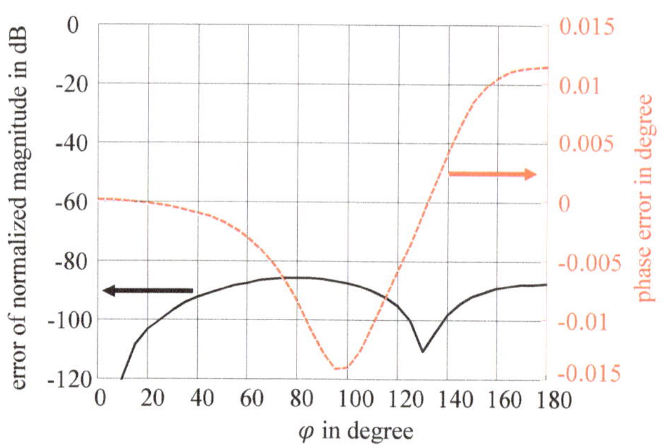

Fig. 12. Errors of normalized magnitude and phase of the transformation result for the spherical near-field measurement.

this seems to be that the measurement was done in a facility that is not appropriate for near-field measurements, so that the exact positioning and orientation of AUT and probe antennas could not be guaranteed. This means that the probe field on the disc which was assumed for the synthesis might have been different from the probe field during the measurement in the anechoic chamber. This also explains the difference between simulated and measured near-field data in Fig. 9.

5 Conclusions

The presented near-field far-field transformation technique based on plane-wave synthesis allows to split the transformation process into two steps. In the first step, the plane wave is synthesized by solving an inverse problem for the filter vector for the virtual probe array, which might be time consuming. In the second step, the transformation of the near-field data can be performed by a faster filtering procedure.

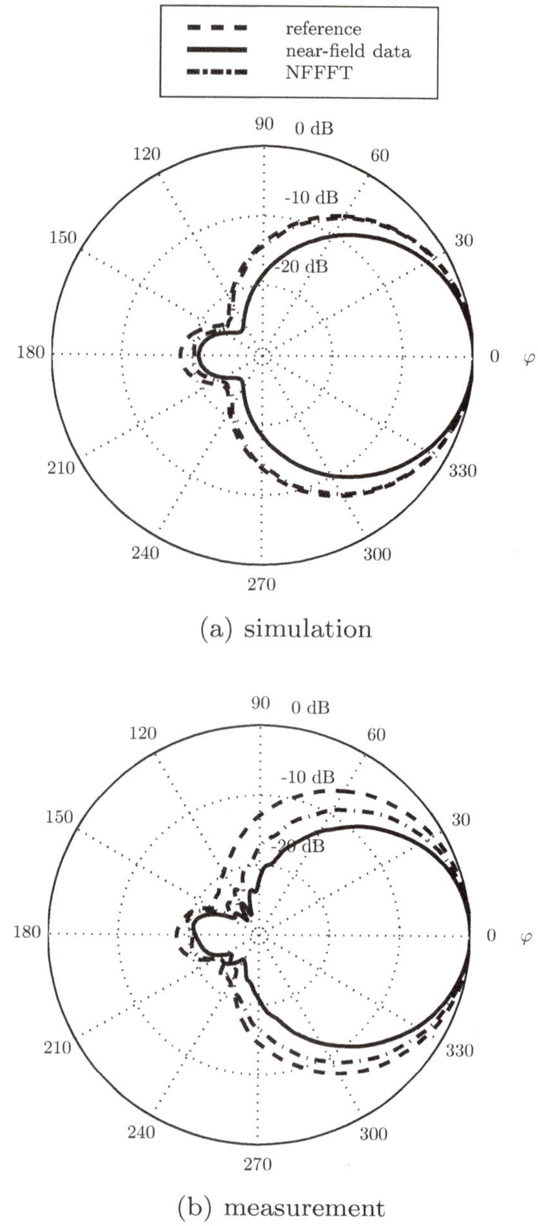

Fig. 13. Results of the NFFF transformation of the simulated and measured near-field data acquired on the measurement circle.

The method was verified by applying it to simulated and real near-field measurement data.

If the computed filters are stored they can be reused for the transformation of near-field measurement data of different AUTs as long as the measurement setup including probe, probe sampling rate, frequency, measurement surface and the shape of the Huygens' surface do not change. The synthesis of arbitrary field distributions inside arbitrarily shaped volumes is an issue of further investigation.

References

Bennett, J. and Schoessow, E.: Antenna near-field/far-field transformation using a plane-wave-synthesis technique, Proceedings of the Institution of Electrical Engineers, 125, 179–184, doi:10.1049/piee.1978.0048, 1978.

EM Software and Systems: FEKO Suite 6.1, http://www.feko.info, 2011.

Hansen, J. E.: Spherical near-field antenna measurements, IEE Electromagnetic Waves Series 26, Peter Peregrinus Ltd., London, UK, 1988.

Harrington, R. F.: Time-Harmonic Electromagnetic Fields, Wiley, J., Weinheim, 2001.

Yaghjian, A.: An overview of near-field antenna measurements, IEEE Transactions on Antennas and Propagation, 34, 30–45, doi:10.1109/TAP.1986.1143727, 1986.

Yamaguchi, R., Kimura, Y., Komiya, K., and Cho, K.: A far-field measurement method for large size antenna by using synthetic aperture antenna, in: 3rd European Conference on Antennas and Propagation, 1730–1733, 2009.

16

Using quasi-guided modes for modeling the transfer behavior of bent dielectric slab waveguides

M. Stallein, B. Bandlow, and R. Schuhmann

Universität Paderborn, Theoretische Elektrotechnik, Warburger Strasse 100, 33098 Paderborn, Germany

Abstract. The connection of two straight dielectric multimode slab waveguides by a circular bent waveguide is analyzed by means of quasi-guided modes. These modes correspond to the well known leaky modes, but own real eigenvalues, thus the mathematical description is simpler. Furthermore they are derived as approximate solutions of the exact theory. This work will first give a brief introduction to the basic theory, followed by a discussion of the properties of quasi-guided modes. After a validation by comparison with a numerical simulation using the Finite Integration Technique, results for the bending loss of multimode waveguides are presented.

1 Introduction

The modeling of bending loss of optical waveguides has been of special interest since the demonstration of the first functional laser in 1960. Many classical publications exist, e.g. Marcatili (1969); Lewin et al. (1977); Marcuse (1982); Snyder and Love (2000), but the presented approaches are mostly approximations for small radiation loss. Often the leaky mode ansatz is presented as exact solution of Maxwell's equations for the cylindrical slab waveguide. This ansatz uses Hankel functions in the outer region in order to satisfy the radiation condition. But it is not the single mode that has to satisfy this condition, it is the sum of all modes. In addition the discrete and limited set of leaky modes can not build a complete set of solutions. In 1990 Morita and Yamata presented a new, initially nonrestrictive theory, but also applied the radiation condition to the individual mode (Morita and Yamada, 1990). Hence it was Kerndlmaier in 1992 who first presented the theory, that is not limited to any special case. Thus he proposed it to be the exact theory (Kerndlmaier, 1992). To the best of the authors knowledge Kerndlmaier did not publish his results apart from his PhD-thesis, thus Sect. 2 deals with the basic ideas of his theory.

The mode spectrum of the bent waveguide is continuous, like the set of radiation modes of the straight slab waveguide. As the primal purpose of our approach is to estimate the radiation loss of bent waveguides connected with straight waveguides, that only carry guided modes at incoming side, it would be advantageous if equivalent modes exist for the bent waveguide. Actually it is possible to isolate a set of quasi guided modes that correspond to the set of leaky modes, but own real eigenvalues. Section 3 characterizes the properties of the quasi-guided modes and describes the connection to straight waveguide elements. For the purpose of validation, results of a numerical simulation using the Finite Integration Technique are presented in Sect. 4. Finally Sect. 5 presents results for the bending loss of multimode slab waveguides involving a 90° bend.

Throughout the following presentation only the case of transverse electric fields is regarded explictly. But all statements also hold for transverse magnetics fields.

Correspondence to: M. Stallein
(stallein@tet.upb.de)

2 Basic theory

The following sections briefly describe the theory of quasi-guided modes of bent dielectric slab waveguides. A detailed presentation including all derivations can be found in Kerndlmaier (1992); Stallein (2010).

2.1 General solution of Maxwell's equations

Consider the waveguide structure shown in Fig. 1. The different sections, divided by the inner radius R^- and outer radius R^+, are free of sources and filled with homogeneous material. The refractive indices in the inner cladding, the

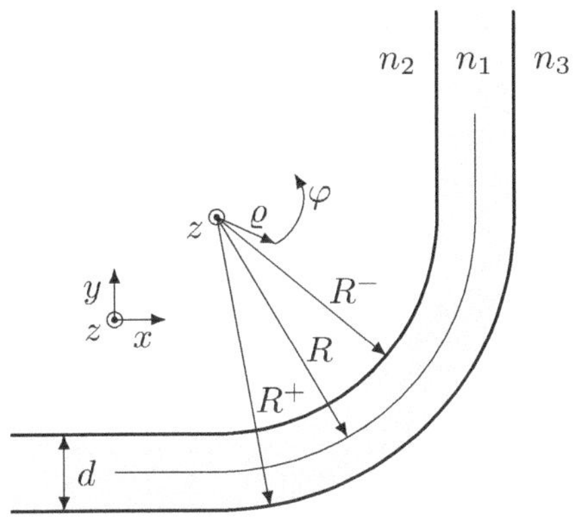

Fig. 1. Connection of two straight waveguides by a bent element.

outer cladding and the core region are n_2, n_3, and n_1, respectively, while the permeability remains constant $\mu = \mu_0$. Using cylindrical coordinates (ϱ, φ, z) the straightforward solution of Maxwell's equations for the bent section in the frequency domain is

$$\underline{\boldsymbol{E}}_\nu(\boldsymbol{r}) = a_\nu \boldsymbol{e}_z \underline{\mathcal{E}}_\nu(\varrho) \exp(-j\nu\varphi) \quad (1)$$

with

$$\underline{\mathcal{E}}_\nu(\varrho) = \begin{cases} J_\nu(k_2\varrho) & ,0 \le \varrho \le R^- \\ A_{1\nu} J_\nu(k_1\varrho) + A_{2\nu} N_\nu(k_1\varrho) & ,R^- \le \varrho \le R^+ \\ A_{3\nu} J_\nu(k_3\varrho) + A_{4\nu} N_\nu(k_3\varrho) & ,R^+ \le \varrho \end{cases} \quad (2)$$

and $k_i = n_i \frac{2\pi}{\lambda}$, $i = 1,2,3$. Here a wave propagation in positive φ-direction has been assumed and ν is called "angular wavenumber" or simply eigenvalue of the mode. J_ν and N_ν are the Bessel and Neumann functions, and λ is the free space wavelength. The magnetic field is given by Faraday's law $\underline{\boldsymbol{H}}_\nu = -1/(j\omega\mu)\nabla \times \underline{\boldsymbol{E}}_\nu$.

Enforcing the boundary conditions in $\varrho = R^-$ and $\varrho = R^+$ results in the following expressions for $A_{1\nu}$–$A_{4\nu}$.

$$A_{1\nu} = \frac{\pi R^-}{2}\Big(k_1 J_\nu(k_2R^-)N'_\nu(k_1R^-) - k_2 J'_\nu(k_2R^-)N_\nu(k_1R^-)\Big) \quad (3)$$

$$A_{2\nu} = \frac{\pi R^-}{2}\Big(k_2 J'_\nu(k_2R^-)J_\nu(k_1R^-) - k_1 J_\nu(k_2R^-)J'_\nu(k_1R^-)\Big) \quad (4)$$

$$A_{3\nu} = \frac{\pi R^+}{2}\Big(-k_1 N_\nu(k_3R^+) \big(A_{1\nu}J'_\nu(k_1R^+) + A_{2\nu}N'_\nu(k_1R^+)\big) + k_3 N'_\nu(k_3R^+) \big(A_{1\nu}J_\nu(k_1R^+) + A_{2\nu}N_\nu(k_1R^+)\big)\Big) \quad (5)$$

$$A_{4\nu} = \frac{\pi R^+}{2}\Big(k_1 J_\nu(k_3R^+) \big(A_{1\nu}J'_\nu(k_1R^+) + A_{2\nu}N'_\nu(k_1R^+)\big) - k_3 J'_\nu(k_3R^+) \big(A_{1\nu}J_\nu(k_1R^+) + A_{2\nu}N_\nu(k_1R^+)\big)\Big) \quad (6)$$

The property a_ν remains as initially undetermined weight function of the continuous mode spectrum of the bent waveguide.

Because of the analogy to the spectrum of radiation modes of the straight waveguide one can assume that purely real eigenvalues ν are adequate to describe any essential physical field problem (Kerndlmaier, 1992), although in general any complex eigenvalue ν is permitted. Therefore in the following only purely real eigenvalues ν are considered and for further characterization of the mode spectrum a scaling of the weight function

$$a_\nu = \sqrt{\frac{2\omega\mu}{A_{3\nu}^2 + A_{4\nu}^2}} \quad (7)$$

is useful.

2.2 Non-orthogonality of the mode spectrum

Testing the orthogonality requires the solution of an integral like

$$\overline{P}'_{\nu\xi} = \frac{\overline{P}_{\nu\xi}}{\Delta z} = \frac{1}{4}\int_0^\infty \Big(\underline{\boldsymbol{E}}_\nu \times \underline{\boldsymbol{H}}^*_\xi + \underline{\boldsymbol{E}}^*_\xi \times \underline{\boldsymbol{H}}_\nu\Big)_\varphi \mathrm{d}\rho \quad (8)$$

where Δz is a distance in z-direction. According to Morita and Yamada (1990) the solution of Eq. (8) is

$$\overline{P}'_{\nu\xi} = \Im\left\{\frac{(A_{3\nu} + jA_{4\nu})}{\sqrt{A_{3\nu}^2 + A_{4\nu}^2}} \frac{(A_{3\xi} - jA_{4\xi})}{\sqrt{A_{3\xi}^2 + A_{4\xi}^2}} \frac{e^{j\frac{\pi}{2}(\nu-\xi)}}{\pi(\nu-\xi)}\right\}. \quad (9)$$

Using the shortcut $\beta_\nu = \frac{A_{3\nu}}{A_{4\nu}}$ it is straightforward to show that no singularity exists for $\nu = \xi$:

$$\overline{P}'_{\nu\nu} = \frac{1}{2} - \frac{1}{\pi}\frac{\mathrm{d}\beta_\nu}{\mathrm{d}\nu}. \quad (10)$$

The expression (9) does not vanish for any $\nu \neq \xi$,

$$\overline{P}'_{\nu\xi} \neq 0, \quad (11)$$

thus the modes of the bent waveguide are nonorthogonal.

2.3 Quasi-guided modes (QGM)

In the continuous mode spectrum of the bent waveguide solutions exist, whose field profiles are focused on the core region. Due to the singularity of the Neumann function N_ν this is the case if

$$A_{3\nu} = 0. \quad (12)$$

If only quasi-guided modes are used in Eq. (9), the expression simplifies with $\beta_\nu = \beta_\xi = 0$ to

$$\overline{P}'_{\nu\xi} = \frac{\sin(\frac{\pi}{2}(\nu-\xi))}{\pi(\nu-\xi)} \qquad \text{if} \quad \nu \neq \xi. \tag{13}$$

As will be shown, in case of quasi-guided modes $\frac{d\beta_\nu}{d\nu}$ is always a large negative quantity, thus

$$\overline{P}'_{\nu\nu} \gg \overline{P}'_{\nu\xi} \tag{14}$$

is valid and the quasi-guided modes are considered as quasi-orthogonal, though the complete mode spectrum is non-orthogonal.

2.4 Physical fields

The modes of the bent waveguide are exited by the fields of a straight waveguide, see Fig. 1. If the fields at the end of the straight waveguide in $\varphi = 0$ are given by $\underline{\mathcal{E}}$ and $\underline{\mathcal{H}}$, then the overlap with the bent waveguide modes results in an integral equation of the form

$$I_\xi = \int_\nu a_\nu \overline{P}'_{\nu\xi} d\nu \tag{15}$$

with

$$I_\xi = \frac{1}{4}\int_0^\infty \left(\underline{\mathcal{E}} \times \underline{H}^*_\xi + \underline{E}^*_\xi \times \underline{\mathcal{H}}\right)_\varphi d\rho. \tag{16}$$

Equation (15) is a Fredholm integral equation of the first kind, that is even numerically difficult to solve.

For the connection of straight and bent waveguides it is expected that primarily the codomain for ν next to the eigenvalues ν_0 of the quasi-guided modes is excited. Thus a linearization of $\beta_\nu = \frac{A_{3\nu}}{A_{4\nu}}$ around the eigenvalue ν_0 of a quasi-guided mode is possible

$$\bar{\beta}_\nu = -k\bar{\nu} \text{ with } \bar{\nu} = \nu - \nu_0 \text{ and } -k = \left.\frac{d\beta_\nu}{d\nu}\right|_{\nu_0}. \tag{17}$$

The quantity k is always positive. Now Eq. (9) becomes

$$\overline{P}'_{\nu\xi} = \frac{1}{\pi}\left(\frac{k}{\sqrt{(1+k^2\bar{\nu}^2)(1+k^2\bar{\xi}^2)}}\cos(\tfrac{\pi}{2}(\bar{\nu}-\bar{\xi})) + \frac{1+k^2\bar{\nu}\bar{\xi}}{\sqrt{(1+k^2\bar{\nu}^2)(1+k^2\bar{\xi}^2)}}\frac{\sin(\frac{\pi}{2}(\bar{\nu}-\bar{\xi}))}{\bar{\nu}-\bar{\xi}}\right). \tag{18}$$

If the kernel (18) is used in the integral equation (15), then the square root of a Lorentz distribution ($\sqrt{\text{Lorentz}}$-Distribution) is an eigensolution of the integral equation

$$\int_\nu \frac{1}{\sqrt{1+k^2\bar{\nu}^2}}\overline{P}'_{\nu\xi} d\nu = \frac{1}{\sqrt{1+k^2\bar{\xi}^2}}. \tag{19}$$

This can be checked by means of the residue theorem.

Now let us assume the fields in the bent waveguide are defined by a weight function of the form of a $\sqrt{\text{Lorentz}}$-distribution, thus

$$\underline{\boldsymbol{E}}(\varrho,\varphi) = \int_\nu a_\nu \boldsymbol{e}_z \underline{\mathcal{E}}_\nu(\varrho)\exp(-j\nu\varphi)d\nu$$

$$\text{with} \quad a_\nu = \frac{1}{\sqrt{1+k^2\bar{\nu}^2}}. \tag{20}$$

When the wave travels through the bent waveguide the shape of the field profile will undergo deformation. As a measure for the deformation after the wave has passed an arbitrary angle φ, the following integral will calculate the overlap with the fields in the plane $\varphi = 0$

$$B(\varphi) = \frac{1}{4}\int_0^\infty \Big(\underline{\boldsymbol{E}}(\varrho,0) \times \underline{\boldsymbol{H}}^*(\varrho,\varphi) + \underline{\boldsymbol{E}}^*(\varrho,\varphi) \times \underline{\boldsymbol{H}}(\varrho,0)\Big)_\varphi d\rho. \tag{21}$$

Using Eqs. (20), (8) and (19) it follows

$$B(\varphi) = \int_\xi \frac{1}{1+k^2\bar{\xi}^2}\exp(j\xi\varphi)d\xi. \tag{22}$$

If the integration interval can be extended from the real axis to the closed contour of the upper half plane, it is possible to find a solution by means of the residue theorem. The normalized result is

$$\frac{B(\varphi)}{B(0)} = \frac{1}{\pi k}\oint_{\substack{\text{upper}\\ \text{half plane}}} \frac{\exp(j\xi\varphi)}{(\bar{\xi}-j/k)(\bar{\xi}+j/k)}d\xi = \exp(-\varphi/k). \tag{23}$$

Therefore, as long as $k \gg \varphi$, the field profile defined by a $\sqrt{\text{Lorentz}}$-distribution propagates almost undistorted through the waveguide.

2.5 Relation to the Leaky Mode Ansatz (LMA)

With

$$A_{3\nu} = jA_{4\nu} \tag{24}$$

the field description in $\varrho > R^+$ can be written in terms of Hankel functions

$$A_{3\nu}J_\nu(k_3\varrho) + A_{4\nu}N_\nu(k_3\varrho) = jA_{4\nu}H^{(2)}_\nu. \tag{25}$$

This is the so called Leaky Mode Ansatz (LMA) and Eq. (24) is the associated eigenvalue equation. The Solutions ν_l of Eq. (24) are always complex valued with $\Re\{\nu_l\} \neq 0$ and $\Im\{\nu_l\} \neq 0$. Hence the real part of the Poynting vector of a leaky mode has a component in ϱ-direction.

If $-1/k$ is taken as imaginary part of the eigenvalue ν_0 of the quasi-guided mode, then this constructed eigenvalue can be checked against the eigenvalues of the leaky mode ansatz.

Table 1. Eigenvalues and attenuation constants of quasi-guided modes (ν_0 and $-1/k$) and leaky modes ($\Re\{\nu_l\}$ and $\Im\{\nu_l\}$). Parameters: $n_1 = 1.57$, $n_2 = n_3 = 1.55$, $d = 10\mu$m, $R = 3$mm and $\lambda = 850$nm.

	$\Re\{\nu\}$		$\Im\{\nu\}$	
Mode	QGM	LMA	QGM	LMA
1	34820.05	34820.05	−6.199e-34	–
2	34772.80	34772.80	−6.427e-28	–
3	34720.68	34720.68	−1.284e-21	–
4	34650.55	34650.55	−4.565e-14	–
5	34561.82	34561.82	−2.961e-6	−2.957e-6
6	34460.38	34460.58	−1.308	−1.263

Table 1 shows the results for a waveguide with six guided modes in case of the straight waveguide, and also six quasi-guided modes in case of the bent-waveguide. Except for the last mode the real parts agree within seven (or more) digits. The real parts of the sixth mode differ slightly but now the attenuation constants, the imaginary parts of the eigenvalues, already take large (negative) values, thus the mode suffers strong attenuation. The absolute value $|1/k|$ is always a little larger then $|\Im\{\nu_l\}|$. But anyway, for the first four modes the absolute values are very small and can be neglected in case of waveguide bends of 90° or less. Effectively the quantity $|\Im\{\nu_l\}|$ is too small to be determined by conventional root solvers and double precision accuracy, e.g. secant methods.

The results of Table 1 are typical and similar conclusions hold for waveguides with a larger number of guided modes. There is always a specific number of modes in dependency on the radius R, whose attenuation constant is negligible. Only a few modes remain with $-1/k$ in a significant range and all other modes can be disregarded at all, because the attenuation is to high. The differences in the eigenvalues of the quasi-guided modes and the leaky modes are always very small. Since the theory of quasi-guided modes owns a real valued eigenvalue equation (12), it is preferable, because the leaky modes afford the solution of a complex valued Eq. (24).

Finally Fig. 2 shows the electric field profile of the fifth and sixth mode of the quasi-guided mode approach, the real and imaginary part of the leaky mode ansatz, and the related profile of the mode of the straight waveguide. All fields are normalized with respect to the power flow in φ-direction. For the fifth mode both bent waveguide solutions approximately coincide and the overlap with the mode of the straight waveguide is high. In case of the sixth mode the leaky mode ansatz exhibits an apparent imaginary part. Obviously this mode cannot be regarded as quasi-guided because the oscillations in the outer cladding region $R^+ < \varrho$ are oversized. Anyway, according to Table 1 the attenuation constant of this mode is very high.

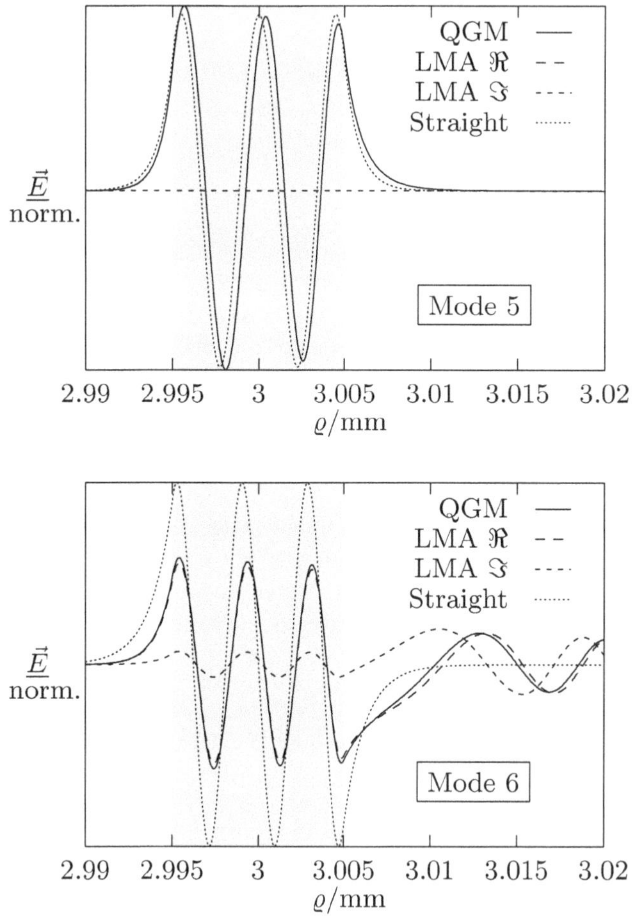

Fig. 2. Comparison of the electric field profile of the fifth and sixth mode of the quasi-guided mode approach (QGM), real and imaginary part of the leaky mode ansatz (LMA $\Re$ and LMA $\Im$), and the related profile of the mode of the straight waveguide (Straight). The waveguide core region is highlighted. Parameters like in Table 1.

3 Calculations on the basis of quasi-guided modes

3.1 Further properties of quasi-guided modes

In Sect. 2.4 the $\sqrt{\text{Lorentz}}$-distribution has been introduced as a physical solution for the weight function a_ν. Because k is a large quantity and $2/k$ is the half width of the Lorentz-distribution, the half width of the first five modes in Table 1 is negligible compared to the distance between the eigenvalues. Even for the sixth mode the distribution concentrates around the quasi eigenvalue. This behavior has been observed for various different waveguide parameters, i.e. different core thicknesses d and radii of curvature R. Therefore it is legitimate to state that the width of the $\sqrt{\text{Lorentz}}$-distribution is infinitesimal small, thus replace it by a Dirac-distribution δ. Then for a multimode waveguide a_ν becomes

$$a_\nu \approx \sum_n C_n \exp(-\alpha_n\varphi)\delta(\nu-\nu_n). \tag{26}$$

Here C_n is the amplitude, ν_n the eigenvalue, and $\alpha_n = 1/k_n$ the attenuation constant of a quasi guided mode. Furthermore the electric field is expressed by

$$\underline{E}(\varrho,\varphi) = \int_{\nu} a_{\nu}\underline{\mathcal{E}}_{\nu}(\varrho)\exp(-j\nu\varphi)d\nu = \sum_n C_n\underline{\mathcal{E}}_n(\varrho)\exp\big(-j(\nu_n - j\alpha_n)\varphi\big). \quad (27)$$

As will be shown later, this is quite a good approximation whenever the waveguide bend is free of loss due to radiation.

The connection of the eigenvalues of the straight and the bent waveguide becomes obvious if the propagation terms are compared:

$$\exp(-jk_z z) \approx \exp(-jk_z R\varphi) \quad \Rightarrow \quad \nu \approx k_z R. \quad (28)$$

Here k_z is the eigenvalue of the straight waveguide. For small radii this relation is just a guesstimate, but it is useful for the definition of a search interval. Based on the well know solution interval of the straight waveguide it follows

$$k_0 n_2 < k_z < k_0 n_1 \quad \Rightarrow \quad k_0 n_2 R^- < \nu < k_0 n_1 R^+. \quad (29)$$

However, the eigenvalue ν of the fundamental mode might slightly exceed the upper boundary in Eq. (29).

3.2 Connection to straight multimode waveguides

If reflected waves can be neglected at the transition from the straight to the bent waveguide only the transverse components of the electric or the magnetic fields must be matched. Thus if the fields in the bent waveguide are given by Eq. (27), then the amplitudes of the quasi-guided modes are

$$C_n^{(b)} \approx \frac{1}{2\overline{P}_{nn}^{\prime(b)}} \int_0^{\infty} \left(\sum_{\mu} C_{\mu}^{(l)}\underline{\mathcal{E}}_{t\mu}^{(l)}\right) \times \underline{\mathcal{H}}_{tn}^{(b)}\,\mathrm{d}\rho. \quad (30)$$

The index (b) indicates the bent waveguide and (l) the straight waveguide. Furthermore the index μ discerns the modes of the straight waveguide and t the transverse components of the fields. In Eq. (30) the orthogonality relation (14) has been applied. Accordingly $\overline{P}_{nn}^{\prime(b)}$ is the normalized power of a quasi-guided mode.

An analog expression to Eq. (30) holds for the second transition after passing through a bent segment defined by the angle φ_0:

$$C_{\mu}^{(r)} \approx \frac{1}{2\overline{P}_{\mu}^{\prime(r)}} \int_0^{\infty} \left(\sum_n C_n^{(b)}\exp\big(-j(\nu_n - j\alpha_n)\varphi_0\big)\underline{\mathcal{E}}_{tn}\right) \times \underline{\mathcal{H}}_{t\mu}^{(r)}\,\mathrm{d}\rho. \quad (31)$$

Here the index (r) indicates the second straight waveguide which follows the bent waveguide.

As mentioned before, the calculation rules (30) and (31) are good approximations for small radiation loss. It is obvious that radiated fields cannot be expressed just in terms of quasi-guided modes. In this case the full mode spectrum of the bent waveguide must be regarded and an integral equation like Eq. (15) has to be solved. But often the exact solution for the radiated fields is not a matter of particular interest. The primary parameter of a transmitting channel is the outgoing power flow. If the angle φ_0 of the bent waveguide is sufficiently large, so that all radiating fields are emitted and just the quasi-guided modes contribute to the fields in the endface of the bent waveguide, then Eqs. (30) and (31) are still capable to calculate the outgoing power flow. In the following this approach is referred to as QGM-Method.

4 Numerical validation

Quasi-guided modes are exact solutions of Maxwell's equations, but they are certainly not a complete set of modes. Whether they are suited to model the power flow like proposed in Sect. 3.2 should be tested by comparison with a numerical simulation. This simulation has been done with CST MICROWAVE STUDIO® (MWS2009), i.e. the Finite Integration Technique. Because the wavelength $\lambda = 1.55\,\mu$m is very small compared to the 90° bent waveguide trajectory, the radius R is decreased to $R = 250\,\mu$m and $R = 125\,\mu$m respectively, in order to reduce the memory requirements. Furthermore no straight waveguides have been considered explicitly, although the field in $\varphi = 0$ is defined by the fundamental mode of a straight waveguide. The thickness of the core is $d = 125\,\mu$m and the refractive indices are $n_1 = 1.1$ and $n_2 = n_3 = 1.0$. For the numerical simulation a discretization with 20 lines per wavelength leads to approximately 60 millions of unknowns. Finally open boundaries have been applied to produce the results for $R = 250\,\mu$m presented on the left side in Fig. 3.

Depicted is the magnitude of the electric field within the ϱ-φ-plane. As is it shown with $R = 250\,\mu$m there is no radiation loss, i.e. there is almost no power flow in radial direction in $R^+ < \varrho$. And as supposed there are only minor differences in the results of the QGM-Method. The graphs on the right in Fig. 3 show the the fields in the input plane $\varphi = 0$ and in the output-plane $\varphi = \pi/2$, respectively, before and after the expansions (30) and (31), denoted by the indices $-/+$ in the keys. For the interpretation it is important to know, that 10 quasi guided modes with $k < 5$ exist in the bent waveguide, compared to 12 guided modes in the straight waveguide. The results demonstrate that the boundary conditions in $\varphi = 0$ and $\varphi = \pi/2$ are approximately fulfilled. Furthermore there is good agreement between the results of QGM-Method and the numerical simulation in $\varphi = \pi/2$.

Figure 4 shows the corresponding results for $R = 125\,\mu$m. Due to the reduced radius of the bent waveguide and since only the fundamental mode of a straight waveguide is excited

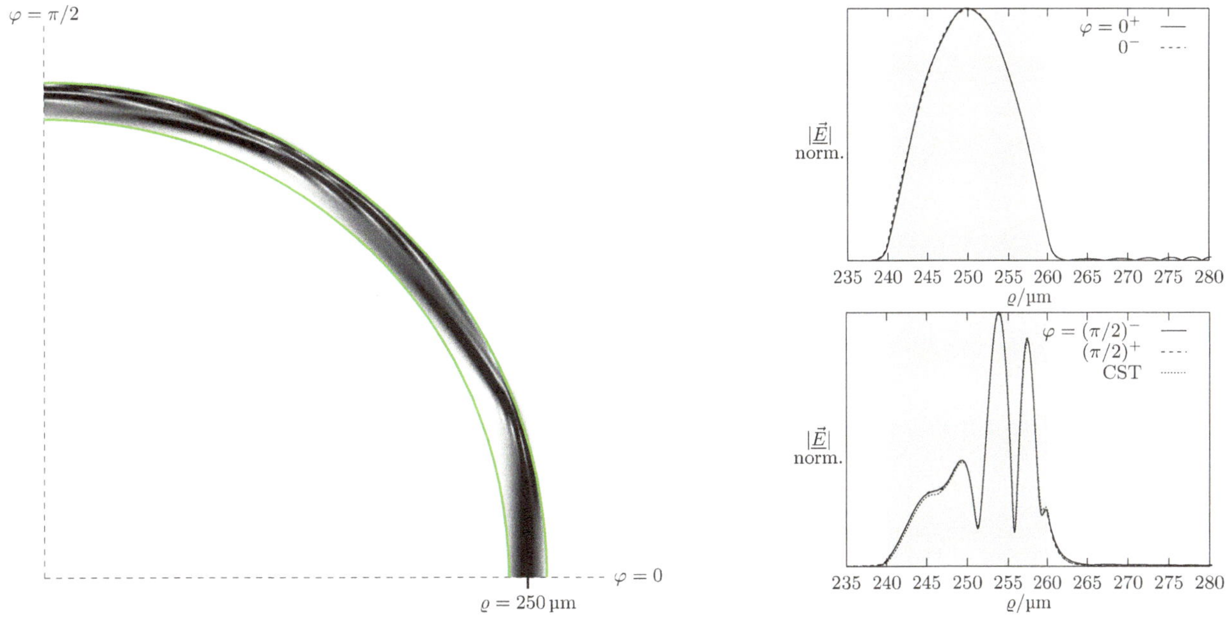

Fig. 3. Left side: Magnitude of the electric field in the ϱ-φ-plane simulated by CST MICROWAVE STUDIO®. Right side: Magnitude of the electric field in the port planes $\varphi = 0$ and $\varphi = \pi/2$ simulated by the QGM-method. The indices $-/+$ in the keys denote whether the fields are calculated before or after the expansions (30) and (31) and the key CST denotes the results of the numerical simulation. The waveguide core region is highlighted.

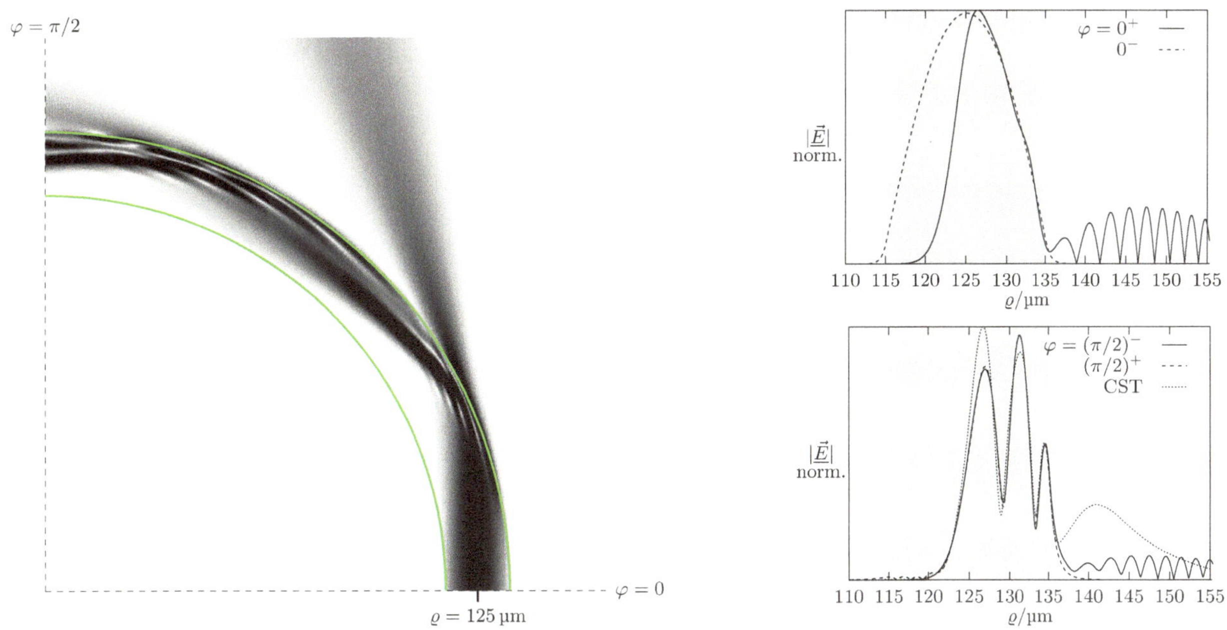

Fig. 4. See caption of Fig. 3, but $R = 125\,\mu m$.

in $\varphi = 0$, mainly at two position in the planes $\varphi \approx 30°$ and $\varphi \approx 80°$ the power radiates in form of a beam. Overall about 50% of the power coupled into the waveguide is radiated away from the core region.

As mentioned before, radiated fields cannot be expressed in terms of quasi-guided modes. Thus the QGM-Method fails especially for small angles φ and consequently the boundary conditions in $\varphi = 0$ are not satisfied. It should be noted that for $R = 125\,\mu m$ only 6 quasi-guided modes with $k < 5$ remain. At the second transition, after passing the bent waveguide section of 90°, one can observe a clearly larger overlap between the different field solutions in the interface.

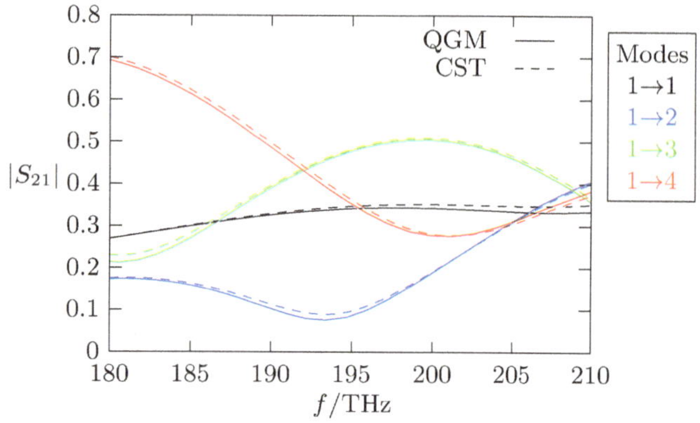

Fig. 5. Comparison of the transmission scattering parameters for the simulation with $R = 250\,\mu$m (Fig. 3). The graphs show the coupling from the fundamental mode of the incoming waveguide to the first four modes of the outgoing waveguide.

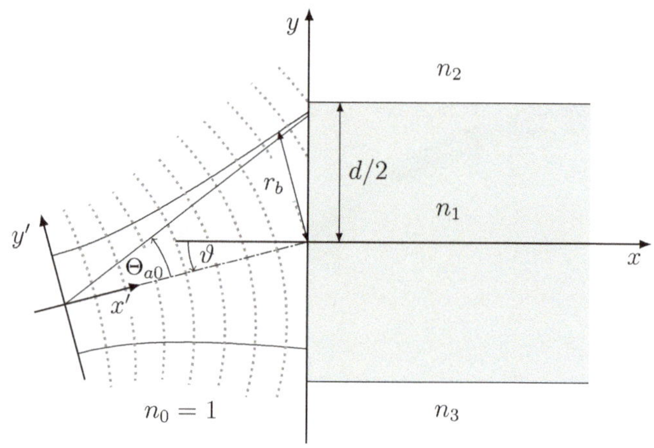

Fig. 6. Coupling conditions in the front face of the incoming straight waveguide.

That is, within the QGM-method the boundary conditions in the core region are approximately fulfilled and the overlap with the numerical solution is large. Obvious differences just remain in the outer cladding region. However, it is expected that these parts are radiated away if the wave propagates a distance within the straight waveguide, because fields in $R^+ < \varrho$ cannot significantly excite guided modes of the straight waveguide. Therefore the presented results for $R = 125\,\mu$m are a further indication for the validity of the QGM-Method.

In addition scattering parameters have been generated from the fields in the port planes, which again show good agreement between the QGM-method and the numerical simulation. Figure 5 shows the results of the simulation with $R = 250\,\mu$m for the first four modes of the outgoing waveguide, when in the incoming waveguide only the fundamental mode is excited. Similar agreement holds for the remaining eight modes of the straight waveguide.

5 Bending loss of multimode waveguides

The QGM-method presented in Sect. 3.2 is an efficient approach to calculate the radiation loss of a circular bended connection of two straight slab waveguides, cf. Fig. 1. In the following some exemplary simulation results for different sets of parameters will be presented. The modes of the incoming straight waveguide are excited by a Gaussian beam with a beam width $b = 2r_b$ and an asymptotic angle of divergence of $\Theta_{a0} = 5°$, see Fig. 6 (Stallein et al., 2004; Stallein, 2008).

In dependency on the angle of incidence of the beam axis ϑ, different excitations of the mode spectrum result and different bending loss is expected. The bending loss $\overline{P}_A$ is defined as the ratio of the power in the guided modes of the outgoing and the incoming straight waveguide, thus the length of the outgoing waveguide is arbitrary. For the results in Fig. 5 the length of the incoming waveguide z_p tends to zero and in Fig. 5 some special values z_p have been used. Further parameters are the ratio of core thickness and beam width $d/b = 1.5$, the wavelength $\lambda = 850$ nm, and the refractive indices $n_1 = 1.57$ and $n_2 = n_3 = 1.55$.

Figure 7 shows results for the bending loss $\overline{P}_A$ in dependency on the radius R for a 90° bend. In Fig. 5 various angles of incidence of the Gaussian beam ϑ have been considered, while the core thickness is constant $d = 75\,\mu$m. It becomes clear that the bending loss usually increases if the coupling conditions are not ideal, i.e. $|\vartheta| > 0°$. An exception is the case of a small rotation of the beam axis against the direction of the bend, which is the negative φ-direction.

Finally Fig. 5 shows some effects of multimode interference. The core thickness is once again $d = 75\,\mu$m. Because of the symmetrical stimulation with $\vartheta = 0°$, the field pattern within the incoming straight waveguide has some distinctive local maxima (Soldano and Pennings, 1995). If the length z_p is chosen such that one, two, or three maxima exist at the end of the incoming waveguide, then the dependency of the bending loss on the radius R is in the form of steps.

6 Conclusions

The theory of quasi-guided modes (QGM) is an important element in the modeling of dielectric bent waveguides. Compared with the well know leaky mode ansatz it is advantageous, because the eigenvalue equation is real valued. Nevertheless the deviations are small, e.g. all results of Sect. 5 can also be generated by means of leaky modes. The analytic calculation with a limited and discrete set of modes is time-efficient in contrast to a numerical simulation, because the optical wavelength is very small compared with the length of the waveguide trajectory. For the purpose of validation a comparison with the results of CST MICROWAVE

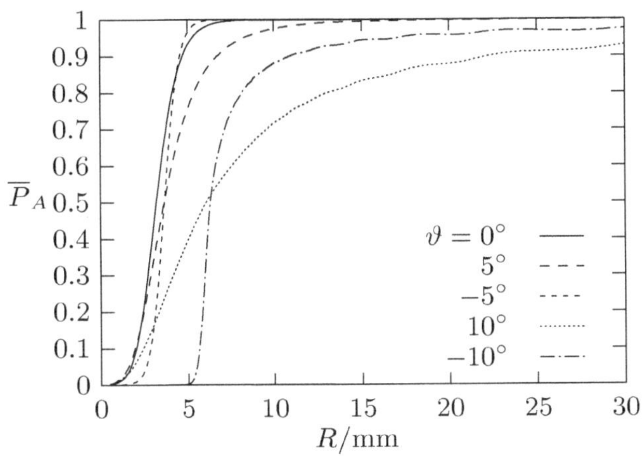

(a) $z_p = 0$

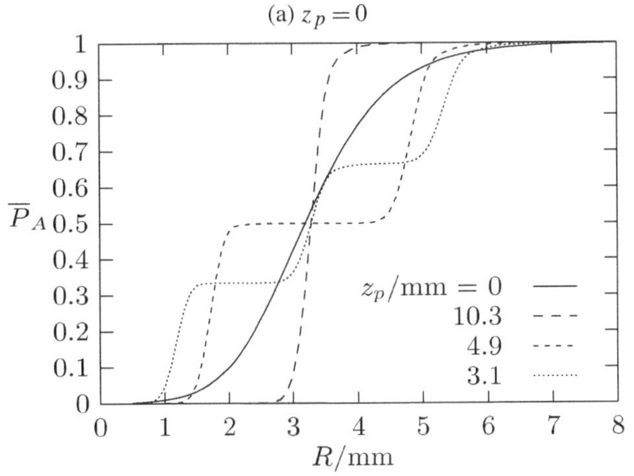

(b) $\vartheta = 0°$

Fig. 7. Normalized output power after passing through a 90° bend: **(a)** for different angles of incidence ϑ; **(b)** for different lengths z_p of the incoming waveguide.

STUDIO® (MWS2009) confirmed the basic suitability of the QGM-method. For large radii of curvature, i.e. small bending loss, the QGM-method is quite a good approximation for the fields in the bent waveguide. And even though the exact solution of a radiation problem of course requires the consideration of the complete continuous mode spectrum, it is possible to calculate the field at the end of the bent section, if the angle of the bend is sufficiently large. In this case all radiating parts of the fields have already passed the core region and only the quasi-guided modes contribute to the fields in the output port.

References

Kerndlmaier, W.: Differentialgeometrische Formulierung der Maxwell-Gleichungen und deren Anwendung auf die Berechnung gekrümmter dielektrischer Wellenleiter, Ph.D. thesis, München, Techn. Univ., 1992.

Lewin, L., Chang, D. C., and Kuester, E. F.: Electromagnetic waves and curved structures, IEE Electromagnetic Waves Series, Volume 2, Peter Peregrinus, Ltd., Stevenage, Herts., England, 1977.

Marcatili, E. A. J.: Bends in optical dielectric guides, The Bell System Technical Journal, 48, 2103–2132, 1969.

Marcuse, D.: Light Transmission Optics, Van Nostrand, New York, 2 edn., 1982.

Morita, N. and Yamada, R.: Electromagnetic fields in circular bends of slab waveguides, IEEE/OSA J. Lightwave Technol., 8, 16–22, 1990.

Snyder, A. W. and Love, J. D.: Optical Waveguide Theory, Kluwer Academic Publishers, Boston, 2000.

Soldano, L. B. and Pennings, E. C. M.: Optical multi-mode interference devices based on self-imaging: principles and applications, IEEE/OSA J. Lightwave Technol., 13, 615–627, 1995.

Stallein, M.: Coupling Efficiency of Gaussian Beams Into Step-Index Waveguides – An Improved Ray-Optical Approach, IEEE/OSA J. Lightwave Technol., 26, 2937–2945, 2008.

Stallein, M.: Einkopplung in multimodale Lichtwellenleiter – Wellentheoretische Analyse und ein Vergleich zu strahlenoptischen Modellierung, Ph.D. thesis, Paberborn Univ., March 2010.

Stallein, M., Kolleck, C., and Mrozynski, G.: Coupling of a Gaussian Beam into a Planar Slab Waveguide using the Mode Matching Method, in: Extended Papers of the Progress in Electromagnetics Research Symposium (PIERS 2004), 309–312, Pisa, Italy, 2004.

17

Phaseless spherical near-field antenna measurements for low and medium gain antennas

C. H. Schmidt[1,2], **S. F. Razavi**[1], **T. F. Eibert**[2], **and Y. Rahmat-Samii**[1]

[1]Department of Electrical Engineering, University of California, Los Angeles, USA
[2]Lehrstuhl für Hochfrequenztechnik, Technische Universität München, Germany

Abstract. The characterisation of antenna radiation patterns by measurements in the near-field and a following near-field far-field transformation require accurate amplitude and phase data. Especially at higher frequencies phase measurements are demanding in terms of instrumentation and measurement accuracy. Phaseless techniques which require amplitude only data on one or more measurement surfaces are therefore of special interest. In this paper a phaseless spherical algorithm based on spherical modal expansion is presented. The algorithm works with amplitude only near-field data measured on two spheres of different radii and is therefore not restricted to certain types of antennas as for planar measurements for example. Simulated as well as measured results of low and medium gain antennas are shown.

1 Introduction

The radiation characteristic of an antenna under test (AUT) is typically defined at far-field distance and is therefore to be measured in the far-field region of the AUT. Two common alternatives to direct measurements in the far-field are compact range measurements (Balanis, 2008) which emulate far-field conditions at close distance to the AUT and secondly near-field measurement techniques. The far-field of an AUT consists of the same modes as the near-field. Therefore it is possible to determine performance of the AUT from a measurement in the radiating near-field. The field modes are determined and processed to the far-field by a post-processing near-field far-field transformation (Yaghjian, 1986). Especially for electrically large AUTs which satisfy far-field conditions at a far distance ($\approx 2D^2/\lambda$, D being the diameter of the minimum sphere enclosing the AUT and λ the wavelength), near-field measurements can be carried out much closer to the AUT in relatively small sized measurement facilities.

A critical requirement for the near-field far-field transformation is the accurate knowledge of the near-fields amplitude and phase. Amplitude measurements are less critical in terms of measurement equipment and accuracy compared to phase measurements, even for high frequencies. This is why phaseless measurement techniques (Yaccarino and Rahmat-Samii, 1999; Pierri et al., 1999; Las-Heras and Sarkar, 2002; Costanzo et al., 2005; Soldovieri et al., 2005; Schejball et al., 2008) have obtained considerable interest in research. Costly vector network analysers are no longer required for phase measurements and also phaseless techniques show some superior robustness for some classes of errors (Razavi and Rahmat-Samii, 2008). The lack of phase information is often compensated either by further amplitude only measurements on additional surfaces or by measurements with additional probes. The Iterative Fourier Technique (IFT) (Razavi and Rahmat-Samii, 2006) is known from planar phaseless measurements and requires amplitude only measurements on two planes. Planar measurement techniques (Kerns, 1981) rely on the finite sized scan plane in front of the AUT and are therefore only suitable for highly directive antennas.

A more common approach for the measurement of medium and low gain antennas is the spherical measurement technique (Hansen, 1988), where the radiated fields of the AUT over a sphere are recorded and the radiation in backward direction is also considered. To utilise the advantages of phaseless measurements, a phaseless spherical transformation algorithm based on a spherical modal expansion has been presented (Schmidt and Rahmat-Samii, 2009) and applied to a low gain patch antenna. The algorithm is particularly suitable for low as well as medium gain antennas and utilises measured amplitude only near-field data on two spheres. A modal expansion representing the radiation characteristic of the AUT is obtained in an iterative manner with some similarities to the IFT.

In this paper the phaseless spherical algorithm is further studied, applied and extended to the needs of medium

Correspondence to: C. H. Schmidt
(carstenschmidt@tum.de)

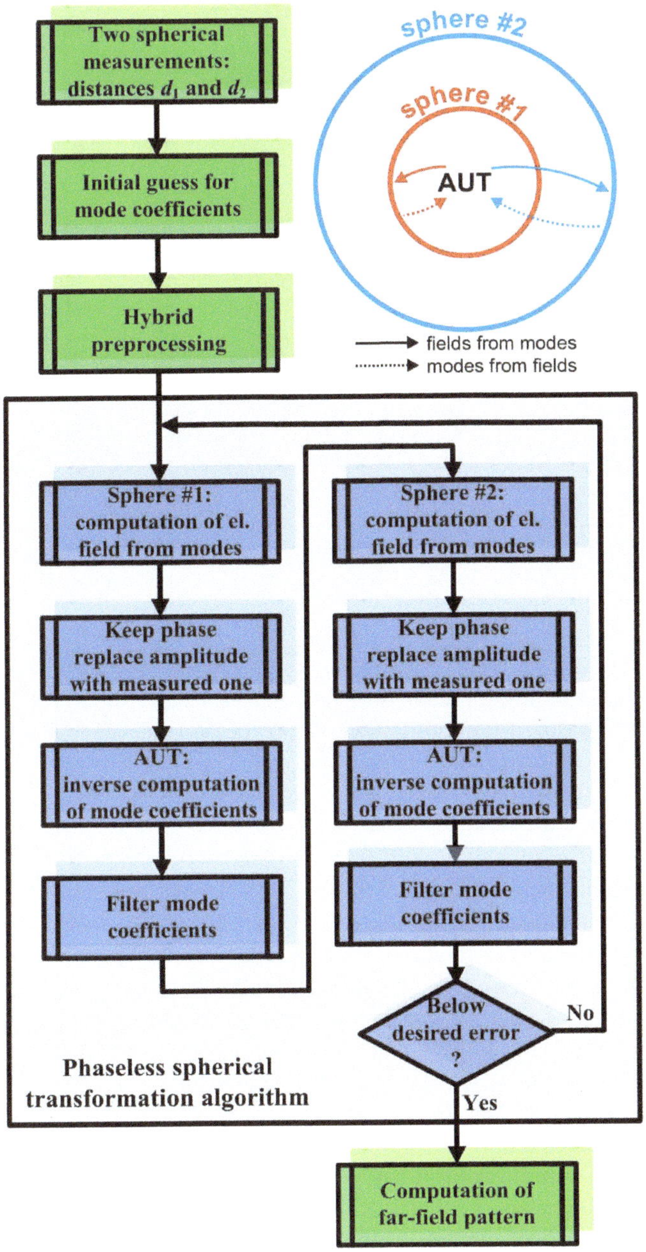

Fig. 1. Flowchart of phaseless spherical transformation algorithm working with amplitude only data on two spheres.

gain antenna measurements. In Sect. 2 a brief fundamental background of spherical near-field measurements is given whereas Sect. 3 introduces the phaseless spherical algorithm. Section 4 shows measured results of a low gain patch antenna and considers simulated near-field scenarios of a medium gain horn antenna. Characteristics of medium gain antennas within the phaseless spherical algorithm are addressed considering the example of a simulated Narda standard gain horn.

2 Spherical near-field far-field transformation

For spherical antenna measurements the radiated field distribution of an AUT

$$\boldsymbol{E}(r,\varphi,\vartheta)=\frac{k}{\sqrt{\eta}}\sum_{s=1}^{2}\sum_{n=1}^{\infty}\sum_{m=-n}^{n}Q_{smn}\boldsymbol{F}_{smn}(r,\varphi,\vartheta) \tag{1}$$

can be expanded in outgoing spherical modes $\boldsymbol{F}_{smn}(r,\varphi,\vartheta)$ (Hansen, 1988). The radiation characteristic of the AUT is obtained by superimposing all considered modes weighted with the mode expansion coefficients Q_{smn} characterising the AUT. k is the wavenumber and η the free space wave admittance. For practical measurements the infinite number of modes required to characterise the AUT is truncated to

$$N=\left\lfloor \pi\frac{D}{\lambda}\right\rfloor+10 \tag{2}$$

depending on the electrical size of the AUT D/λ (Balanis, 2008).

$$\{Q\}=[\mathbf{C}]^{-1}\cdot\{\boldsymbol{E}\} \tag{3}$$

are obtained by inverse solution of the field problem. $[\mathbf{C}]$ is the coupling matrix, containing discrete values of the spherical modes for discrete measurement points. $\{\boldsymbol{E}\}$ is the vector containing the amplitude and phase near-field values measured on a sphere. Once the mode expansion coefficients are known, the desired far-field pattern can be computed using asymptotic expressions for the spherical modes (Hansen, 1988).

In general a probe correction is required to compensate for the field probe effects on a near-field measurement. For spherical measurements and in particular if the probe is not too close to the AUT, probe effects might be sufficiently small and might even be neglected (Hindman and Fooshe, 1998). Therefore the presented phaseless spherical algorithm neglects the probe correction for simplicity. Nevertheless it is possible to adapt the technique to one of the several probe corrected spherical near-field algorithms, e.g. Hansen (1988); Laitinen et al. (2005); Laitinen and Breinbjerg (2008); Leibfritz and Landstorfer (2006).

3 Phaseless spherical algorithm

The phaseless spherical algorithm determines a modal expansion of the AUT fields from amplitude only near-field measurements on two spheres. The individual steps are shown in the flowchart in Fig. 1. The algorithm starts with an initial guess for the mode expansion coefficients $\{Q^{\mathrm{IG}}\}$. The initial guess is important for the convergence of the algorithm and different choices are introduced and presented for the near-field scenarios in Sect. 4. Amplitude and phase near-field values $\{E_1^{\mathrm{P}}\}$ are computed from the initial guess $\{Q^{\mathrm{IG}}\}$ according to Eq. (1). The propagated phase information on

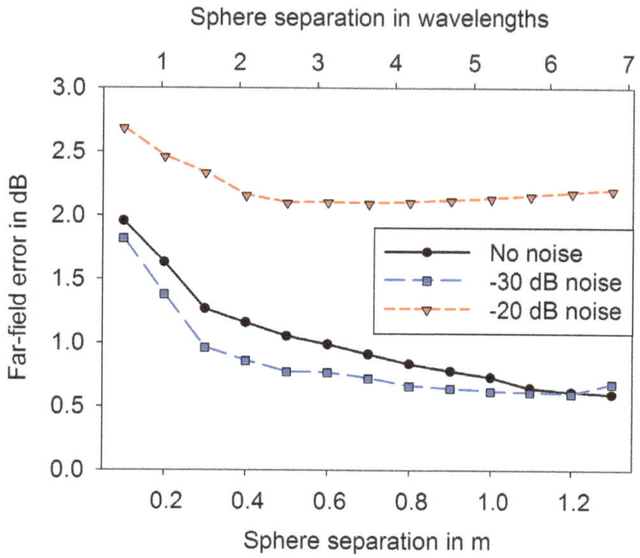

Fig. 2. Deviation of far-field pattern obtained by phaseless spherical algorithm from MoM reference pattern, unweighted average of all points on the sphere (step size 1° in ϑ and φ). Near-field data with and without noise considered.

the first sphere is retained while the field amplitude is replaced by the measured amplitude $\{E_1^{\mathrm{m}}\}$ resulting in $\{E_1^{\mathrm{mp}}\}$. A new set of mode coefficients $\{Q\}$ is computed from $\{E_1^{\mathrm{mp}}\}$ according to Eq. (3). The same procedure is repeated for the second measurement sphere. During the iterations the error norm

$$\epsilon = \frac{\sum_{i=1}^{S}\left\{\left(\sqrt{\left|E_{\varphi,i}^{\mathrm{p}}\right|^2+\left|E_{\vartheta,i}^{\mathrm{p}}\right|^2}-\sqrt{\left|E_{\varphi,i}^{\mathrm{m}}\right|^2+\left|E_{\vartheta,i}^{\mathrm{m}}\right|^2}\right)^2\right\}}{\sum_{i=1}^{S}\left\{\left|E_{\varphi,i}^{\mathrm{m}}\right|^2+\left|E_{\vartheta,i}^{\mathrm{m}}\right|^2\right\}} \tag{4}$$

of the total field amplitude on the first measurement sphere is monitored as convergence metric. The subscripts φ und ϑ denote the two polarisation components of the measured near-fields and S is the total number of near-field samples on the sphere.

The mode expansion coefficients of an AUT are known to decrease rapidly. Filtering the higher order mode expansion coefficients and setting them to zero during the iterations helps the algorithm to find a modal expansion of the AUT, which incorporates this behaviour. So undesired rapid oscillations in the far-field pattern can be reduced. An appropriate filter index i_{filter} is obtained by a hybrid preprocessing method (Razavi and Rahmat-Samii, 2006) which evaluates the main algorithm for a few iterations and various filter indices. i_{filter} is chosen to minimise the error norm ϵ.

An important parameter of the phaseless spherical algorithm is the separation of the measurement spheres. The lack of phase information in the transformation is overcomed by additional amplitude information. Therefore it is required, that the measured amplitude pattern on the second sphere contains additional field information. This can be achieved by placing the first measurement sphere close to the AUT and the second sphere with a preferable high separation to the first one. To evaluate the minimum required sphere separation, a patch antenna (as shown in Sect. 4.1) is considered and electric field values are simulated for various sphere separations using the field simulator FEKO (http://www.feko.info). For the phaseless spherical algorithm, the first sphere remains constant whereas the radius of the second sphere is further increased. Figure 2 shows the deviation from the method of moments (MoM) reference far-field pattern, averaged over all points in the pattern. Also transformations have been carried out with additive noise added to the amplitude only near-field data. A high sphere separation is preferable in order to gain further field information on the second sphere with increasing measurement distance. According to this simplified analysis, a minimum sphere separation of at least two wavelengths is recommended.

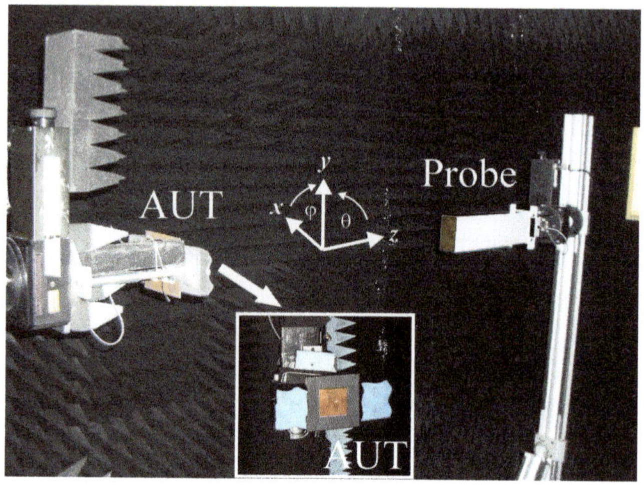

Fig. 3. Measurement setup inside the spherical near-field chamber at the Electrical Engineering Department, UCLA, with patch antenna as AUT measured at 1.57 GHz and open ended waveguide probe.

4 Results

4.1 Patch antenna measurement

In the first example, the phaseless spherical algorithm is tested for a low gain patch antenna operating at 1.57 GHz. The measurement setup with positioner, AUT and open ended waveguide probe is shown in Fig. 3. The field distribution has been measured on two spheres of radii 1.7 m and 2.2 m employing a WR430 open-ended waveguide probe and a spherical NSI (Nearfield Systems Inc.) near-field scanner at UCLA. The initial guess $\{Q^{\mathrm{IG}}\}$ is obtained according

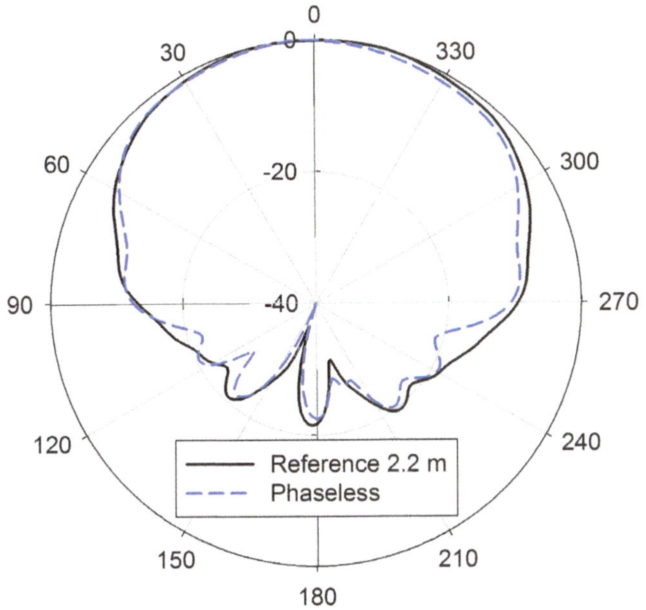

$\varphi = 0°$ cut.

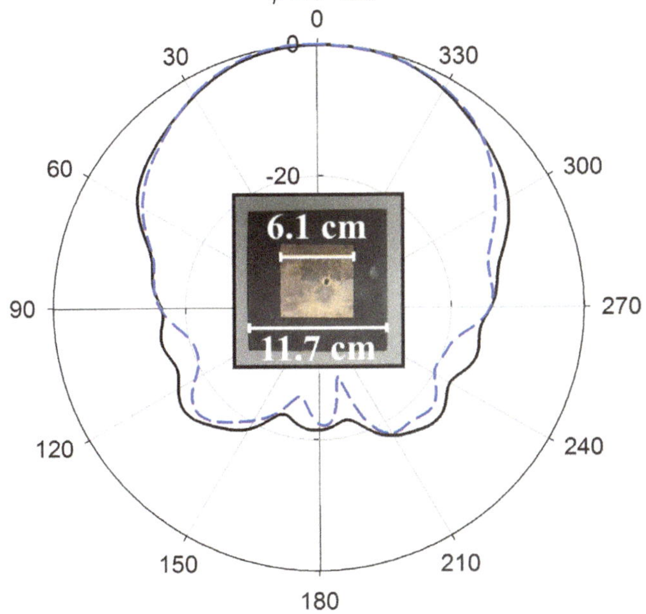

$\varphi = 90°$ cut.

Fig. 4. Comparison of patch antenna far-field patterns obtained by phaseless spherical algorithm from amplitude only data on two spheres with NSI reference pattern obtained from amplitude and phase data on a single sphere.

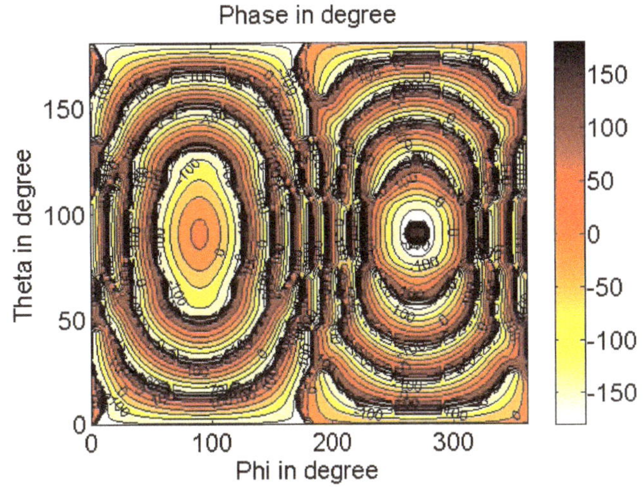

Fig. 5. Simulated near-field phase of horn antenna on sphere of radius 1.75 m.

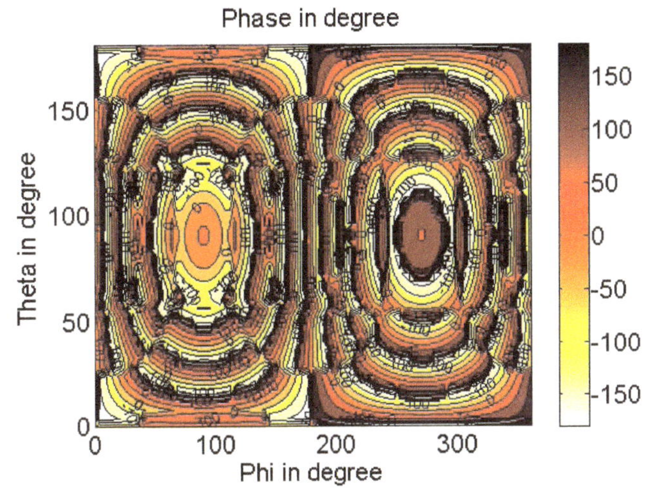

Fig. 6. Near-field phase of electric/magnetic aperture current distribution on sphere of radius 1.75 m, used to compute the initial guess together with simulated/measured amplitude on second sphere.

to Eq. (3) for the amplitude only data $\{E_2^{\mathrm{m}}\}$ measured on the second sphere. In order to have a reference, amplitude and phase have been measured and the far-field pattern has been computed with the NSI2000 software. Figure 4 shows the far-field pattern obtained by the phaseless spherical algorithm together with the NSI reference in the principal cuts. The patterns show an acceptable agreement to the reference if one keeps in mind that accurate amplitude and phase measurements usually give more accurate results due to the complete description of the fields also by phase information.

During the measurement, the AUT positioner moves between AUT and probe while measuring the radiation in backward direction. This should be considered when evaluating measured patterns in backward direction of the AUT.

4.2 Horn antenna simulation

The second example considers a medium gain horn antenna at 6 GHz. A Narda standard gain horn model 642 is therefore simulated with FEKO. Near-field values have been computed on spheres of 1.27 m and 1.75 m radii. For the horn antenna two different initial guesses $\{Q^{\mathrm{IG}}\}$ are considered. The first one follows the same procedure as in the previous example.

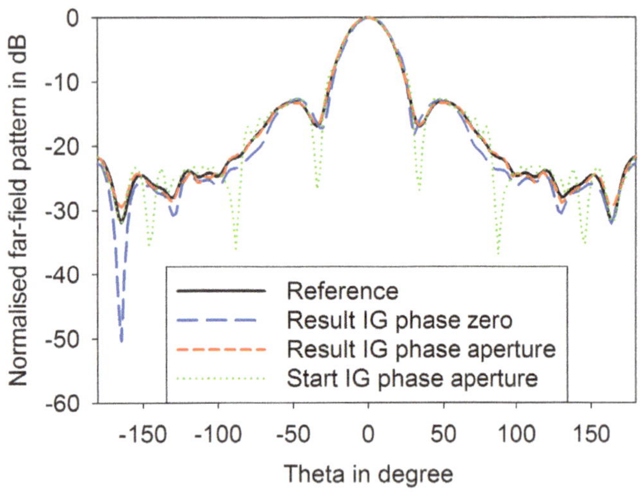

$\varphi = 0°$ cut.

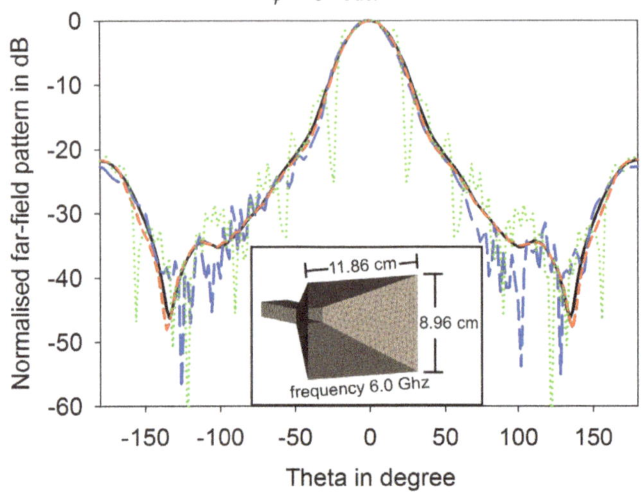

$\varphi = 90°$ cut.

Fig. 7. Comparison of horn antenna far-field patterns obtained by phaseless spherical algorithm from amplitude only data on two spheres with MoM reference pattern.

The initial guess for the mode expansion coefficients is obtained from the amplitude only data on the second sphere (referred to as "result IG phase zero" in the figures). The second initial guess computes the field distribution of an electric/magnetic current distribution in the aperture of the horn. A uniform amplitude and phase distribution is assumed for the currents. The obtained phase on the second measurement sphere together with the simulated/measured amplitude is used to compute the initial guess (referred to as "result IG phase aperture" in the figures). Figure 5 shows the phase of the horn antenna on the second measurement sphere, whereas Fig. 6 shows the phase of the aperture current distribution. The aperture phase shows already close agreement with the real phase and so it is suited to compute the initial guess. The far-field pattern obtained by the phaseless spherical algorithm is shown in Fig. 7 in the principal cuts. The pattern

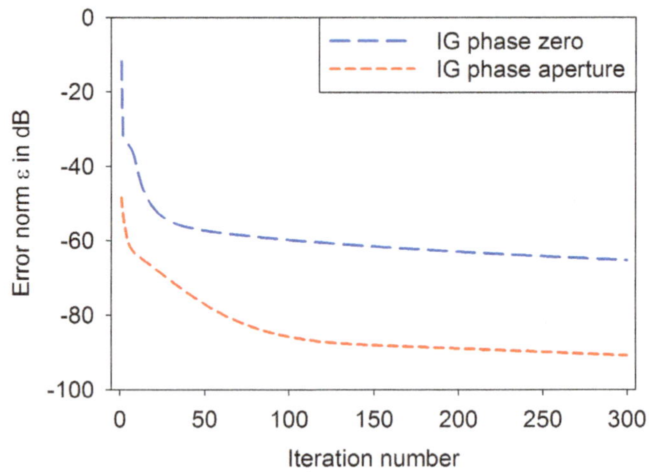

Fig. 8. Error norm ϵ of field amplitude error on first measurement sphere for phaseless spherical algorithm utilising simulated amplitude only near-field values on two spheres.

obtained by the second initial guess shows a much better performance than the first one. The green dotted curve shows the far-field pattern computed from the initial guess $\{Q^{\mathrm{IG}}\}$ employing the aperture phase distribution and demonstrates, how the far-field pattern is finally retrieved from the initial guess by the phaseless spherical algorithm resulting in the red short dashed curve. The error norm ϵ is plotted in Fig. 8 and shows a better performance for the initial guess employing the aperture phase distribution. This was expected since more a priori information of the AUT, in form of a simplified but good approximation of the phase distribution, was used to generate the initial guess.

5 Conclusions

A phaseless spherical near-field transformation algorithm has been presented. The algorithm determines the far-field pattern of an antenna from two amplitude only measurements on spheres of different radii in an iterative manner. Due to the spherical measurement configuration, the algorithm is not restricted to a particular class of antennas. Two different initial guesses for the mode expansion coefficients as starting point for the algorithm have been considered and evaluated. Simulated and measured results of a low gain patch antenna as well as a medium gain horn antenna have been presented, showing the applicability of the phaseless spherical algorithm for these types of antennas.

Acknowledgements. The authors are grateful to Shenheng Xu, Electrical Engineering Department UCLA, for fruitful discussions.

This work was supported in part by a Ph.D. research fellowship of the German Academic Exchange Service (DAAD).

References

Balanis, C. A.: Modern Antenna Handbook, John Wiley & Sons, Inc., 2008.

Costanzo, S., Massa, G. D., and Migliore, M. D.: A Novel Hybrid Approach for Far-Field Characterization From Near-Field Amplitude-Only Measurements on Arbitrary Scanning Surfaces, IEEE Trans. Antennas Propag., 53(6), 1866–1874, 2005.

FEKO Suite 5.5, http://www.feko.info, access: 2010.

Hansen, J. E.: Spherical Near-Field Antenna Measurements, IEE Electromagnetic Wave Series 26, Exeter, UK, 1988.

Hindman, G. and Fooshe, D. S.: Probe Correction Effects on Planar, Cylindrical and Spherical Near-Field Measurements, Antenna Measurement Techniques Association Conference, 1998.

Kerns, D. M.: Plane-Wave Scattering-Matrix Theory of Antennas and Antenna-Antenna Interactions, National Bureau of Standards, Boulder CO, 1981.

Laitinen, T. and Breinbjerg, O.: A First/Third-Order Probe Correction Technique for Spherical Near-Field Antenna Measurements Using Three Probe Orientations, IEEE Trans. Antennas Propag., 56(5), 1259–1268, 2008.

Laitinen, T., Pivnenko, S., and Breinbjerg, O.: Iterative Probe Correction Technique for Spherical Near-Field Antenna Measurements, IEEE Antennas Wireless Propag. Lett., 4, 221–223, 2005.

Las-Heras, F. and Sarkar, T.: A Direct Optimization Approach for Source Reconstruction and NF-FF Transformation Using Amplitude-Only Data, IEEE Trans. Antennas Propag., 50(4), 500–510, 2002.

Leibfritz, M. and Landstorfer, F.: Full Probe Correction for Near-Field Antenna Measurements, Proc. IEEE AP-S Int. Symp., Albuquerque, NM, 2006.

Nearfield Systems Inc.: http://www.nearfield.com, access: 2010.

Pierri, R., D'Elia, G., and Soldovieri, F.: A Two Probes Scanning Phaseless Near-Field Far-Field Transformation Technique, IEEE Trans. Antennas Propag., 47(5), 792–802, 1999.

Razavi, S. F. and Rahmat-Samii, Y.: Phaseless Planar Near Field Measurements for Scanned Beams: Difficulties, a Hybrid Solution and Measured Results, Proc. IEEE AP-S Int. Symp., Albuquerque, NM, 2006.

Razavi, S. F. and Rahmat-Samii, Y.: On The Robustness Of Planar Phaseless Near-Field Measurements To Probe Positioning Errors, AMTA, Boston, MA, 2008.

Schejbal, V., Kovarik, V., and Cermak, D.: Synthesized-Reference-Wave Holography for Determining Antenna Radiation Characteristics, IEEE Antenn. Propag. M., 50(5), 71–83, 2008.

Schmidt, C. H. and Rahmat-Samii, Y.: Phaseless Spherical Near-Field Antenna Measurements: Concept, Algorithm and Simulation, Proc. IEEE AP-S Int. Symp., Charleston, SC, 2009.

Soldovieri, F., Liseno, A., D'Elia, G., and Pierri, R.: Global Convergence of Phase Retrieval by Quadratic Approach, IEEE Trans. Antennas Propag., 53(10), 3135–3141, 2005.

Yaccarino, R. and Rahmat-Samii, Y.: Phaseless Bi-Polar Planar Near-Field Measurements and Diagnostics of Array Antennas, IEEE Trans. Antennas Propag., 47(3), 574–583, 1999.

Yaghjian, A. D.: An Overview of Near-Field Antenna Measurements, IEEE Trans. Antennas Propag., 34(1), 30–45, 1986.

18

High resolution time of arrival estimation for a cooperative sensor system

C. Morhart and E. M. Biebl

Fachgebiet Höchstfrequenztechnik, Technische Universität München, Munich, Germany

Abstract. Distance resolution of cooperative sensors is limited by the signal bandwidth. For the transmission mainly lower frequency bands are used which are more narrowband than classical radar frequencies. To compensate this resolution problem the combination of a pseudo-noise coded pulse compression system with superresolution time of arrival estimation is proposed. Coded pulsecompression allows secure and fast distance measurement in multi-user scenarios which can easily be adapted for data transmission purposes (Morhart and Biebl, 2009). Due to the lack of available signal bandwidth the measurement accuracy degrades especially in multipath scenarios. Superresolution time of arrival algorithms can improve this behaviour by estimating the channel impulse response out of a band-limited channel view. For the given test system the implementation of a MUSIC algorithm permitted a two times better distance resolution as the standard pulse compression.

1 Introduction

For security systems and a number of logistic applications the combination of clear identification and exact localization is a vital task. Actions and reactions can only be taken if the classification of the target is assured. Cooperative sensors are therefore a key technology as the use of active target transponders offers an easy identification by the transponder signal. Furthermore, the communication between cooperative sensors can even be established in difficult communication scenarios when there's no line of sight. This property is dependent on the transmission frequency and it was shown (Fackelmeier et al., 2008) that lower frequencies are advantageous for the detection of visually hidden objects. Nevertheless, bandwidth is limited for lower frequencies like ISM bands, which are additionaly used by other applications like data transmission. In Morhart and Biebl (2009) a distance measurement system was presented for the protection of vulnerable road users in city scenarios. With this approach in the 2.4 GHz band highly accurate results for line of sight scenarios could be achieved. On the other side, this system offered a path resolution of only 5 m unidirectionally and respectively 2.5 m bidirectionally. For the time of arrival (ToA) estimation, the fastest path is also the most important one. Thus, in multipath scenarios with limited resolution a distance error can result because of unresolvable signal arrivals. This is especially disadvantageous in non line of sight scenarios where several comparable strong transmission links occur. A possibility to improve this bandwidth problem is with superresolution ToA estimation (Krim and Viberg, 1996). Here, the complete channel impulse response is predicted out of a bandlimited channel view by superimposing a specific channel model. It has been shown in (Manolakis et al., 2000) that especially eigenstructure based methods like MUSIC and ESPRIT show superior performance. In combination with biphase modulated pulse compression the approach of (Dumont et al., 1994) is advantageous. Therein a MUSIC implementation for a CDMA based system is described. This method could easily be adapted for the cooperative distance measurement system from Morhart and Biebl (2009).

2 System model

The base system for the implementation of the MUSIC superresolution algorithm is the pedestrian protection system of Morhart and Biebl (2009). This approach uses a Round Trip Time of Flight principle in combination with bi-phase modulated pulse compression for improved measurement accuracy. For a better understanding of this method it is reasonable to repeat the main principles.

Correspondence to: C. Morhart
(morhart@tum.de)

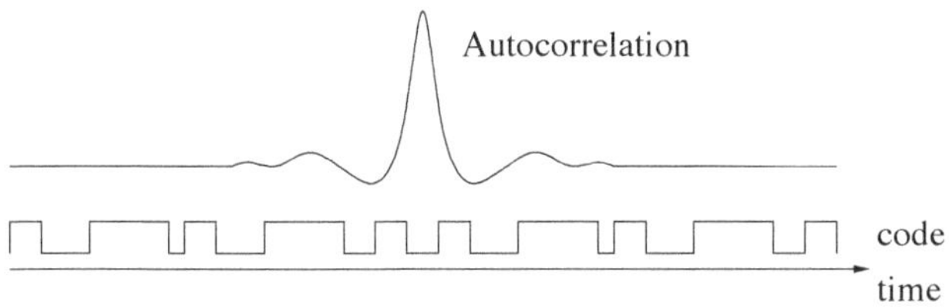

Fig. 1. Schematic capture of biphase modulated pulse compression.

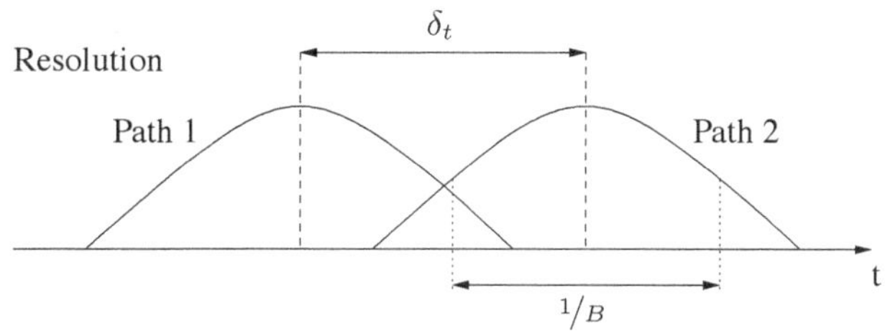

Fig. 2. Resolution between 2 multipath with different time delay.

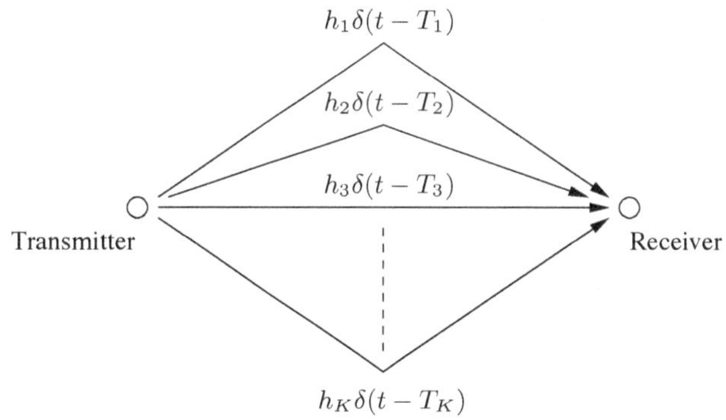

Fig. 3. Channel model.

2.1 Round trip time of flight

The round trip time of flight (RToF) method resembles the classical radar principle. A signal transmitted by the car is received at the pedestrian side and sent back after a finite waiting time T_w. By knowing the waiting time the car can compute the distance Δs out of the time T_p passed since the start of transmission.

$$\Delta s = \frac{\Delta T}{2} c_0 = \frac{T_\mathrm{p} - T_\mathrm{w}}{2} c_0 \tag{1}$$

The advantage of this method is that no absolute time synchonization is needed but only a relative one to determine the exact delay time.

2.2 Pulse compression

High precision distance measurement and insensitivity to noise require a high Signal to Noise Ratio (SNR). With limited output power a possible way to increase this ratio is by introducing pulse compression. This method raises the pulse energy by spreading signal information in time domain at constant amplitude and bandwidth (Skolnik, 2001). The energy-gain is due to a larger time bandwidth product that results equivalently in a slower measurement process compared to a single pulse system. An additional advantage of pseudo noise coded pulse compression is a differentiation of multi-user by applying different transmit codes. The ToA is then determined by correlating the received signal with the appropriate code (cf. Fig. 1). In that way the energy of the code sequence is compressed onto the autocorrelation peak, leading to a SNR gain in the ratio of the compression factor.

2.3 Multipath resolution

A key property of cooperative sensor technology is that target interference can be avoided by using signal multiplexing techniques like time division multiple access or frequency division multiple access. Nevertheless, each signal of an active transponder can still generate multipath interference due to reflections and diffractions on obstacles in the transmission path. The superposition of these signals leads to different correlation maxima in a pulse compression. The minimum distance between two signals which can still be distinguished is called multipath resolution. In the classical approach the ToA is defined by the correlation peak. Thus, the width of correlation maximum is proportional to the resolution (cf. Fig. 2). For the case of pulse modulated signal transmission, this resolution can be estimated by the pulse bandwidth (Haykin, 1995):

$$\delta_\mathrm{t} = \frac{1}{B} \tag{2}$$

As signal bandwidth is limited for lower frequency bands, classical multipath resolution is comparatively large.

2.4 Channel model

For signal transmission in city scenarios, multipath propagation because of reflection and diffraction on obstacles has to be regarded (cf. Fig. 3). In dependency on the surroundings there are K multipaths having different time delay $T_k \in \mathbb{R}_0^+$ and different path attenuation $h_k \in \mathbb{C}$. Therefore, the received signal consists of K replicas of the same signal. Accordingly, the continuous channel impulse response is defined by:

$$h(t) = \sum_{k=0}^{K} h_k \delta(t - T_k) \tag{3}$$

The complex channel impulse factor h_k incorporates the channel attenuation and the channel phase variation, which can be assumed randomly distributed between $[0, 2\pi]$.
This channel transfer function can be described as a sum of complex exponentials in frequency domain:

$$H(f) = \sum_{k=0}^{K} |h_k| \mathrm{e}^{-j(2\pi f T_k + \angle h_k)} \tag{4}$$

Different path delays T_k correspond to different oscillation periods, whereas the oscillation phase is randomly distributed. With this channel model a time domain peak search can be changed into the determination of the oscillation period in frequency domain. For this task several superresolution algorithms are available (Krim and Viberg, 1996). For the pedestrian protection system the MUSIC algorithm was chosen because of its superior performance even with low SNR.

2.5 MUSIC Algorithm

The MUSIC Algorithm is based on a signal model consisting of a sum of harmonic signals and noise (Schmidt, 1986).

$$x[n] = \sum_{k=0}^{K} \alpha_k e^{j\omega_k n} + \eta[n] \tag{5}$$

Each harmonic function is characterized by a complex amplitude $\alpha_k \in \mathbb{C}$ and an oscillation frequency $\omega_k \in \mathbb{R}$. $\eta[n]$ corresponds to the additional noise. This relation can be expressed in matrix-vector notation as:

$$\boldsymbol{x} = \sum_{k=0}^{K} \alpha_k \boldsymbol{v}(\omega_k) + \boldsymbol{\eta} \tag{6}$$

$\boldsymbol{x} = [x[0], x[1], \ldots, x[N-1]]^T$ is the $N \times 1$ data vector, $\boldsymbol{v}(\omega) = [1, e^{j\omega}, \ldots, e^{j(N-1)\omega}]^T$ the $N \times 1$ complex phase vector and $\boldsymbol{\eta}$ the noise vector. By building the autocorrelation matrix $\mathbf{R}_{xx}$ the correlation result can be divided into the information part $\mathbf{R}_{ss}$ and the noise $\mathbf{R}_{\eta}$.

$$\mathbf{R}_{xx} = \mathbf{R}_{ss} + \mathbf{R}_{\eta} \tag{7}$$

$$\mathbf{R}_{ss} = \sum_{k=0}^{K} |h_k|^2 \boldsymbol{v}(\omega_k) \boldsymbol{v}^H(\omega_k) \tag{8}$$

The matrix dimension for $\mathbf{R}_{xx}$ is $P \times P$ with $P \leq N$. $P = N$ means maximum resolution, whereas $P \ll N$ means more redundancy and less computational costs. The noise correlation matrix in the case of white noise is a diagonal-matrix with

$$\mathbf{R}_{\eta} = \sigma_{\eta}^2 \mathbf{E}. \tag{9}$$

In the case of $K < P$, The eigenvalue decomposition results in K signal and noise eigenvalues and $P - K$ pure noise eigenvalues.

$$\lambda_p = P|h_p|^2 + \sigma_{\eta}^2 \qquad \text{for} \quad p \leq K \tag{10}$$

$$\lambda_p = \sigma_{\eta}^2 \qquad \text{for} \quad K < p \leq P \tag{11}$$

According to these eigenvalues there are P signal- and $P-K$ noise-eigenvectors.

$$\mathbf{Q}_s = [\boldsymbol{q}_1, \boldsymbol{q}_2, \ldots, \boldsymbol{q}_K]^T \tag{12}$$

$$\mathbf{Q}_{\eta} = [\boldsymbol{q}_{K+1}, \ldots, \boldsymbol{q}_P]^T \tag{13}$$

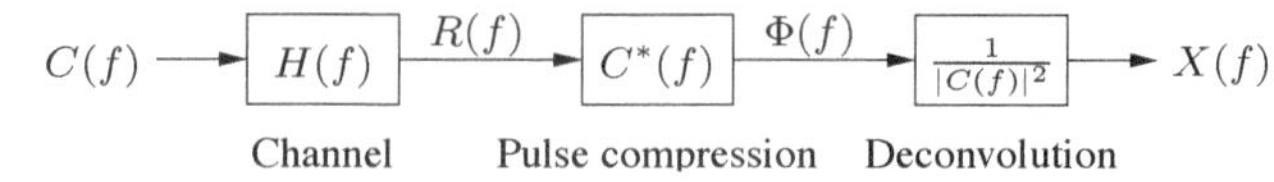

Fig. 4. Idealized schematic of signal processing chain.

Because of the conjugate transposed structure of the correlation matrix all eigenvectors are orthogonal, i.e. $\mathbf{Q}_s \perp \mathbf{Q}_{\eta}$. Therefore signal information $h_k \boldsymbol{v}(\omega_k)$ is a linear combination of $\mathbf{Q}_s$ and orthogonal to $\mathbf{Q}_{\eta}$. This property can be used to determine the signal frequencies ω_k. The magnitude of the scalar product with a noise eigenvector $|\boldsymbol{v}(\omega) \cdot \boldsymbol{q}_{\eta}|$ is minimum if $\omega = \omega_k$ and $\boldsymbol{v}(\omega_k)$ a signal vector. Thus, the MUSIC pseudospectrum can be defined as the reciprocal of the sum of scalar products with all noise eigenvectors.

$$R(\omega) = \frac{1}{\sum_{p=K+1}^{P} |\boldsymbol{v}(\omega) \cdot \boldsymbol{q}_p|^2} \tag{14}$$

3 System implementation

Harmonic functions plus noise are the initial point of the MUSIC algorithm. This model matches with the channel transfer function in frequency domain. Thus, for the implementation of this superresolution algorithm a reconstruction of the channel transfer function is necessary. In a bi-phase modulated pulse compression system this is possible by deconvolving with the autocorrelation function. Figure 4 shows an illustration for a baseband transmission without nonlinearities and model errors. Following, the stages of ToA determination will be explained.

3.1 Correlation

Pulse compression or equivalently signal correlation can be carried out in time or frequency domain. In time domain correlation can be achieved with a FIR Filter structure, in frequency domain with a scalar product of frequency vectors.

$$\varphi[n] = \sum_{N} c[n] r^*[n+N] \tag{15}$$

$$\Phi[m] = C[m] R^*[m] \tag{16}$$

The latter one is computational cost-saving especially for long code sequences. The SNR of the correlation output is also dependent on the code length L_c.

$$\mathrm{SNR}_k = \frac{2|h_k|^2 L_c E_\mathrm{p}}{N_0} \tag{17}$$

E_p is the pulse energy and N_0 the noise power density. For maximum target dynamic it is important to choose codes with minimum correlation side lobes like m-sequences and Barker codes.

3.2 Deconvolution

For optimum performance of superresolution algorithms a maximum similarity between real communication and the chosen signal model is needed. The pure correlation result is therefore not suited as additional components like correlation pulse form, spectral filter characteristics or hardware nonlinearities are also included. The goal is to cancel those parts out of the received signal. This method is called Deconvolution and carried out by a division with the correlation power density.

$$X[m] = \frac{\Phi[m]}{|C[m]|^2} \approx H[m] \qquad (18)$$

With the neglect of model nonidealities, this result matches with the channel impulse response. In a real implementation not only the correlation pulse form but also spectral filter charactersitics should be compensated. In this context, digital filters are advantegeous as the filter coefficients can be exactly known on transmitter and receiver side.

3.3 MUSIC implementation

For the computation of the MUSIC pseudospectrum the parameter P and K have to be determined. P describes the dimension of $\mathbf{R}_{xx}$ and is therefore a degree for the frequency aperture of the algorithm (Manolakis et al., 2000). Thus larger values of P have better resolution capabilities whereas smaller values result in more signal averaging and less variations in the determination of noise eigenvalues. Another convenient characteristic of smaller values of P is a decreasing computational effort. In the literature there's no optimal solution for the relation of P to N with values ranging from $P \in [N/3, 3N/5]$ (Li and Pahlavan, 2004). With regard to signal processing costs for mobile transponder $P = N/3$ was chosen.

Another important point is the estimation of the included signal components K. This can be done with the result of the eigenvalue decomposition. In the case of an adequate SNR ratio, signal eigenvalues outrange noise eigenvalues. Hence, there's a detectable step between signal and noise subspace. For the following section, especially the case $K = 2$ was examined.

4 Verification

The benefit of superresolution algorithms for cooperative sensor technology lies in an increase of ToA resolution. This improvement is possible because of the reconstruction of a limited channel view with an assumed signal model. Nevertheless the detected channel information and thus the signal bandwidth has still a deciding influence on the quality of this method.

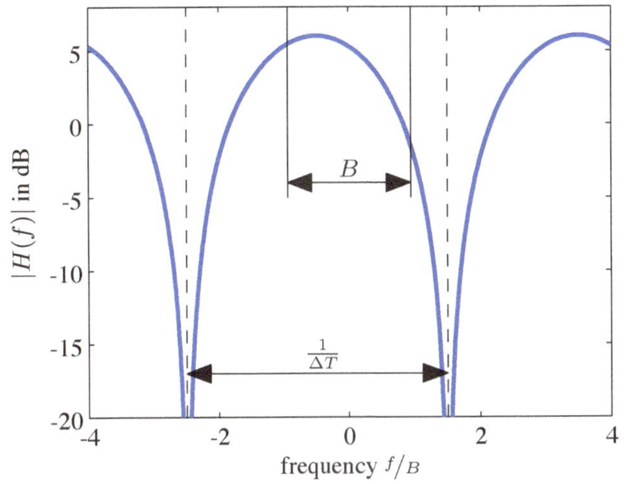

Fig. 5. Magnitude of $|H(f)|$ for time difference $\Delta T = 1/4\delta_t$ and $\Delta\varphi = -\pi/4$.

4.1 Channel information

At first the available channel information should be analyzed. For superresolution ToA determination only scenarios with at least two transmission paths and time difference ΔT smaller than the classical resolution δ_t are interesting. Figure 5 shows the spectral magnitude of a channel transfer function for the case of a two path transmission with a time difference of $\Delta T = 1/4\delta_t$ and a phase difference of $\Delta\varphi = -\pi/4$. The detectable channel spectrum is limited by the transmission bandwidth B, so that only a small section of the channel information is visible. This section is furthermore dependent on the phase difference which can be assumed randomly distributed between $\Delta\varphi \in [0, 2\pi]$. As a consequence the superresolution results are connected to the position of the channel extract. If concentrated at a fading point, the SNR and thus the quality of channel reconstruction is degrading. Hence, the influence of the SNR on the MUSIC resolution was investigated.

4.2 Simulation of multipath resolution

For the determination of the resolution boundary a simulation with a two path transmission with variable time difference and SNR was executed. Transmission parameters were chosen according to the proposed cooperative sensor system from Morhart and Biebl (2009) with 256 Bit long m-sequences and a two times oversampling receiver. The phase difference was assumed randomly distributed between $[0, 2\pi]$. Figure 6 shows the resolution boundary for the MUSIC and the correlation ToA determination. The correlation boundary is according to Eq. (2) at $\Delta T = \delta_t$ with two resolved paths for $\Delta T > \delta_t$.

The situation is different for the MUSIC algorithm. Simulation results for the resolution boundary show a relation

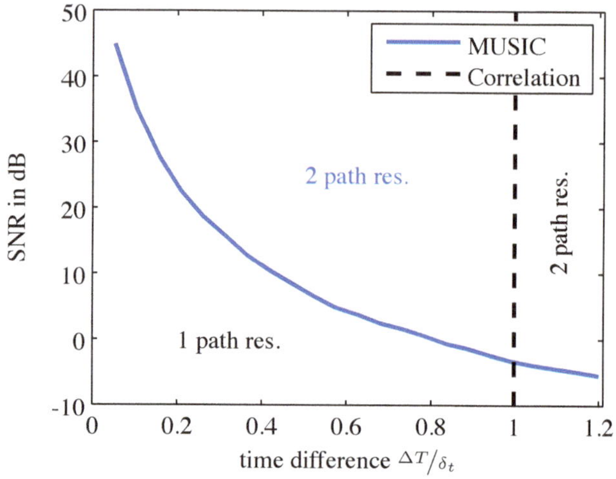

Fig. 6. Simulation of resolution boundary for two path transmission with variable time difference ΔT and SNR.

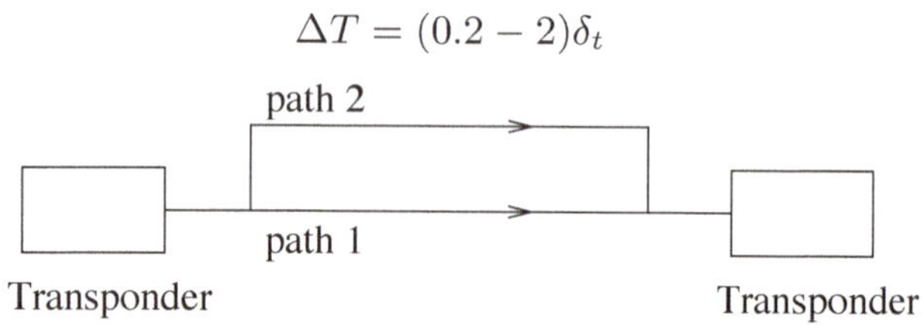

Fig. 7. Measurement assembly for the determination of resolution boundary.

between the time difference and the SNR. Summarized, 2 paths can even be resolved below the classical resolution boundary if the SNR is sufficient. Unfortunately, for too small time differences huge SNR values are necessary, what makes a resolution unlikely. The big advantage for a pulse compression system is that ToA resolution together with the SNR can be adapted with the length of the code sequence.

4.3 Test measurement

To check out the performance of the MUSIC algorithm for a cooperative sensor system test measurements were done with a prototype hardware. Two transponders were used which were connected via coaxial cable (cf. Fig. 7). The signal was divided into two transmission paths with variable time difference ΔT. This factor was modified by changing the cable length difference. Furthermore, to equal amplitude levels additional attenuation elements were used into the shorter path. On the receiver side the correlation result was passed onto a PC to compute the superresolution method. This is necessary because the MUSIC algorithm was not implemented on the receiver FPGA yet. In Fig. 8 the measured time difference between the two paths is illustrated. If they can not be distinguished the difference drops to zero. It can be seen, that MUSIC post processing offers a more than two times better path resolution compared with the correlation result. It is also remarkable that correlation does not reach the theoretical resolution limit. This is due to pulse widening in the transmission process. Another problem of the actual prototype are nonidealities and model errors because of uncompensated filter and amplifier charateristics. Even though superresolution post processing is able to significantly improve the ToA resolution.

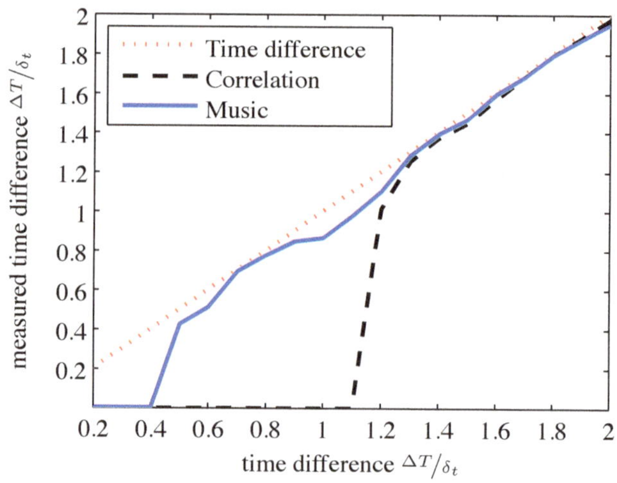

Fig. 8. Measurement of signal resolution for two path transmission with variable time difference.

5 Conclusions

Cooperative sensor systems have to use the same signal bands as data communication systems. Bandwidth in these bands is tightly limited resulting equivalently in a low path resolution. The application of superresolution methods is therefore a suitable alternative for the improvement of ToA determination. In the simulation it could be shown that the resolution of the MUSIC algorithm is limited by the SNR and the time difference. For the proof of this fact a MUSIC ToA algorithm was implemented into a existing cooperative sensor system. Measurement results demonstrated a more than two times higher path resolution compared with the standard pulse compression although a non ideal sensor hardware has been used. The next steps are therefore a minimization of model errors by improving hardware linearity and an implementation of the MUSIC algorithm on the transponder FPGA.

References

Dumont, L., Fattouche, M., and Morrison, G.: Super-resolution of multipath channels in a spread spectrum location system, Electronics Lett., 30, 1583–1584, 1994.

Fackelmeier, A., Morhart, C., and Biebl, E.: Evaluation of Diffraction Effects for Identifying Hidden Targets, in: GeMIC 2008 ITG-Fachbericht Band 206, 2008.

Haykin, S.: Adaptive Filter Theory, Prentice Hall, 1995.

Krim, H. and Viberg, M.: Two decades of array signal processing research: the parametric approach, Signal Processing Magazine, IEEE, 13, 67–94, 1996.

Li, X. and Pahlavan, K.: Super-resolution TOA estimation with diversity for indoor geolocation, Wireless Communications, IEEE Transactions on, 3, 224–234, 2004.

Manolakis, D. G., Ingle, V. K., and Kogon, S. M.: Statistical and Adaptive Signal Processing: Spectral Estimation, Signal Modeling, Adaptive Filtering, and Array Processing, McGraw Hill, New York, 2000.

Morhart, C. and Biebl, E. M.: Cooperative multi-user detection and ranging based on pseudo-random codes, Adv. Radio Sci., 7, 55–59, 2009, http://www.adv-radio-sci.net/7/55/2009/.

Schmidt, R.: Multiple emitter location and signal parameter estimation, Antennas and Propagation, IEEE Transactions on, 34, 276–280, 1986.

Skolnik, M. I.: Introduction to RADAR systems, chap. 6, pp. 317–331, McGraw-Hill, New York, 3. edn., 2001.

Sensitivity analysis of waveguide eigenvalue problems

N. Burschäpers[1], S. Fiege[2], R. Schuhmann[1], and A. Walther[2]

[1]Universität Paderborn, EIM, Fachgebiet Theoretische Elektrotechnik, Germany
[2]Universität Paderborn, EIM, Lehrstuhl für Mathematik und ihre Anwendungen, Germany

Abstract. We analyze the sensitivity of dielectric waveguides with respect to design parameters such as permittivity values or simple geometric dependencies. Based on a discretization using the Finite Integration Technique the eigenvalue problem for the wave number is shown to be non-Hermitian with possibly complex solutions even in the lossless case. Nevertheless, the sensitivity can be obtained with negligible numerical effort. Numerical examples demonstrate the validity of the approach.

1 Introduction

Numerical simulations of waveguide structures by finite methods have been used for many years, and a number of eigenvalue formulations are available, based, e.g., on the Finite Integration Technique (FIT, Weiland, 1977), or the Finite Element method (FE, e.g. Farle et al., 2004). If the calculations aim mainly at the analysis of port planes of originally 3-D setups, the number of unknowns in the 2-D waveguide model is typically low, and results are available within seconds. However, the situation changes for optical applications such as fibers or integrated waveguides, where due to the small wavelengths in the optical regime the problem sizes (of the two-dimensional discrete model) sometimes exceed several 10^6 unknowns. If additionally the dependencies of the waveguide modes w.r.t. design parameters such as geometric dimensions or material parameters are searched for, it is desirable to have sophisticated approaches for fast parameter sweeps at hand.

Several such approaches have been reported in the context of Model Order Reduction (MOR) techniques. In most cases they are based on projections of the system matrices by low-order projection matrices, with the frequency as the main design parameter. Recently, some effort has been taken to extend them to the so-called multi-variate case (Farle et al., 2006; Stavrakakis et al., 2009). The application to 2-D eigenvalue problems seems to be straight-forward, although only a few such attempts have been published (Polstyanko et al., 1997; Schultschik et al., 2008).

A different approach, the so-called sensitivity analysis of electromagnetic systems using adjoint techniques (Nikolova et al., 2004), has recently gained large interest. Starting with analytical differentiations of the algebraic matrix equations, compact formulas can be derived for the sensitivities of output quantities w.r.t. an arbitrary number of design parameters. Adjoint techniques have been applied to various formulations in electromagnetic modeling, but to our knowledge not yet to 2-D waveguide eigenvalue problems.

In this paper we apply a classical sensitivity analysis to the eigenvalue problem arising from a 2-D FIT-discretization of inhomogeneous dielectric waveguides. In Sect. 2 we show that the system matrix is non-Hermitian, supporting complex modes also in the lossless case. This has some consequences for the adjoint technique to be applied which will be addressed in Sect. 3, where we also discuss an efficient Algorithmic Differentiation (AD) method for the matrix derivatives. Some numerical examples in Sect. 4 demonstrate the validity and efficiency of the approach.

2 Waveguide eigenvalue problem using FIT

We consider a cross section of a dielectric waveguide (e.g., an optical fiber) and use the Finite Integration Technique, FIT, Weiland, 1977, 1996, for the discretization of Maxwell's Equations in frequency domain. For sake of simplicity, a standard Cartesian mesh is used and PEC boundary conditions are imposed at a distance far enough from the core region. The derivation of the resulting eigenvalue problem has been described in detail in Weiland (1977); Schuhmann and Weiland (2001) and is only briefly revisited here.

The state variables of FIT are integral quantities which are defined on edges $L_i, \widetilde{L}_i$ and facets $A_i, \widetilde{A}_i$ of the primary grid

Correspondence to: N. Burschäpers
(burschaepers@tet.upb.de)

G and the dual grid $\widetilde{G}$, respectively. These are the grid voltages

$$\widehat{e}_i = \int_{L_i} \boldsymbol{E}(\boldsymbol{r},t) \cdot d\boldsymbol{s}, \quad \widehat{h}_i = \int_{\widetilde{L}_i} \boldsymbol{H}(\boldsymbol{r},t) \cdot d\boldsymbol{s}, \tag{1}$$

and the grid fluxes (omitting currents)

$$\widehat{\widehat{d}}_i = \int_{\widetilde{A}_i} \boldsymbol{D}(\boldsymbol{r},t) \cdot d\boldsymbol{A}, \quad \widehat{\widehat{b}}_i = \int_{A_i} \boldsymbol{B}(\boldsymbol{r},t) \cdot d\boldsymbol{A}. \tag{2}$$

Using these definitions, Maxwell's equations (neglecting currents and charges) are transformed into a set of matrix-vector-equations for the component vectors $\widehat{\mathbf{e}}$, $\widehat{\widehat{\mathbf{d}}}$, $\widehat{\mathbf{h}}$, $\widehat{\widehat{\mathbf{b}}}$,

$$\mathbf{C}\,\widehat{\mathbf{e}} = -j\omega\,\widehat{\widehat{\mathbf{b}}}, \qquad \widetilde{\mathbf{C}}\,\widehat{\mathbf{h}} = j\omega\,\widehat{\widehat{\mathbf{d}}}, \tag{3}$$

$$\mathbf{S}\,\widehat{\widehat{\mathbf{b}}} = \mathbf{0}, \qquad \widetilde{\mathbf{S}}\,\widehat{\widehat{\mathbf{d}}} = \mathbf{0}, \tag{4}$$

which are referred to as *Maxwell's Grid Equations*. The matrix $\mathbf{C}$ is the discrete curl-operator, $\mathbf{S}$ is the discrete div-operator of the primary grid, and $\widetilde{\mathbf{S}}$ and $\widetilde{\mathbf{C}}$ are the corresponding operators for the dual grid. In Cartesian grid systems with N_{P} primary nodes these are structured matrices, e.g.

$$\mathbf{C} = \begin{pmatrix} \mathbf{0} & -\mathbf{P}_{\mathrm{z}} & \mathbf{P}_{\mathrm{y}} \\ \mathbf{P}_{\mathrm{z}} & \mathbf{0} & -\mathbf{P}_{\mathrm{x}} \\ -\mathbf{P}_{\mathrm{y}} & \mathbf{P}_{\mathrm{x}} & \mathbf{0} \end{pmatrix}, \quad \widetilde{\mathbf{S}} = \left(-\mathbf{P}_{\mathrm{x}}^T \; -\mathbf{P}_{\mathrm{y}}^T \; -\mathbf{P}_{\mathrm{z}}^T\right), \tag{5}$$

and the $N_{\mathrm{P}} \times N_{\mathrm{P}}$-blocks $\mathbf{P}_{\mathrm{x}}$, $\mathbf{P}_{\mathrm{y}}$, $\mathbf{P}_{\mathrm{z}}$ can be identified as discrete partial differentiation operators (Weiland, 1996). From grid topology we find the relations $\mathbf{SC} = \mathbf{0}$, $\widetilde{\mathbf{S}}\widetilde{\mathbf{C}} = \mathbf{0}$ and $\widetilde{\mathbf{C}} = \mathbf{C}^T$. The formulation is completed by the *material relations* (for linear media)

$$\widehat{\widehat{\mathbf{d}}} = \mathbf{M}_{\varepsilon}\,\widehat{\mathbf{e}}, \qquad \widehat{\mathbf{h}} = \mathbf{M}_{\mu}^{-1}\,\widehat{\widehat{\mathbf{b}}}. \tag{6}$$

Both material matrices $\mathbf{M}_{\varepsilon}$ and $\mathbf{M}_{\mu}^{-1}$ are diagonal and may be complex (in frequency domain) to account for dielectric or magnetic losses. Their entries contain the locally averaged permittivity and permeability distribution as well as the metrics of the grid.

For the discretization of waveguide cross sections we use a two-dimensional Cartesian grid with $N_{\mathrm{P}} = N_{\mathrm{x}} \cdot N_{\mathrm{y}}$ primary nodes and assume a wave propagation $\boldsymbol{E}, \boldsymbol{H} \sim e^{-jk_z z}$ in z-direction, with k_z the (unknown) wave number. Thus, the longitudinal differential operator becomes

$$\mathbf{P}_{\mathrm{z}} = -jk_z\mathbf{I} \tag{7}$$

with the identity matrix $\mathbf{I}$. Note, that applying Eq. (7) to (5) results in complex-valued matrices, which are only valid for one distinct k_z. Besides, the transposed sub-matrices in $\widetilde{\mathbf{C}}$ and $\widetilde{\mathbf{S}}$ have to be replaced by the Hermitian expression $\mathbf{P}_{\mathrm{z}}^{\mathrm{H}} = +jk_z\mathbf{I}$.

In order to derive an eigenvalue formulation for the modes in such waveguide cross sections, we start with the discrete curl-curl eigenproblem and the divergence-free condition of the fields

$$\mathbf{C}^T\mathbf{M}_{\mu}^{-1}\mathbf{C}\,\widehat{\mathbf{e}} = \omega^2\mathbf{M}_{\varepsilon}\,\widehat{\mathbf{e}} \tag{8}$$

$$\mathbf{0} = \widetilde{\mathbf{S}}\,\widehat{\widehat{\mathbf{d}}} = -\mathbf{P}_{\mathrm{x}}^T\,\widehat{\widehat{\mathbf{d}}}_{\mathrm{x}} - \mathbf{P}_{\mathrm{y}}^T\,\widehat{\widehat{\mathbf{d}}}_{\mathrm{y}} - jk_z\,\widehat{\widehat{\mathbf{d}}}_{\mathrm{z}} \tag{9}$$

Equation (9) allows to eliminate the longitudinal components from the eigenvalue equation, since the expression (7) as well as the material matrices can be easily inverted:

$$\widehat{\mathbf{e}}_{\mathrm{z}} = \frac{1}{jk_{\mathrm{z}}}\mathbf{M}_{\varepsilon z}^{-1}(-\mathbf{P}_{\mathrm{x}}^T\mathbf{M}_{\varepsilon x}\,\widehat{\mathbf{e}}_{\mathrm{x}} - \mathbf{P}_{\mathrm{y}}^T\mathbf{M}_{\varepsilon y}\,\widehat{\mathbf{e}}_{\mathrm{y}}). \tag{10}$$

Finally a $2N_{\mathrm{P}} \times 2N_{\mathrm{P}}$-eigenvalue problem can be derived:

$$(\mathbf{A}_{\mathrm{E,2\text{-}D}} - \omega^2\mathbf{B}_{\mathrm{E,2\text{-}D}} + k_{\mathrm{z}}^2\mathbf{I})\begin{pmatrix}\widehat{\mathbf{e}}_{\mathrm{x}} \\ \widehat{\mathbf{e}}_{\mathrm{y}}\end{pmatrix} = \mathbf{0}, \tag{11}$$

$$\mathbf{A}_{\mathrm{E,2\text{-}D}} = \begin{pmatrix}\mathbf{A}_{11} & \mathbf{A}_{12} \\ \mathbf{A}_{21} & \mathbf{A}_{22}\end{pmatrix}$$

$$\begin{aligned} \text{with} \quad \mathbf{A}_{11} &= \mathbf{P}_{\mathrm{x}}\mathbf{M}_{\varepsilon z}^{-1}\mathbf{P}_{\mathrm{x}}^T\mathbf{M}_{\varepsilon x} + \mathbf{M}_{\mu y}\mathbf{P}_{\mathrm{y}}^T\mathbf{M}_{\mu z}^{-1}\mathbf{P}_{\mathrm{y}}, \\ \mathbf{A}_{12} &= \mathbf{P}_{\mathrm{x}}\mathbf{M}_{\varepsilon z}^{-1}\mathbf{P}_{\mathrm{y}}^T\mathbf{M}_{\varepsilon y} - \mathbf{M}_{\mu y}\mathbf{P}_{\mathrm{y}}^T\mathbf{M}_{\mu z}^{-1}\mathbf{P}_{\mathrm{x}}, \\ \mathbf{A}_{21} &= \mathbf{P}_{\mathrm{y}}\mathbf{M}_{\varepsilon z}^{-1}\mathbf{P}_{\mathrm{x}}^T\mathbf{M}_{\varepsilon x} - \mathbf{M}_{\mu x}\mathbf{P}_{\mathrm{x}}^T\mathbf{M}_{\mu z}^{-1}\mathbf{P}_{\mathrm{y}}, \\ \mathbf{A}_{22} &= \mathbf{P}_{\mathrm{y}}\mathbf{M}_{\varepsilon z}^{-1}\mathbf{P}_{\mathrm{y}}^T\mathbf{M}_{\varepsilon y} + \mathbf{M}_{\mu x}\mathbf{P}_{\mathrm{x}}^T\mathbf{M}_{\mu z}^{-1}\mathbf{P}_{\mathrm{x}}, \end{aligned} \tag{12}$$

$$\mathbf{B}_{\mathrm{E,2\text{-}D}} = \begin{pmatrix}\mathbf{M}_{\varepsilon x}\mathbf{M}_{\mu y} & \mathbf{0} \\ \mathbf{0} & \mathbf{M}_{\varepsilon y}\mathbf{M}_{\mu x}\end{pmatrix} \quad \text{(diagonal).} \tag{13}$$

For a fixed angular frequency ω we obtain a simple, non-symmetric eigenproblem $(\mathbf{A} - \lambda\mathbf{I})\mathbf{x} = \mathbf{0}$ with the system matrix $\mathbf{A} = \mathbf{A}_{\mathrm{E,2\text{-}D}} - \omega^2\mathbf{B}_{\mathrm{E,2\text{-}D}}$ and the eigenpair

$$\lambda = -k_{\mathrm{z}}^2, \qquad \mathbf{x} = \begin{pmatrix}\widehat{\mathbf{e}}_{\mathrm{x}} \\ \widehat{\mathbf{e}}_{\mathrm{y}}\end{pmatrix}. \tag{14}$$

Note that $\mathbf{A}$ is non-symmetric also in the lossless case: This is not a property of our specific formulation but rather of the physical setting itself, since it is well-known that dielectric waveguides may support so-called complex modes. In the lossy case (with complex material matrices) the system is non-Hermitian and supports complex wave numbers $k_{\mathrm{z}} = \beta - j\alpha$. Actually, dielectric losses can be allowed in the following without any restriction (except for some more effort in the algebraic solver for the eigenproblem).

3 Sensitivity analysis

3.1 Eigenvalue perturbation theory

We are interested in the sensitivity of the eigensolutions with respect to a number of design parameters such as geometric dimensions or permittivity values. For simplicity of notation we restrict here to one single parameter p and calculate the derivatives

$$\lambda' = \frac{d\lambda}{dp}, \qquad \mathbf{x}' = \frac{d\mathbf{x}}{dp}. \tag{15}$$

The derivation makes use of the left-eigenvectors $\mathbf{y}$ of the system (the eigenvector of the Hermitian matrix $\mathbf{A}^{\mathrm{H}}$, hence: *adjoint technique*) with

$$\mathbf{y}^{\mathrm{H}}(\mathbf{A} - \lambda\mathbf{I}) = \mathbf{0} \tag{16}$$

and the orthonormality condition of right- and left-eigenvectors (i, j denoting the index number of the eigensolution)

$$\mathbf{y}^{(j)\mathrm{H}}\mathbf{x}^{(i)} = \delta_{ij} = \begin{cases} 1 & (i = j), \\ 0 & (i \neq j). \end{cases} \tag{17}$$

Following the standard perturbation theory for eigenproblems (e.g., Nelson, 1976) we build the derivative

$$\frac{\mathrm{d}}{\mathrm{d}p}\{(\mathbf{A}-\lambda\mathbf{I})\mathbf{x}=\mathbf{0}\}$$
$$\Rightarrow \quad (\mathbf{A}'-\lambda'\mathbf{I})\mathbf{x}+(\mathbf{A}-\lambda\mathbf{I})\mathbf{x}'=\mathbf{0}. \tag{18}$$

Multiplying from the left by $\mathbf{y}^{\mathrm{H}}$ yields, together with Eqs. (16) and (17), the desired eigenvalue sensitivity:

$$\lambda' = \mathbf{y}^{\mathrm{H}}\mathbf{A}'\mathbf{x}. \tag{19}$$

Once λ' has been calculated, Eq. (18) defines a linear system for the eigenvector derivative $\mathbf{x}'$. Its singularity can be easily removed using a normalization condition (Nelson, 1976), here defined again by $\mathbf{y}^{\mathrm{H}}\mathbf{x} = 1$.

To summarize, the ingredients to calculate the sensitivities are

1. the eigensolution $\{\mathbf{x}, \lambda\}$ itself, properly normalized,
2. the corresponding left-eigenvector $\mathbf{y}$,
3. the derivative of the system matrix $\mathbf{A}'$.

Note that some care must be taken in the previous steps in case of multiple eigensolutions (degenerated modes).

3.2 Calculation of the left-eigenvector

Unfortunately, in a non-Hermitian system as given here, the left- and right eigenvectors are not identical. A symmetrization of the system (like, e.g., in lossless 3-D eigenproblems) is generally not available, since this loss of symmetry is due to the physics of inhomogeneous waveguides rather than an inefficient formulation.

However, physical considerations again can help to avoid an additional solver step as shown in the following. The orthogonality property (17) between the searched left eigenvector and the original right eigenvector (the transversal electric field) suggest that $\mathbf{y}$ may be related to the magnetic field in the guide which fulfills the continuous orthogonality relation

$$\int_A \boldsymbol{E}^{(i)} \times \boldsymbol{H}^{(j)} \cdot \boldsymbol{n}_z \,\mathrm{d}A = \delta_{ij} \tag{20}$$

for the fields of two modes i and j. It has already been shown previously (Schuhmann and Weiland, 2001) that this type of orthogonality can be reproduced within the discrete setting by

$$\sum_n (\widehat{e}^{(i)}_{x,n}\widehat{h}^{(j)}_{y,n} - \widehat{e}^{(i)}_{y,n}\widehat{h}^{(j)}_{x,n})$$
$$= \begin{pmatrix} \widehat{\mathbf{h}}^{(j)}_{y} \\ -\widehat{\mathbf{h}}^{(j)}_{x} \end{pmatrix}^T \begin{pmatrix} \widehat{\mathbf{e}}^{(i)}_{x} \\ \widehat{\mathbf{e}}^{(i)}_{y} \end{pmatrix} = \delta_{ij}. \tag{21}$$

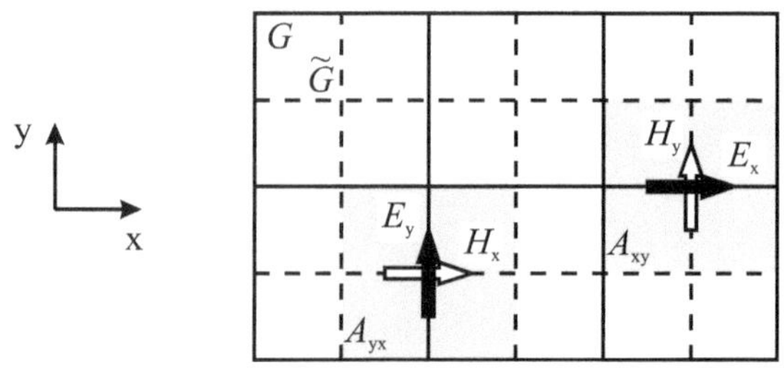

Fig. 1. Discrete evaluation of the $\boldsymbol{E} \times \boldsymbol{H}$-based orthogonality relation in the waveguide cross section. The voltages $\widehat{e}_x \approx E_x\Delta x$, $\widehat{e}_y \approx E_y\Delta y$ and $\widehat{h}_y \approx H_y\widetilde{\Delta y}$, $\widehat{h}_x \approx H_x\widetilde{\Delta x}$ include the lengths which define the integration areas A_{yx}, A_{yx} (shaded gray).

As shown in Fig. 1 this formula directly rebuilds the desired cross product in a Cartesian system, and the area integration is implicitly included due to the integral character of the grid voltages $\widehat{e}$ and $\widehat{h}$.

Obviously, the vector (now omitting the mode number j in the notation)

$$\mathbf{y} = \begin{pmatrix} \widehat{\mathbf{h}}^*_y \\ -\widehat{\mathbf{h}}^*_x \end{pmatrix} = \begin{pmatrix} \mathbf{0} & \mathbf{I} \\ -\mathbf{I} & \mathbf{0} \end{pmatrix}\begin{pmatrix} \widehat{\mathbf{h}}^*_x \\ \widehat{\mathbf{h}}^*_y \end{pmatrix} = \mathbf{K}\begin{pmatrix} \widehat{\mathbf{h}}^*_x \\ \widehat{\mathbf{h}}^*_y \end{pmatrix} \tag{22}$$

seems to be the left-eigenvector we are looking for. To prove this assumption we can derive the eigenvalue formulation for the magnetic field components in a similar way than above for the electric field: Starting with the curl-curl problem

$$\mathbf{C}\mathbf{M}_\varepsilon^{-1}\widetilde{\mathbf{C}}\,\widehat{\mathbf{h}} = \omega^2\mathbf{M}_\mu\,\widehat{\mathbf{h}} \tag{23}$$

we can use the divergence-free condition for the magnetic fluxes $\mathbf{S}\,\widehat{\widehat{\mathbf{b}}} = \mathbf{S}\mathbf{M}_\mu\,\widehat{\mathbf{h}} = \mathbf{0}$ together with Eq. (7) to eliminate the longitudinal $\widehat{\mathbf{h}}_z$ components. After some calculation and a rearrangement of the matrix blocks according to Eq. (22) we find

$$(\mathbf{A}_{\mathrm{H,2\text{-}D}} - \omega^2\mathbf{B}_{\mathrm{H,2\text{-}D}} + k_z^2\mathbf{I})\,\mathbf{y} = \mathbf{0}$$
$$\text{with } \mathbf{A}_{\mathrm{H,2\text{-}D}} = \mathbf{A}^{\mathrm{H}}_{\mathrm{E,2\text{-}D}} \quad \text{and} \quad \mathbf{B}_{\mathrm{H,2\text{-}D}} = \mathbf{B}^{\mathrm{H}}_{\mathrm{H,2\text{-}D}}, \tag{24}$$

which proves that $\mathbf{y}$ is the desired left-eigenvector. Besides the nice physical interpretation of this result (the transversal magnetic and electric fields of the modes constitute the left- and right-eigenvectors in this formulation), there is an even more important algebraic consequence: Although the matrix is non-symmetric, the left-eigenvector can be easily calculated without any additional solver step, simply by applying Faraday's law:

$$\widehat{\mathbf{h}} = \frac{1}{-j\omega}\mathbf{M}_\mu^{-1}\mathbf{C}\,\widehat{\mathbf{e}}, \qquad \mathbf{y} = \mathbf{K}\begin{pmatrix} \widehat{\mathbf{h}}^*_x \\ \widehat{\mathbf{h}}^*_y \end{pmatrix}. \tag{25}$$

Note that the z-components in $\widehat{\mathbf{e}}$ have to be calculated first from the eigenvector $\mathbf{x}$ by applying Eq. (10), and this step as well as the $\mathbf{C}$-matrix in Eq. (25) contain the (square-root of the) eigenvalue k_z.

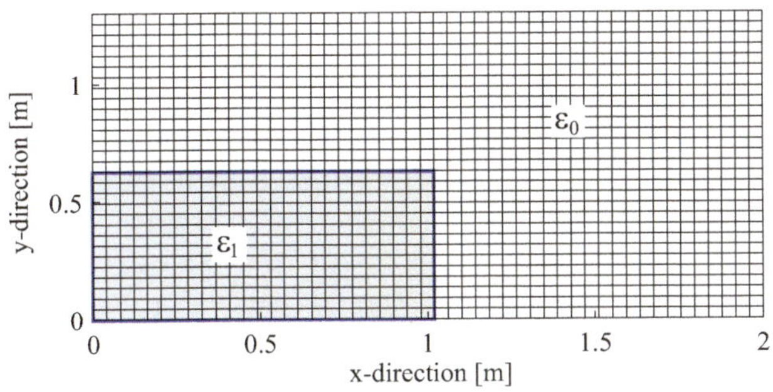

Fig. 2. Geometric data of a waveguide with dielectric inset.

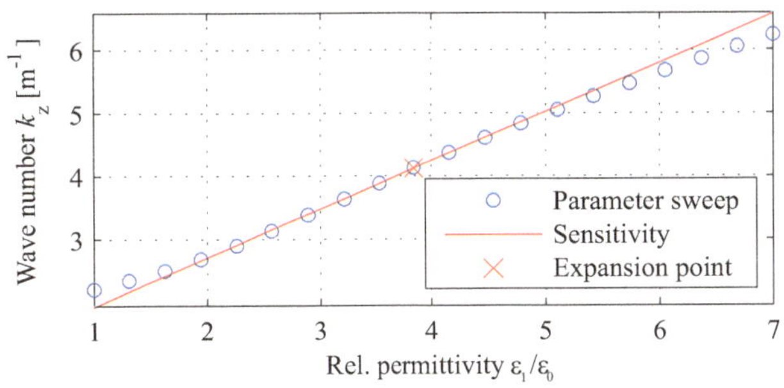

Fig. 3. Sensitivity analysis with regard to the permittivity.

3.3 Calculation of the matrix derivative

The next step is the calculation of the matrix derivative $\mathbf{A}' = \frac{\partial}{\partial p}\mathbf{A}$ in Eq. (19). We have implemented two variants: The first one depends on the type of parameter p and may only be available for simple constellations. If, e.g., the design parameter is related to the permittivity within the waveguide, the dependency of the material operator $\mathbf{M}_{\varepsilon}(p)$ in Eqs. (12) and (13) can be analyzed by analytical differentiation, and the main task is to provide for an efficient implementation.

As a more general approach we also apply an Algorithmic Differentiation (AD) algorithm as described in Griewank and Walther (2008) and implemented in the ADOL-C package (ADOL-C). It enables the differentiation w.r.t. arbitrary parameters at low computational cost, once the original function has been coded in a computer program. As expected, the results are identical up to the level of numerical noise.

Note that AD is only applied to the matrix assembly in Eqs. (11)–(13). The parameter dependency in these formulas is considered as a function $\mathbf{F}: \mathbb{R}^n \mapsto \mathbb{R}^m$, $\mathbf{y} = \mathbf{F}(\mathbf{x})$ describing an algebraic mapping from $\mathbb{R}^n$ to $\mathbb{R}^m$ and should be defined by an evaluation procedure in a high-level computer language like Fortran or C. The technique of algorithmic differentiation, also called automatic differentiation, provides derivative information of arbitrary order for the code segment in the computer that evaluates $\mathbf{F}(\mathbf{x})$ within working accuracy. For this purpose, the basic differentiation rules such as, the product rule, the quotient rule etc., are applied to each statement of the given code segment. This local derivative information is then combined by the chain rule to calculate the overall derivatives. Hence the code is decomposed into a long sequence of simple evaluations, e.g., additions, multiplications, and calls to elementary functions such as $\sin(x)$ or $\exp(x)$, the derivatives of which can be easily calculated. Exploiting the chain rule yields the derivatives of the whole sequence of statements with respect to the input variables.

Over the past decades, extensive research activities led to a thorough understanding of AD and its two basic modes, the forward and the reverse mode. The complexity estimates for these approaches to compute derivative information are based on the operation count $O_{\mathbf{F}}$, i.e., the number of floating point operations required to evaluate $\mathbf{F}(\mathbf{x})$. Using the forward mode, one computes the required derivatives together with the function evaluation in one sweep as illustrated above. The forward mode yields one *column* of the Jacobian $\nabla\mathbf{F}$ at no more than three times $O_{\mathbf{F}}$ (Griewank and Walther, 2008). One *row* of $\nabla\mathbf{F}$, e.g., the gradient of a scalar-valued component function of $\mathbf{F}$, is obtained using the reverse mode in its basic form also at no more than four times $O_{\mathbf{F}}$ (Griewank and Walther, 2008). It is important to note that this bound for the reverse mode is completely independent of the number n of input variables.

For the problem formulation considered here, it suffices to apply the forward mode. The more sophisticated reverse mode will be required for more detailed problem descriptions to be considered in future work.

4 Numerical examples

We apply our algorithm to the fundamental modes in two inhomogeneous dielectric waveguides, where the permittivity distributions are given by simple one-dimensional parametrized models. Goal quantities are the wave number k_z, or the effective refractive index $n_{\mathrm{eff}} = k_z/k_0$, respectively. In both cases their sensitivities have to be derived by a simple post-processing step from the derivative of the eigenvalue $\lambda = -k_z^2$. Parameter sweeps serve as a reference for the sensitivities. Note that the dimension of the system matrices has no influence on the performance and accuracy of the sensitivity analysis.

4.1 Waveguide with dielectric inset

Figure 2 shows the FIT 2-D computational grid and the geometry of a waveguide with dielectric inset. The parameter of the model is the relative permittivity $p = \varepsilon_1/\varepsilon_0$ of the dielectric inset, and we calculate the wave number $k_z(p)$. Figure 3 shows the results from a parameter sweep together with a tangent on this curve which visualizes the first order derivative of the eigenvalue at the expansion point. Note that this sensitivity is available after only one run of the eigensolver and negligible additional cost.

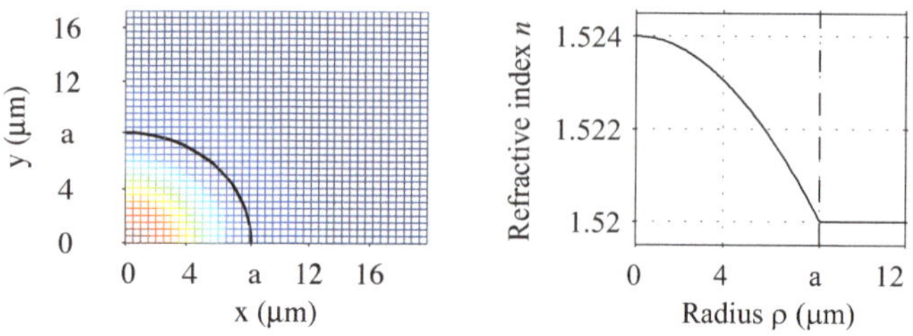

Fig. 4. Grid model and index profile of the graded index waveguide.

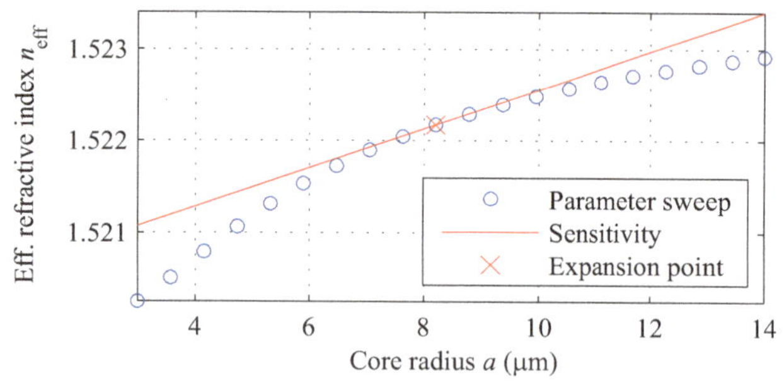

Fig. 5. Sensitivity analysis of $n_{eff}(a)$ in the graded index guide.

4.2 Graded index waveguide

The second example is a graded index waveguide with parabolic index profile. Figure 4 shows the 2-D computational grid, where due to the twofold symmetry only a quarter of the waveguide has to be discretized. Also shown is the profile of the refractive index $n = \sqrt{\varepsilon_r}$ as a function of the radius ρ. The design parameter $p = a$ is the core size of the waveguide, i.e. a geometrical parameter, which can of course easily be mapped on the permittivity distribution in each cell. Thus, the existence of the first derivatives of the matrices and the goal function w.r.t. the design parameter is assured.

Figure 5 shows again the reference results together with a tangent using the first order derivative from the sensitivity analysis. The results agree very nicely.

5 Conclusions

The numerical examples show that the accurate calculation of the first derivative of the eigenvalue w.r.t. some design parameters can be performed at negligible computational cost. Although the original formulation is structurally non-Hermitian, the required left-eigenvector is available from the magnetic field of the modes without a further solving step. The main implementation effort has to be spent on providing a matrix derivative, where the AD approach promises to be both numerical efficient and generally applicable.

Some issues to be addressed in future work are the following: Although the result curves shown here are quite smooth (suggesting that the wave number depends only weakly on the design parameters), it might be desirable to have also second order derivatives. Following the ideas presented above, their calculation is straight-forward, provided that the differentiability of the operators involved is assured. The required second order matrix derivatives can again be obtained from the AD approach. Either second order derivatives or multiple first order derivatives at different expansion points may be used to attain broadband estimations of the parameter dependencies. Especially if applied to the frequency as parameter. this clearly is another link to related MOR approaches and the discussion on single- or multipoint approximations.

Finally, the sensitivity results can be very useful in optimization processes. Especially when derived for more than a single design parameter, the existing implementation will benefit from the generality of the AD approach.

References

ADOL-C: An open-source package for the automatic differentiation of C and C++ programs, https://projects.coin-or.org/ADOL-C, maintained by A. Walther, last access: 14 December 2010.

Farle, O., Hill, V., and Dyczij-Edlinger, R.: Finite-Element Waveguide Solvers Revisited, IEEE Trans. Mag., 40, 1468–1471, 2004.

Farle, O., Hill, V., Nickel, P., and Dyczij-Edlinger, R.: Multivariate Finite Element Model Order Reduction for Permittivity or Permeability Estimation, IEEE Trans. Mag., 42, 623–626, 2006.

Griewank, A. and Walther, A.: Evaluating Derivatives: Principles and Techniques of Algorithmic Differentiation, 2nd Ed., SIAM, 2008.

Nelson, R.: Simplified Calculation of Eigenvector Derivatives, AIAA Journal, 14, 1201–1205, 1976.

Nikolova, N. K., Bandler, J. W., and Bakr, M. H.: Adjoint Techniques for Sensitivity Analysis in High-Frequency Structure CAD, IEEE Trans. MTT, 52, 403–419, 2004.

Polstyanko, S. V., Dyczij-Edlinger, R., and Lee, J.-F.: Fast Frequency Sweep Technique for the Efficient Analysis of Dielectric Waveguides, IEEE Trans. MTT, 45, 1118–1126, 1997.

Schuhmann, R. and Weiland, T.: Conservation of Discrete Energy and Related Laws in the Finite Integration Technique, in: Geometric Methods for Computational Electromagnetics, edited by: Teixeira, F. L., vol. 32 of PIER, pp. 301–316, 2001.

Schultschik, A., Farle, O., and Dyczij-Edlinger, R.: A Model Order Reduction Method for the Finite-Element Simulation of Inhomogeneous Waveguides, IEEE Trans. Mag., 44, 1394–1397, 2008.

Stavrakakis, K., Wittig, T., Ackermann, W., and Weiland, T.: Linearization of Parametric FIT-Discretized Systems for Model Order Reduction, IEEE Trans. Mag., 45, 1380–1383, 2009.

Weiland, T.: A Numerical Method for the Solution of the Eigenwave Problem of Longitudinally Homogeneous Waveguides, Electronics and Communication, 31, 308–314, 1977.

Weiland, T.: Time Domain Electromagnetic Field Computation with Finite Difference Methods, Int. Journal of Numerical Modelling: Electr. Networks, Devices and Fields, 9, 295–319, 1996.

Automotive radar – investigation of mutual interference mechanisms

M. Goppelt[1], **H.-L. Blöcher**[2], **and W. Menzel**[1]

[1]Institute of Micowave Techniques, University of Ulm, 89081 Ulm, Germany
[2]Group Research & Advanced Engineering, Daimler AG, 89081 Ulm, Germany

Abstract. In the past mutual interference between automotive radar sensors has not been regarded as a major problem. With an increasing number of such systems, however, this topic is receiving more and more attention. The investigation of mutual interference and countermeasures is therefore one topic of the joint project "Radar on Chip for Cars" (RoCC) funded by the German Federal Ministry of Education and Research (BMBF). RoCC's goal is to pave the way for the development of high-performance, low-cost 79 GHz radar sensors based on Silicon-Germanium (SiGe) Monolithic Microwave Integrated Circuits (MMICs).

This paper will present some generic interference scenarios and report on the current status of the analysis of interference mechanisms.

1 Introduction

In the automotive field, radar sensors are key components for comfort and safety functions, for example adaptive cruise control (ACC) or collision mitigation systems (CMS). With an increasing number of automotive radar sensors operated close to each other at the same time, radar sensors may receive signals from other radar sensors. The reception of foreign signals (interference) can lead to problems such as ghost targets or a reduced signal-to-noise ratio. Figure 1 shows such a simple automotive interference scenario with direct interference from an oncoming vehicle.

Up to now, interference has not been considered as a major problem because the percentage of vehicles equipped with radar sensors and therefore the probability of interference was low, and the sensors were used mainly for comfort functions. In this case it may be sufficient to detect interference and turn off the function for the duration of the interference. On the contrary, safety functions of future systems require very low failure rates. So in spite of a predicted higher number of radar systems, the probability of interference-induced problems has to be reduced considerably. Therefore effective countermeasures have to be introduced to minimize mutual interference even with high traffic density (e.g. in large cities) and a rising percentage of vehicles equipped with radar sensors.

In Germany currently three frequency bands are assigned to automotive radar applications:

- 24 GHz ultra-wideband (UWB)
- 76–77 GHz
- 77–81 GHz

In addition, some systems also are operated in the 24 GHz ISM band. Detailed information about the frequency bands can be found in Table 1.

New vehicles can be equipped with 24 GHz UWB sensors only until 2013. From then on, the 79 GHz frequency band with up to 4 GHz bandwidth will replace the 24 GHz band.

From 2004 to 2007, the Daimler AG took part in the German Federal Ministry of Education and Research (BMBF) funded joint project KOKON. Aim of the project was to demonstrate the feasibility of 79 GHz automotive radar sensors with SiGe-based MMICs. Presently, the Daimler AG contributes to the successor project "Radar on Chip for Cars" (RoCC) running from 2008–2011.

Special emphasis of this projects lies on the following items:

- reducing cost and size
- improving the radio-frequency (RF) packaging technology
- improving sensor performance
- improving sensor reliability

The last item includes the investigation of mutual interference and interference minimization techniques (countermeasures). The specific tasks are

Correspondence to: M. Goppelt
(markus.goppelt@uni-ulm.de)

Table 1. Frequency bands and equivalent isotropically radiated power (EIRP) for automotive radar applications in Germany.

	24 GHz UWB short-range	24 GHz ISM short-range	77 GHz long-range	79 GHz short-range
frequency range (GHz)	21.650–26.650	24.000–24.250	76.000–77.000	77.000–81.000
center frequency (GHz)	24.150	24.125	76.500	79.000
max. average power (EIRP) (dBm/MHz)	−41.3			−9
max. average power (EIRP) (dBm)			50	
max. peak power (EIRP) (dBm)		20	55	55

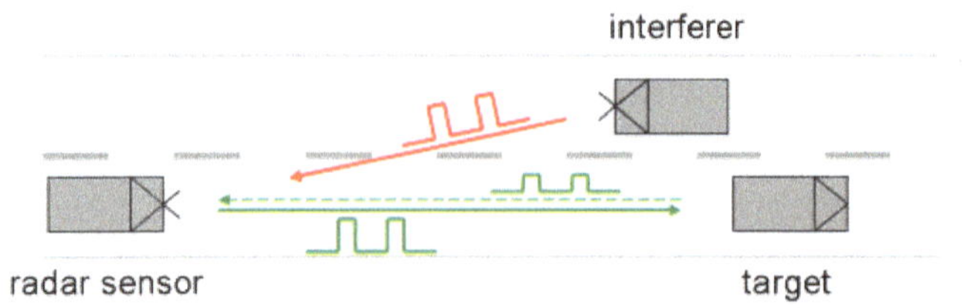

Fig. 1. Simple automotive interference scenario with one target and one interferer.

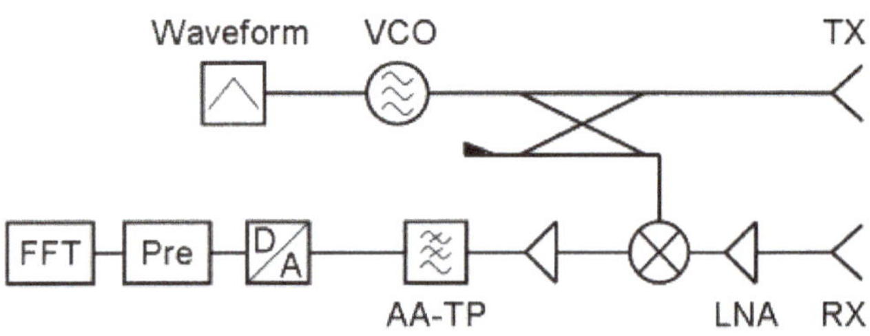

Fig. 2. Block diagram of a simple generic FMCW radar.

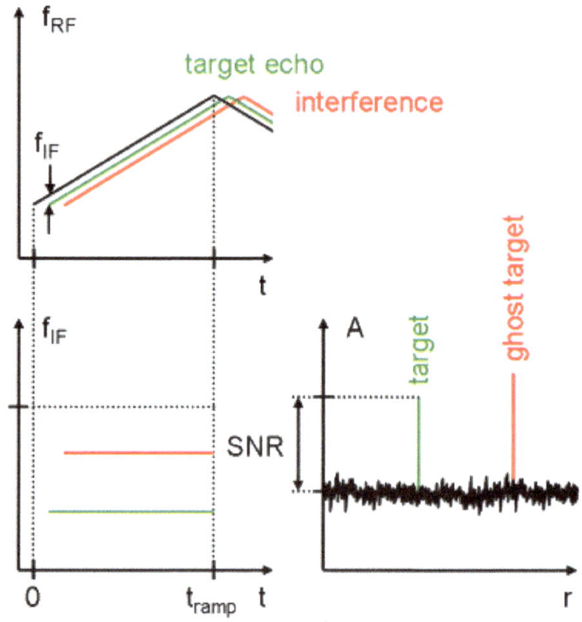

Fig. 3. Diagrams illustrating a case where an FMCW radar interferes with another FMCW radar generating a ghost target. The upper diagram shows the frequency course of the local oscillator signal (black) and the signals received by the receive (RX) antenna. The lower left-hand diagram shows the down-converted IF frequencies as a function of time, the lower right-hand diagram the Fourier transform of the IF signal.

- to identify interference mechanisms,
- to find efficient countermeasures and
- to test countermeasures in the laboratory and with radar sensors in vehicles.

At the beginning of Sect. 2 of this paper, two radar architectures are defined, and the analysis approach to interference mechanisms between different types of radar sensors is outlined. Sections 2.1 to 2.4 illustrate the interference mechanisms. Target and interference power levels are analyzed in Sect. 2.5 for the simple interference scenario in Fig. 1.

2 Interference mechanisms

Since there are many different radar sensors on the automotive market, it was decided to start the analysis of interference mechanisms with a simple generic frequency-modulated continuous wave (FMCW) radar (Fig. 2) and a simple generic pulsed radar with correlation receiver (Fig. 10).

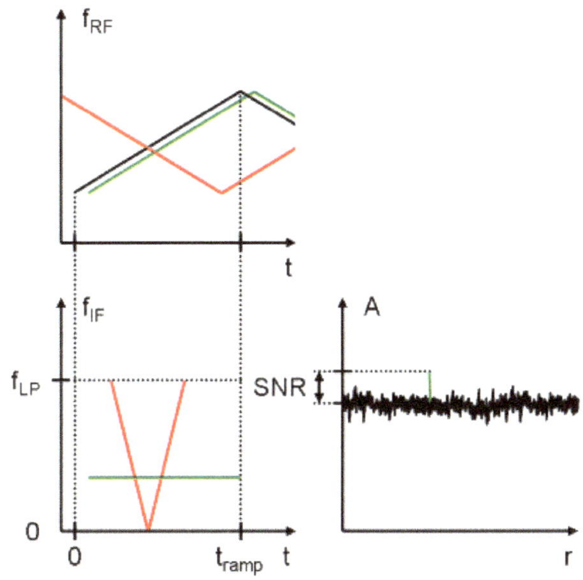

Fig. 4. Diagrams illustrating a case with interference between two FMCW radars causing a uniform increase of the noise floor.

In a first step the interference mechanisms between the different types of radar were roughly identified, followed by simulations and measurements to verify the approximations and to develop a more precise understanding of the interference mechanisms. The last item, however, has not yet been completed. Previous research on interference mechanisms had been conducted by

- Brooker (2007): Investigation of the interference between an FMCW radar and a pulsed radar and propositions for interference minimization techniques.
- Oprisan and Rohling (2005): Analysis of interference effects between continuous wave (CW), frequency shift keying (FSK), frequency-modulated continuous wave (FMCW), frequency-modulated shift keying (FMSK) and pulsed radars. Some interference reduction techniques are proposed.
- Tullsson (1997): Investigation of the interference sensitivity of an automotive FMCW radar and analysis of interference elimination techniques.

2.1 FMCW radar with FMCW interference

To illustrate the effects of interference from an FMCW radar on an FMCW radar, two cases were selected. Both radars have the same triangular FMCW modulation (same ramp duration and sweep bandwidth). In the first case (Fig. 3) the interference leads to a constant intermediate frequency (IF) generating a second line in the radar range profile. The second line then is interpreted as an additional target (ghost target).

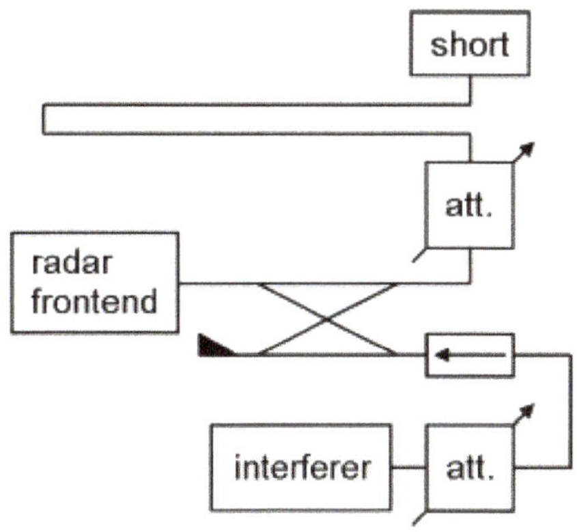

Fig. 5. Block diagram of the waveguide-based measurement setup to demonstrate the effects of interference on an FMCW radar.

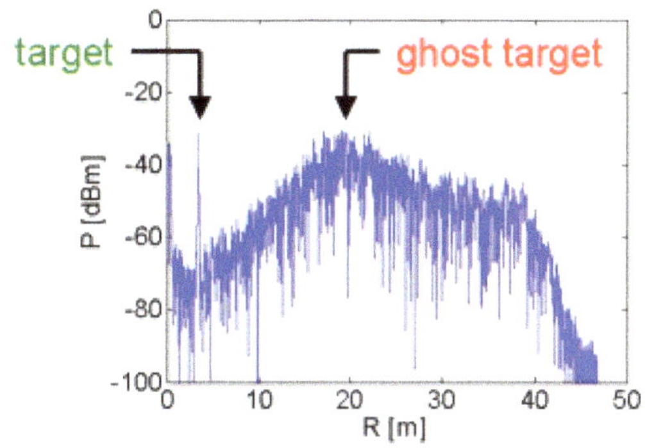

Fig. 6. Measured FMCW radar range profile with a ghost target generated by interference from another FMCW radar.

In the second case (Fig. 4) the frequency ramp of the interfering signal starts earlier in time, consequently the crossing of the ramps lead to an IF signal with fast V-shaped frequency variation over time, resulting in a uniform increase of the noise floor over the whole IF-band. Thus, the target signal-to-noise-ratio can be reduced considerably.

Both cases were demonstrated in the laboratory using a waveguide-based measurement setup (Fig. 5). The target was simulated using an approximately three meters long waveguide terminated with a short circuit. The interfering signal was added to the reflected signal using a directional coupler. Figure 6 shows the result for the first case (parallel frequency ramps). The ghost target's broad spectrum originates from the fact that there is no correlation between the phase noise of the interfering signal and of the radar's local oscillator signal.

For the second case, a CW signal was coupled into the receiver of the FMCW radar. In the frequency domain, the CW signal intersects the ramps of the FMCW radar just like the interfering FMCW signal in Fig. 4. The upper diagram of Fig. 7 shows the resulting IF signal in the time domain. The IF signal generated by the interfering CW signal is short in relation to the ramp duration . This is due to the fact that the

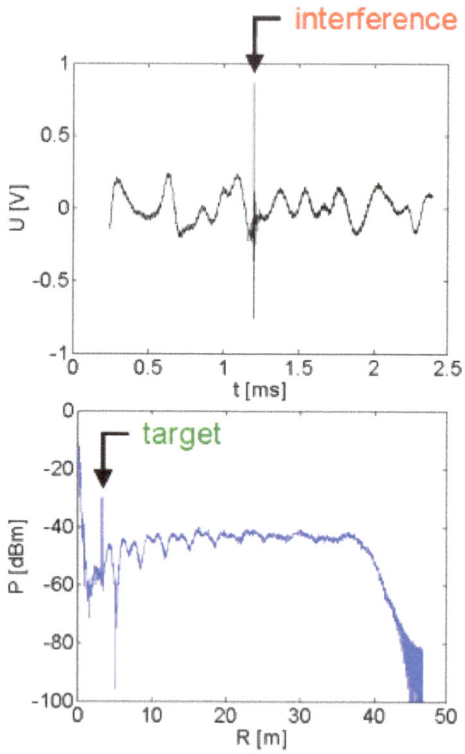

Fig. 7. Measured FMCW radar IF signal and range profile with CW interference.

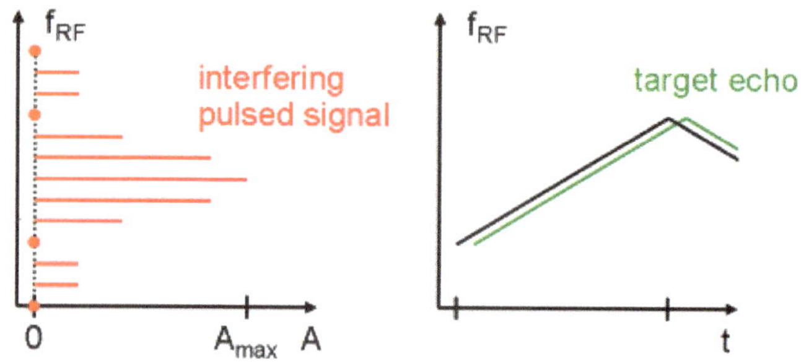

Fig. 8. Diagrams illustrating the case of pulsed signals interfering with an FMCW radar. The left-hand diagram shows the spectrum of a pulsed signal, the right-hand diagram shows the FMCW radar signals. Some of the equidistant spectral lines of the interfering pulsed signal intersect the FMCW radar ramp.

radar's IF bandwidth is much smaller than the radar's frequency deviation (e.g. by a factor of 1000). The IF noise floor of the radar without interference is below −60 dBm, the noise floor increases to about −45 dBm due to the interfering CW signal as can be seen in the lower diagram of Fig. 7.

2.2 FMCW radar with pulsed interference

A similar situation results, when a pulse modulated signal interferes with an FMCW radar. The spectrum of a pulsed signal consists of equidistant lines with a $\sin(x)/x$ envelope for rectangular pulses. This is depicted in the left part of Fig. 8. Some lines intersect the FMCW radar ramp leading to a short IF signal with V-shaped frequency course similar to the case shown in Fig. 4.

An exemplary measurement with pulsed signal interference is shown in Fig. 9. It can be seen that the resulting IF signal (upper diagram) is wider than in the CW case (upper diagram of Fig. 7). The resulting increase in the noise floor is not as uniform as in the CW case (lower diagrams of Figs. 7 and 9).

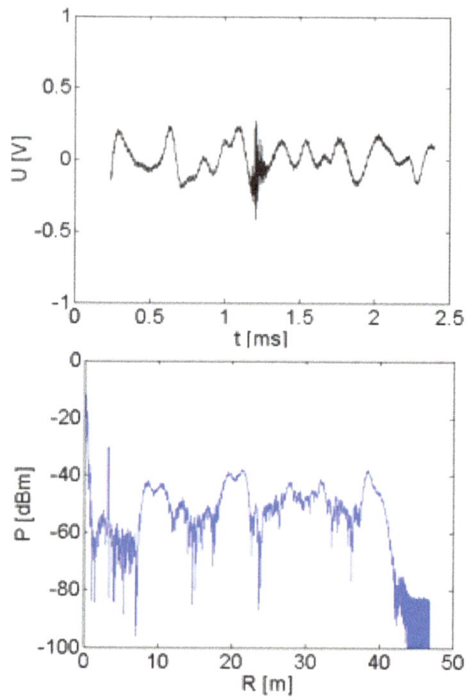

Fig. 9. Measured FMCW radar IF signal and range profile with pulsed signal interference.

2.3 Pulsed radar with FMCW interference

Pulsed radars with correlation receiver, as typically used for automotive applications and also selected for this investigation, work similar to sampling oscilloscopes. The time delay t_D (Fig. 10) is successively increased until the maximum signal flight time of interest is reached. An IF signal (in principal a correlation function) appears at the mixer output if a radar signal delay and t_D are equal. This is done in such a way that every range cell is sampled several times, and the resulting output signals are integrated in a lowpass filter where the resulting amplitude is detected. Following this, the sampling process starts again with the lowest time delay. In practice, the mixer in Fig. 10 has to be substituted by a pair of mixers to get the quadrature signals. A pulsed radar sensor with correlation receiver is described in detail in a publication by Gresham et al. (2004).

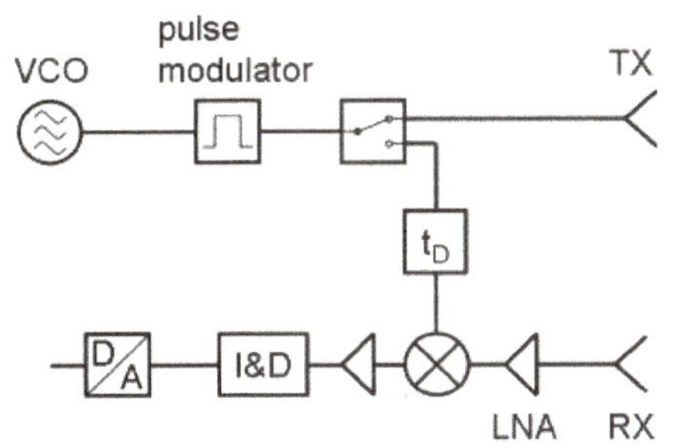

Fig. 10. Block diagram of a simple generic (coherent) pulsed radar with correlation receiver.

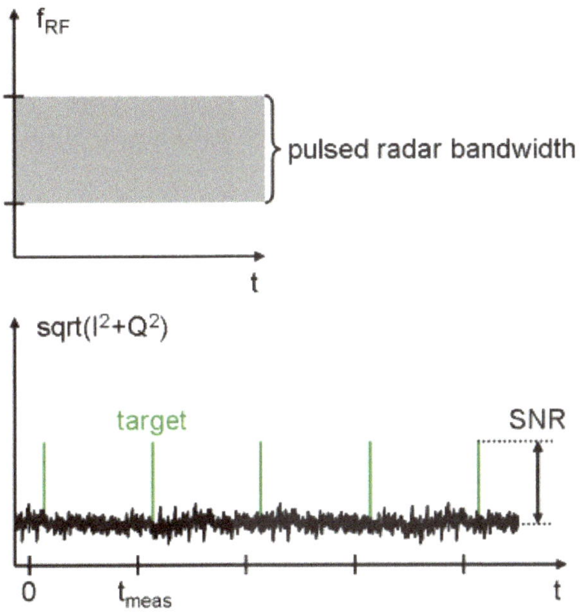

Fig. 11. Reception bandwidth and range profile of the pulsed radar without interference.

Figure 11 shows the receiver bandwidth and the range profile of the pulsed radar without interference. The amplitudes in the lower diagram are the absolute values of the complex I-Q output of the mixer.

The effect of an interference from a narrowband FMCW radar on this type of radar is a uniform increase in the noise floor (lower diagram of Fig. 12). The target signal-to-noise ratio decreases accordingly.

2.4 Pulsed radar with pulsed interference

Interference from an identical pulsed radar with the same pulse repetition frequency (PRF) generates a ghost target at a constant "distance" (Fig. 13).

A slightly different PRF results in a moving ghost target since the time difference between the transmitted pulse and the interfering pulse increases or decreases from pulse repetition period to pulse repetition period. Figure 14 displays this

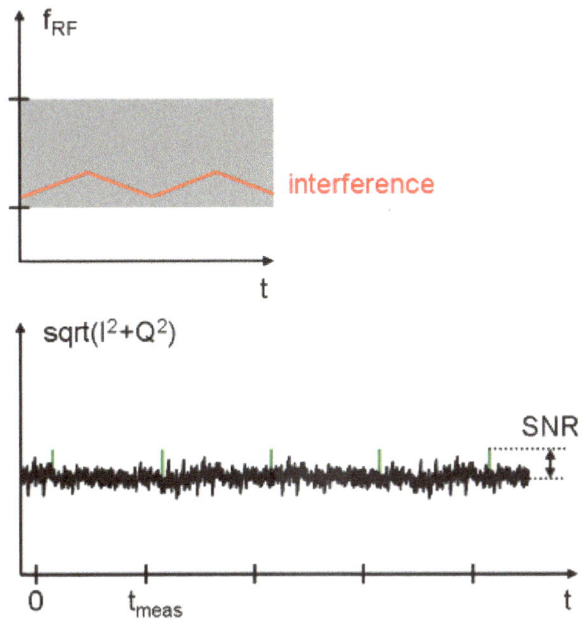

Fig. 12. Reception bandwidth and range profile of a pulsed radar with interference by a narrowband FMCW signal causing a uniform increase of the noise floor and therefore a reduction of the target signal-to-noise-ratio (SNR).

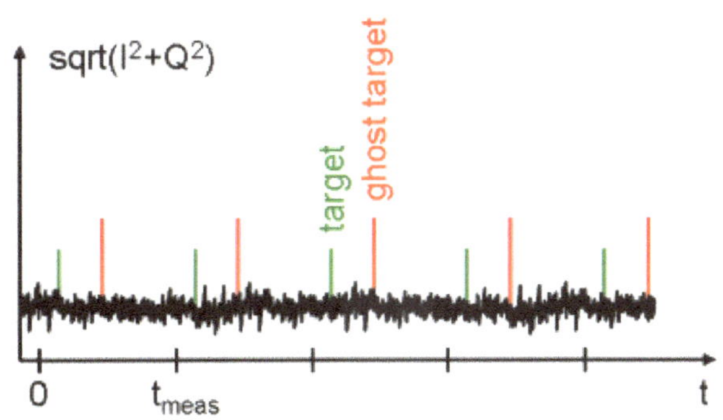

Fig. 13. Range profile of a pulsed radar with interference from an identical pulsed radar with the same PRF generating a stationary ghost target.

situation for a slightly lower PRF resulting in a ghost target with increasing distance. This diagram, however, does not yet take into account the effect of integration in the correlation receiver. Depending on the actual difference in PRF, less interference pulses may be integrated, leading to a possibly improved situation.

2.5 Consideration of the interference power level

So far, the power levels of the target echo and the interfering signal have not been regarded in detail. To start with an analysis of power levels, the simple interference scenario in Fig. 1 has been considered. Simplified radiation diagrams with constant antenna gain in a given sector were used as

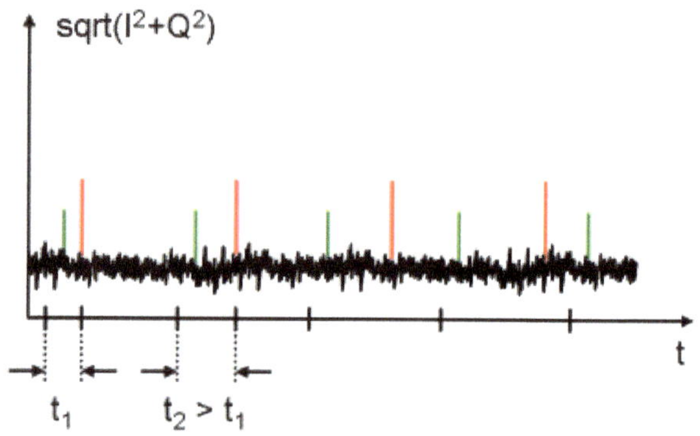

Fig. 14. Range profile of a pulsed radar with interference from an identical pulsed radar with slightly lower PRF generating a ghost target with increasing distance.

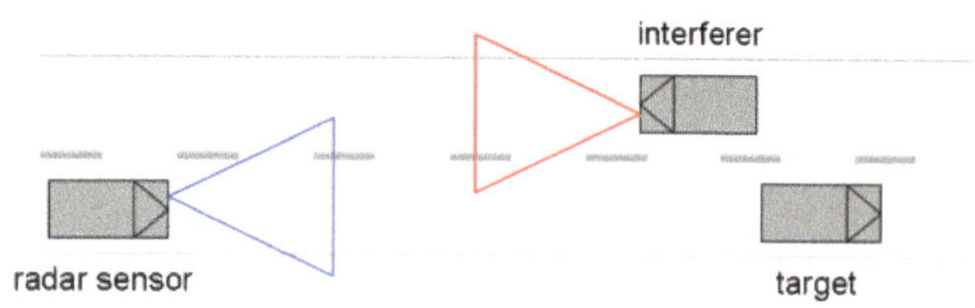

Fig. 15. Simple automotive interference scenario with simplified radiation diagrams. The interferer radiates directly into the radar sensor's receiver.

depicted in Fig. 15. The positions of the radar sensors on the vehicles as well as widths and orientations of the antenna beams are just one example. Power levels then are calculated using Friis' formula.

Figure 16 shows the simulated power levels at the antenna port of the receiving radar. The interference lasts as long as the radar sensors "see" each other. The target level is constant since it is assumed that the target vehicle is at a constant distance. As the interfering radar radiates directly into the receiver, the interference level can be much higher than the target level (more than 60 dB above the target level). This high interference level may have different effects. On the one hand, it may drive receiver components into saturation or into their nonlinear regime, resulting in unwanted harmonics. On the other hand, the interference signals will be further processed and can appear as ghost targets or as an increased noise level, reducing the desired SNR of the receiving radar. Some further investigations are necessary, however, including the respective analogue and digital signal processing; this may reduce the influence of the interference to some extent.

3 Conclusion

Mutual interference between automotive radar sensors is expected to become an increasingly important topic. This is due the fact that, on the one hand, automotive radar sensors

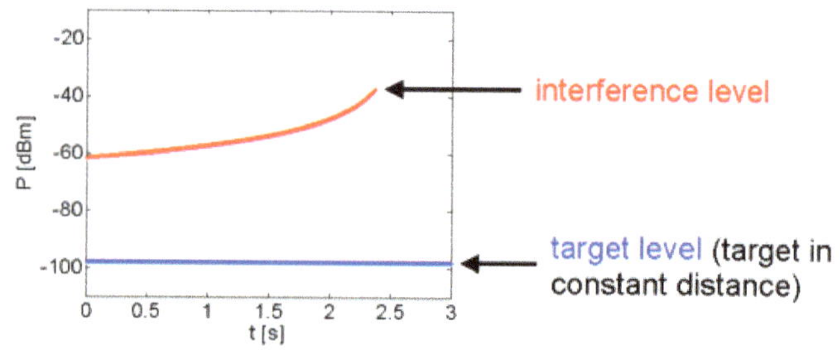

Fig. 16. Target and interference power level at the radar sensor's receive (RX) antenna port for the scenario in Fig. 15.

are used increasingly for safety functions. On the other hand, traffic density and percentage of vehicles equipped with radar sensors will increase considerably. Interference mechanisms have to be precisely understood in order to design and to verify the effectiveness of countermeasures to minimize mutual interference.

This paper has described some fundamental interference effects appearing in a scenario with FMCW and pulsed radar sensors, partly by a simple analysis using generic radar sensors, partly by first experiments using radar frontends built by the University of Ulm.

The two major effects detected so far are the appearance of ghost targets and an increase of the noise or interference level in the radar receiver. Concerning the interference levels, very high signal strengths can occur if two sensors directly face each other.

Acknowledgements. The authors wish to thank the German Federal Ministry of Education and Research (BMBF) for funding this work within the RoCC project.

References

Brooker, G.: Mutual Interference of Millimeter-Wave Radar Systems, IEEE Transactions on Electromagnetic Compatibility, 49, 170–181, 2007.

Oprisan, D. and Rohling, H.: Analysis of Mutual Interference between Automotive Radar Systems, International Radar Symposium, Berlin, Germany, Session: Automotive Radar I, 2005.

Tullsson, B.: Topics in FMCW Radar Disturbance Suppression, Radar 97, Edinburgh, United Kingdom, 1–5, 1997.

Gresham, I., Jenkins, A., Egri, R., et al.: Ultra-Wideband Radar Sensors for Short-Range Vehicular Applications, IEEE Transactions on Microwave Theory and Techniques, 52, 2105–2122, 2004.

Improved measurement surface for MEG using magnetic-dipole sources and a spherical-multipole expansion

F. Argin, H. Ahrens, and L. Klinkenbusch

Institut für Elektrotechnik und Informationstechnik, Christian-Albrechts-Universität zu Kiel, Germany

Correspondence to: F. Argin (far@tf.uni-kiel.de)

Abstract. The multipole representation of Magnetoencephalography (MEG) signals is known as a useful tool for distinguishing between magnetic fields arising from the brain and external disturbances. In this contribution we extend this concept and show that a closed double-layer surface with magnetometer probes is better suited to determine the corresponding multipole amplitudes α_{lm} than a conventional single-layer surface with gradiometers and magnetometer probes. For two different source configurations we show that the α_{lm} rapidly converge to the exact values. This proof of concept motivates to further optimize the geometry of the double-layer surface and the sensors' positions.

1 Introduction

Neuronal activity of the human brain involves electrical currents which give rise to a magnetic field outside of the brain. Since the typical intensity of the field is in the femtotesla range its measurement requires a lot of effort. Current MEG techniques involve magnetically shielded environments and *S*uperconducting *Qu*antum *I*nterference *D*evices (SQUIDs). From the measured magnetic field the field sources can be estimated by different algorithms. Obviously the reduction of noise and the elimination of external disturbances improve the estimation of the field sources. Assuming divergence-free current sources only, the magnetic field outside the brain can be described by a scalar magnetic potential which will be represented by a spherical-multipole expansion. It has been shown that this technique allows to separate external disturbances from internal sources (Signal Space Separation) (Taulu et al., 2004). We extend this idea and introduce a closed double-layered measurement surface. The field within this source-free enclosure can be uniquely determined if Neumann or Dirichlet boundary data are given on its surface. From the knowledge of its boundary values the magnetic field is uniquely determined also outside of the enclosure, particularly also in the magnetically homogeneous brain outside of the sources. The field measured on this new measurement surface is represented by a spherical-multipole expansion which conveniently can be used to interpret the biomagnetic field as well as to optimize the shape of the measurement surface and the sensors' positions.

In Sect. 2 we demonstrate that the neuronal activity, that is, the divergence-free currents which produce the biomagnetic field can be described by magnetic dipoles which leads us to a most simple characterization of the forward problem. Since the domain outside the brain is free of sources, we deduce a spherical-multipole expansion for that curl-free magnetic field arising from sources in the brain. In Sect. 3 we show how the measurement surface has been designed to ensure that we are able to determine the multipole-amplitudes uniquely. Section 4 illustrates how the single-layer and double-layer surfaces were implemented, as well as the set-up of the system of linear equations for numerically determining the spherical-multipole amplitudes α_{lm}.

Numerical results for a single-layer measurement surface with symmetric axial first order gradiometer as well as magnetometer data, and for a closed double-layer surface using magnetometer data only are finally presented in Sect. 5.

2 Forward problem

2.1 Magnetic dipole moments

While the brain is processing information, small electric currents flow inside the neural system producing a magnetic field outside of the head. This extremely weak magnetic field can be recorded by suitable MEG devices (Hämäläinen et al., 1993). Because of the complexity of the neuronal system, the exact current flows are difficult to describe. Without going into detail of the brain physiology, we shall point out a few of the relevant aspects leading to MEG signals. It is

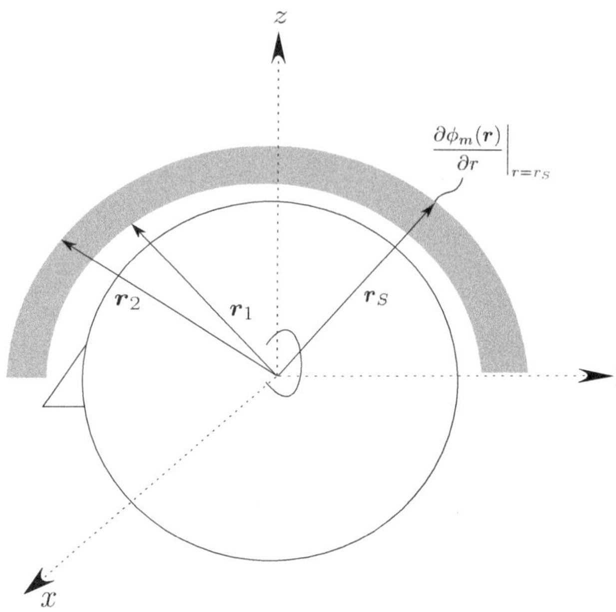

Fig. 1. Schematic view of a simplified head model and a closed double-layer surface. The radius of the first layer is $\boldsymbol{r}_1$ and the radius of the second one is $\boldsymbol{r}_2$. The measurement surface is the entire (closed) surface of the gray domain. The vector $\boldsymbol{r}_s$ points on that surface where $B_r \sim \left.\frac{\partial \phi_m(\boldsymbol{r})}{\partial r}\right|_{r=r_S}$ is measured.

assumed that the main source of MEG signals are *e*xcitatory *p*ost*s*ynaptic *p*otentials (EPSPs) (Nunez, 1981) in the *input* part of the neurons. The EPSPs cause current flows inside the excited neurons (Gloor, 1985), which – due to the commonly presumed conservation of charge at each time – must be balanced by an electrical counterflow outside of the neuron (Baillet et al., 2001). Usually, the intracellular current density $\boldsymbol{J}_p(\boldsymbol{r})$ is called primary current density while the extracellular part $\boldsymbol{J}_V(\boldsymbol{r})$ is referred to as the volume current density.

The total electric-current density $\boldsymbol{J}(\boldsymbol{r})$ is

$$\boldsymbol{J}(\boldsymbol{r}) = \boldsymbol{J}_p(\boldsymbol{r}) + \boldsymbol{J}_V(\boldsymbol{r}). \tag{1}$$

It is commonly presumed that at any time there is no accumulation of charge in the neuronal system, hence the total current must be divergence-free according to

$$\nabla \cdot \boldsymbol{J}(\boldsymbol{r}) = 0 \tag{2}$$

which we may electrically interpret as a closed current circuit. Hence it is possible to represent $\boldsymbol{J}(\boldsymbol{r})$ by a suitable number of located magnetic dipoles described by corresponding dipole moments $\boldsymbol{m}$. For a given current density located in ΔV it holds that

$$\boldsymbol{m}|_{\Delta V} = \frac{\mu_0}{2} \int_{\Delta V} \boldsymbol{r} \times \boldsymbol{J}(\boldsymbol{r})\, \mathrm{d}^3 r \tag{3}$$

which in case of a single closed current circuit with a (directed) area $\boldsymbol{F}$ and a current strength I has the solution

$$\boldsymbol{m} = \mu_0 \boldsymbol{F} I. \tag{4}$$

Commonly tens of thousands of neurons have to be activated at the same time to have measurable MEG signals, therefore the magnetic dipole moment can be understood as an equivalent source of the net effect of current circuits covering several cubic centimeters of activated brain tissue.

In general, the frequencies for brain signals are in the range of or below 100 Hz, hence we can neglect the time derivatives $\sigma|\partial \boldsymbol{E}/\partial t|$ and $|\partial \boldsymbol{B}/\partial t|$ in Maxwell's equations (Hämäläinen et al., 1993). This leads us to the following quasi-static approximation of Maxwell's equations for the magnetic field produced by a magnetic dipole moment $\boldsymbol{m}$ located at $\boldsymbol{r}_0$

$$\boldsymbol{B}(\boldsymbol{r}) = \frac{\mu_0}{4\pi} \frac{3(\boldsymbol{r}-\boldsymbol{r}_0)(\boldsymbol{m}\cdot(\boldsymbol{r}-\boldsymbol{r}_0))}{|\boldsymbol{r}-\boldsymbol{r}_0|^5} - \frac{\boldsymbol{m}}{|\boldsymbol{r}-\boldsymbol{r}_0|^3}. \tag{5}$$

2.2 Multipole expansion

Assuming that the domain outside of the head is homogeneous and free of sources, Ampère's law in the magneto-quasi-static approximation reads

$$\nabla \times \boldsymbol{B}(\boldsymbol{r}) = 0. \tag{6}$$

Outside of the head the magnetic field produced by closed-current circuits (or by magnetic dipoles) in the head consists of curl-free fields only, hence it is always possible to deduce that magnetic field from a scalar magnetic potential $\phi_m(\boldsymbol{r})$ according to

$$\boldsymbol{B}(\boldsymbol{r}) = -\mu_0 \nabla \phi_m(\boldsymbol{r}). \tag{7}$$

With $\nabla \cdot \boldsymbol{B} = 0$ we easily derive the Laplace equation

$$\Delta \phi_m(\boldsymbol{r}) = 0. \tag{8}$$

In spherical coordinates (r, ϑ, φ) the general solution can be found as a spherical-multipole expansion

$$\phi_m(\boldsymbol{r}) = \sum_{l=0}^{\infty} \sum_{m=-l}^{l} \left[\alpha_{\mathrm{lm}} \frac{Y_{\mathrm{lm}}(\theta,\phi)}{r^{l+1}} + \beta_{\mathrm{lm}} r^l Y_{\mathrm{lm}}(\theta,\phi) \right] \tag{9}$$

with multipole-amplitudes (expansion coefficients) α_{lm} and β_{lm}. The surface-spherical harmonics are related to associated Legendre functions and harmonic functions by

$$Y_{\mathrm{lm}}(\theta,\phi) = \sqrt{\frac{2l+1}{4\pi} \frac{(l+1)!}{(l-1)!}} P_l^m(\cos\theta) e^{jm\phi}. \tag{10}$$

As can be seen from Eq. (9) only the part in the multipole expansion belonging to α_{lm} remains regular for $r \to \infty$. Consequently, that part can be interpreted as the magnetic field arising from magnetic sources inside the head. On the other hand, only the part in the multipole expansion belonging to

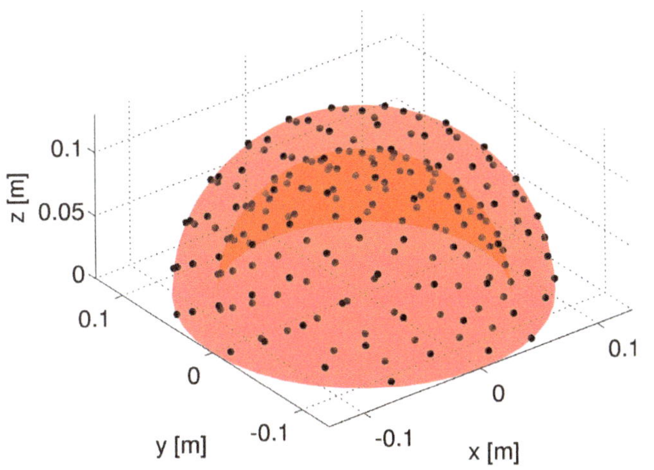

Fig. 2. Closed double-layer surface with isotropically distributed pointlike magnetometers (black dots). The radius of the first layer is $r_1 = 10\,\text{cm}$ and the radius of the second layer is $r_2 = 13\,\text{cm}$.

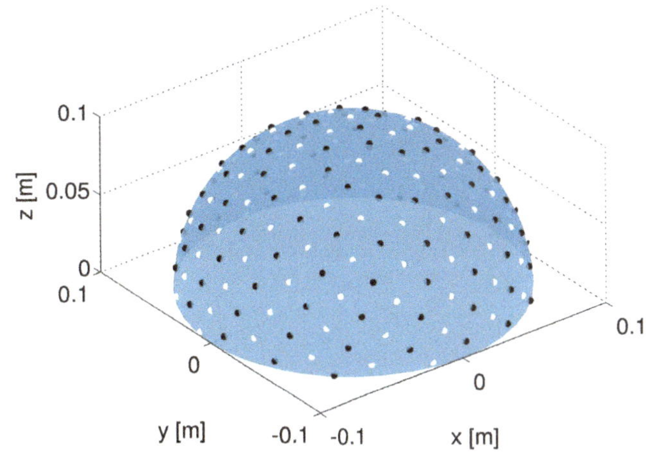

Fig. 3. Single-layer surface with isotropically distributed pointlike magnetometers (black dots) and pointlike gradiometers (white dots). The units of the axes are in meters.

β_{lm} remains finite for $r \to 0$, therefore this part represents the magnetic field arising from sources outside the head.

In this paper we consider sources inside the head only and neglect external (interfering) sources, hence only the field part with the α_{lm} is relevant.

3 Implementation of the measurement surfaces

The scalar potential $\phi(\boldsymbol{r})$ within the closed measurement surface (gray domain in Fig. 1) can be determined uniquely, if the corresponding Neumann $(\partial\phi(\boldsymbol{r})/\partial n|_{\boldsymbol{r}=\boldsymbol{r}_S})$ or Dirichlet $(\phi(\boldsymbol{r})|_{\boldsymbol{r}=\boldsymbol{r}_S})$ boundary data are known. In this paper, the measurement surface consists of two connected hemispheres with different radii (Fig. 2). From Eq. (7) we obtain

$$B_r(\boldsymbol{r})|_{\boldsymbol{r}=\boldsymbol{r}_S} = -\mu_0 \left.\frac{\partial\phi_m(\boldsymbol{r})}{\partial r}\right|_{\boldsymbol{r}=\boldsymbol{r}_S}, \tag{11}$$

on the hemispheres, that is, the solution of the Neumann problem is equivalent to finding the normal component of the magnetic field of the measurement surface. Practically, we measure the radial component at isotropically distributed points on a closed double-layer surface (Fig. 2). The results will be compared to those ones obtained from a single measurement surface (Fig. 3) on which we assume given Cauchy data, i.e., $\boldsymbol{B}_r(\boldsymbol{r})|_{\boldsymbol{r}=\boldsymbol{r}_S}$ and simultaneously $(\partial\boldsymbol{B}_r(\boldsymbol{r})/\partial r)|_{\boldsymbol{r}=\boldsymbol{r}_S}$. The latter configuration is the standard procedure in MEG measurement, where magnetometers as well as gradiometers are taken into account.

To compare the results obtained from the different measurement surfaces, we solve the forward problem analytically from Eq. (5) for two simple source configurations. For the single-layer configuration the positions of the magnetometers and gradiometers are alternating. The radius of the single-layer surface as well as of the first layer of the closed surface is $r_1 = 10\,\text{cm}$. For the second layer of the closed double-layer surface the radius has been chosen to $r_2 = 13\,\text{cm}$.

Next, the magnetic field $\boldsymbol{B}(\boldsymbol{r})$ of the mentioned simple source configuration is expressed as a spherical-multipole expansion Eq. (9) with multipole amplitudes α_{lm} only. For each of the field values at the probes' positions we set up a corresponding spherical-multipole expansion. Doing this for each probe, we end up with a system of linear equations. Its solution yields the multipole amplitudes of the reconstructed magnetic field, which then is compared to the original (true) field value at the probes' positions.

The first of the source configurations consists of a magnetic dipole located at the origin of the coordinate system and polarized in the z-direction, while for the second one we place the magnetic dipole at $\boldsymbol{r} = 5\,\text{cm}\,\hat{z}$ polarized in the x-direction. Arranging the multipole expansions for each of the measurement points in rows, the system of linear equations reads as

$$\underbrace{\begin{pmatrix} \Omega^1_{1,-1} & \Omega^1_{1,0} & \cdots & \Omega^1_{l,l} \\ \Omega^2_{1,-1} & \Omega^2_{1,0} & \cdots & \Omega^2_{l,l} \\ \vdots & \ddots & \ddots & \vdots \\ \Omega^N_{1,-1} & \Omega^N_{1,0} & \cdots & \Omega^N_{l,l} \end{pmatrix}}_{=\mathbf{A}} \underbrace{\begin{pmatrix} \alpha_{1,-1} \\ \alpha_{1,0} \\ \vdots \\ \alpha_{l,l} \end{pmatrix}}_{=\boldsymbol{\alpha}} = \underbrace{\begin{pmatrix} B_r(\boldsymbol{r}_1) \\ B_r(\boldsymbol{r}_2) \\ \vdots \\ B_r(\boldsymbol{r}_N) \end{pmatrix}}_{=\boldsymbol{b}}. \tag{12}$$

Here, $\Omega^k_{l,m} = \frac{(l+1)Y_{\text{lm}}(\theta_k,\phi_k)}{r_k^{l+2}}$, α_{lm} are the spherical-multipole amplitudes while $B_r(\boldsymbol{r}_k)$ denotes the radial component of the measured magnetic field at the k-th measurement point.

The exact solution of the system of linear equations might be obtained only if it is at least squared and of full rank. Because the number of columns of $\mathbf{A}$ are depending on the multipole order l, the number of necessary rows (measurement points) is given by $N = (l+1)^2 - 1$.

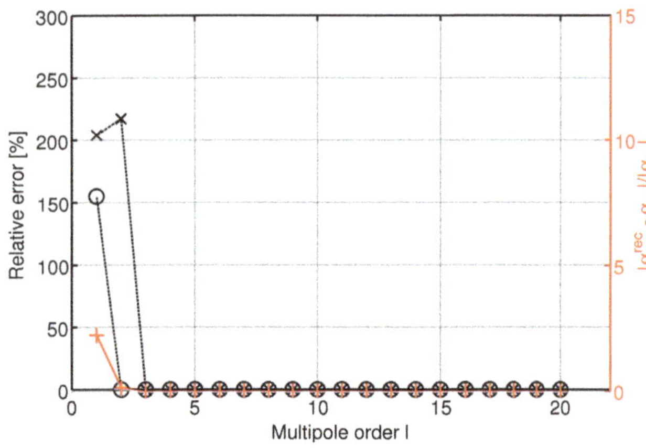

Fig. 4. Relative reconstruction error of the components B_r (black ∘) and B_θ (black ×) on a closed double-layer surface and of the original multipole expansion (red) as a function of the order l of the multipole expansion. Source: Dipole at the origin, polarized in z-direction.

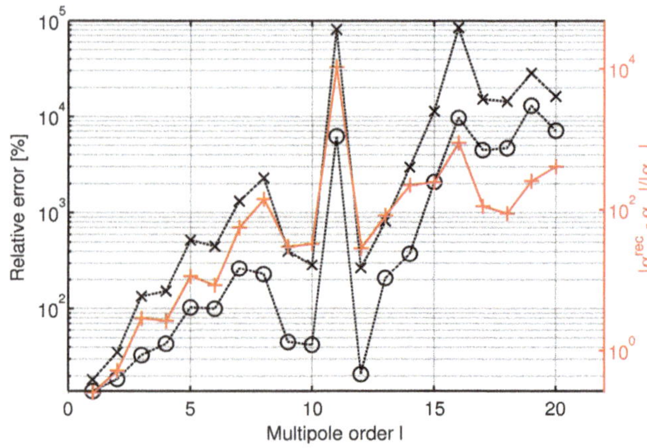

Fig. 5. Relative reconstruction error of the components B_r (black ∘) and B_θ (black ×) on a single-layer surface and of the original multipole expansion (red) as a function of the order l of the multipole expansion. Source: Dipole at the origin, polarized in z-direction.

For the double-layer surface we have to combine the data of the inner and the outer layer. The single-layer surface consists of magnetometers and (symmetrically arranged) gradiometers. A gradiometer provides a first order derivation of $\boldsymbol{B}(\boldsymbol{r})$ in radial direction at the measurement point, thus the elements of the coefficient matrix of the gradiometer probe are

$$\tilde{\Omega}^k_{l,m} = \frac{\frac{(l+1)Y_{lm}(\theta_k,\phi_k)}{r^{l+2}_{g_2,k}} - \frac{(l+1)Y_{lm}(\theta_k,\phi_k)}{r^{l+2}_{g_1,k}}}{r_{g_2,k} - r_{g_1,k}} = \frac{\Omega^{g_2,k}_{l,m} - \Omega^{g_1,k}_{l,m}}{\Delta r_{g,k}} \tag{13}$$

while the elements of the right hand side read as

$$\tilde{B}_r(\boldsymbol{r}_k) = \frac{B_r(\boldsymbol{r}_{g_2,k}) - B_r(\boldsymbol{r}_{g_1,k})}{\Delta r_{g,k}}. \tag{14}$$

Here, $r_{g_1,k}$ denotes the distance of the first coil and $r_{g_2,k}$ of the second one of the axial gradiometer. Thus the combination of magnetometer and gradiometer probes leads us to the following system of linear equations

$$\begin{pmatrix} \tilde{\Omega}^1_{1,-1} & \tilde{\Omega}^1_{1,0} & \cdots & \tilde{\Omega}^1_{l,l} \\ \Omega^2_{1,-1} & \Omega^2_{1,0} & \cdots & \Omega^2_{l,l} \\ \vdots & \ddots & \ddots & \vdots \\ \tilde{\Omega}^{N-1}_{1,-1} & \tilde{\Omega}^{N-1}_{1,0} & \cdots & \tilde{\Omega}^{N-1}_{l,l} \\ \Omega^N_{1,-1} & \Omega^N_{1,0} & \cdots & \Omega^N_{l,l} \end{pmatrix} \begin{pmatrix} \alpha_{1,-1} \\ \alpha_{1,0} \\ \vdots \\ \alpha_{l,l} \end{pmatrix} = \begin{pmatrix} \tilde{B}_r(\boldsymbol{r}_1) \\ B_r(\boldsymbol{r}_2) \\ \vdots \\ \tilde{B}_r(\boldsymbol{r}_{N-1}) \\ B_r(\boldsymbol{r}_N) \end{pmatrix}. \tag{15}$$

For estimating the quality of the reconstructed fields we define a relative error according to

$$\mathrm{err}_\psi = \frac{\sum_{k=1}^N |B^k_\psi - B^{k(rec)}_\psi|}{\sum_{k=1}^N |B^k_\psi|} \quad , \qquad \psi \in \{\mathrm{r}, \theta, \phi\} \tag{16}$$

4 Numerical results

The first example concerns the magnetic field produced by a magnetic dipole moment $\boldsymbol{m} = m\hat{z}$ located at the origin of the coordinate system. For this configuration B_ϕ vanishes, and we are investigating the relative errors of the B_r and B_θ components only. Moreover, in this case the only non-zero multipole amplitude is $\alpha_{1,0}$. In the following text we will denote the analytically obtained (exact) multipole amplitudes as α_{lm} while the multipole amplitudes obtained from the reconstruction procedure are referred to as $\alpha^{\mathrm{rec}}_{\mathrm{lm}}$.

The black curves in Fig. 4 show that the measurement on the closed double-layer surface yields an excellent reconstruction of the magnetic field at the measurement points with a few multipole amplitudes considered. The magnetic-field error which can be estimated by the summed difference between corresponding multipole amplitudes of the reconstructed field and the original field, also shows an excellent and stable behavior when a few multipole amplitudes are considered. In contrast, for the single-layer measurement surface the deviation between reconstructed and original fields even diverge with an increasing order of the multipole expansion as we can deduce from the results shown in Fig. 5.

The second example was performed with a magnetic dipole moment $\boldsymbol{m} = m\hat{x}$ located at $x = 0$, $y = 0$, $z = 5$cm. Since the number of relevant multipole amplitudes increases if the sources are located off the origin, also the numbers of multipole amplitudes needed for the reconstruction increase. However, we again observe from Fig. 6 that the magnetic field reconstructed from the measurement on the closed double-layer surface constantly converges to the original field.

Note that reconstruction of the tangential components $B_\theta(\boldsymbol{r})$ and $B_\phi(\boldsymbol{r})$ also is converging, however, not as fast

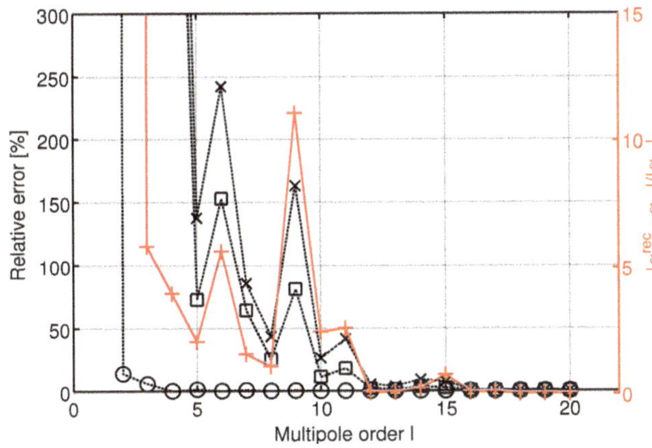

Fig. 6. Relative reconstruction error of the components B_r (black o), B_θ (black ×) , and B_ϕ (black □) on a closed double-layer surface and of the original multipole expansion (red) as a function of the order l of the multipole expansion. Source: Dipole at $\boldsymbol{r} = 5$ cm $\hat{z}$, polarized in x-direction.

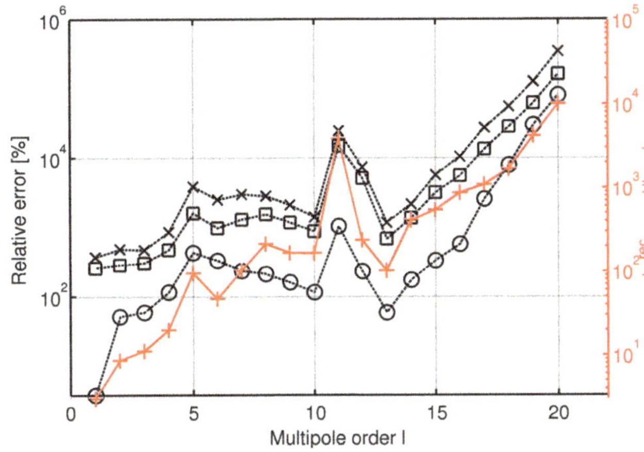

Fig. 7. Relative reconstruction error of the components B_r (black o), B_θ (black ×) , and B_ϕ (black □) on a single-layer surface and of the original multipole expansion (red) as a function of the order l of the multipole expansion. Source: Dipole at $\boldsymbol{r} = 5$ cm $\hat{z}$, polarized in x-direction.

as for the radial component. The reason for this behavior can simply be found in the underlying numerical procedure, as the corresponding system of linear equations has been solved with respect to the radial component at the measurement points. Again, the reconstruction of the magnetic field from measurements on a single-layer measurement surface fails (see Fig. 7), i.e., it diverges with an increasing number of multipole amplitudes considered.

5 Conclusions

It has been shown that the use of a closed double-layer measurement surface is excellently suited to reconstruct the spherical-multipole expansion of a divergence-free quasi-static magnetic field. In contrast, the same method equivalently applied to the conventionally used single-layer surface with magnetometers and gradiometers showed no convergence. The method can be applied to Magnetoencephalography as well as to Magnetocardiography. Particularly, it is best suited for the currently developed new type of magnetic field sensors based on magnetoelectric composite materials.

Acknowledgements. The work was supported by the Deutsche Forschungsgemeinschaft within the SFB855: *Magnetoelectric composite materials – Future biomagnetic interface.*

References

Baillet, S., Mosher, J. C., and Leahy, R. M: Electromagnetic brain mapping, IEEE Signal. Proc. Mag., 18, 14—30, 2001.

Gloor, P.: Neuronal generators and the problem of localization in electro-encephalography: Application of volume conductor theory to electroencephalography, J. Clin. Neurophysiol., 2, 327–354, 1985.

Hämäläinen, M., Hari, R., Ilmoniemi, R. J., Knuutila, J., and Lounasmaa, O. V.: Magnetoencephalography – Theory, instrumentation, and applications to noninvasive studies of the working human brain, Rev. Mod. Phys., 65, 413–497, 1993

Nunez, P. L.: Electric Fields of the Brain, New York:Oxford, USA, 1981.

Taulu, S. and Matti, K.: Presentation of electromagnetic multichannel data: The signal space separation method, J. Appl. Phys., 97, 124905–124905-10, 2005.

Taulu, S., Kajola, M., and Simola, J.: Suppression of Interference and Artifacts by the Signal Space Separation Method, Brain Topogr., 16, 269–275, 2004.

22

Properties of axial surface waves along dielectrically coated conducting cylinders

U. Siart, S. Adrian, and T. Eibert

Lehrstuhl fur Hochfrequenztechnik, Technische Universität München, 80290 München, Germany

Correspondence to: U. Siart (uwe.siart@tum.de)

Abstract. In this paper the fundamental properties of surface waves along conducting cylinders with and without dielectric coating are investigated for cylinder diameters in the centimeter range and frequencies in the gigahertz range and higher. Analytical results for the phase constant and attenuation versus the cylinder radius are derived and cutoff frequencies of various TE, TM, and hybrid waves are computed. The radial power distribution is computed in order to investigate the spatial extent of the wave fields.

1 Introduction

Electromagnetic wave propagation along cylinders has been studied in the past. Pioneering work was done by Sommerfeld, Hondros, and Goubau from the beginning to mid of last century. In recently emerging application fields for high-frequency measurement and monitoring systems the properties of surface waves along cylinders with dielectric coating are of particular significance. Exceptionally because of the availability of frequencies for industrial, scientific, and medical applications (ISM) in the gigahertz range a study of cylindrical surface waves in the high-frequency limit where the diameter of the guiding structure is no longer small compared to wavelength is eligible. An exemplary application is the monitoring of surface condition and detracting deposits in ductwork and piping under harsh conditions where ultrasound or optical techniques are inapplicable. A possible configuration for monitoring the surface condition of metal pipes by launching surface waves is suggested in Fig. 1. Conical horns may be used as wave launchers as already presented by Goubau (1951). In case there is a metal wall dielectric windows are suitable to provide through-wall transition. A second horn at the end of the pipe receives the wave. From a transmission and reflection measurement the propagation parameters of the surface wave along the pipe can be reconstructed. This in turn may allow conclusions on the surface condition like deposits or disruptions.

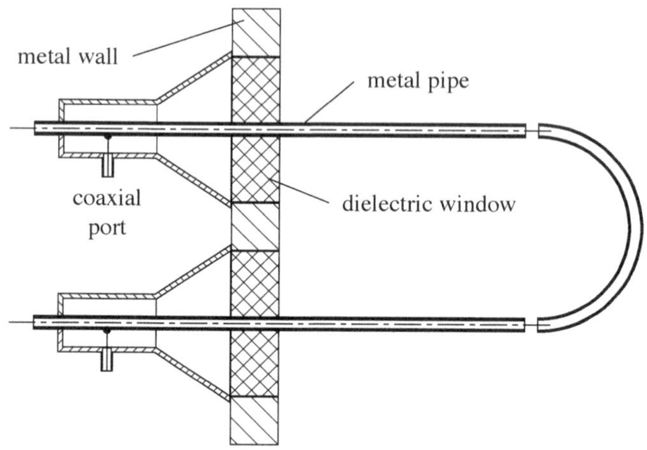

Fig. 1. Example configuration: monitoring the surface condition of metal pipes by launching surface waves.

In this paper we present analytical solutions for the characteristics of electromagnetic axial surface waves along dielectrically coated conducting cylinders. Dispersion characteristics are derived for TM, TE, and hybrid waves. The effect of losses due to finite conductivity and dielectric losses is particularly considered. Emphasis is placed on gigahertz frequencies and cylinder diameters and layer thicknesses in the centimeter range. As a result we present analytical solutions for dispersion characteristics and attenuation of different wave modes versus frequency and permittivity with a focus on the fundamental TM_{00} mode which is suitable for wave excitation through a coaxial waveguide structure. Finally a study of the radial power distribution as a function of material parameters is also presented.

Table 1. Classification of complex waves after Ishimaru (1991).

β_z	α_z	β_x	α_x	the term used
proper waves				
+	0	+	0	Fast wave (waveguide modes)
+	−	+	+	Backward leaky wave
+	0	0	+	Trapped surface wave (slow wave)
+	+	−	+	Zenneck wave
+	0	−	0	Plane-wave incidence
improper waves				
+	−	−	−	
+	0	0	−	Untrapped surface wave
+	+	+	−	Forward leaky wave

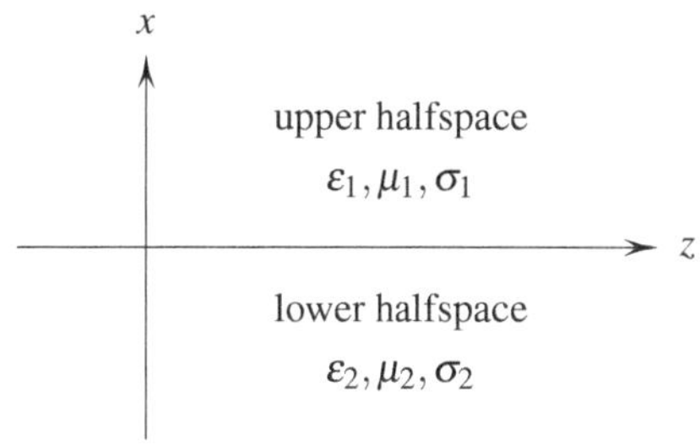

Fig. 2. Zenneck wave guiding geometry.

2 Review of complex waves

As an introduction to the topic of surface waves we give a short review of the properties of complex waves as they were discussed by Ishimaru (1991). We consider a 2-D geometry with infinite extent in cartesian coordinates x, y, and z. There are two media in the half spaces $x < 0$ and $x > 0$, respectively. Directions x and z are considered directions of wave propagation. The problem shell not depend on y, i.e. $\partial/\partial y = 0$ for all quantities. We obtain the following wave equation for a scalar quantity $u(x,z)$:

$$\left(\frac{\partial^2}{\partial x^2} + \frac{\partial^2}{\partial z^2} + k^2\right) u(x,z) = 0. \tag{1}$$

It's general solution is given by

$$u(x,z) = \mathrm{e}^{-\mathrm{j}k_x x - \mathrm{j}k_z z} \tag{2}$$

where k_x and k_z are wave numbers and $k_{x,z} = \beta_{x,z} - \mathrm{j}\alpha_{x,z}$. From $k_x^2 + k_z^2 = k^2 = \omega^2\varepsilon\mu - \mathrm{j}\omega\mu\sigma$ and by comparing real parts and imaginary parts we get the relations

$$\begin{aligned} &\beta_x^2 - \alpha_x^2 + \beta_z^2 - \alpha_z^2 = \omega^2\mu\varepsilon, \\ &\beta_x\alpha_x + \beta_z\alpha_z = \omega\mu\sigma\,. \end{aligned} \tag{3}$$

With $\boldsymbol{\alpha} = (\alpha_x\ \alpha_z)^{\mathrm{T}}$ and $\boldsymbol{\beta} = (\beta_x\ \beta_z)^{\mathrm{T}}$ it follows that in the lossless case ($\sigma = 0$) it is $\boldsymbol{\beta}\cdot\boldsymbol{\alpha} = 0$. Several solutions for wave numbers that satisfy the boundary conditions in each point of the interface plane $x = 0$ can be found. They may be classified depending on the values of $\alpha_{x,z}$ and $\beta_{x,z}$. An overview is given in Table 1. Solutions where $\alpha_x < 0$ will increase exponentially and are therefore unphysical. They are termed improper waves. Even though they are unphysical they yield practically meaningful wave types when the extent of the interface is not infinite but finite.

Let us summarize some of the solutions in Table 1 in more detail starting with the *Zenneck wave*. It can exist when at least one of the two media is lossy. In this case the dispersion equation may be solved exact and analytically. For a circular cylindrical interface the solution can only be found numerically. Also for cylinders the solution will depend what is the outer and what is the inner medium. A trivial example is the different nature of a conducting wire and a hollow waveguide. However, even for a plane interface it is found that there exists no solution of TEM type.

Assuming a configuration as shown in Fig. 2 and restricting to TM solutions the following differential equation for H_y is found:

$$\frac{1}{\sigma + \mathrm{j}\varepsilon\omega}\frac{\partial^2 H_y}{\partial x^2} - \left(k_z^2 \frac{1}{\sigma + \mathrm{j}\varepsilon\omega} + \mathrm{j}\mu\omega\right) H_y = 0. \tag{4}$$

It is solved in both half-spaces by the ansatz $H_y = C\mathrm{e}^{-\mathrm{j}k_x x}$. By further investigating the solution it is readily found that the wave numbers k_z, $k_{x,1}$ and $k_{x,2}$ correspond to the case of TM incidence at Brewster's angle. In this case Brewster's angle appears to be complex due to the losses in the medium. Hence there is no coincidence between planes of constant amplitude and planes of constant phase planes i.e. the solution represents an inhomogeneous TEM wave.

The *trapped surface wave* is a solution that is obtained by the assumption of loss-free propagation in z direction. Its phase velocity is given by

$$v_{\mathrm{ph}} = \frac{\omega}{\beta_z} = \frac{\omega}{k_z} = \frac{\omega}{k}. \tag{5}$$

Since the phase velocity of a trapped surface wave is smaller than in free space it belongs to the class of slow wave solutions. From

$$k_x^2 + \beta_z^2 = k^2 \tag{6}$$

it follows that $k_x^2 = k^2 - \beta_z^2 < 0$. Furthermore, if k and β_z are real it is necessary that $\mathrm{Re}\{k_x\} = 0$ i.e.

$$\alpha_x \neq 0 \quad \beta_x = 0. \tag{7}$$

An example geometry where trapped surface waves can exist is shown in Fig. 3. For both TE and TM waves the same dispersion relation

$$\frac{k_{x,1}}{\varepsilon_1}\tan(k_{x,1}d) = \mathrm{j}\frac{k_{x,0}}{\varepsilon_0} \tag{8}$$

$$k_{x,0/1} = \sqrt{k_{0,1}^2 - k_z^2} \tag{9}$$

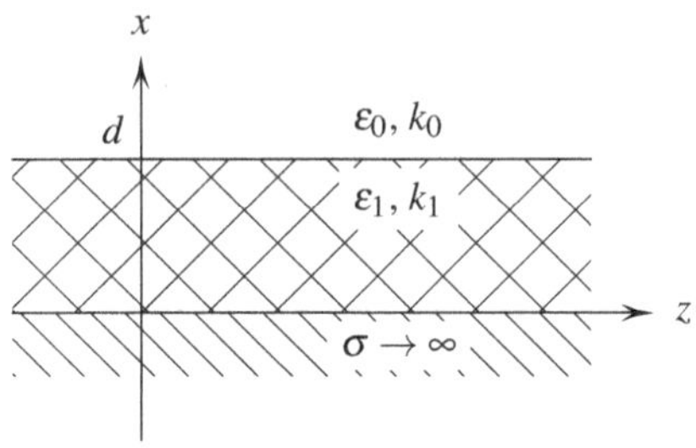

Fig. 3. Slow-wave guiding geometry.

holds. In this case the dispersion relation is transcendental and may be solved e.g. graphically.

3 Waves along conducting cylinders

Let us review surface waves in cylindrical structures. To start we discuss conducting cylinders without dielectric coating first. An efficient way to a general solution of the wave equation in cylindrical coordinates is to introduce the Hertz's vector $\boldsymbol{\pi}(\boldsymbol{r},t)$. It must satisfy the Helmholtz equation

$$\Delta\boldsymbol{\pi}+k^2\boldsymbol{\pi}=0. \tag{10}$$

A practical ansatz in frequency domain is given by

$$\boldsymbol{\pi}(\boldsymbol{r},t)=\boldsymbol{\pi}(\boldsymbol{r})\mathrm{e}^{\mathrm{j}\omega t}. \tag{11}$$

Solutions are obtained by introducing the electric and the magnetic Hertz's vector in the forms

$$\boldsymbol{\pi}_{\mathrm{m}}=\pi_{\mathrm{m}}(r,\varphi,z)\boldsymbol{e}_z \quad \boldsymbol{\pi}_{\mathrm{e}}=\pi_{\mathrm{e}}(r,\varphi,z)\boldsymbol{e}_z. \tag{12}$$

It can be shown that the choice of one common direction is sufficient for a complete and general solution. The choice of the direction $\boldsymbol{e}_z$ along the cylinder axis is advantageous in this case. By separation of variables, i.e. $\pi_{\mathrm{m,e}}=R(r)\Phi(\varphi)Z(z)$ solutions of the form

$$\pi_{\mathrm{e},z}=a_n\cdot Z_n\left(r\sqrt{k^2-k_z^2}\right)\mathrm{e}^{\pm\mathrm{j}n\varphi}\mathrm{e}^{-\mathrm{j}k_z z}\mathrm{e}^{\mathrm{j}\omega t}\quad \text{TM waves} \tag{13}$$

$$\pi_{\mathrm{m},z}=b_n\cdot Z_n\left(r\sqrt{k^2-k_z^2}\right)\mathrm{e}^{\pm\mathrm{j}n\varphi}\mathrm{e}^{-\mathrm{j}k_z z}\mathrm{e}^{\mathrm{j}\omega t}\quad \text{TE waves} \tag{14}$$

are found. Here $Z_n(x)$ is composed from two particulate solutions of the Bessel differential equation. Considering TM waves along a cylinder with finite conductivity the continuity of fields from (o)utside to (i)nside yields the dispersion equation

$$\frac{k_\mathrm{o}^2}{\mu_\mathrm{o}s_\mathrm{o}}\frac{\mathrm{H}_0^{(1)\prime}(s_\mathrm{o}r_0)}{\mathrm{H}_0^{(1)}(s_\mathrm{o}r_0)}=\frac{k_\mathrm{i}^2}{\mu_\mathrm{i}s_\mathrm{i}}\frac{\mathrm{J}_0'(s_\mathrm{i}r_0)}{\mathrm{J}_0(s_\mathrm{i}r_0)};\quad s_{\mathrm{o,i}}=\sqrt{k_{\mathrm{o,i}}^2-k_z^2}. \tag{15}$$

Since we expect solutions with $k_z^2\approx k_{\mathrm{out}}^2$ we assume that s_o is very small, i.e. $s_\mathrm{o}\to 0$. Inside the conducting cylinder the

Table 2. TM dominant wave: phase constant versus conductivity ($r=2.5$ cm, $f=2.41$ GHz).

Conductivity σ_i (in S m^{-1})	phase constant β_z (in m^{-1})	phase velocity (in % of c_0)
10^1	51.0368	1.0326
10^2	50.6889	0.3533
10^3	50.5577	0.0946
10^4	50.5227	0.0254
10^5	50.5133	0.0069
10^6	50.5108	0.0019
10^7	50.5101	0.0005

wavenumber is predominantly determined by its conductivity σ_i. Hence, $\mathrm{Re}\{s_\mathrm{i}\}$ will be a large value and $\mathrm{Im}\{s_\mathrm{i}\}$ will be a large and negative value. Therefore the approximations

$$\mathrm{J}_0'(u)/\mathrm{J}_0(u)=\mathrm{j};\quad \mathrm{H}_0(v)=\frac{2\mathrm{j}}{\pi}\ln\frac{\gamma v}{2\mathrm{j}};\quad \gamma=1.781072\ldots. \tag{16}$$

are justified. They allow for an iterative numerical solution algorithm. An exemplary result for a non-magnetic cylinder in vacuum with radius $r=2.5$ cm at signal frequency $f=2.41$ GHz is $\alpha=2.84$ dB km^{-1}.

In the case of TE waves along a cylinder with finite conductivity one obtains the dispersion relation

$$\frac{\mu_\mathrm{i}}{s_\mathrm{i}}\frac{\mathrm{J}_0'(s_\mathrm{i}r_0)}{\mathrm{J}_0(s_\mathrm{i}r_0)}=\frac{\mu_\mathrm{o}}{s_\mathrm{o}}\frac{\mathrm{H}_0^{(1)\prime}(s_\mathrm{o}r_0)}{\mathrm{H}_0^{(1)}(s_\mathrm{o}r_0)}. \tag{17}$$

In contrast to the TM case the conditions for dominant wave with low attenuation lead to inconsistencies. Hence no TM dominant waves exist. For the properties of possible subsidiary waves with high attenuation it can be said that if $u=r_0\sqrt{k_\mathrm{i}^2-k_z^2}$ is large then $\mathrm{Im}\{u\}$ must also be large and negative. Then, also $\mathrm{Im}\left\{v=r_0\sqrt{k_\mathrm{o}^2-k_z^2}\right\}<0$. This indicates exponentially growing fields in radial direction. Therefore it can be concluded that leaky waves exist for sufficiently large values u. Tables 2 and 3 show some numerical results for the phase constant of the TM dominant wave for a cylinder with radius $r=2.5$ cm and signal frequency $f=2.41$ GHz.

Figure 4 shows the numerical result for the attenuation of the TM dominant wave versus frequency for three different radii. The conductivity is $\sigma=10^7$ S m^{-1}.

4 Waves along dielectrically coated cylinders

Finally, we consider a perfectly conducting circular cylinder coated with a dielectric material. This structure has been studied in the past for larger ratios between wavelength and wire diameter. It is well known as Harms-Goubeau line and it is shown in Fig. 5 together with conical horns as wave

Table 3. TM dominant wave: phase constant versus radius ($f =$ 2.41 GHz).

Radius r (in cm)	phase constant β_z (in m^{-1})	phase velocity (in % of c_0)
0.1	50.51525	0.01066
0.5	50.51107	0.00240
0.9	50.51057	0.00140
1.3	50.51037	0.00100
1.7	50.51026	0.00078
2.1	50.51019	0.00064
2.5	50.51014	0.00055

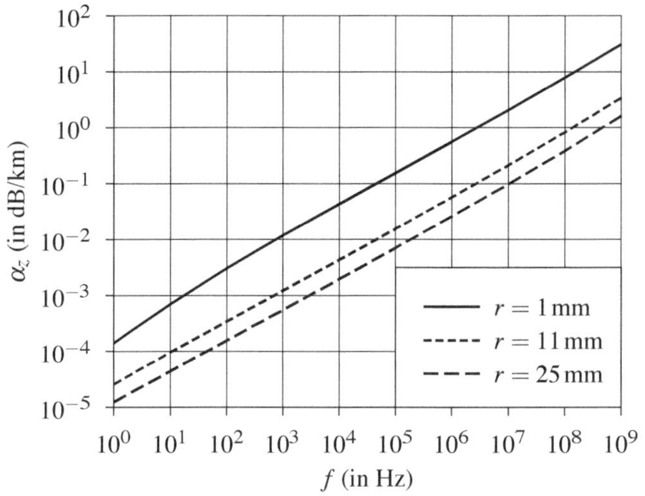

Fig. 4. TM dominant wave: attenuation for $\sigma = 10^7\,\mathrm{S\,m^{-1}}$.

launchers. We can use the same ansatz as in Sect. 3. Again, we skip the lengthy derivation and refer to the derivation of Harms (1907). Because of the more intricate boundary conditions, the eigenvalue equation reads

$$\left[\frac{k_\mathrm{o}^2}{\mu_\mathrm{o} s_\mathrm{o}} \frac{\mathrm{H}_n^{(1)\prime}(s_\mathrm{o} r_1)}{\mathrm{H}_n^{(1)}(s_\mathrm{o} r_1)} - \frac{k_\mathrm{d}^2}{\mu_\mathrm{d} s_\mathrm{d}} \frac{\mathrm{J}_n'(s_\mathrm{d} r_1)\mathrm{N}_n(s_\mathrm{d} r_0) - \mathrm{N}_n'(s_\mathrm{d} r_1)\mathrm{J}_n(s_\mathrm{d} r_0)}{\mathrm{J}_n(s_\mathrm{d} r_1)\mathrm{N}_n(s_\mathrm{d} r_0) - \mathrm{N}_n(s_\mathrm{d} r_1)\mathrm{J}_n(s_\mathrm{d} r_0)}\right]$$

$$\cdot \left[\frac{\mu_\mathrm{o}}{s_\mathrm{o}} \frac{\mathrm{H}_n^{(1)\prime}(s_\mathrm{o} r_1)}{\mathrm{H}_n^{(1)}(s_\mathrm{o} r_1)} - \frac{\mu_\mathrm{d}}{s_\mathrm{d}} \frac{\mathrm{J}_n'(s_\mathrm{d} r_1)\mathrm{N}_n'(s_\mathrm{d} r_0) - \mathrm{N}_n'(s_\mathrm{d} r_1)\mathrm{J}_n'(s_\mathrm{d} r_0)}{\mathrm{J}_n(s_\mathrm{d} r_1)\mathrm{N}_n'(s_\mathrm{d} r_0) - \mathrm{N}_n(s_\mathrm{d} r_1)\mathrm{J}_n'(s_\mathrm{d} r_0)}\right]$$

$$= \frac{k_z^2 n^2}{r_1^2} \left(\frac{1}{s_\mathrm{d}^2} - \frac{1}{s_\mathrm{o}^2}\right)^2. \tag{18}$$

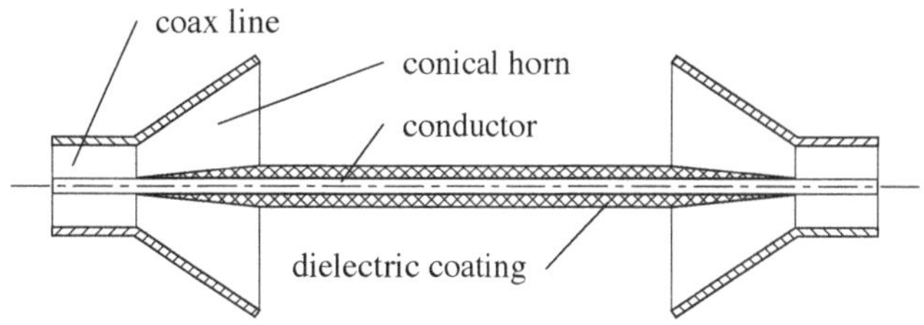

Fig. 5. Structure of a Harms-Goubeau line.

In general, the wave number k_z can be complex, which renders it difficult to find a solution. However, if we consider a loss-free dielectric material, then we know that k_z is a real number. In addition, we know that the wave number is bounded by $k_\mathrm{free} < k_z < k_\mathrm{d}$. Bessel functions of first and second kind are real for real-valued arguments. The Hankel function of first kind and even orders $n = 0, 2, 4, \ldots$ will give for an argument which has only an imaginary component a number with only an imaginary component. For odd orders $n = 1, 3, 5, \ldots$ the result is purely real. As s_o has only an imaginary component, we know that the contribution of the Hankel functions will be real-valued, like the contribution of the Bessel functions. Hence, to find solutions of the eigenvalue equation, we can just consider the real part of eigenvalue equation.

Furthermore, we see that for $n = 0$ the equation decouples into two separate equations, where the first factor of Eq. (18) will result in TM waves and the second factor in TE waves. In general, however, we have hybrid waves. The question is, whether we can find a fundamental wave, i.e. a wave which has no cutoff frequency.

As shown by Fikioris and Roumeliotis (1979), Eq. (18) can be simplified and we obtain formulas for different values of n. For $n = 0$ or $n = 1$, we have TM_{0m} waves

$$\mathrm{J}_0(s_\mathrm{d} r_1)\mathrm{N}_0(s_\mathrm{d} r_0) - \mathrm{N}_0(s_\mathrm{d} r_1)\mathrm{J}_0(s_\mathrm{d} r_0) = 0, \tag{19}$$

TE_{0m} waves

$$\mathrm{J}_0(s_\mathrm{d} r_1)\mathrm{N}_1(s_\mathrm{d} r_0) - \mathrm{N}_0(s_\mathrm{d} r_1)\mathrm{J}_1(s_\mathrm{d} r_0), \tag{20}$$

hybrid waves, HE_{1m}, where $m > 1$

$$M_1 = \mathrm{J}_1(s_\mathrm{d} r_1)\mathrm{N}_1(s_\mathrm{d} r_0) - \mathrm{N}_1(s_\mathrm{d} r_1)\mathrm{J}_1(s_\mathrm{d} r_0), \tag{21}$$

and hybrid waves, EH_{1m}, where $m \geqslant 1$

$$L_1 = \mathrm{J}_1(s_\mathrm{d} r_1)\mathrm{N}_1'(s_\mathrm{d} r_0) - \mathrm{N}_1(s_\mathrm{d} r_1)\mathrm{J}_1'(s_\mathrm{d} r_0). \tag{22}$$

For finding the cutoff frequencies, we have to find the zeros of these equations. Using the formula

$$s_\mathrm{d} = \omega_\mathrm{cutoff} \frac{\sqrt{\varepsilon_\mathrm{r,d}\mu_\mathrm{r,d} - \varepsilon_\mathrm{r,o}\mu_\mathrm{r,o}}}{c}, \tag{23}$$

we obtain the cutoff frequencies. The constant c is the speed of light in vacuum. For the TM and TE waves, we can also state a closed form expression. When the frequency is

Table 4. TM waves: some cutoff frequencies.

$\varepsilon_{r,d}$	thickness $d = 1$ mm			thickness $d = 5$ mm			thickness $d = 20$ mm		
	f_{01} (in GHz)	f_{02} (in GHz)	f_{03} (in GHz)	f_{01} (in GHz)	f_{02} (in GHz)	f_{03} (in GHz)	f_{01} (in GHz)	f_{02} (in GHz)	f_{03} (in GHz)
2	149.90	299.79	449.68	29.97	59.95	89.93	7.49	14.98	22.48
4	86.54	173.08	259.62	17.30	34.61	51.92	4.32	8.65	12.98
6	67.03	134.07	201.10	13.40	26.81	40.22	3.35	6.70	10.05
8	56.65	113.31	169.96	11.33	22.66	33.99	2.83	5.66	8.49

Table 5. TE waves: some cutoff frequencies.

$\varepsilon_{r,d}$	thickness $d = 1$ mm			thickness $d = 5$ mm			thickness $d = 20$ mm		
	f_{01} (in GHz)	f_{02} (in GHz)	f_{03} (in GHz)	f_{01} (in GHz)	f_{02} (in GHz)	f_{03} (in GHz)	f_{01} (in GHz)	f_{02} (in GHz)	f_{03} (in GHz)
2	74.94	224.84	374.74	14.98	44.96	74.94	3.74	11.24	18.73
4	43.27	129.81	216.35	8.65	25.96	43.27	2.16	6.49	10.81
6	33.51	100.55	167.58	6.70	20.11	33.51	1.67	5.02	8.37
8	28.32	84.98	141.63	5.66	16.99	28.32	1.41	4.24	7.08

large enough, we can use the asymptotic approximation of the Bessel functions. Therefore, we obtain

$$\omega_{m,\mathrm{cutoff}} = m \cdot \frac{\pi}{d} \frac{c}{\sqrt{\varepsilon_{r,d}\mu_{r,d} - \varepsilon_{r,o}\mu_{r,o}}} \tag{24}$$

and

$$\omega_{m,\mathrm{cutoff}} = \frac{1}{d}\left(m + \frac{1}{2}\right)\pi \cdot \frac{c}{\sqrt{\varepsilon_{r,d}\mu_{r,d} - \varepsilon_{r,o}\mu_{r,o}}} \tag{25}$$

for the TM and TE case, respectively, where $d = r_1 - r_0$ is the thickness of the dielectric coating.

Regarding Eq. (19), we find that $s_d = 0$ satisfies this equation. Hence, there is a fundamental TM wave, similar to the one we had in the case of the uncoated conducting cylinder. However, by deriving the Eqs. (19) through (22), it appears that for $n = 1$ again $s_d = 0$ is a valid solution (Fikioris and Roumeliotis, 1979). Therefore, there is also a fundamental hybrid wave HE_{11} so that on a dielectric coated conducting cylinder two fundamental waves exist. The cutoff frequencies for different configurations are presented in Tables 4, 5 and 6.

Although more waves can exist than just the fundamental TM_{00} wave, we are interested in its properties as we can easily excite this type of wave with, e.g. a horn structure. In Fig. 6 we can see the dispersion behavior of the fundamental TM_{00} wave. This wave is propagating on a cylindrical conductor with radius $r = 2.5$ cm and which has a dielectric coating with relative permittivity $\varepsilon_r = 8$ and thickness $d = 5$ mm. We can observe that for the lower frequencies the $k_z \approx k_o$, while for high frequencies $k_z \approx k_d$. This indicates that for high frequencies the wave is guided closer to the conductor, i.e. the wave is contained mostly in the dielectric.

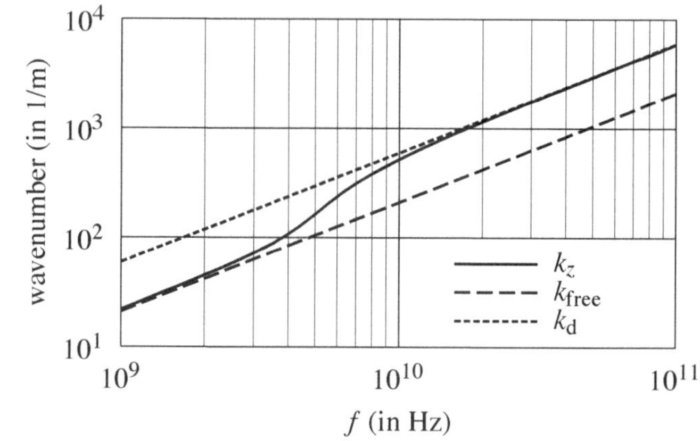

Fig. 6. Dispersion diagram of the TM_{00} fundamental mode.

The power transported in the dielectric is

$$P_d = 2\pi \int_{r_0}^{r_1} \frac{1}{2} E_r H_\varphi r \,\mathrm{d}r = \frac{k_z}{4\pi\omega\varepsilon_d} \ln\frac{r_1}{r_0} |I|^2, \tag{26}$$

while the power transported in the exterior is

$$P_a = \pi |a_0^a|^2 \sqrt{\frac{\varepsilon_a}{\mu_a}} |s_a|^2 k_z k_a \frac{1}{2\varsigma_a{}^2} \cdot [F(\varsigma_a r) - F(\varsigma_a r_1)], \tag{27}$$

where

$$F(\varsigma_a r) = (\varsigma_a r)^2 \left\{ \frac{2}{\varsigma_a r} \mathrm{j} \mathrm{H}_0^{(1)}(\mathrm{j}\varsigma_a r) \mathrm{H}_1^{(1)}(\mathrm{j}\varsigma_a r) + [\mathrm{H}_0^{(1)}(\mathrm{j}\varsigma_a r)]^2 + [\mathrm{H}_1^{(1)}(\mathrm{j}\varsigma_a r)]^2 \right\}, \tag{28}$$

Table 6. Hybrid waves: some cutoff frequencies.

$\varepsilon_{r,d}$	thickness $d = 1$ mm			thickness $d = 5$ mm			thickness $d = 20$ mm		
	f_{12} (in GHz)	f_{13} (in GHz)	f_{14} (in GHz)	f_{12} (in GHz)	f_{13} (in GHz)	f_{14} (in GHz)	f_{12} (in GHz)	f_{13} (in GHz)	f_{14} (in GHz)
2	149.90	299.79	449.69	30.01	59.97	89.95	7.59	15.03	22.51
4	86.54	173.08	259.62	17.33	34.62	51.93	4.38	8.68	13.00
6	67.03	134.07	201.10	13.42	26.82	40.22	3.39	6.72	10.07
8	56.65	113.31	169.96	11.34	22.66	33.99	2.86	5.68	8.51

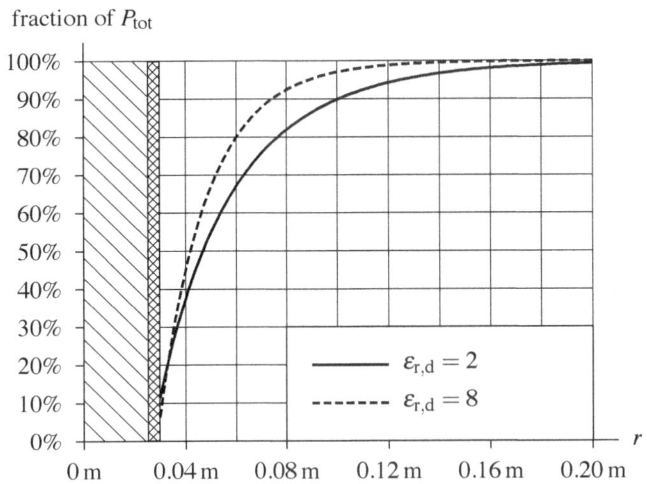

Fig. 7. Spatial power distribution of the TM^{00} fundamental mode, shown is the fraction of the total power P_{tot} that is guided inside a cylindrical surface with radius r. The thickness of the dielectric layer is $d = 5$ mm.

and $\varsigma_a = \sqrt{k_z^2 - k_a^2}$ as given by Goubau (1950). The spatial power distribution for a conductor with radius 2.5 cm and dielectric coating with thickness $d = 5$ mm and relative permittivities $\varepsilon_r = 2$ and $\varepsilon_r = 8$, respectively, is shown in Fig. 7. More than 75 % of the power is contained within a 5 cm radius when the relative permittivity is $\varepsilon_r = 8$.

5 Conclusions

In this paper the fundamental properties of surface waves along conducting cylinders have been investigated for cylinder diameters in the centimeter range and frequencies in the gigahertz range and higher. For cylinders without additional dielectric coating the TM_{00} dominant wave (axial Sommerfeld wave) has been considered. Its dispersion and attenuation characteristics have been derived and the effect of cylinder radius and conductivity on the phase constant has been studied. As an exemplary result on the guiding properties it can be concluded that for a radius $r = 2.5$ cm, frequency $f = 2.41$ GHz, and conductivity $\sigma = 10^7\ \mathrm{S\,m^{-1}}$ 75 % of the total power are carried within a radius of $r = 1$ m.

For the case of conducting cylinders with additional dielectric coating on their surface TM_{0m}, TE_{0m}, and hybrid waves have been studied. We computed cutoff frequencies for various material parameters and derived and approximate formula for the wave attenuation assuming perfect conductor and dielectric losses. Again the effect of cylinder radius and permittivity has been studied. As an exemplary result on the guiding properties with additional dielectric coating it can be said that for a cylinder radius $r = 2.5$ cm, frequency $f = 2.41$ GHz, and dielectric coating with relative permittivity $\varepsilon_r = 8$ and thickness $d = 1$ mm 75 % of the total power are carried within a radius of $r = 10$ cm.

References

Fikioris, J. G. and Roumeliotis, J. A.: Cutoff Wavenumbers of Goubau Lines, IEEE T. Microw. Theory, 27, 570–573, 1979.

Goubau, G.: Surface Waves and Their Application to Transmission Lines, J. Appl. Phys., 21, 1119–1128, 1950.

Goubau, G.: Single-Conductor Surface-Wave Transmission Lines, Proceedings of the IRE, 39, 619–625, 1951.

Goubau, G.: Cylindrical Dielectric Waveguide Modes, IRE Transactions on Antennas and Propagation, 7, 140–146, 1959.

Harms, F.: Elektromagnetische Wellen an einem Draht mit isolierender zylindrischer Hülle, Ann. Phys.-Berlin, 328, 44–60, 1907.

Ishimaru, A.: Electromagnetic Wave Propagation, Radiation and Scattering, Prentice Hall, Englewood Cliffs, New Jersey, 10th Edn., 1991.

Roberts Jr., T. E.: Theory of the Single-Wire Transmission Line, J. Appl. Phys., 24, 57–67, 1953.

Schelkunoff, S. A.: Anatomy of Surface waves, IEEE Transactions on Antennas and Propagation, 7, 133–139, 1959.

Snitzer, E.: Cylindrical Dielectric Waveguide Modes, J. Opt. Soc. Am. (1917–1983), 51, 491–498, 1961.

Scattering centers modeling of non-anechoic measurement environments

K. A. Yinusa, C. H. Schmidt, and T. F. Eibert

Lehrstuhl für Hochfrequenztechnik, Technische Universität München, Arcisstr. 21, 80333 Munich, Germany

Correspondence to: K. A. Yinusa (k.y@tum.de)

Abstract. Near-field measurements are established techniques to obtain the far-field radiation pattern of an Antenna Under Test via near-field measurements and subsequent near-field far-field transformation. For measurements acquired in echoic environments, additional post-processing is required to eliminate the effects of multipath signals in the resulting far-field pattern. One of such methods models the measurement environment as a multiple source scenario whereby the collected near-field data is attributed to the AUT and some scattering centers in the vicinity of the AUT. In this way, the contributions of the AUT at the probe can be separated from those of the disturbers during the near-field far-field transformation if the disturber locations are known. In this paper, we present ways of modeling the scattering centers on equivalent surfaces such that echo suppression is possible with only partial or no information about the geometry of the scatterers.

1 Introduction

In order to determine the far-field radiation pattern of an Antenna Under Test (AUT), measurement of the field distribution can be carried out directly in the far-field. The far-field is the region surrounding the AUT from a radial distance of approximately $2D^2/\lambda$, where D is the largest dimension of the aperture and λ is the wavelength. For electrically small antennas i.e. antennas with small physical size relative to the wavelength, the far-field criteria is met at relatively small distance from the AUT and a controlled environment can be easily provided for direct far-field measurement. A controlled environment is one that averts threats due to weather, security, electromagnetic interference, multipath propagation, etc. Such an environment is often achieved in an anechoic chamber, which is a room shielded to prevent electromagnetic waves penetrating from outside and interior lined with radiation absorbent material (RAM) to prevent reflections off the ceiling and floor of the room. In this way, a free space environment is provided for the AUT and the probe antenna so that only the mutual interactions between the AUT and the probe antenna have influence on the measurement. But for electrically large antennas, the far-field characteristics are satisfied at a relatively large distance from the AUT and indoor measurement will require an anechoic chamber of equally large size. An alternative is to measure the antenna in the radiating near-field; a region extending from a distance of about a wavelength from the aperture to the far-field region (Yaghjian, 1986). Measurement in the near-field reduces the required dimension of the anechoic chamber and the amount of absorbers needed to cover the walls. A near-field far-field transformation (NFFFT) algorithm is then applied on the measured near-field data to obtain the far-field pattern. As the dimension of the AUT becomes larger relative to the wavelength, providing an anechoic chamber becomes expensive even for near-field measurements and it is desirable to have techniques that allow for semi-anechoic or outdoor measurements in exchange for some echo suppression post-processing. Techniques based on the matrix pencil (Sarkar and Pereira, 1995) and fast Fourier transform (FFT) techniques attempt to identify the line-of-sight (LOS) components from a broadband measurement (Loredo et al., 2004). The requirement for a broadband measurement however means that these techniques are not applicable for narrowband antennas and require a substantial increase in the measurement time for broadband antennas. Modal filtering techniques, applicable to measurement on canonical geometries i.e. planar, cylindrical, and spherical scan geometries, such as MARS (Gregson et al., 2009) and IsoFilter™ (Hess, 2011) modify the measurement setup such that higher order modes corresponding to the multipath contributions are excited and identifiable. In source reconstruction techniques (González and Álvarez López, 2011; Cano et al., 2010; Araque Quijano et al., 2011), equivalent current sources over fictitious surfaces enclosing the echo sources are computed alongside those representing the AUT

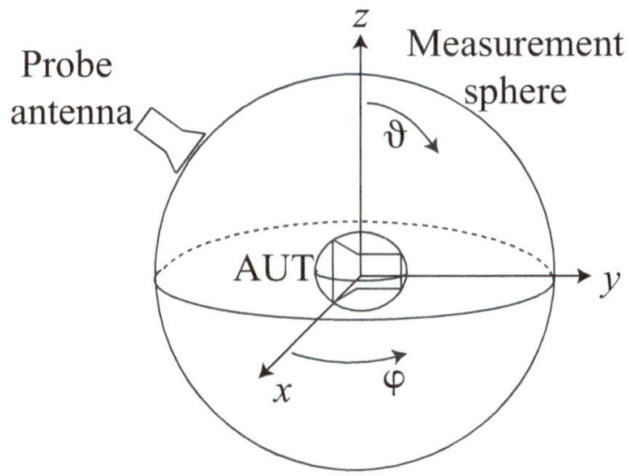

Fig. 1. Measurement setup.

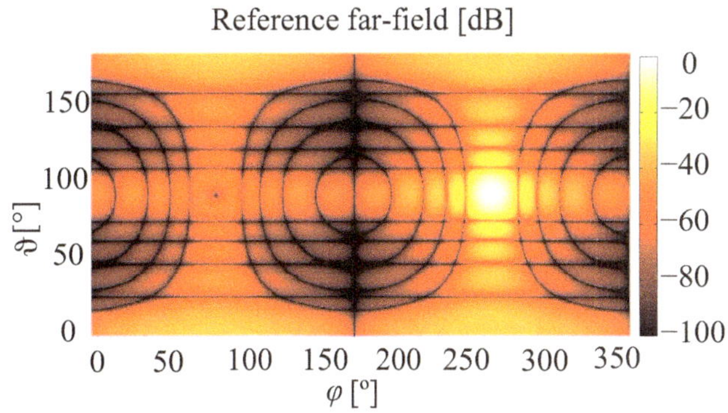

Fig. 2. Reference far-field of the AUT.

such that the AUT far-field can be computed with the isolated AUT equivalent current sources. In the scattering center representation (Schmidt and Eibert, 2009), the echo sources are treated as independent sources contributing to the measured near-field data and with the knowledge of the echo locations, these sources can be integrated into the NFFF transformation algorithm and their contributions can be separated from that of the AUT. The need for echo locations however can be impractical for situations involving several disturbers. In this paper, we present ways of modeling the contributing echo sources such that knowledge of their locations is not necessary for echo suppression. The scattering center representation as implemented in this paper is integrated with a plane wave based near-field far-field transformation algorithm described in Schmidt et al. (2008). A brief description of this algorithm is given in Sect. 2.1 for self-containment. In Sects. 3, 4, and 5, we present scattering centers modeling and results for localized, partially localized, and non-localized echo scenarios, respectively.

2 Preliminaries

2.1 Plane wave based near-field far-field transformation

We consider a spherical near-field measurement setup as illustrated in Fig. 1, where a set of plane waves in all directions $(\bar{\mathbf{I}} - \hat{k}\hat{k}) \cdot \tilde{\boldsymbol{J}}(\hat{k})$ with respect to the origin of the coordinate system is used to describe the radiation behavior of the AUT (Schmidt et al., 2008). The recorded probe signal

$$\boldsymbol{U}(\boldsymbol{r}_\mathrm{M}) = -j\frac{\omega\mu}{4\pi}\sum_{\hat{k}} T_L(\hat{k},\boldsymbol{r}_\mathrm{M})\bar{\mathbf{P}}(\hat{k},\hat{r}_\mathrm{M}) \cdot (\bar{\mathbf{I}} - \hat{k}\hat{k}) \cdot \tilde{\boldsymbol{J}}(\hat{k}) \quad (1)$$

at a measurement point $\boldsymbol{r}_\mathrm{M}$ is reconstructed by translating all the plane waves from the AUT to the point $\boldsymbol{r}_\mathrm{M}$. $\bar{\mathbf{P}}(\hat{k},\hat{r}_\mathrm{M})$ is the probe correction coefficient used to weight the plane waves to compensate for the probe influence and

$$T_L(\hat{k},\boldsymbol{r}_\mathrm{M}) = -\frac{jk}{4\pi}\sum_{l=0}^{L}(-j)^l(2l+1)h_l^{(2)}(kr_\mathrm{M})P_l(\hat{k}\cdot\hat{r}_\mathrm{M}) \quad (2)$$

is the diagonalized translation operator derived from the fast multipole method (FMM) (Coifman et al., 1993) and it converts radiated plane waves from the AUT into incident plane waves at the probe. $h_l^{(2)}$ and P_l are the spherical Hankel function of the second kind and a Legendre polynomial, respectively and k is the wavenumber. The plane wave translation is done in this way to all the measurement points and a linear system of equations

$$\boldsymbol{U} = -j\frac{\omega\mu}{4\pi}\|\mathbf{C}\| \cdot \tilde{\boldsymbol{J}} \quad (3)$$

is formulated with the number of equations and unknowns equaling the number of measurement points and number of plane waves for each polarization, respectively. The system of equations is solved with an iterative solver such as GMRES (Saad and Schultz, 1986) to retrieve the AUT plane wave coefficients.

2.2 Measurement setup

The AUT in the following sections is a 5 GHz square horn antenna with an aperture size of 0.4245 m and a symmetric amplitude and phase excitation. The AUT and echo sources are simulated using dipole distributions and the echoic data is generated by evaluating and superimposing the dyadic Green's function of free space for the individual dipoles at the measurement points. Data is generated for the spherical measurement geometry with a spacing of 3° along ϑ and φ directions at a distance of 0.9 m with the AUT located at the origin of the coordinate system and oriented as shown in Fig. 1. The echo sources are point sources distributed randomly outside the measurement sphere. The reference far-field and far-field with echoes are shown in Figs. 2, and 3, respectively.

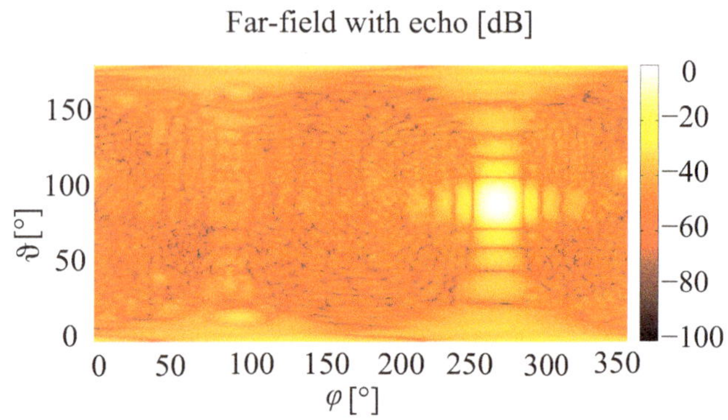

Fig. 3. Far-field with echo: obtained far-field for echoic near-field data without a subsequent echo suppression.

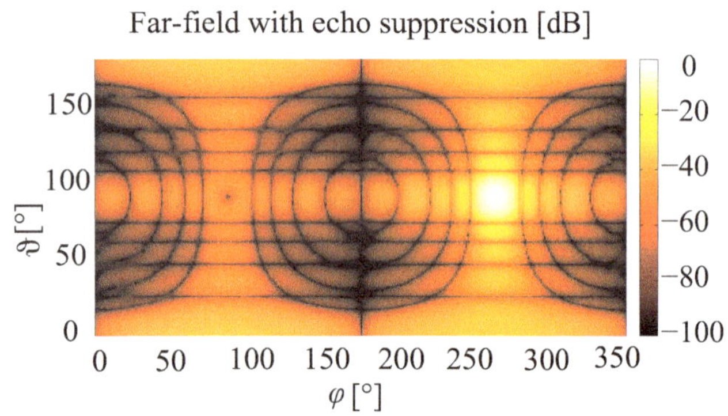

Fig. 4. Far-field after echo suppression with localized echo sources: echo sources are replaced with auxiliary sources at the same locations during the NFFFT.

3 Scattering centers modeling of localized echo sources

The recorded near-field data is a superposition of the field emanating directly from the AUT and multipath signals. The linear system of equations in Eq. (3) can then be rewritten with increased number of unknowns as follows:

$$\boldsymbol{U}' = -j\frac{\omega\mu}{4\pi}\|\mathbf{C}_{\mathrm{AUT}}\| \cdot \tilde{\boldsymbol{J}} - j\frac{\omega\mu}{4\pi}\sum_{i=1}^{N_{\mathrm{SC}}}\|\mathbf{C}_{\mathrm{SC}i}\| \cdot \tilde{\boldsymbol{J}}_{\mathrm{SC}i} \tag{4}$$

where $\mathbf{C}_{\mathrm{AUT}}$ and $\mathbf{C}_{\mathrm{SC}i}$ represent the coupling matrices for the AUT and the i-th scattering center, respectively. N_{SC} is the number of scattering centers employed. When the echo locations are known, $\mathbf{C}_{\mathrm{SC}}$ can be computed and Eq. (4) is solved to retrieve the plane wave coefficients $\tilde{\boldsymbol{J}}$ of the AUT and $\tilde{\boldsymbol{J}}_{\mathrm{SC}i}$ of the i-th scattering center. Figure 4 shows the far-field pattern after echo suppression.

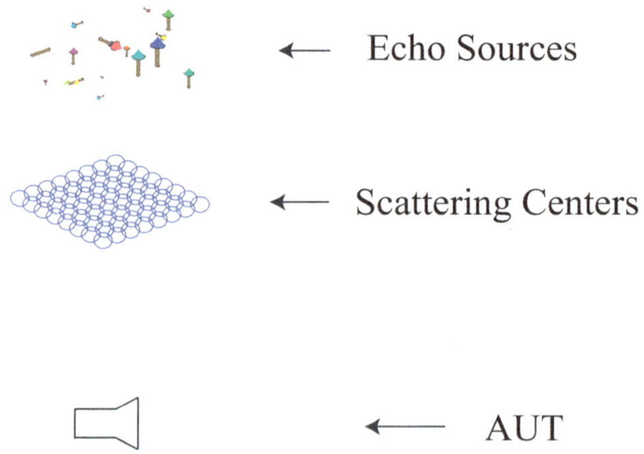

Fig. 5. SCs placed to block the LOS between the AUT and the echo sources.

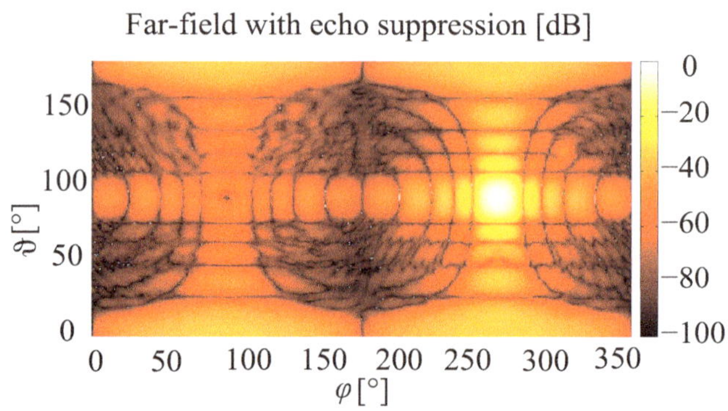

Fig. 6. Far-field after echo suppression with partial knowledge of echo locations.

4 Scattering centers modeling of partially localized echo sources

Often, precise information regarding echo sources is not available and when it is, manual input of large numbers of locations makes it more efficient to combine neighboring echo sources. In these cases, we place the scattering centers along the line-of-sight (LOS) between the AUT and the echo sources such that the measurement geometry is shielded from the target disturbers as shown in Fig. 5. This requires only the knowledge of the direction from which we expect to receive the multipath signals. The result with this representation is shown in Fig. 6 and the difference between the reference and the obtained far-fields before and after echo suppression for $\vartheta = 90°$ cut is plotted in Fig. 7.

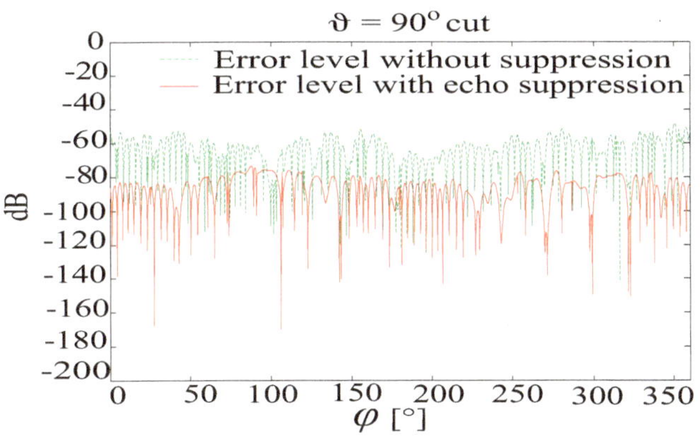

Fig. 7. Error level before and after echo suppression for the representation in Fig. 5.

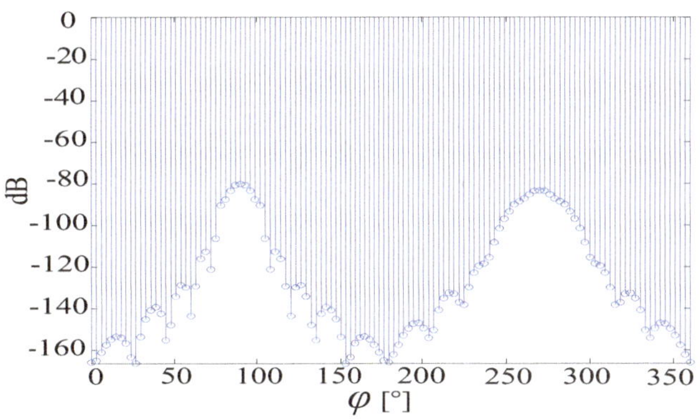

Fig. 10. Search for locations with least error penalty: placing scattering centers at locations where there are no echo sources leads to error in the obtained far-field. Plotted is the energy erroneously attributed to a test scattering center at different locations relative to the AUT total energy.

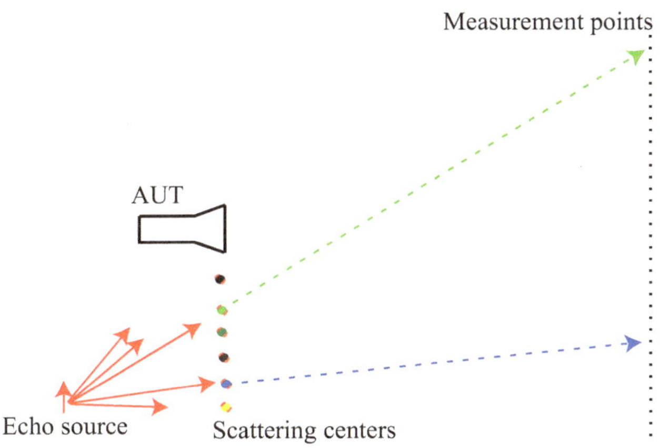

Fig. 8. Several scattering centers can account for contributions of an echo source at different measurement points.

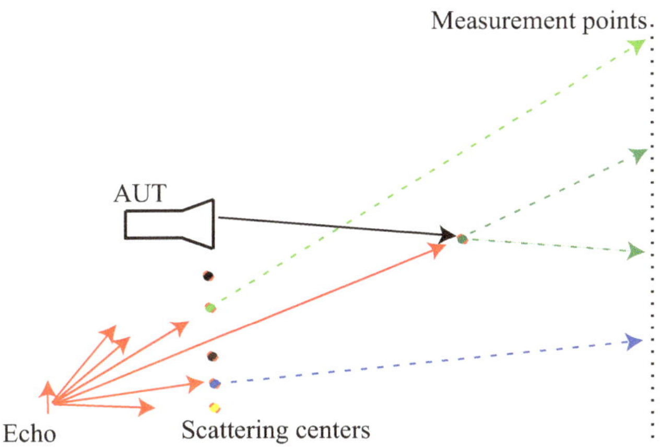

Fig. 9. Scattering centers can also contribute errors in the far-field by replacing components of the AUT signal.

5 Scattering centers modeling without echo localization

Without knowledge of the echo locations, $\mathbf{C}_{SC}$, the coupling matrix that describes how an individual echo source contributes to the measured near-field data, cannot be computed. We therefore place the scattering centers around the AUT since, as shown in Fig. 8, such scattering centers can potentially represent components of the echo contributions. In the same way however, these scattering centers can also introduce errors into the obtained far-field pattern as part of the plane waves originating from the AUT might be attributed to the scattering centers during the NFFFT. This scenario is depicted in Fig. 9. To improve the achievable echo suppression without knowledge of the echo locations, we optimize the scattering centers placement by searching for locations where the least amount of error is introduced to the obtained AUT pattern. To this end, an echo free measurement is made and a single scattering center is placed at a particular location during the NFFFT. The NFFFT is repeated several times with the scattering center at varying locations and the attributed energy is recorded. An ideal scattering center location in this scenario will attribute an energy of $-\infty$ dB to the signal originating from the scattering center since the measurement is echo free. The search is done for ϑ fixed at 90° and φ from 0 to 360° with a step of 3°. The result in Fig. 10 shows that the least error is introduced around the null axis of the AUT i.e. along the AUT extended aperture. As a result of this observation, the scattering centers are placed along the plane containing the AUT aperture as shown in Fig. 11 such that no scattering center is placed within the measurement sphere. In this way, the scattering centers replace only the echo sources and components of the AUT signal are not attributed to the scattering centers. The result with this representation is shown in Fig. 12 and the error levels before and after echo suppression in Fig. 13.

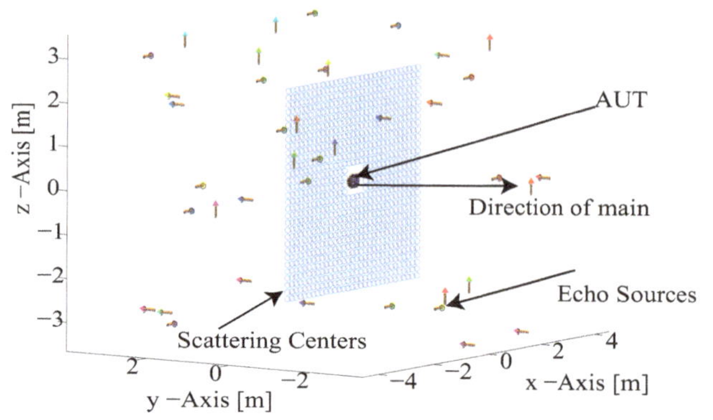

Fig. 11. Scattering centers are placed outside the measurement sphere along AUT extended aperture.

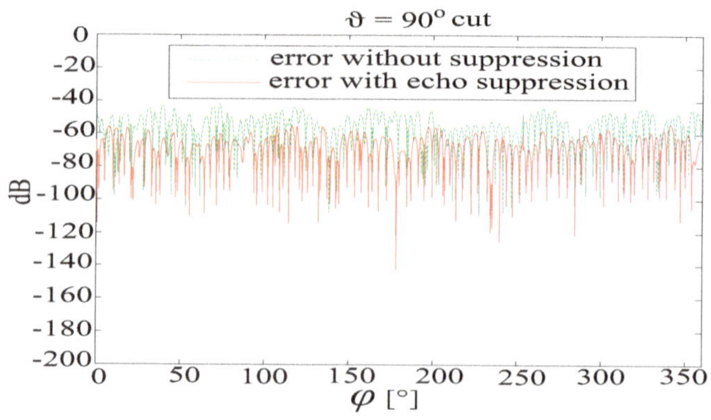

Fig. 13. Error levels before and after echo suppression for the case where the echo locations are unknown.

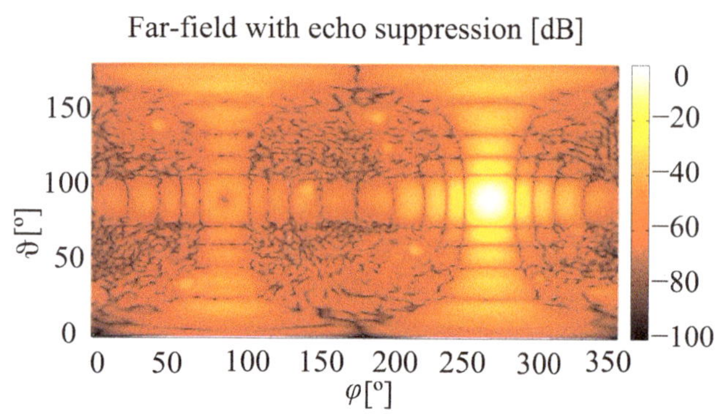

Fig. 12. Far-field after echo suppression without echo localization.

6 Conclusions

We presented methods of modeling echoic measurement environments with auxiliary sources by utilizing the available knowledge about the measurement facility. In situations where the echo locations are known exactly or partially, we simply replace them with scattering centers so that their contributions are integrated within the NFFFT. When the locations are unknown, we optimize the scattering centers placement such that the echo contributions are removed considerably while at the same time keeping the AUT contribution isolated.

References

Araque Quijano, J., Vecchi, G., Sabbadini, M., Scialacqua, L., Bencivenga, B., Mioc, F., and Foged, L.: Source reconstruction in advanced processing of antenna measurements, in: Proceedings of the 5th European Conference on Antennas and Propagation (EUCAP), 3875–3879, 2011.

Cano, F., Sierra-Castaner, M., Burgos, S., and Besada, J. L.: Applications of sources reconstruction techniques: Theory and practical results, in: Proceedings of the Fourth European Conference on Antennas and Propagation (EuCAP), 1–5, 2010.

Coifman, R., Rokhlin, V., and Wandzura, S.: The fast multipole method for the wave equation: a pedestrian prescription, IEEE Antenn. Propag. M., 35, 7–12, doi:10.1109/74.250128, 1993.

González, C. G. and Álvarez López, Y.: Characterization of antenna interaction with scatterers by means of equivalent currents, Pr. Electromagn. Res., 116, 185–202, 2011.

Gregson, S., Newell, A., and Hindman, G.: Reflection Suppression in Cylindrical Near-field Antenna Measuremennt Systems, Cylindrical MARS, AMTA 31st Annual Meeting and Symposium, Salt Lake City, 2009.

Hess, D.: Spherical modal filtering of antenna patterns augmented by translation of coordinate origin, in: IEEE International Symposium on Antennas and Propagation (APSURSI), 1632–1635, doi:10.1109/APS.2011.5996615, 2011.

Loredo, S., Pino, M., Las-Heras, F., and Sarkar, T.: Echo Identification and Cancellation Techniques for Antenna Measurement in Non-anechoic Test Sites, IEEE Antenn. Propag. M., 46, 100–107, doi:10.1109/MAP.2004.1296154, 2004.

Saad, Y. and Schultz, M.: GMRES: A generalized minimal residual algorithm for solving nonsymmetric linear systems, SIAM J. Sci. Stat. Comp., 7, 856–869, 1986.

Sarkar, T. and Pereira, O.: Using the Matrix Pencil Method to Estimate the Parameters of a Sum of Complex Exponentials, IEEE Antenn. Propag. M., 37, 48–55, doi:10.1109/74.370583, 1995.

Schmidt, C. and Eibert, T.: Near-field far-field transformation in echoic measurement environments employing scattering center representations, in: 3rd European Conference on Antennas and Propagation, EuCAP 2009, 3370–3374, 2009.

Schmidt, C., Leibfritz, M., and Eibert, T.: Fully Probe-Corrected Near-Field Far-Field Transformation Employing Plane Wave Expansion and Diagonal Translation Operators, IEEE T. Antenn. Propag., 56, 737–746, doi:10.1109/TAP.2008.916975, 2008.

Yaghjian, A.: An overview of near-field antenna measurements, IEEE T. Antenn. Propag., 34, 30–45, doi:10.1109/TAP.1986.1143727, 1986.

Generation of postured voxel-based human models for the study of step voltage excited by lightning current

J. Gao[1,2], I. Munteanu[1,3], W. F. O. Müller[2], and T. Weiland[1,2,3]

[1]Graduate School of Computational Engineering, Technische Universität Darmstadt, Dolivostraße 15, 64293 Darmstadt, Germany

[2]Institut für Theorie Elektromagnetischer Felder, Technische Universität Darmstadt, Schloßgartenstraße 8, 64289 Darmstadt, Germany

[3]CST – Computer Simulation Technology AG, Bad Nauheimer Straße 19, 64289 Darmstadt, Germany

Abstract. With the development of medical technique and computational electromagnetics, high resolution anatomic human models have already been widely developed and used in computation of electromagnetic fields induced in human body. Although these so called voxel-based human models are powerful tools for research on electromagnetic safety, their unchangeable standing posture makes it impossible to simulate a realistic scenario in which people have a lot of different postures. This paper describes a poser program package which was developed as an improved version of the free-from deformation technique to overcome this problem. It can set rotation angles of different human joints and then deform the original human model to make it have different postures. The original whole-body human model can be deformed smoothly, continuity of internal tissues and organs is maintained and the mass of different tissues and organs can be conserved in a reasonable level.

As a typical application of the postured human models, this paper also studies the effect of the step voltage due to a lightning strike on the human body. Two voxel-based human body models with standing and walking posture were developed and integrated into simulation models to compute the current density distribution in the human body shocked by the step voltage. In order to speed up the transient simulation, the reduced c technique was used, leading to a speedup factor of around 20. The error introduced by the reduced c technique is discussed and simulation results are presented in detail.

Correspondence to: J. Gao
(gao@gsc.tu-darmstadt.de)

1 Introduction

Since coming into being in the last century, medical imaging technique has developed rapidly. Nowadays, it allows the construction of digital three-dimensional computational human models based on actual anatomy of individual humans by scanning a real person by means of computer tomography (CT) or magnetic resonance imaging (MRI) techniques. The usefulness of these so-called voxel-based human body models lies in their faithful representation of human anatomy. Until now there are already more than twenty voxel-based human models reported including different ethnicities, ages, sizes, heights and genders (Caon, 2004).

However, most of these voxel-based human models only have their initial standing posture, which significantly limits related research activities including a realistic scenario where postured human bodies must be considered. In order to solve this problem, several deformed voxel-based human models have already been developed, such as the sitting human models developed by Dawson et al. (1999), Allen et al. (2003, 2005), Findlay and Dimbylow (2005, 2006) et al.; the deformed human model with outstretched or forward stretched arms developed by Dawson et al. (1999, 2002), Findlay and Dimbylow (2005); and the deformed human models with several different postures developed by Nagaoka and Watanabe (2008). Until now the most popular method used to develop a postured human model is realized by cutting, rotating and re-attaching the limb to the trunk, and then adding manual editing around the joint attachment areas to ensure tissue continuity. However, developing a deformed human model from the heterogeneous voxel model with a manual editing procedure is very time-consuming. Thus it's necessary to develop a software tool which can be used to generate different kinds of postured human models conveniently.

In this paper, we describe a poser software package which was developed to posture the most popular voxel-based

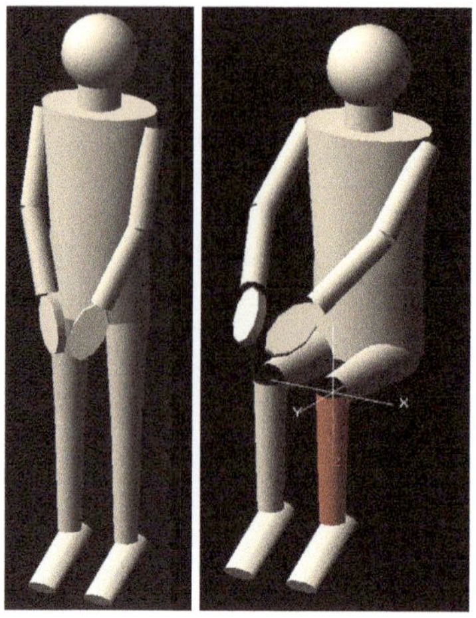

Fig. 1. Simplified human model with initial standing posture and deformed sitting posture, which was designed to define rotation angles of different joints of the human body.

human model built from the Visible Human Project®. Compared with other voxel-based human models, this model is even more challenging, because its two arms are originally bent in front of the abdomen and touch the trunk. A segmentation approach was developed to separate arm voxels away from the trunk, which avoids the possible artifacts caused by the simple cutting method. Apart from that deformation of the knee joint was dealt with separately to avoid unreasonable deformation of the femur and tibia, and to rotate the patella into its right position when the knee joint is bent.

2 Generation of postured voxel-based human models

2.1 Voxel-based human model

The voxel-based human model studied in this paper named "Hugo" (Dipp GmbH, 2010) was built based on the most popular human model from the Visible Human Project® (VHP) funded by the United States National Library of Medicine (Ackerman, 1998). It has already been used for different kinds of applications especially in electromagnetics (Gjonaj et al., 2002). This human model offers a variable voxel resolution ranging from $8 \times 8 \times 8\,mm^3$ to $1 \times 1 \times 1\,mm^3$. Each voxel is assigned a unique tissue type, with a total of 31 different tissue types. In the finest resolution, the model consists of approximately 380 million voxels. It represents a 38 years old, male person of 187 cm height and an approximate weight of 113 kg. Its original posture was lying on an experimental table, similar to an upright standing posture.

2.2 Workflow of the developed poser program

The deformation technique used in the poser program was developed based on the free-form deformation technique (FFD) introduced by Sederberg and Parry in 1986 (Sederberg and Parry, 1986). The FFD technique is a very versatile tool to arbitrarily deform three-dimensional solid geometric models. Through a lattice of movable control points, the FFD technique can deform any kinds of complex geometric model in a free form style. At the same time, the continuity and the topology of the original model can be conserved.

In order to deform the original voxel-based human model as conveniently as possible, a graphical user interface was developed, whose main functionality is described below.

The very first step of the poser program is to import the original voxel dataset file and to display the three-dimensional human models based on different tissues and organs. The rendering function of the poser program was realized by the marching cubes algorithm (Lorensen and Cline, 1987), which can extract outer surfaces of different tissues and organs of the voxel dataset. After checking the original human model, the user can switch to a simplified human model (Fig. 1) in order to set the rotation angles of different joints, and, at the same time, to get a general impression about the deformed shape of the postured human model.

After the rotation angles of all considered joints are defined satisfactorily, the deformation calculation can be performed to posture the original human model. Two special issues related to this step are described in more detail in Sects. 2.4 and 2.5. At any point, the user can go back to the simplified human model mentioned above to redefine the rotation angles of some joints and perform deformation calculation again, until a satisfactory result is achieved. In the last step, the postured human body model is meshed and exported as new voxel dataset file.

The outer surface and the internal structure of a sitting human model generated by the poser program are shown in Fig. 2. It can be seen that the outer surface of the whole human body is deformed smoothly, and the continuity of all internal tissues and organs is conserved.

2.3 Mass conservation

An important issue for the deformation of biological models is the conservation of their mass. This is especially important for simulations involving the calculation of the Specific Absorption Rate (SAR), whose accuracy depends in an essential way on the tissue mass.

In our poser program, for the sitting human model with 1 mm resolution, the relative error of the mass of any tissue or organ is below 3.5% as compared to the original model. The relative error of the whole body mass, which is summation of all tissues and organs, is even below 1%. This error level is acceptable for most simulations involving human models.

The mass error decreases with increasing resolution Δx of the human body model, as shown in Fig. 3. The numerical experiments indicate that the convergence of the whole body mass is of the order $O(\Delta x^{1/2})$.

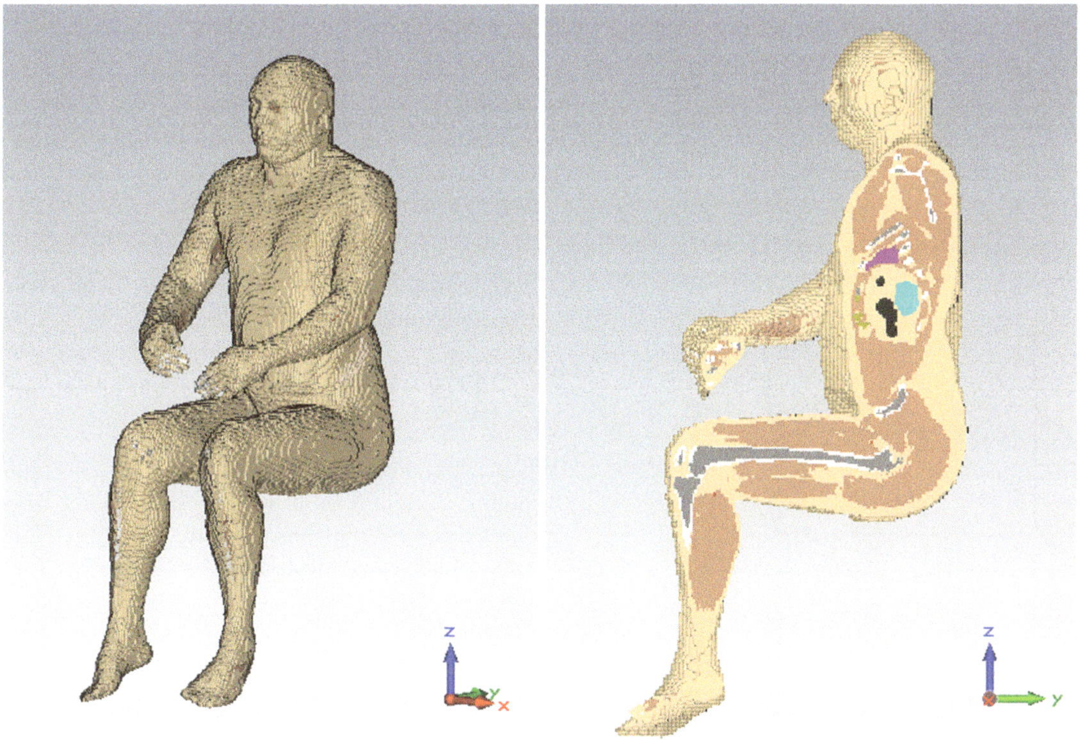

Fig. 2. Outer surface and internal structure of the sitting human model generated from the poser program.

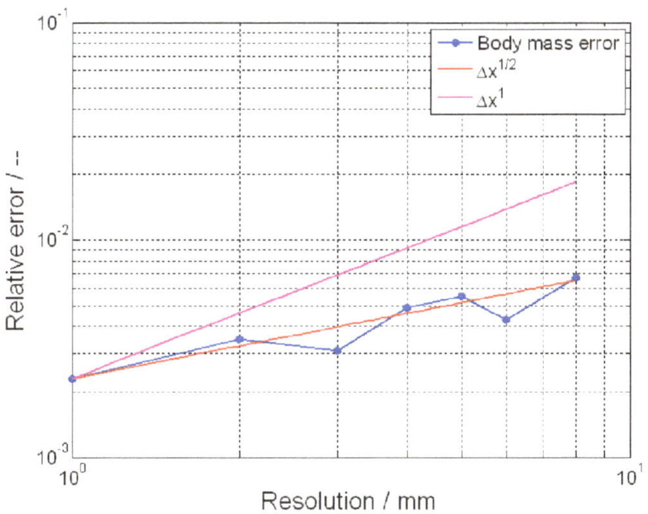

Fig. 3. Relative error of the total body mass, after the movement of the human model to a sitting position. The curves Δx^1 and $\Delta x^{1/2}$ are also displayed for reference.

2.4 Movement of the shoulder joint

Most voxel-based human models have the arms well separated from the trunk. However, even for this situation a simple cutting procedure must be performed to separate the arm voxels from those of the chest near the armpit, a procedure which is recognized as error-prone (Nagaoka and Watanabe, 2008).

The configuration of the voxel-based human model studied in this paper is even more complicated. Originally its arms are bent in front of the abdomen and they touch the trunk, as shown in Fig. 4. To be able to correctly deform the arm model, all voxels which belong to the arm must be separated from the trunk voxels. Fortunately, in the finest resolution of the voxel dataset, a continuous layer of skin voxels exists around the whole human body model, which can be used for the arm segmentation. The complexity of the human body model near the arm required that the whole arm model be divided into six parts which are dealt with separately in order to get an accurate enough result (Fig. 4).

As an example, the first part of the right arm, shown in Fig. 5a, placed below the right shoulder, can be checked from the center towards the right side as the red arrow in Fig. 5a shows. All voxels found after the skin voxel can be set as arm voxels. The other five parts of the right arm model and their checking directions are listed in respective sub-figures of Fig. 5. They can be dealt with in a similar manner, until all voxels belonging to the arm are separated away from trunk.

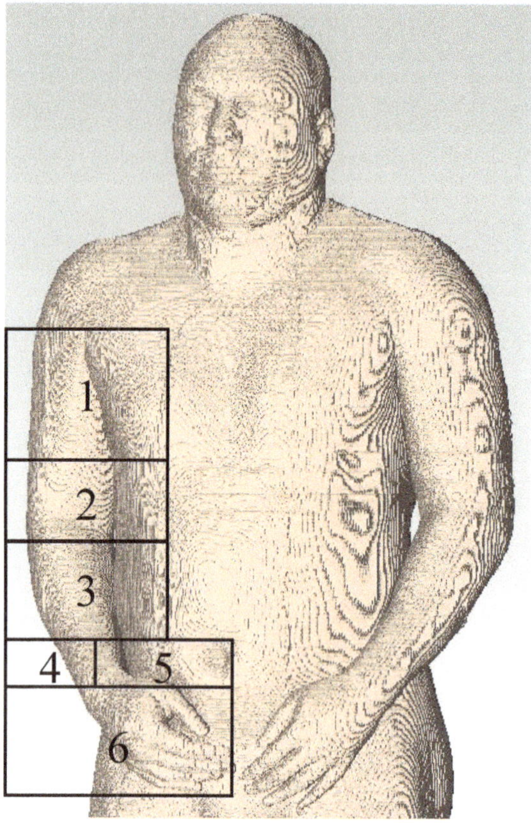

Fig. 4. Initial configuration of the original human model and six parts of the right arm.

2.5 Movement of the knee joint

The main advantage of the FFD technique in deforming the voxel-based human model lies in maintaining the continuity of internal tissues and organs, due to the intrinsic "flexible" deformation ensured by the FFD.

However, the anatomically correct movement is complicated in some special joints such as the knee joint. The movement of the knee joint includes not only flexible deformations of the fat, muscle, skin etc., but also rigid body movement and rotation of the skeleton. When the knee joint is bent, the tibia rotates around the bottom surface of the femur, while at the same time, the relative position of the patella with respect to the femur changes. Straightforward application of the FFD technique would also lead to a deformation of the bones and would maintain the patella in an incorrect position. This disadvantage is well known and was previously mentioned in the literature (Nagaoka and Watanabe, 2008).

To circumvent the above mentioned problems, some special measures were taken. They are based on the observation that rigid body movement can also be achieved with FFD, if the FFD control lattice itself is not deformed.

In order to realize all movements of the knee joint mentioned above, three sets of FFD control lattices were defined around the knee joint. The first set is designed to deform the flexible materials such as fat, muscle, skin etc, while the second and third set control lattice are used to rotate the femur, tibia and patella as a whole. The final deformation of the knee joint of the sitting human model generated from the poser program can be seen in Fig. 2, right and consistent with the real anatomic structure.

3 Application: study on step voltage with postured human models

3.1 Introduction to the step voltage caused by lightning current

When a lightning flash strikes the earth or buildings, the lightning current can flow into the ground through any grounded conductive object, and the discharge current in the ground will cause an electric field in the near area. If a human happens to stand or walk in this region, an electric potential difference between his or her legs exists – the so-called step voltage, whose value is an important reference parameter that must be considered in designing, manufacturing and optimizing lightning protection equipment.

For analysis purposes, the International Electro-technical Commission (IEC) suggests to use following formula to express current wave shape of the short lightning flash stroke (IEC, 2003)

$$i(t) = \frac{I}{k} \cdot \frac{(t/\tau_1)^{10}}{1+(t/\tau_1)^{10}} \cdot e^{(-t/\tau_2)}. \qquad (1)$$

Here, $I = 200\,\text{kA}$ is the peak current value, $k = 0.93$ is a correction factor for the peak value, $\tau_1 = 19\,\mu\text{s}$ and $\tau_2 = 485\,\mu\text{s}$ are front and tail time constants. The duration of the lightning current signal is defined as $350\,\mu\text{s}$.

This time-domain current signal was used as excitation source for all simulations in this section.

3.2 Simulation models

In order to study the influence of the step voltage on human's health and to consider the effects of human's posture on the results, two simulation models, one standing and one walking human body model, were used. The original voxel-based human model, with 4 mm resolution, was deformed into two these postures with the poser program. A step voltage of around 240 kV peak value was used to excite two perfect electric conductive patches placed below the human's feet, which models a human standing around 11 meters away from the point where the lightning strikes. The simulation setup including the walking and the standing human body model is shown in Fig. 6.

A fast Fourier transformation applied to the lightning current signal expressed by Eq. (1) shows that most part of the frequency spectrum of this current signal is below 100 kHz. Because of the very low frequency spectrum of the excitation

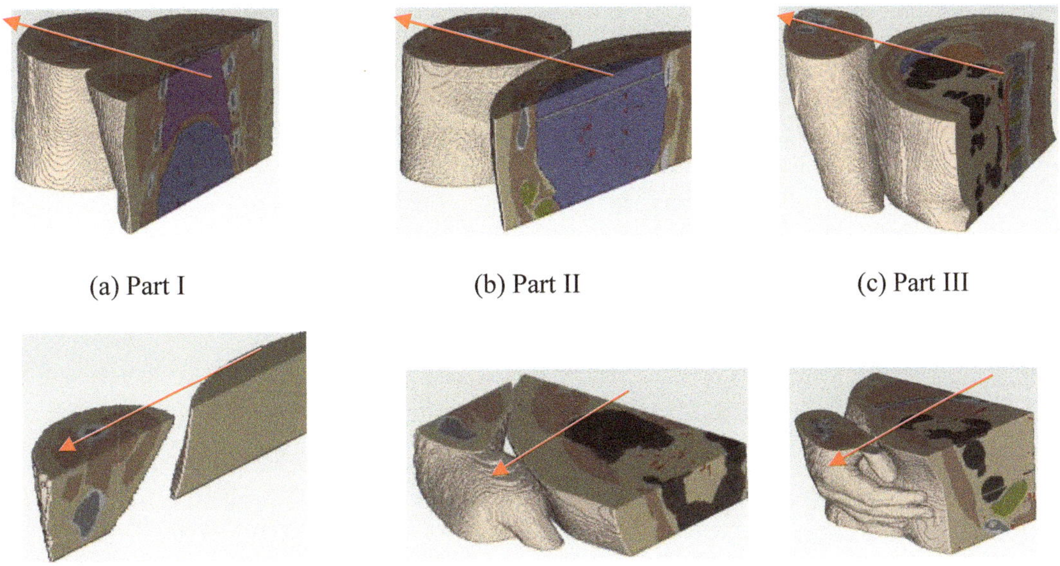

(a) Part I (b) Part II (c) Part III

(d) Part IV (e) Part V (f) Part VI

Fig. 5. Six parts of the right arm and their checking directions used to separate arm voxels away from the trunk.

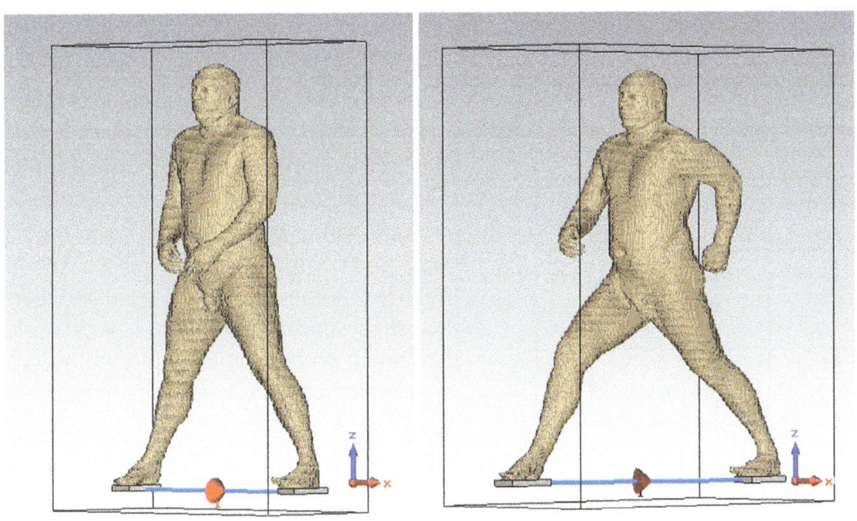

Fig. 6. Simulation models of the step voltage excited by lighting current including human models with standing and walking posture.

signal, the simulation studied here is in principle an electro-quasistatic problem. However, in order to take into account the dispersive characteristics of the biological materials, we used a full-wave transient simulator (CST AG, 2010), which can deal with such materials.

The direct use of this explicit full-wave transient simulator for the quasistatic simulation results in a huge number of time steps (around 10^8) that need to be simulated and therefore a very long calculation time. The reduced c technique, first introduced by Holland in 1995 to solve a magneto-quasistatic problem (Holland, 1995), was studied to be applied in this simulation. To our knowledge, it has not been previously applied to electro-quasistatic simulations, in which the relative permeability, rather than the relative permittivity, can be increased to mathematically reduce the speed of light, and thus to obtain a much larger stable time step than the value specified by the Courant limit. A detailed error analysis showed that increasing the relative permeability by a factor of 400 leads to a speedup of the simulation around 20 times, while the relative error thus introduced is below 5% before the peak value of the excitation signal at $t = 30\,\mu$s, and below 0.5% near and after the peak value.

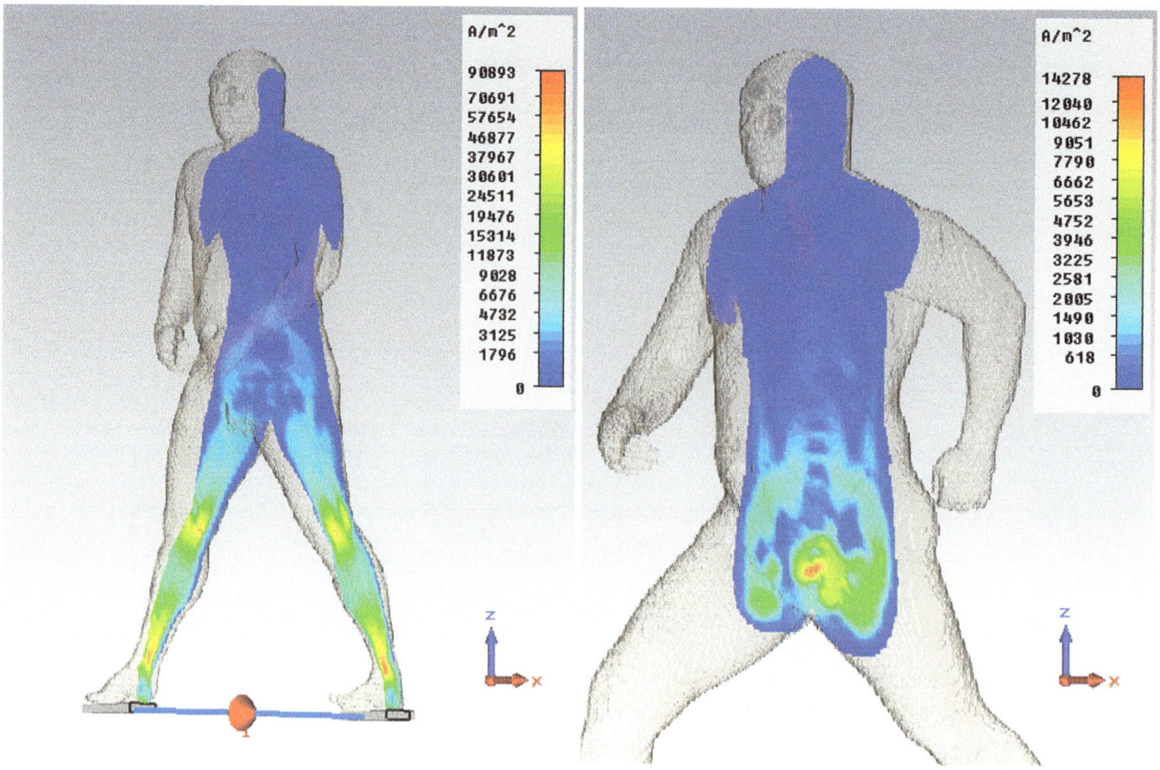

Fig. 7. Current density distribution in the standing and walking human body shocked by the step voltage.

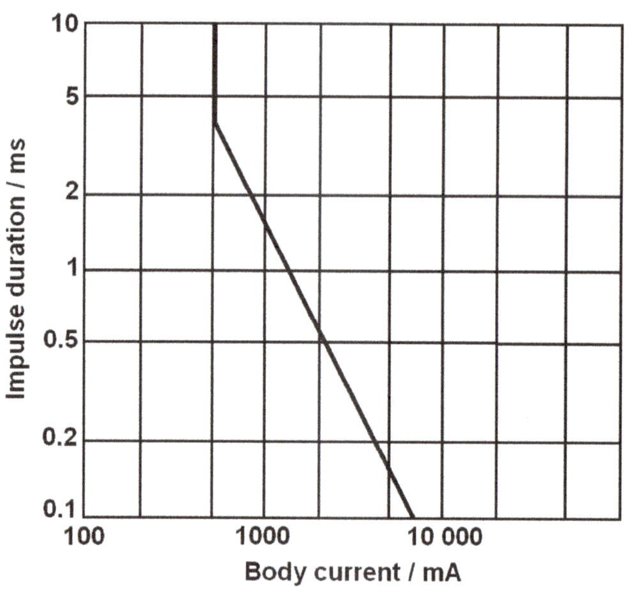

Fig. 8. Threshold of ventricular fibrillation with different body current and impulse duration (IEC, 2006).

3.3 Simulation results

Figure 7 shows the current density distribution in the standing and walking human model at $t = 30\,\mu$s, which is near the peak value of the lightning current signal. The largest current density appears in the leg muscles, especially near the knee and ankle joints, as well as in the bladder. The current density is comparatively much weaker in bones, due to the much lower conductivity of bone and marrow with respect to that of the muscle, blood and bladder. In the upper part of the human body, which is quite far away from the excitation sources, the current density is lower as well.

As a very important organ in the whole human body, the safety of the heart is always of great concern. The standard 60479 of the International Electrotechnical Commission (IEC) (IEC, 2005) defines the whole body current threshold of ventricular fibrillation of the heart for different excitation impulse durations, as Fig. 8 shows for a signal type which is very similar to a lightning strike. The curve in the figure indicates the probability of fibrillation risks for current flowing along the path from left hand to feet. For a path from left foot to right foot, the value should be divided by a heart current correction factor $F = 0.04$ (IEC, 2006). The measurement results from Fig. 8 show that the threshold of the ventricular fibrillation is around $4000\,\text{mA}/0.04 = 100\,\text{A}$ effective value for impulse duration $350\,\mu\text{s} = 0.35\,\text{ms}$ along the path from one foot to the other. Our simulation results here show that the *peak* value of the whole-body current can reach around 100 A, thus its effective value is much smaller than this value. This means that, due to the very short duration of the excitation signal and to the comparatively long distance between the heart and the excitation sources, under the simulation condition above, the whole body current is below the threshold of ventricular fibrillation prescribed by the IEC standard.

Apart from that, the total body current is the almost the same for the standing and the walking human model, which means that the human's posture does not have a big influence on the human body's impedance in frequency range below 100 kHz.

4 Conclusions

In this paper, we describe a poser program to deform one of the most popular voxel-based human models from the Visible Human Project®. The program is based on the free form deformation technique, a versatile and powerful deformation technique in computer graphics. New techniques were developed to segment two arms away from the trunk and for separately dealing with body parts requiring rigid body motion, such as the knee joint, in order to get an anatomically correct postured human body model. The poser program allows the convenient and quick movement of the voxel-based human model, with a minimum of user interaction. The outer surface of the whole human body can be deformed smoothly, and the continuity of internal tissues and organs is ensured. In addition, the relative error of the mass of every tissue and organ between the original and the postured human model can be conserved in a reasonable level.

As an application example, we studied the internal current density distribution in the human body shocked by the step voltage, by using two postured voxel-based human body models. Time-domain calculations were all speeded up around 20 times with low error by using the reduced c technique. Simulation results have shown all dangerous areas of the human body with comparatively higher current density values, which are valuable references for lightning protection engineering.

Acknowledgements. This work is supported by the 'Excellence Initiative' of the German Federal and State Governments and the Graduate School of Computational Engineering at Technische Universität Darmstadt.

References

Ackerman, M. J.: The visible human project, Proceedings of the IEEE, 86(3), 504–511, 1998.

Allen, S. J., Adair, E. R., Mylacraine, K. S., Hurt, W., and Ziriax, J.: Empirical and theoretical dosimetry in support of whole body resonant RF exposure (100 MHz) in human volunteers, Bioelectromagnetics, 24(7), 502–509, 2003.

Allen, S. J., Adair, E. R., Mylacraine, K. S., Hurt, W., and Ziriax, J.: Empirical and theoretical dosimetry in support of whole body radio frequency (RF) exposure in seated human volunteers at 220 MHz, Bioelectromagnetics, 26(6), 440–447, 2005.

Caon, M.: Voxel-based computational models of real human anatomy: a review, Radiat. Environ. Bioph., 42(4), 229–235, 2004.

CST AG: CST – Computer Simulatin Technology, online available at: http://www.cst.com/, accessed: 20 August 2010.

Dawson, T. W., Caputa, K., and Stuchly, M. A.: Numerical evaluation of 60 Hz magnetic induction in the human body in complex occupational environments, Phys. Med. Biol., 44, 1025–1040, 1999.

Dawson, T. W., Caputa, K., and Stuchly, M. A.: Magnetic field exposures for UK live-line workers, Phys. Med. Biol., 47, 995–1012, 2002.

Dipp GmbH: Anatomical volume data sets, online available at: http://www.vr-laboratory.com/, accessed: 8 November 2010.

Findlay, R. P. and Dimbylow, P. J.: Effects of posture on FDTD calculations of specific absorption rate in a voxel model of the human body, Phys. Med. Biol., 50(16), 3825–3835, 2005.

Findlay, R. P. and Dimbylow, P. J.: FDTD calculations of specific energy absorption rate in a seated voxel model of the human body from 10 MHz to 3 GHz, Phys. Med. Biol., 51(9), 2339–2352, 2006.

Gjonaj, E., Bartsch, M., Clemens, M., Schupp, S., and Weiland, T.: High-resolution human anatomy models for advanced electromagnetic field computations, IEEE Transactions on Magnetics, 38(2), 357–360, 2002.

Holland, R.: FDTD analysis of nonlinear magnetic diffusion by reduced c, IEEE Transactions on Antennas and Propagation, 43(7), 653–659, 1995.

IEC: IEC 62305-1: Protection against lighting – Part 1: General principles, International Electrotechnical Commission, 2003.

IEC: IEC 60479-1: Effects of current on human beings and livestock – Part 1: General aspects, International Electrotechnical Commission, 2005.

IEC: IEC 60479-2: Effects of current passing through the human body – Part 2: Special aspects, International Electrotechnical Commission, 2006.

Lorensen, W. E. and Cline, H. E.: Marching cubes: A high resolution 3D surface construction algorithm, in: Proceedings of the 14th annual conference on Computer graphics and interactive techniques (ACM Siggraph 87), 24, 163–169, 1987.

Nagaoka, T. and Watanabe, S.: Postured voxel-based human models for electromagnetic dosimetry, Phys. Med. Biol., 53(24), 7047–7061, 2008.

Sederberg, T. W. and Parry, S. R.: Free-form deformation of solid geometric models, Proceedings of the 13th annual conference on computer graphics and interactive techniques (ACM Siggraph 86), 20(4), 151–160, 1986.

A comparison of minimum norm and MUSIC for a combined MEG/EEG sensor array

H. Ahrens, F. Argin, and L. Klinkenbusch

Institut für Elektrotechnik und Informationstechnik, Christian-Albrechts-Universität zu Kiel, Germany

Correspondence to: H. Ahrens (ha@tf.uni-kiel.de)

Abstract. Many different algorithms for imaging neuronal activity with magnetoencephalography (MEG) or electroencephalography (EEG) have been developed so far. We validate the result of other authors that a combined MEG/EEG sensor array provides smaller source localisation errors than a single MEG or single EEG sensor array for the same total number of sensors. We show that Multiple Signal Classification (MUSIC) provides smaller localisation errors than an unweighted minimum norm method for activity located in the cortical sulcus regions. This is important for many medical applications, e.g. the localisation of the origin of epileptic seizures (focal epilepsy) that can be located very deep in the cortical sulcus.

1 Introduction

Imaging of neuronal activity can be very helpful in medical applications, e.g. before a resection surgery of brain areas including the origin of epileptic seizures (Pataraia et al., 2002). Functional Magnetic Resonance Imaging (fMRI) measures the oxygen consumption due to synaptic activity and provides a high spatial resolution of the order of a few millimeters but due to the slow haemodynamic response a relatively low temporal resolution of the order of one second (Baillet et al., 2001). In contrast MEG or EEG directly measure the magnetic field or electrostatic scalp potential produced by neuronal currents outside the brain and provide a much higher temporal resolution of a few milliseconds but a lower spatial resolution of a few centimeters. Today in MEG the magnetic fields of the order of a few femtotesla are measured with Superconducting Quantum Interference Devices (SQUIDs) and in EEG the potential is measured with electrodes directly at the scalp. With future uncooled and unshielded magnetoelectric sensors developed by the research project SFB 855 of the Deutsche Forschungsgemeinschaft maybe also the magnetic field can be measured with sensors directly at the scalp. This could provide on one hand a higher SNR, on the other hand a better alignment of the head to the sensor array for applications in MEG/EEG video monitoring of epileptic seizures. It has been shown before (Sharon et al., 2007; Liu et al., 2002) that a combined SQUID-MEG/EEG sensor array provides smaller source localisation errors than single SQUID-MEG or single EEG sensor arrays for the same total number of sensors. In Sect. 5 we affirm these results with a simulation for a combined MEG/EEG sensor array with MEG magnetometers located directly at the scalp. EEG electrodes and MEG sensors contain different information of the primary currents and for a spherical conductor model even complementary information (Dassios et al., 2007; Dassios, 2008). Due to the non-uniqueness of the solution of the inverse problem (Fokas et al., 2004) lots of different algorithms for the localisation of neuronal activity have been developed in the last decades (Baillet et al., 2001). In Sect. 6 we compare the localisation accuracy of an unweighted minimum norm method and the MUSIC (Mosher et al., 1992) beamforming technique for a focal activity located at the cortical gyrus and in the cortical sulcus region of the brain. We show that the localisation error for the cortical sulcus activity is smaller for the MUSIC algorithm than for the unweighted minimum norm method.

2 Realistic head model

In the simulations of this work we use a realistic head model with the realistic head layers cortex, innerskull, outerskull and scalp extracted from an MRI-Scan of the patients head. The layers are tesselated into a triangular mesh with a vertex

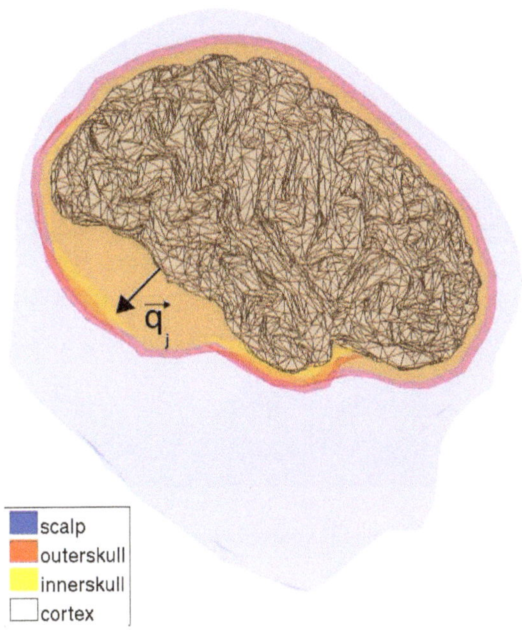

Fig. 1. Realistic head geometry extracted from MRI. Number of dipoles $\boldsymbol{q}_j$ on the cortical mesh $n = 15\,010$.

to vertex distance of the order of 1 mm. Figure 1 shows the used layers. The cortical mesh contains 15 010, the innerskull surface 642, the outerskull surface 642 and the scalp surface 1082 vertices. The MRI geometries were provided by Sabine Meunier (Hôpital de la Salpêtrière, Paris) as part of the tutorial dataset distributed with the BrainStorm program (Tadel et al., 2011).

3 Forward model

The total current density $\boldsymbol{j}$ present in the brain can be separated into a primary current density $\boldsymbol{j}_p$ that corresponds to the synaptic activity and a volume current density $\boldsymbol{j}_V$ that corresponds to the passive response currents to fulfil the charge conservation $\nabla \cdot \boldsymbol{j} = 0$:

$$\boldsymbol{j} = \boldsymbol{j}_p + \boldsymbol{j}_V \tag{1}$$

To calculate the magnetic field and the electrostatic potential at the sensor positions we use current dipoles $\boldsymbol{q}_j$ fixed at the vertices $\boldsymbol{r}_j$ of the cortical mesh to model the primary current density (distributed model):

$$\boldsymbol{j}_p(\boldsymbol{r}) = \sum_{j=1}^{n} \boldsymbol{q}_j \delta(\boldsymbol{r} - \boldsymbol{r}_j) \tag{2}$$

We choose dipole orientations normal to the cortical surface (Baillet et al., 2001, p. 16) and consider n dipoles on the cortical mesh and p sensors. We use a piecewise homogeneous and isotropic conductor model with the 4 conductivity domains brain, skull, scalp and air separated by the 3 layers innerskull, outerskull and scalp as shown in Fig. 1. We use the default conductivity parameters of the BrainStorm program 1 (brain), 0.0125 (skull), 1 (scalp) and 0 (air). The magnetic field and electrostatic potential at the sensor positions is calculated with the Boundary Element Method (BEM) taking into account the contributions of both primary and volume currents (Mosher et al., 1999, p. 246). For the numerical solution of the BEM we use the forward modeling program OpenMEEG (Gramfort et al., 2010; Kybic et al., 2005). The signal of the MEG sensor is proportional to the magnetic flux through the sensor that can be calculated with the superposition principle:

$$\varphi_i = \sum_{j=1}^{n} G_{B,ij} q_j \tag{3}$$

For a data time series $t_1, \ldots, t_T$ the time dependent dipole amplitude matrix $\mathbf{Q}$ is defined by

$$\mathbf{Q} = \begin{pmatrix} q_1(t_1) & \ldots & q_1(t_T) \\ \vdots & \ddots & \vdots \\ q_n(t_1) & \ldots & q_n(t_T) \end{pmatrix} \tag{4}$$

and the time dependent data matrix $\boldsymbol{\phi}_B$ is defined by

$$\boldsymbol{\phi}_B = \begin{pmatrix} \varphi_1(t_1) & \ldots & \varphi_1(t_T) \\ \vdots & \ddots & \vdots \\ \varphi_p(t_1) & \ldots & \varphi_p(t_T) \end{pmatrix} \tag{5}$$

Inserting Eqs. (4) and (5) into Eq. (3) yields a system of linear equations

$$\boldsymbol{\phi}_B = \mathbf{G}_B \mathbf{Q} \tag{6}$$

with a $(p \times n)$ gain matrix $\mathbf{G}_B$.

4 Forward model for combined MEG/EEG

Using the superposition principle we can derive an analogue system of linear equations for the EEG signal:

$$\boldsymbol{\phi}_V = \mathbf{G}_V \mathbf{Q} \tag{7}$$

where V denotes the electrostatic scalp potential and an analogue system of linear equations for a combined MEG/EEG data matrix:

$$\boldsymbol{\phi} = \begin{pmatrix} \boldsymbol{\phi}_B \\ \boldsymbol{\phi}_V \end{pmatrix} = \begin{pmatrix} \mathbf{G}_B \\ \mathbf{G}_V \end{pmatrix} Q = GQ \tag{8}$$

where $\mathbf{G}_B$ denotes the gain matrix for the MEG signal and $\mathbf{G}_V$ denotes the gain matrix for the EEG signal.

Fig. 2. Combined MEG/EEG sensor array with 177 normal MEG magnetometers and 177 EEG electrodes. In the single MEG simulation the EEG electrodes are substituted by MEG magnetometers, in the single EEG simulation the MEG magnetometers are substituted by EEG electrodes.

5 Comparison of the crosstalk errors for a single EEG, single MEG and combined MEG/EEG sensor array

5.1 Sensor arrays

Figure 2 shows the combined MEG/EEG sensor array that is used in the simulations. The MEG magnetometers (red) measure just the component of the magnetic field normal to the scalp, the EEG electrodes (green) measure the electrostatic potential at the scalp. Both MEG and EEG sensors are considered as point sensors without noise.

5.2 Crosstalk error

The most simple solution of the inverse problem is the unweighted minimum norm solution

$$\hat{\mathbf{Q}} = \mathbf{G}^{\dagger}\boldsymbol{\phi} \tag{9}$$

giving the solution with the smallest norm $||\hat{\mathbf{Q}}||$. In a forward-inverse-simulation we can insert $\boldsymbol{\phi} = \mathbf{GQ}$ into Eq. (9) and get $\hat{\mathbf{Q}} = \mathbf{G}^{\dagger}\mathbf{GQ}$. If q_j denotes the given amplitude of the j-th dipole, $\hat{q}_i = \sum_j (\mathbf{G}^{\dagger}\mathbf{G})_{ij} q_j$ is the reconstructed amplitude of the i-th dipole. The crosstalk error (Liu et al., 2002)

$$C(i) = \frac{1}{n}\sum_{j=1}^{n} \frac{|(\mathbf{G}^{\dagger}\mathbf{G})_{ij}|^2}{|(\mathbf{G}^{\dagger}\mathbf{G})_{ii}|^2} \tag{10}$$

measures how the reconstructed amplitude of the dipole at position i depends on the given dipole amplitudes at other positions j. So the crosstalk error quantifies the mislocalisation of activity. The cortical average of the crosstalk error is given by

$$\overline{C} = \frac{1}{n}\sum_{i=1}^{n} C(i) \tag{11}$$

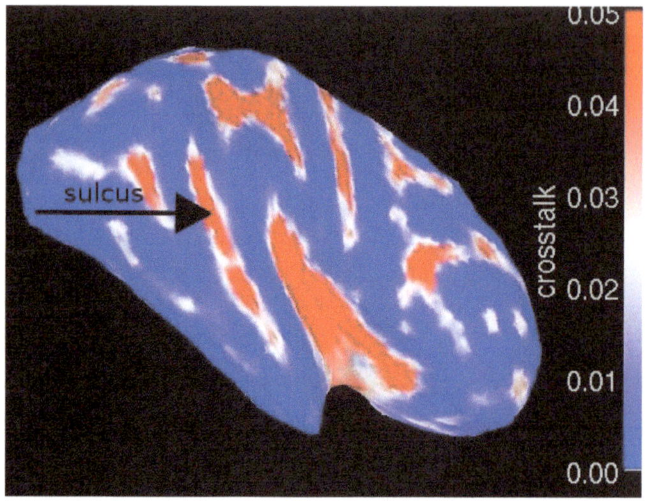

Fig. 3. Crosstalk error map for 354 EEG sensors. $\overline{C} = 0.032$.

5.3 Results

Figures 3, 4 and 5 show the crosstalk localisation error plotted on a flat representation of the right cortex for a single EEG, single MEG and a combined MEG/EEG sensor array, respectively. The localisation errors are smallest for the combined MEG/EEG sensor array, especially in the cortical sulcus regions (black arrow). Also the averaged crosstalk $\overline{C}$ is smallest in the case of the combined MEG/EEG sensor array.

6 Comparison of minimum norm and MUSIC

6.1 MUSIC

The minimum norm solution of Eq. (9) for a distributed model reconstructs the amplitudes of a huge number of dipoles ($n = 15\,010$). In contrast the MUSIC-algorithm tries to represent the measured fields with a smaller number of dipoles $r \ll n$, namely the number r of active areas in the brain. Assuming that the time series and the field topologies of the r active areas are linearly independent the number of

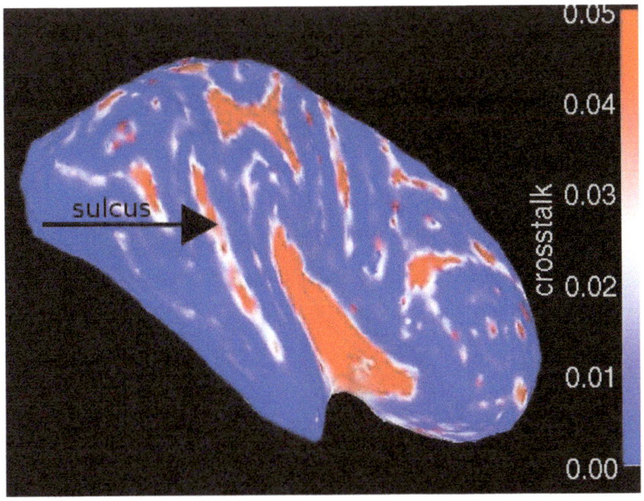

Fig. 4. Crosstalk error map for 354 MEG sensors. $\overline{C}=0.038$.

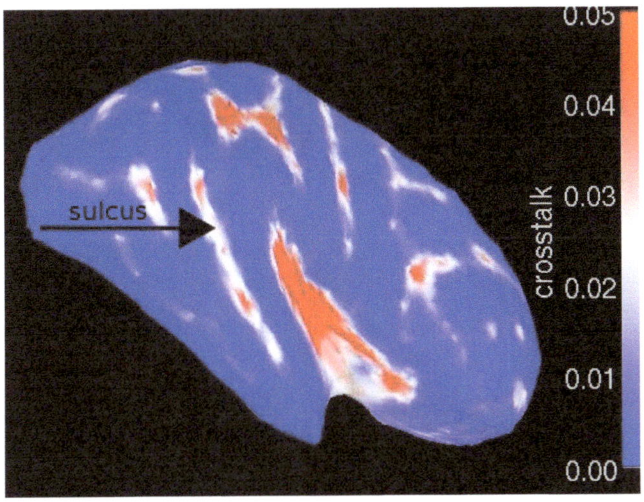

Fig. 5. Crosstalk error map for 177 MEG and 177 EEG sensors. $\overline{C}=0.020$.

active areas r can be estimated from the rank of the data matrix $r \approx \text{rank}(\phi)$ (Mosher and Leahy, 1998). The positions $\boldsymbol{l}_i$ of the r active areas can be found as the r zero points of a cost function u:

$$u(\boldsymbol{l}_i) = \frac{||\mathbf{P}^{\perp}\boldsymbol{g}(\boldsymbol{l}_i)||^2}{||\boldsymbol{g}(\boldsymbol{l}_i)||^2} = 0 \quad \text{for} \quad i = 1, \ldots, r \tag{12}$$

where $\boldsymbol{g}(\boldsymbol{l}_i)$ are the column vectors of $\mathbf{G}$ and $\mathbf{P}^{\perp}$ is a rank $(p-r)$ orthogonal projector that can be calculated with a Singular Value Decomposition (SVD) of the data matrix $\boldsymbol{\phi}$ (Mosher et al., 1992). In praxis the reciprocal of the cost function u is plotted on the cortex and the positions of the r active areas can be found as r sharp peaks in the MUSIC-map. In our simulations we consider the number of active areas to be known as one.

6.2 Results

6.2.1 Activation in cortical sulcus

Figures 6 and 7 show the results of the simulation in the combined MEG/EEG sensor array with 177 MEG and 177 EEG sensors for the minimum norm and MUSIC, respectively. The given activity is represented by a cortical patch of 8 dipoles activated in the cortical sulcus normally to the cortex with maximum activity at the white dot in the middle. The localisation error is shown with a white double-arrow and is defined as the distance between the maximum of the given activity and the maximum of the reconstructed activity. The localisation error of the minimum norm solution is 1.8 cm (Fig. 6), the localisation error of the MUSIC-map is 0.4 cm (Fig. 7). The MUSIC-map provides a smaller localisation error than the minimum norm solution.

6.2.2 Activation in cortical gyrus

Figures 8 and 9 show the results of the simulation in the combined MEG/EEG sensor array with 177 MEG and 177 EEG sensors for the minimum norm and MUSIC, respectively. The given activity is located at the cortical gyrus. The localisation error of the minimum norm solution is 0.5 cm (Fig. 8), the localisation error of the MUSIC-map is 0.7 cm (Fig. 9). The difference between the localisation errors of the minimum norm and the MUSIC solution for a cortical gyrus activation is smaller than for a cortical sulcus activation. For a cortical gyrus activation minimum norm and MUSIC provide comparable localisation errors.

7 Conclusions

In Sect. 5 we showed that a combined MEG/EEG sensor array provides smaller crosstalk localisation errors than a single EEG and a single MEG sensor array for the same total number of sensors. In Sect. 6 we compared the unweighted minimum norm solution and the MUSIC solution of the inverse problem for a given one source activation in the cortical sulcus and at the cortical gyrus. For the cortical sulcus activation the localisation error of the minimum norm solution is 1.4 cm bigger than for the MUSIC solution. In the case of the cortical gyrus activation the minimum norm solution and the MUSIC solution are more comparable: the localisation error of the MUSIC solution is 0.2 cm bigger than the localisation error of the minimum norm solution. This motivates the application of MUSIC beamforming techniques for the localisation of deep one source activation (e.g. focal epilepsy) in the brain.

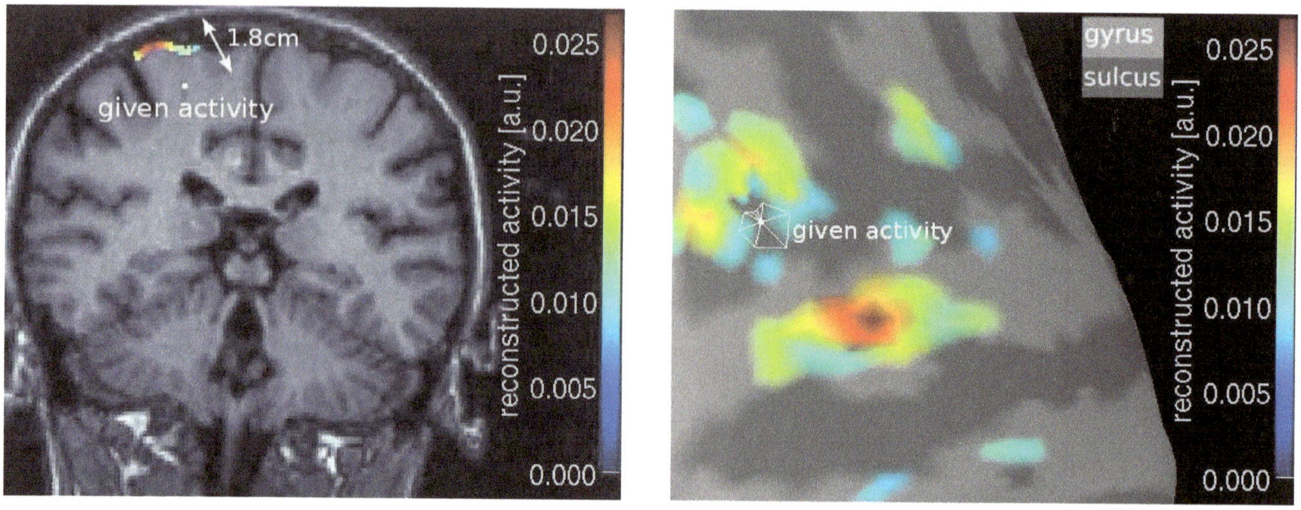

Fig. 6. Minimum norm solution (colored) for a given activity in the cortical sulcus and the combined MEG/EEG sensor array from Fig. 2. Left: reconstructed dipole amplitudes plotted in MRI-Scan. Right: activity plotted on a flat representation of the cortex.

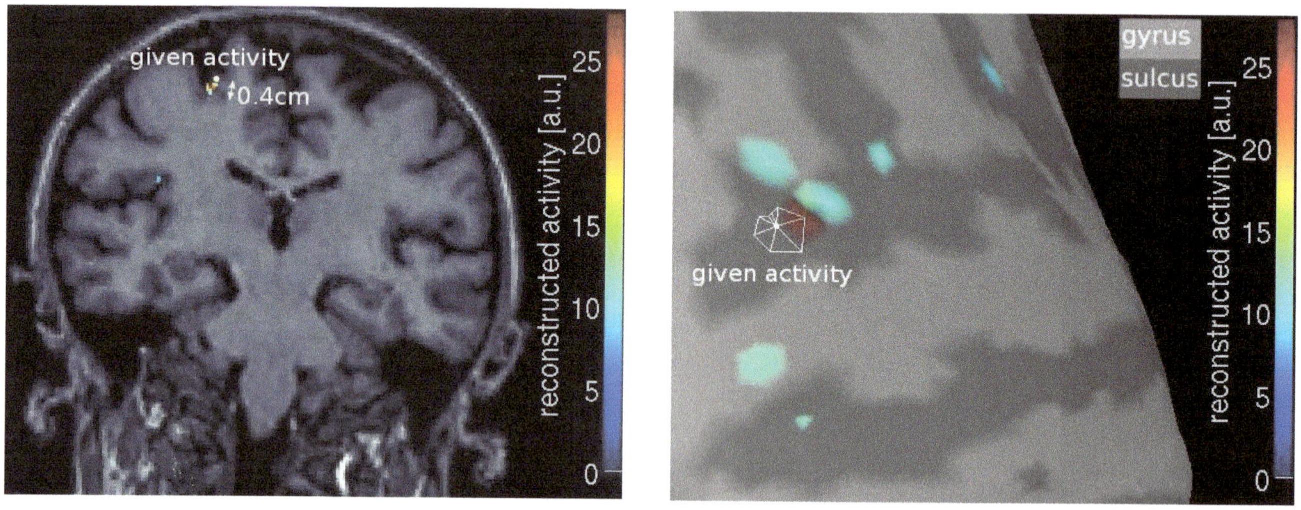

Fig. 7. MUSIC-map (colored) for a given activity in the cortical sulcus and the combined MEG/EEG sensor array from Fig. 2. Left: MUSIC-map plotted in MRI-Scan. Right: activity plotted on a flat representation of the cortex.

Acknowledgements. This work was supported by the SFB 855 *Magnetoelectric Composites – Future Biomagnetic Interfaces* of the Deutsche Forschungsgemeinschaft.

References

Baillet, S., Mosher, J. C., and Leahy, R. M.: Electromagnetic brain mapping, IEEE Signal Proc. Mag., 18, 14–30, 2001.

Dassios, G.: Neuronal currents and EEG-MEG fields, Math. Med. Biol., 25, 133–139, 2008.

Dassios, G., Fokas, A. S., and Hadjiloizi, D.: On the complementarity of electroencephalography and magnetoencephalography, IOP Science Inverse Problems, 23, 2541–2549, 2007.

Fokas, A. S., Kurylev, Y., and Marinakis, V.: The unique determination of neuronal currents in the brain via magnetoencephalography, Institute of Physics Publishing, Inverse Probl., 20, 1067–1082, 2004.

Gramfort, A., Papadopoulo, T., Olivi, E., and Clerc, M.: OpenMEEG: opensource software for quasistatic bioelectromagnetics, Biomed. Eng. Online, 9, 45, doi:10.1186/1475-925X-9-45, 2010.

Kybic, J., Clerc, M., Abboud, T., Faugeras, O., Keriven, R., and Papadopoulo, T.: A common formalism for the integral formulations of the forward EEG problem, IEEE T. Med. Imaging, 24, 12–28, 2005.

Liu, A. K., Dale, A. M., and Belliveau, J. W.: Monte Carlo Simulation Studies of EEG and MEG Localization Accuracy, Hum. Brain Mapp., 16, 47–62, 2002.

Mosher, J. C. and Leahy, R. M.: Recursive MUSIC: A framework for EEG and MEG source localization, IEEE T. Bio-Med. Eng., 45, 1342–1354, 1998.

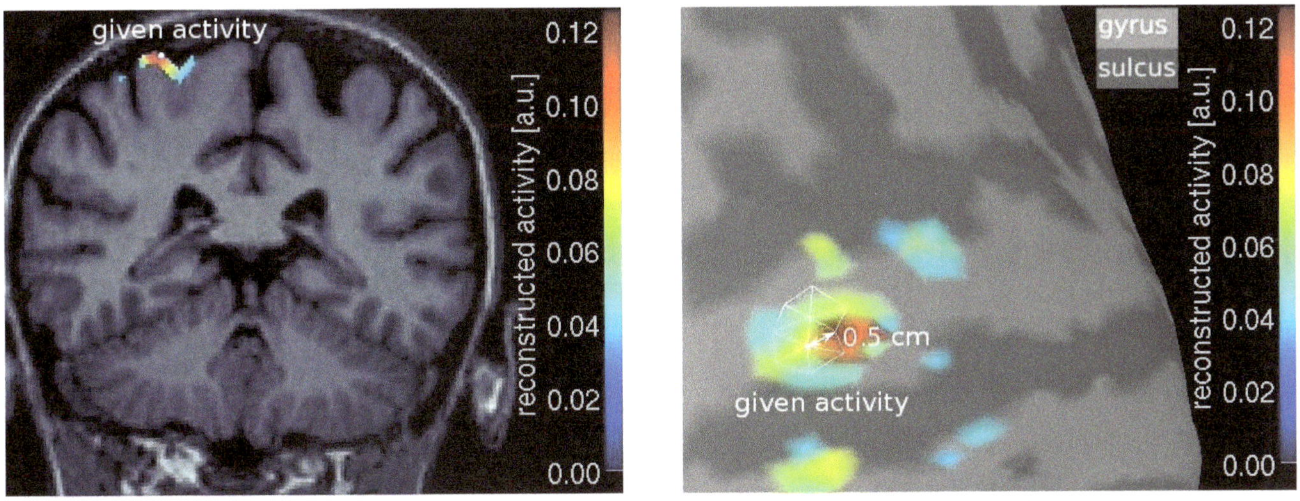

Fig. 8. Minimum norm solution (colored) for a given activity at the cortical gyrus and the combined MEG/EEG sensor array from Fig. 2. Left: reconstructed dipole amplitudes plotted in MRI-Scan. Right: activity plotted on a flat representation of the cortex.

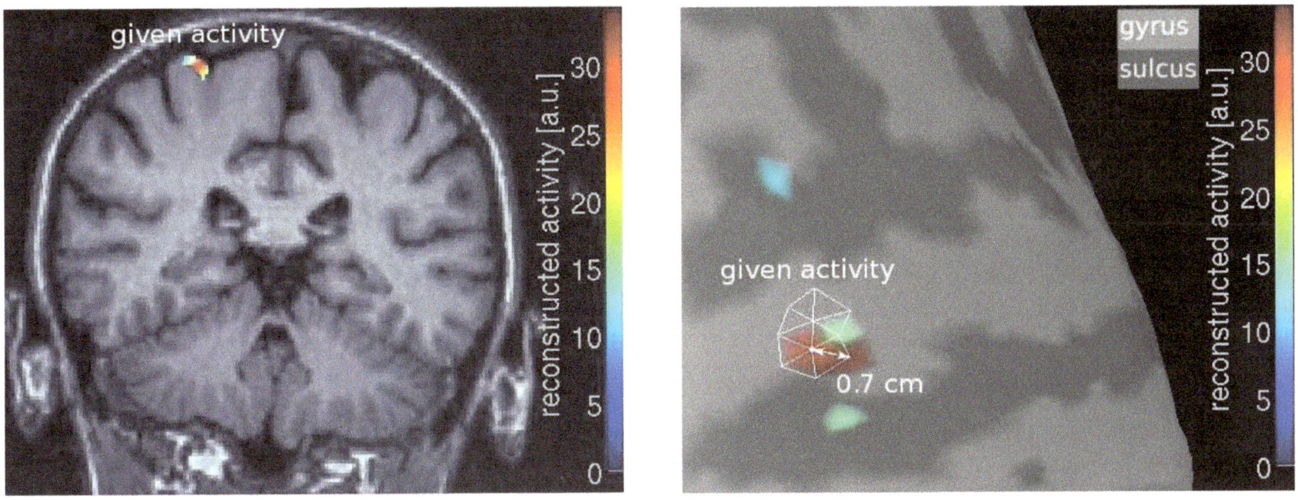

Fig. 9. MUSIC-map (colored) for a given activity at the cortical gyrus and the combined MEG/EEG sensor array from Fig. 2. Left: MUSIC-map plotted in MRI-Scan. Right: activity plotted on a flat representation of the cortex.

Mosher, J. C., Lewis, P. S., and Leahy, R. M.: Multiple Dipole Modeling and Localization from Spatio-Temporal MEG Data, IEEE T. Bio-Med. Eng., 39, 541–557, 1992.

Mosher, J. C., Leahy, R. M., and Lewis, P. S.: EEG and MEG: Forward Solutions for Inverse Methods, IEEE T. Bio-Med. Eng., 46, 245–259, 1999.

Pataraia, E., Baumgartner, C., Lindinger, G., and Deecke, L.: Magnetoencephalography in presurgical epilepsy evaluation, Neurosurg. Rev., 25, 141–159, 2002.

Sharon, D., Hämäläinen, M. S., Tootell, R. B., Halgren, E., and Belliveau, J. W.: The advantage of combining MEG and EEG: comparison to fMRI in focally stimulated visual cortex, Neuroimage, 36, 1225–1235, 2007.

Tadel, F., Baillet, S., Mosher, J. C., Pantazis, D., and Leahy, R. M.: BrainStorm: A User-Friendly Application for MEG/EEG Analysis, Computational Intelligence and Neuroscience, 2011, 13 pp., 879716, 2011.

26

Simulations of magnetocardiographic signals using realistic geometry models of the heart and torso

C. V. Motrescu and L. Klinkenbusch

Institut für Elektrotechnik und Informationstechnik, Christian-Albrechts-Universität zu Kiel, Germany

Correspondence to: L. Klinkenbusch (lbk@tf.uni-kiel.de)

Abstract. Although the first measurement of the cardiac magnetic field was reported almost half a century ago magnetocardiography (MCG) is not yet widely used as a clinical diagnostic technique. With the development of a new generation of magnetoelectric sensors it is believed that MCG will become widely accepted in the clinical diagnosis. Our goal is to build a computer-based tool for medical diagnosis and to use it for the clarification of open electro-physiological questions. Here we present results from modelling of the cardiac electrical activity and computation of the generated magnetic field. For the simulations we use MRT-based anatomical models of the human atria and ventricles where the shape of the action potential is determined by ionic currents passing through the cardiac cell membranes. The monodomain reaction-diffusion equation is chosen for the description of the heart's electrical activity. This equation is solved for the transmembrane voltage which is in turn used to calculate current densities at discrete time instants. In subsequent simulations these current densities represent primary sources of magnetostatic fields arising from a volume conduction problem. In these simulations the heart is placed in a realistic torso model where the lungs are also considered. Both, the volume conduction problem as well as the reaction-diffusion problem are modelled using Finite-Element techniques.

1 Introduction

The human body can be seen as a volume conductor where the electrical activity of the heart gives rise to small electric currents in the body with associated potential distributions. The electrocardiography (ECG) records the variation in time of the electrical potential produced by the heart at the surface of the body. The graphical representation of the recorded potentials depicts the well known electrocardiogram. However, the same currents in the body give also rise to weak magnetic fields. The magnetocardiography (MCG) records the variation in time of the magnetic fields produced by the heart outside the body and the graphical representation of the recorded magnetic signals is called magnetocardiogram. Whereas the ECG is a widely used technique in the clinical diagnosis, the MCG is not yet. The reasoning for this may lay in the fact that currently used magnetic sensors (Superconducting QUantum Interference Devices – SQUIDs) require extremely low temperatures as well as magnetically shielded environments. Therefore, electrocardiography (ECG) is considerably more attractive for clinicians, though MCG and ECG supposedly supply disjunctive information and MCG offers several advantages (e.g. no electrodes needed, magnetic transparency of tissue). Efforts are nowadays directed to the overcoming of the above mentioned disadvantages of MCG by using magnetoelectric composite materials (Jahns et al., 2011) with giant magnetoelectric effect. In the same time computational tools for modelling the electrical activity of the heart as well as for solving the inverse problem of magnetocardiography are developed as a framework for studies on open electrophysiological questions such as the non-response of some patients to bi-ventricular stimulation.

In this paper we present a modelling of the electrical activity of the heart at both, cellular and organ level and a method to compute the associated magnetic field produced by a healthy heart.

2 Biological and electro-physiological background

The heart is a hollow muscular organ being structured on four chambers i.e. two atria and two ventricles. The function of the heart is to pump blood. This function is achieved by contractions of the muscles composing the atria and ventricles (working myocardium). The contraction of the cardiac muscles is electrically controlled by the excitation-conduction system of the heart. This system is made of cells that are able to fire impulses – a property called autorhythmicity.

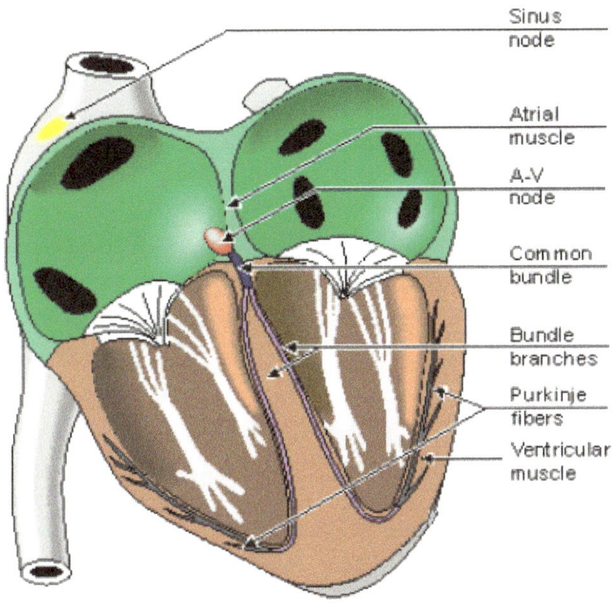

Fig. 1. Schematic representation of a section through the human heart showing the excitation-conduction system. Picture adapted from (Malmivuo and Plonsey, 1995).

Electrical impulses result from ionic activity at the cellular level. At rest, the cell membrane of the cardiac muscle cells (myocytes) composing the atria and ventricles, is electrically polarized in the sense that the inner side of the membrane is more electronegative than the outer side of the membrane. This produces the so called transmembrane voltage. When a strong enough stimulus is applied to the membrane, certain protein channels will open letting particular ionic species to cross the membrane along their electro-chemical gradients. This depolarization of the membrane is followed by a repolarization which brings the myocyte in the resting state. In contrast to the atrial and ventricular myocytes (of the working myocardium), the cells belonging to the excitation-conduction system of the heart do not need an external stimulus for the depolarization because they are able to depolarize themselves initiating thus electrical impulses (pacemaker cells). The transmembrane voltage propagates from a myocyte to the neighboring myocytes in the form of an action potential. The spread of the action potential is supported by some rapid intercellular channels called gap junctions which couple the interiors of adjacent cells. Thus, at the organ level the excitation propagates in the form of isochronal surfaces also called wave-fronts. Electrically active cells of the heart are located within these wave-fronts and they are usually associated to current dipoles. It is these moving current dipoles with their accompanying conduction currents that gives rise to the magnetic field of the heart.

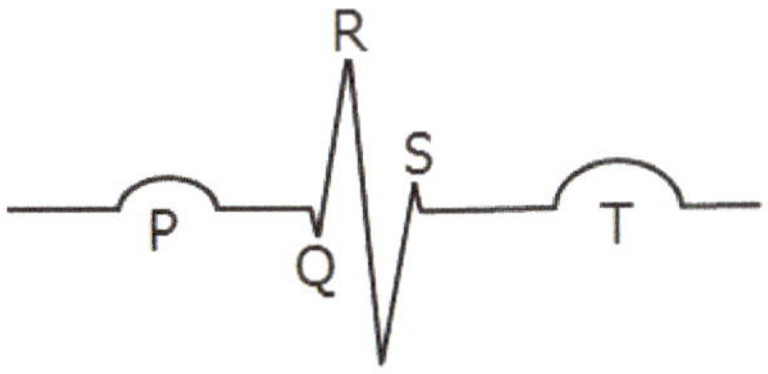

Fig. 2. Stylized form of the normal electrocardiogram.

3 The normal electrocardiogram

The excitation-conduction system of the heart (see Fig. 1) consists of the sinoatrial (SA) node, the atrioventricular (AV) node, the common bundle also called the bundle of His, the left and right bundle branches and the Purkinje fibres.

Under normal conditions, the SA node generates action potentials at the rate of about 70 pulses/min (Malmivuo and Plonsey, 1995) which spread through the right and left atria depolarizing them and determining the occurrence of the so called P-wave in the recorded electrocardiogram (see Fig. 2). The only conduction path between the atria and the ventricles is provided by the AV node which has a low conduction velocity and consequently introduces a delay in the propagation of the excitation that corresponds to the PQ-segment of the electrocardiogram. Once the AV node has been passed, the excitation propagates through a high conduction-velocity system (the His-Purkinje system) to the inner sides of the ventricular wall. The ventricular depolarization determines the QRS complex on the electrocardiogram. After the complete ventricular depolarization, it follows a plateau phase which corresponds to the ST-segment on the electrocardiogram and the repolarization of the ventricles that determines the T-wave.

The atrial repolarization takes place during the QRS-complex but it is masked by the much stronger process of ventricular depolarization.

4 Mathematical modelling at the cellular level

As already mentioned in Sect. 2 the electrical activity of the cardiac myocyte is mainly the result of the ionic currents that are crossing the cell membrane through ion channels and transporters. These currents determine the shape of the action potential of a cell. The available cellular ionic models at present can be divided into three classes (Clayton and Panfilov, 2008):

- simplified two-variable models that reduce the behavior of all the ionic currents to two variables that describe excitation and recovery;

- biophysically detailed models that use experimental information about the voltage and time dependence of ion

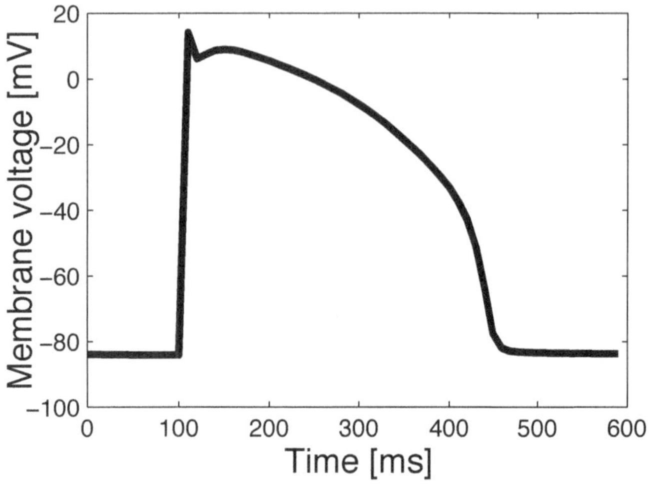

Fig. 3. Action potential shape of the Luo-Rudy phase 1 cellular ionic model.

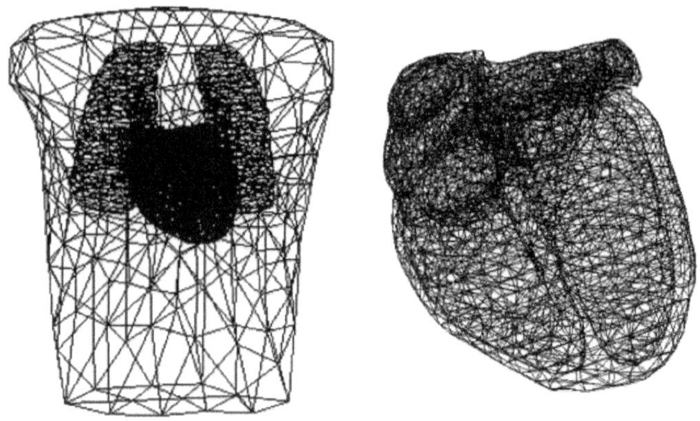

Fig. 4. MRT-based anatomical mesh models provided with the freely available software code ECGsim (van Oosterom, 2004). Left: heart and lungs inside the torso. Right: detailed view of the heart.

channel conductances (first generation) as well as intracellular ionic concentrations (second generation) in order to reproduce the action potential;

- reduced cardiac models that simplify the underlying electrophysiology to a minimum number of currents for which the activation and inactivation is controlled by some empirical variables.

Choosing the appropriate cellular model for a certain problem is a matter of balancing its benefits and limitations. For this work we have chosen the Luo-Rudy phase 1 model of mammalian myocytes (Luo and Rudy, 1991). This cellular model belongs to the first generation of the biophysically detailed models and it is considered that this model offers a good balance between electrophysiological details and computational efficiency. Within this model the Hodgkin-Huxley type of formalism is used to describe the action potential. The rate of change of the membrane voltage V_m is given by:

$$C_m \frac{dV_m}{dt} = -(I_{ion} + I_{st}) \tag{1}$$

where C_m is the membrane capacitance, I_{st} is an applied stimulus current and I_{ion} is the sum of six ionic currents: a fast sodium current (I_{Na}), a slow inward current (I_{si}), a time-dependent potassium current (I_k), a time-independent potassium current (I_{k1}), a plateau potassium current (I_{kp}) and a time-independent background current (I_b). In general, each current I_S carried by an ionic species "S" through an ion channel is expressed as:

$$I_S = \overline{G}_S + x_s^1 x_s^2 ... x_s^n (V_m - V_S) \tag{2}$$

with $\overline{G}_S$ being the maximal conductance of the ion channel, x_s being gating variables with values varying in the interval [0, 1] and describing the activation, inactivation and recovery of the ion channel and V_S being the membrane voltage of the species "S" alone. The gating variables are obtained by solving a system of eight nonlinear ordinary differential equations (ODEs) of the general form:

$$\frac{dx_S}{dt} = \frac{x_{S\infty}(V_m) - x_S}{\tau_S(V_m)} \tag{3}$$

where $x_{S\infty}$ is the steady-state value of a single gating variable x_S and τ_S is the time constant for the return of the variable x to its steady-state value after a voltage perturbation. The voltage dependencies of $x_{S\infty}$ and τ_S are experimentally determined.

The shape of the action potential as given by the Luo-Rudy phase 1 cellular model and used in this work is shown in Fig. 3.

5 Mathematical modelling at the tissue level

At the tissue level the granular structure may be neglected and the cardiac muscle may be modelled as a functional syncytium in which the membrane voltage propagates smoothly. Under this assumption the action potential can be considered to propagate by diffusion in an excitable medium that consists of one or two phases. The bi-domain model considers the cardiac muscle to be a two-phase medium consisting of intracellular and extracellular spaces separated at each point in space by the cellular membrane. Although this is a very detailed model, it is also computationally expensive. The simpler monodomain model is based on a single phase description of the cardiac muscle but it is still able to reproduce many of the experimentally observed phenomena with a much higher numerical efficiency. The simulation of a problem with the monodomain model can be by a factor of ten or more faster than the same problem simulated

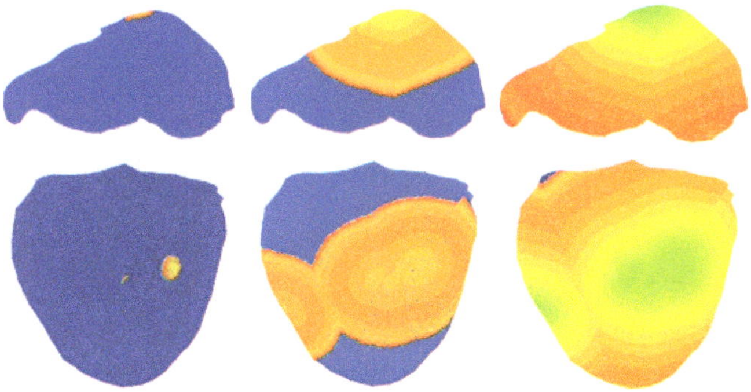

Fig. 5. The spread of electrical excitation in the heart during the atrial depolarization (top row), and ventricular depolarization (bottom row).

with the bi-domain model (Clayton and Panfilov, 2008). In this work we consider that the action potential propagates by diffusion in the monodomain model of the cardiac tissue with local excitation of the membrane voltage. The so called monodomain reaction-diffusion model consists of a parabolic partial differential equation (PDE) coupled to a system of ODEs (Whiteley, 2006):

$$\nabla \cdot (\sigma \nabla V_{\mathrm{m}}) = \beta \left[C_{\mathrm{m}} \frac{\partial V_{\mathrm{m}}}{\partial t} + I_{\mathrm{ion}} + I_{\mathrm{st}} \right] \tag{4}$$

where σ (Sm^{-1}) is the electrical conductivity tensor, V_{m} (V) is the transmembrane voltage, β (m^{-1}) is the surface to volume ratio of the cardiac cells, C_{m} (Fm^{-2}) is the specific membrane capacitance, I_{ion} (Am^{-2}) is the current flow through unit area of the cell membrane and I_{st} (Am^{-2}) is an applied stimulation current. The net membrane current

$$I_{\mathrm{m}} = \left[C_{\mathrm{m}} \frac{\partial V_{\mathrm{m}}}{\partial t} + I_{\mathrm{ion}} \right] \tag{5}$$

is given by the cellular ionic models such as the one described in Sect. 4.

6 Anatomical models

Anatomical models of the heart, lungs and torso used in this work are those provided with the freely available software code for simulations of electrocardiograms called "ECGsim" (van Oosterom, 2004). The models are based on magnetic resonance tomography (MRT) data of a healthy young male of 22 yr. They are available in the form of triangulated surfaces as shown in Fig. 4.

7 Method

The procedure adopted by us for computing the magnetic field produced by the heart consists of three subsequent steps where the output of one step is used as an input for the next step. The steps are also differentiated by the involved time and space scales.

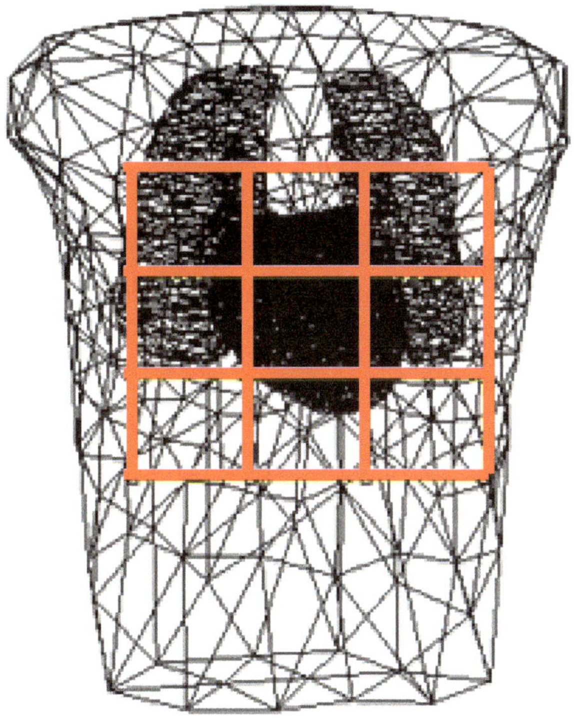

Fig. 6. Position of the magnetic sensor array (red color) relative to the body.

In the first step we simulate the propagation of electrical excitation in the isolated heart shown in the right-hand side of Fig. 4. As a result we obtain the distribution of the transmembrane voltage in the heart at discrete time instants during a complete heart beat.

In the second step a volume conduction problem is solved where the driving force is represented by impressed current densities in the heart which are computed using the transmembrane voltage resulted at each time instant considered in the simulations of the previous (first) step. The computational domain consists of the heart and lungs inside the torso as shown on the left-hand side of Fig. 4. The output of these simulations is represented by the total (impressed and conduction) current density in the entire torso.

In the third step the magnetic field outside the body produced by the total current density in the torso, is computed as a magnetostatic problem at each time instant considered in the previous (second) step.

Thus slowly varying magnetocardiographic signals with time are obtained from a sequence of stationary computations.

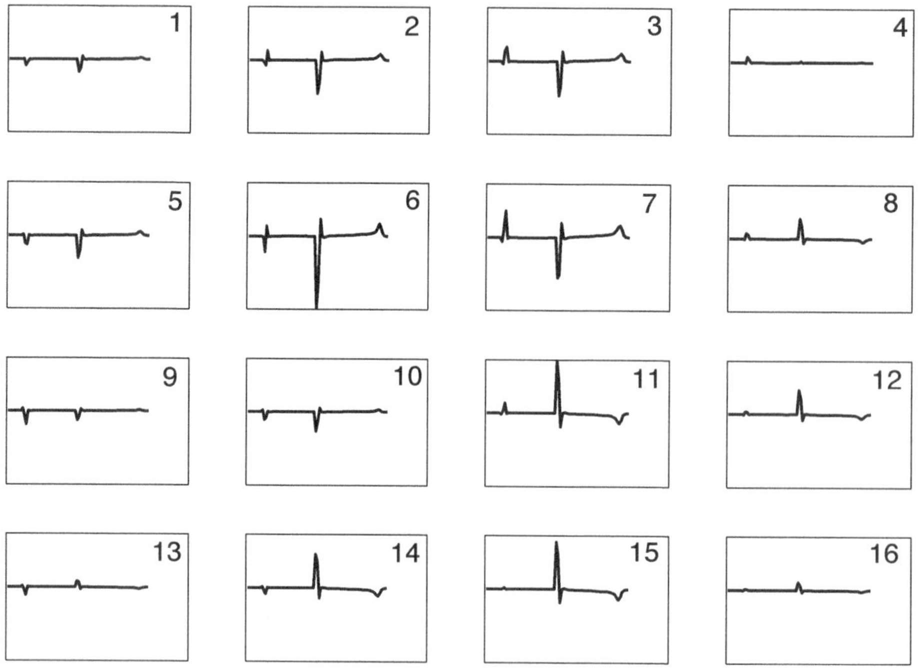

Fig. 7. Magnetic flux density component perpendicular on the chest recorded as a function of time at the position of each node of the sensor array shown in Fig. 6.

8 Simulation of the excitation spreading in the heart

The spread of excitation in the heart is simulated with Finite-Element methods. For this purpose the surface meshes of the atria and ventricles composing the heart shown on the right-hand side of the Fig. 4 are transformed to tetrahedral volume meshes by using the "Tetgen" – a freely available software code for research and non-commercial purposes (Si, 2006). The atria are discretized with 1.23 millions mesh nodes with an average edge length of 0.8 mm and the ventricles contain 1.43 millions mesh nodes with an average edge length of 0.95 mm.

The simulations are based on the monodomain reaction-diffusion model described by Eq. (4). Each node of the computational domain contains the Luo-Rudy ionic cellular model described in Sect. 4. This cellular model with constant action potential duration (APD) is used for both, atria and ventricles. The cardiac muscles are assumed to be homogeneous and isotropic with the scalar electrical conductivity $\sigma = 0.25\,\mathrm{Sm}^{-1}$. Other parameter values used in the simulation are: $C_\mathrm{m} = 1\,\mu\mathrm{Fcm}^{-2}$ and $\beta = 1400\,\mathrm{cm}^{-1}$.

The simulations are performed within "CHASTE" – the "Cancer, Heart And Soft-Tissue Environment" – a multi-purpose software library for simulations in the area of biology (Pitt-Francis et al., 2009) provided under the LGPL license by a team from the Computer Science Department of the University of Oxford, the UK.

Examples showing the membrane voltage distribution at three different time instants during the atrial and ventricular depolarization are provided in Fig. 5. Due to the constant action potential durations considered in the simulations, the repolarizations follow a similar time course. Rectangular current pulses of 2 ms duration are used for the initial stimulations. The areas of initial stimulations are taken from the ECGsim (van Oosterom, 2004) software where depolarization isochrones on the surfaces enveloping the atria and ventricles are obtained from the potentials on the surface of the body by following an inverse procedure. The atria are stimulated in an area corresponding to the position of the SA-node. In the case of the ventricles, three areas inside the left ventricle and one area inside the right ventricle are initially stimulated with some small time delays between them. Animated representations of the spread of electrical excitation during the depolarization and repolarization of the atria and ventricles are also provided in the form of short films.

9 Computation of currents in the body

The electrical excitation in the heart gives rise to currents in the entire body. Impressed current densities in the heart can be computed as $\boldsymbol{J}_\mathrm{e} = -\sigma \nabla V_\mathrm{m}$ (Aoki et al., 1987) using the membrane voltage V_m already computed at discrete time instants during the spread of excitation in the heart. The associated conduction currents $\boldsymbol{J}_\mathrm{c}$ can then be determined by

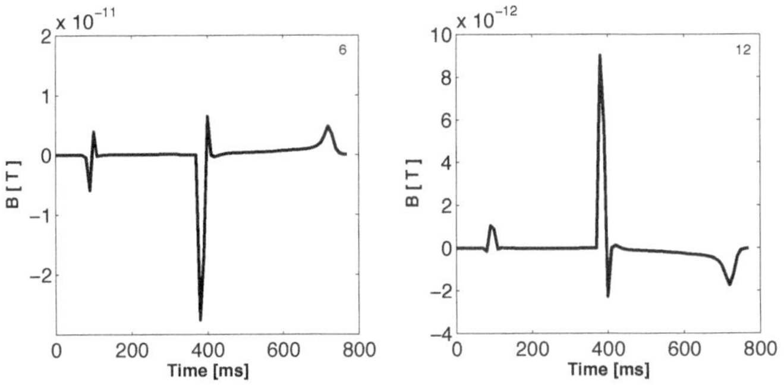

Fig. 8. Detailed view of the magnetocardiographic signals recorded at the positions of the sensors denoted by 6 (left) and 12 (right) in Fig. 7.

solving a stationary volume conduction problem described by the equation $\nabla \boldsymbol{J}_\mathrm{t} = 0$ where $\boldsymbol{J}_\mathrm{t} = \boldsymbol{J}_\mathrm{e} + \boldsymbol{J}_\mathrm{c}$ is the total current density. We compute the total current density in the entire torso shown on the left-hand side of Fig. 4 by using the commercially available software for Finite-Element analysis called COMSOL Multiphysics (COMSOL Multiphysics v. 4.2, 2011).

The electrical conductivity values used for the torso and the lungs are $\sigma_\mathrm{torso} = 0.2\,\mathrm{Sm}^{-1}$ and $\sigma_\mathrm{lung} = 0.04\,\mathrm{Sm}^{-1}$, respectively.

10 Computation of magnetic fields (MCG signals)

The currents in the torso produce magnetic fields inside and outside the body. We compute these magnetic fields using Ampere's law $\nabla \times \boldsymbol{H} = \boldsymbol{J}_\mathrm{t}$ at each time instant previously considered in the computation of the total current density in the torso $\boldsymbol{J}_\mathrm{t}$. These simulations are coupled in COMSOL Multiphysics (COMSOL Multiphysics v. 4.2, 2011) with those for the solution of the volume conduction problems. All biological tissues are considered to be non-magnetic with the relative magnetic permeability $\mu_\mathrm{r} = 1$.

For the results we consider a four by four magnetic sensor array positioned just in front of the torso as shown in Fig. 6. The nodes of the magnetic sensor array are spaced 8.2 cm and 7.2 cm apart from each other on the horizontal and vertical directions, respectively.

The variation in time of the magnetic field component perpendicular to the chest computed in each node of the magnetic sensor array is shown in Fig. 7. In this figure all the magnetocardiographic signals are represented on the same scale. Detailed views of the signals recorded by the sensors 6 and 12 are given in Fig. 8. When plotting the signals recorded by all the sensors in a single figure, the so called butterfly plot is obtained as shown in Fig. 9.

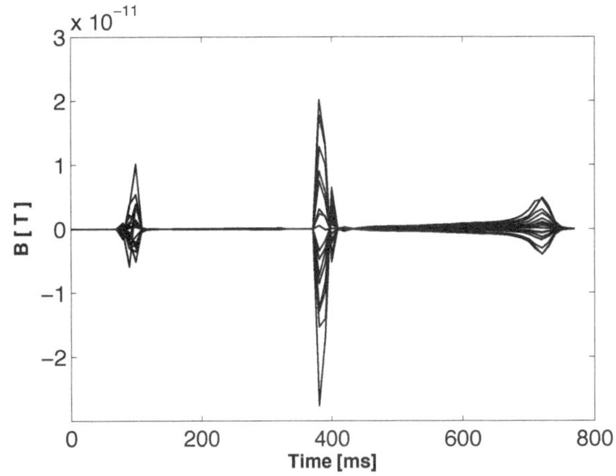

Fig. 9. Butterfly plot of the magnetic flux density component perpendicular on the chest recorded at the position of each point sensor.

11 Conclusions

The method used in this work proved to be able to provide realistic values for the computed magnetic fields of the heart. We coupled different state-of-the-art existent software codes and realistic anatomical models in order to compute the magnetic field produced by the heart. We used the magneto-static approximation to compute slowly varying magnetic field with time at successive time instants. The Finite-Element method (FEM) with tetrahedral elements was used for computing conduction currents in the inhomogeneous and irregularly shaped torso. Future work will focus on the reduction of the number of unknowns in order to reduce the computation time as well as for more conveniently solving the corresponding inverse problem.

Acknowledgements. We are thankful to all the persons who contributed to the freely available software codes (ECGsim, TetGen, CHASTE) used in this work. This work was supported by the Deutsche Forschungsgemeinschaft (DFG) within the SFB 855: *Magnetoelectric composite material – Future biomagnetic interfaces*.

References

Aoki, M., Okamoto, Y., Musha, T., and Harumi, K.-I.: Three-Dimensional Simulation of the Ventricular Depolarisation and Repolarisation Processes and Body Surface Potentials: Normal Heart and Bundle Branch Block, IEEE T. Bio-Med. Eng., 34, 454–462, 1987.

Clayton, R. H. and Panfilov, A. V.: A guide to modelling cardiac electrical activity in anatomically detailed ventricles, Prog. Biophys. Mol. Bio., 96, 19–43, 2008.

COMSOL AB: COMSOL Multiphysics® v. 4.2, available at: http://www.comsol.com, last access date: July 7, 2012.

Dössel, O.: Inverse Problem of Electro- and Magnetocardiography: Review and Recent Progress, International Journal of Bioelectromagnetism, vol. 2, no. 2. Online available at: http://ijbem.k.hosei.ac.jp/volume2/number2/toc.htm (last access date: July 7, 2012), 2000

Jahns, R., Greve, H., Woltermann, E., Quandt, E., and Knöchel, R.H. : Noise performance of magnetometers with resonant thin-film magnetoelectric sensors, IEEE T. Instrum. Meas., 60, 2995–3001, 2011.

Luo, C. and Rudy, Y.: A Model of the Ventricular Cardiac Action Potential – Depolarisation, Repolarisation and Their Interaction, Circ. Res., 68, 1501–1526, 1991.

Malmivuo, J. and Plonsey, R.: Bioelectromagnetism – Principles and Applications of Bioelectric and Biomagnetic Fields, Web-edition, Oxford University Press, New York, 1995.

Nakaya, Y. and Mori, H.: Magnetocardiography, Clin. Phys. Physiol. M., 13, 191–229, 1992.

Pitt-Francis, J., Pathmanathan, P., Bernabeu, M. O., Bordas, R., Cooper, J., Fletcher, A. G., Mirams, G. R., Murray, P. J., Osbourne, J., Walter, A., Chapman, S. J., Garny, A., van Leeuwen, I. M. M., Maini, P. K., Rodriguez, B., Waters, S. L., Whiteley, J. P., Byrne, H. M., and Gavaghan, D. J.: Chaste: A test-driven approach to software development for biological modelling, Comput. Phys. Commun., 180, 2452–2471, 2009.

Sato, M., Terada, Y., Mitsui,T., Miyashita,T., Kandori, A., and Tsukada K.: Visualization of atrial excitation by magnetocardiogram, The International Journal of Cardiovascular Imaging, 18, 305–312, 2002

Si, B. H.: TetGen, A Quality Tetrahedral Mesh Generator and Three-Dimensional Delaunay Triangulator, User's manual v1.4, 18 January 2006, available at: http://tetgen.berlios.de(last access date: July 7, 2012) 2006.

Smith, F. E., Langley, P., van Leeuwen, P., Hailer, B., Trahms, L., Steinhoff, U., Bourke, J.P., and Murray, A.: Comparison of magnetocardiography and electrocardiography: a study of automatic measurement of dispersion of ventricular repolarization, Europace Advance Access, Vol. 8 (10), 887-893, 2006.

van Oosterom, A. and Oostendorp, T.: ECGSIM; an interactive tool for studying the genesis of QRST waveforms, Heart, available at: http://www.ecgsim.org, 90, 165–168, 2004.

Whiteley, J. P.: An efficient numerical technique for the solution of the monodomain and bidomain equations, IEEE T. Bio-Med. Eng., 53, 2139–2147, 2006.

Zaza, A., Wilders, R., and Opthof, T.: Cellular electrophysiology, in: Comprehensive Electrocardiography, 2nd Edn., edited by: Macfarlane, P.W., van Oosterom, A., Pahlm, O., Kligfield, P., Janse, M., and Camm, J., Springer-Verlag London, 107–144, 2011.

Performance analysis of a low cost wireless indoor positioning system with distributed antennas

A. T. Ott[1], M. Shalaby[1], U. Siart[1], E. Kaliyaperumal[1], T. F. Eibert[1], J. Engelbrecht[2], and R. Collmann[2]

[1]Lehrstuhl für Hochfrequenztechnik, Technische Universität München, 80290 Munich, Germany
[2]Professur für Kommunikationstechnik/Funksysteme, Hochschule für Technik und Wirtschaft Dresden, 01069 Dresden, Germany

Abstract. In this paper an approach for indoor localization based on already installed communication transceiver infrastructure is presented. Only marginal additional hardware is required to implement the localization functionality. The idea is to use a power level detection evaluation of the received signals in communication receivers to localize participants. In the planned configuration the power transmitted from mobile terminals is received by a leaky wave cable. Its length and coupling properties are dimensioned according to the dynamic range of two access points connected to the two cable heads. In order to adapt the problem to circuit-based analysis techniques the leaky wave cable is represented by a distributed antenna system. Thereby the considered design consists of a cascade of coupling structures with broadband Vivaldi antennas connected to the coupling ports. An experimental system has been built and also tested using an automated frequency domain measurement setup. Measurement results relating the system bandwidth to positioning accuracy show good agreement with theoretical investigations.

1 Introduction

In recent years indoor positioning systems (IPS) have gained importance in research as commercially available satellite based positioning systems become inaccurate or even drop out completely inside buildings (Gu et al., 2009). Ultra-wideband (UWB) or active radio frequency identification (RFID) systems are a state-of-the-art replacement (Bill et al., 2004). They can provide accuracy down to the centimeter range.

However, these systems are very costly due to complex digital signal processing. A very cost efficient approach is to deploy the already installed network infrastructure. Thus it can be trusted that only little additional hardware is needed for the purpose of localization. One possible approach is to combine the observed power levels at the access points of the system with a-priori knowledge about the properties of the antenna system (Saha et al., 2003) into an estimator for the positions of the client terminals.

A serious disadvantage of this positioning approach is the fact that amplitude detection is rather sensitive with respect to the inherent amplitude fading effects. Hence a larger number of access points (observations) is necessary to reach the required accuracy. A refinement of the positioning resolution can be reached by placing distributed antenna systems between the access points (Saleh et al., 1987).

A novel approach is to achieve localization via power level detection by using leaky wave cables (Engelbrecht et al., 2010). Furthermore communication systems for tunnels and indoor environments based on leaky wave cables are well investigated (Delogne and Deryck, 1980; Motley and Palmer, 1983). However, the dimensioning of a leaky wave cable is very challenging as it is difficult to allocate a radiation pattern to a radiating slot (Walter, 1965). Furthermore any modification of the cable performance involves a redesign of the entire cable structure. To derive design rules a distributed antenna system is perfectly suited as it is possible to change single components with well known properties individually.

In this paper the evaluation of the localization performance of a distributed antenna system related to the properties of a leaky wave cable is reported. Section 2 comprises the basic principles of the localization system followed by a description of the developed antenna architecture in Sect. 3. The key features of the implemented simulation tool is described in Sect. 4. In Sect. 5 the performance of the wireless indoor positioning system is analyzed by simulations and measurements before the paper is concluded in Sect. 6.

Correspondence to: A. T. Ott
(hft@ei.tum.de)

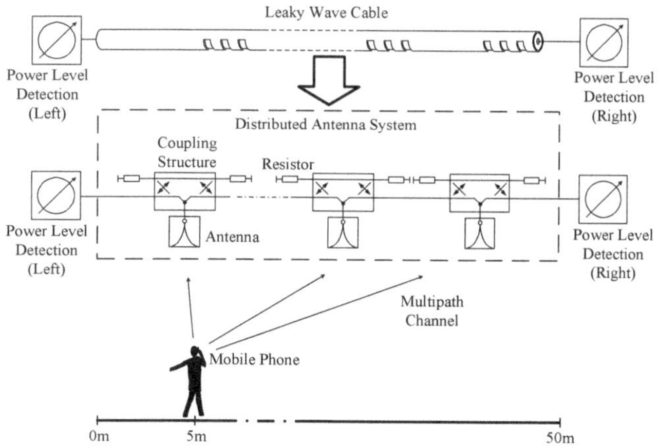

Fig. 1. Schematical diagram of an indoor wireless coverage system and the positioning approach under study. The leaky wave cable is represented by an equivalent distributed antenna system with appropriate coupling factors.

2 Functional principle

The functional principle of the wireless indoor positioning system is depicted in Fig. 1. A leaky wave cable is used to realize the radio link of a participant with a mobile. The position of the mobile should be estimated by means of a comparison of the output power levels at the cable heads. The output power level difference

$$\frac{\Delta P}{\mathrm{dB}} = \frac{P_{\mathrm{left}}}{\mathrm{dBm}} - \frac{P_{\mathrm{right}}}{\mathrm{dBm}} \tag{1}$$

is detected by subtracting the logarithmic power levels P_{left} and P_{right} at the cable heads from each other. By comparing the result with an a-priori known reference characteristic the power levels are mapped to an estimated position in the room (Weber et al., 2009). This so-called RF fingerprinting technique has the advantage that the estimate is based on the level difference ΔP rather than absolute levels. Thus the system is very robust against motions orthogonal to the cable as well as polarization loss. The system requirements have been chosen to achieve a linear curve progression with respect to position and level difference. Thus the location error of simulation and measurement is referred to the deviation from a least mean square error (LMSE) algorithm (Hagan et al., 1996).

The slots of the leaky wave cable have been replaced by a cascade of coupling structures and antenna elements. Consequently the design complexity reduces significantly as coupling between the radiating elements can be neglected. Furthermore the leaky wave cable has to be considered as an array of single radiating elements. The radiation behavior of one slot is not that easy to determine and to control whereas the radiation pattern of a single antenna element is well known.

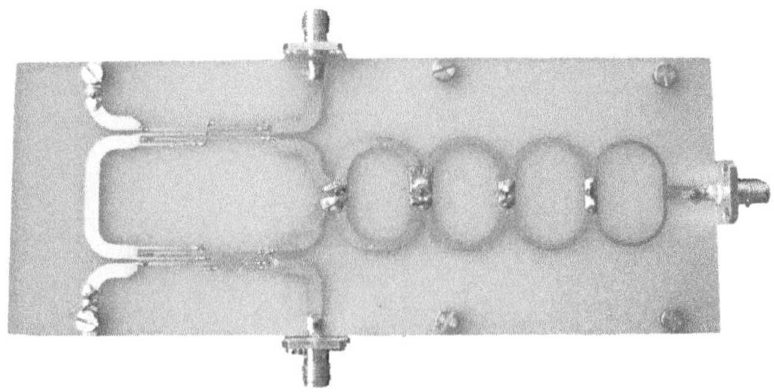

Fig. 2. Photo of the developed coupling structure prototype.

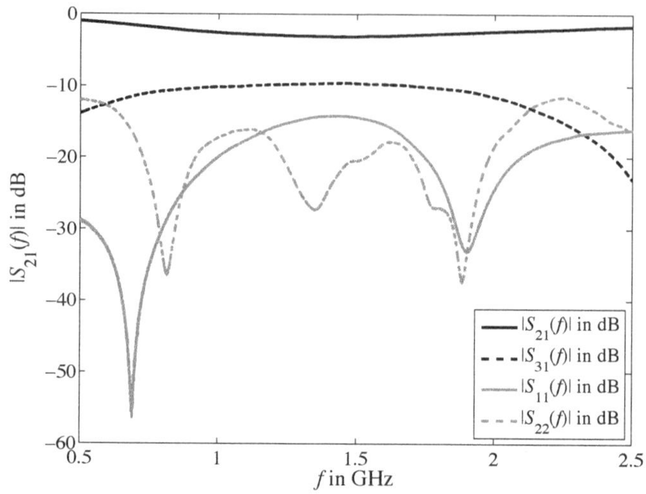

Fig. 3. Scattering parameters of the developed coupling structure (measurement).

3 Antenna prototype

The experimental antenna system has been built-up as a chain of identical individual parts including a coupling structure and an antenna (see Fig. 1). This elementary cell is supposed to represent a radiating slot (or a group of slots) on the shielding of the leaky wave cable. Thus the coupling structure of the elementary cell has been dimensioned in such way that only a small portion of the guided energy from the backbone cable is radiated by each antenna element. A prototype of a single coupling structure is depicted in Fig. 2. The structure consists of two mirrored Lange couplers (Lange, 1969) and a four stage broadband Wilkinson divider (Wilkinson, 1960). Similar to a single radiating slot it implements a weak coupling with no directivity. A drawback of this concept is that almost the complete received energy of the antennas is dissipated in the load elements. The scattering parameters of the coupling structures are illustrated in Fig. 3. The signal from the antenna is attenuated by 9 dB and the signal which passes the coupling structure is attenuated by 3 dB.

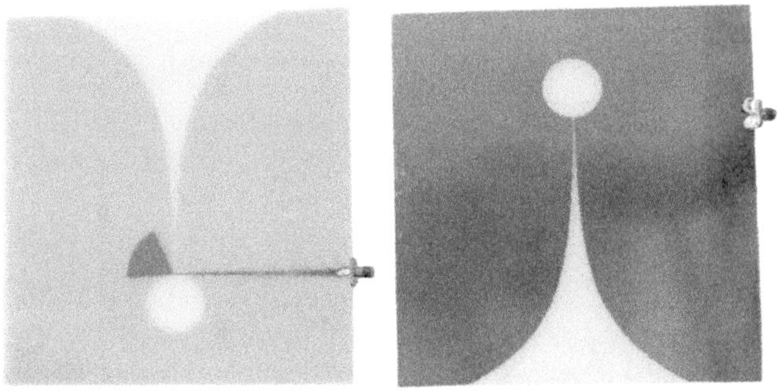

Fig. 4. Photograph of the developed Vivaldi antenna prototypes.

Based on the simulation results we found that Vivaldi antennas (Gibson, 1979) are perfectly suited as receiving antennas. Two antenna prototypes with rear and front view are depicted in Fig. 4. By feeding the antennas with a balun a symmetrical radiation pattern can be achieved. Furthermore the antenna is low directive within the frequency range from 800 MHz to 2 GHz and thus field distributions from different angle of arrivals are equally weighted.

4 System simulation tool

A detailed comparison between various scenarios within different environments provides a deep insight into fundamental limitations of positioning accuracy and reliability. Especially inside large buildings like concert halls or train stations prototypes are costly as many elementary cells are needed and the timbering can be difficult. Hence a simulation tool which enables to plan the entire system has been developed. An overview of the system simulation approach is given in Fig. 5. According to the individual system requirements the number of antenna elements M and the spacing between the antenna elements can be variably chosen. The entire system of cascaded couplers is modeled by a scattering matrix which enables to compute the power waves at the output ports $M+1$ and $M+2$ from the input waves at the M antenna ports. They are computed from the incoming rays of a ray tracing process weighted with the antenna properties.

To estimate the maximum fidelity which can be achieved by the antenna system we identified the impact of reflections in the environment and on the antenna system itself as essential effects for signal distortions. Thus a ray tracer based on a shooting and bouncing ray (SBR) algorithm similar to (Buddendick and Eibert, 2010) has been applied. As ray finding in complex scenarios can be very time intensive SBR algorithms are a good choice to characterize complex scenarios (Ling et al., 1989). Thereby the electrical field at the receiver is computed from the arriving rays on a sphere around the nominal receiver position. Furthermore Computer Aided Design (CAD) data can be applied and simulated.

To obtain accurate and reliable simulation results the receiving antenna elements were characterized by angle and frequency dependent transfer functions similar to Kunisch (2007), Farr and Baum (1998). Consequently this approach also enables to test the system with different antenna properties. The antenna transfer functions have been chosen in a way that the electrical field in free space is related with the power waves at the output ports of the antennas. Given the wave $a(\omega)$ incident at the antenna port and the normalized vector valued antenna transfer function $\boldsymbol{H}_{\mathrm{TX,N}}(\omega,\varphi,\vartheta)$ the electrical field launched by the TX antenna (mobile phone) can be expressed by

$$\boldsymbol{E}_{\mathrm{TX}}(\omega,\varphi,\vartheta)=\frac{\mathrm{j}\omega Z_{\mathrm{F0}}}{2\pi c_0}\boldsymbol{H}_{\mathrm{TX,N}}(\omega,\varphi,\vartheta) \cdot\frac{\sqrt{Z_{\mathrm{L}}}}{Z_{\mathrm{A}}(\omega)+Z_{\mathrm{L}}}a(\omega) \quad (2)$$

with c_0 the speed of light, $Z_{\mathrm{A}}(\omega)$ the antenna terminal impedance, Z_{F0} the intrinsic impedance of free space and Z_{L} the reference impedance of the transmission lines, respectively. Thereby the normalized antenna transfer function is related to the antenna transfer function in the receiving case. Assuming that the transmit antenna and receive antenna m are separated by a distance of R_m in free space, the field incident at antenna m is given by

$$\boldsymbol{E}_{\mathrm{RX},m}(\omega,\varphi,\vartheta)=\boldsymbol{E}_{\mathrm{TX}}(\omega,\varphi,\vartheta)\frac{\mathrm{e}^{-\mathrm{j}\omega\tau_m}}{R_m} \quad (3)$$

with the path delay $\tau_m=R_m/c_0$. In the receiving case K different rays $\boldsymbol{E}_{\mathrm{RX},m}(\omega,\varphi_k,\vartheta_k)$ arrive from arbitrary directions (φ_k,ϑ_k) at the surface of the antenna. Thus the outgoing wave at the antenna port m is related to the sum of all incoming electrical field contributions and the antenna transfer function in the receiving case by

$$b_{1,m}(\omega)=\frac{\sqrt{Z_{\mathrm{L}}}}{Z_{\mathrm{A},m}(\omega)+Z_{\mathrm{L}}} \cdot\sum_{k=1}^{K}\boldsymbol{H}_{\mathrm{RX,N},m}(\omega,\varphi_k,\vartheta_k)\boldsymbol{E}_{\mathrm{RX},m}(\omega,\varphi_k,\vartheta_k). \quad (4)$$

According to the reciprocity theorem of antennas the afore stated ansatz leads to the equivalence of the normalized transmitting and receiving antenna transfer functions

$$\boldsymbol{H}_{\mathrm{TX,N}}(\omega,\varphi,\vartheta)=\boldsymbol{H}_{\mathrm{RX,N}}(\omega,\varphi,\vartheta). \quad (5)$$

Furthermore both transmitting and receiving transfer functions can be related to the classical properties of antennas. Thus the magnitude of the transfer function in the transmitting case is related to the antenna gain $G_{\mathrm{TX}}(\omega,\varphi,\vartheta)$ according to

$$|\boldsymbol{H}_{\mathrm{TX,N}}(\omega,\varphi,\vartheta)|^2=\frac{4\pi c_0^2 G_{\mathrm{TX}}(\omega,\varphi,\vartheta)\mathrm{Re}\{Z_{\mathrm{A}}(\omega)\}}{\omega^2 Z_{\mathrm{F0}}}. \quad (6)$$

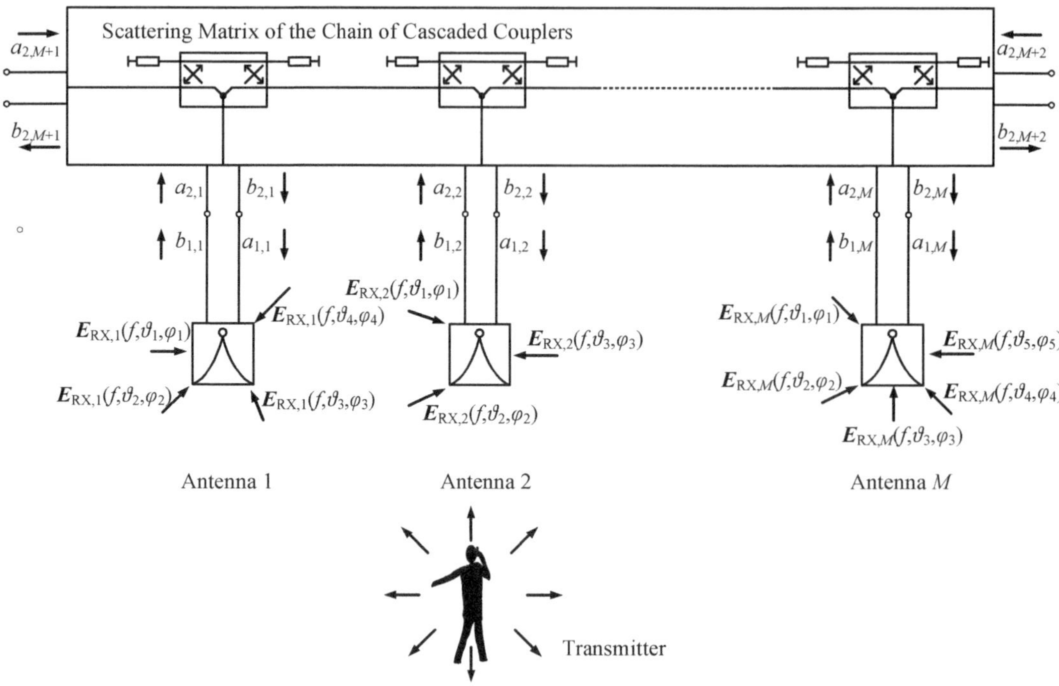

Fig. 5. Architecture of the system simulation.

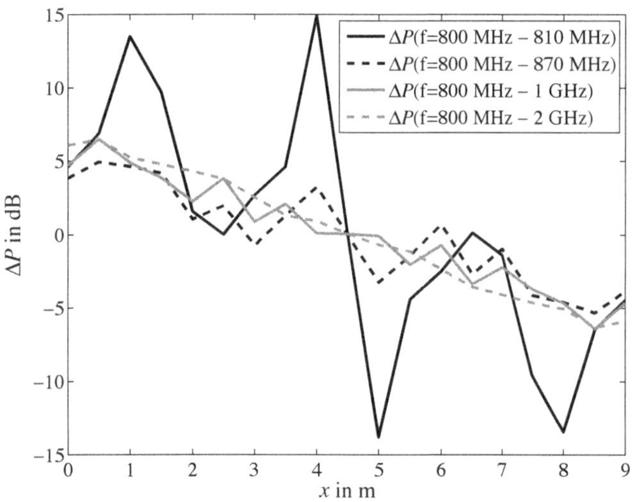

Fig. 6. Simulated level difference at the output ports of a 9 m long distributed antenna system with four equally spaced antenna elements.

Similarly, the receiving transfer function

$$|\boldsymbol{H}_{\mathrm{RX,N}}(\omega,\varphi,\vartheta)|^2 = \frac{4A_{\mathrm{e}}(\omega,\varphi,\vartheta)\mathrm{Re}\{Z_{\mathrm{A}}(\omega)\}}{Z_{\mathrm{F0}}} \tag{7}$$

is related to the effective area $A_{\mathrm{e}}(\omega,\varphi,\vartheta)$. By considering a transmission between two antennas in free space it is possible to extract the antenna properties from the transmission scattering parameter $S_{21}(\omega)$. Thus the antenna transfer function can either be deduced from simulations or from measurements in an anechoic chamber.

Furthermore the waves $b_{1,m}(\omega)$ are used as input waves into the coupling structures. To obtain accurate results all coupling structures and transmission lines are cascaded and mathematically described within one scattering matrix. To obtain the system matrix an efficient algorithm according to Simpson (1981) is applied. To accelerate the computation time the entire scattering matrix of the system is computed in a preprocessing step before the simulation is started. The power level difference of the system is determined according to

$$\frac{\Delta P}{\mathrm{dB}} = 10\cdot\log_{10}\left[\int_{f_c-B/2}^{f_c+B/2}\frac{|b_{2,M+1}(2\pi f)|^2}{2B}\mathrm{d}f\right] - 10\cdot\log_{10}\left[\int_{f_c-B/2}^{f_c+B/2}\frac{|b_{2,M+2}(2\pi f)|^2}{2B}\mathrm{d}f\right]. \tag{8}$$

Thereby the influence of the signal baseband bandwidth B is considered by computing the mean value of the output waves at the center frequency f_c.

5 Performance analysis

A prototype with four antenna elements was tested in a 6 m high building with a length of 12 m and a width of 10 m. The spacings of the single antenna elements was chosen to 3 m. In this paper exemplary results of stationary measurements at

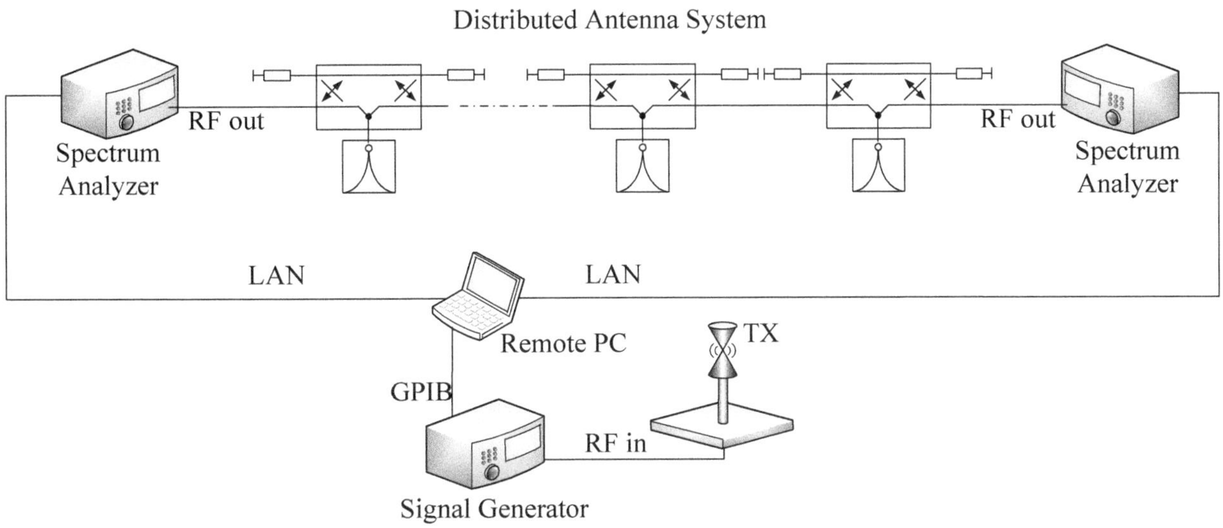

Fig. 7. Illustration of the automated frequency domain measurement setup.

different transmitter positions are presented. Starting under the first antenna element of the system measurements ranging from 0 m to 9 m with an incremental step of 0.5 m were carried out. To identify the potential for commercial use the impact of the system bandwidth was analyzed. Therefore continuous wave (CW) sweeps over the entire frequency range from 800 MHz to 2 GHz at equidistant frequency steps were performed. Averaging the received level differences at different frequency ranges yields an estimation of the localization accuracy. To achieve comparable results equivalent conditions were considered in the simulations.

5.1 Simulations

In the simulations the ray tracing process was started for discrete frequencies with a step width of 1 MHz over the entire frequency range at each transmitter position. In Fig. 6 simulated level differences at varying system bandwidths are presented. It can be observed that with increasing bandwidth the curves become smoother with less variations and therefore less ambiguities in the estimation of position x from power difference ΔP. An almost linear behavior is reached when the signal baseband bandwidth exceeds 70 MHz.

5.2 Measurements

Measurements were carried out using an automated measurement setup including two spectrum analyzers and a signal generator. An important advantage of frequency domain analysis is that the measured data can be used flexibly for evaluation of different system bandwidths. A schematical drawing of all components of the measurement setup is depicted in Fig. 7. The transmitting part consists of a signal generator connected to an omnidirectional low dispersive biconical antenna (Ott and Eibert, 2010). This setup enables to test the system under ideal conditions and makes the measurement results comparable to the simulations with an isotropic transmitting antenna. Instead of root mean square (RMS) level detection two spectrum analyzers were connected to the output ports of the distributed antenna system. A remote personal computer is used to provide communication between the devices over Local Area Network (LAN) and General Purpose Interface Bus (GPIB) connections. This setup enables to characterize the entire frequency range of the signal by applying a CW sweep. Thus predictions within the entire system bandwidth are possible without filtering the signal. Furthermore a larger dynamic range can be investigated compared to measurements with network analyzers.

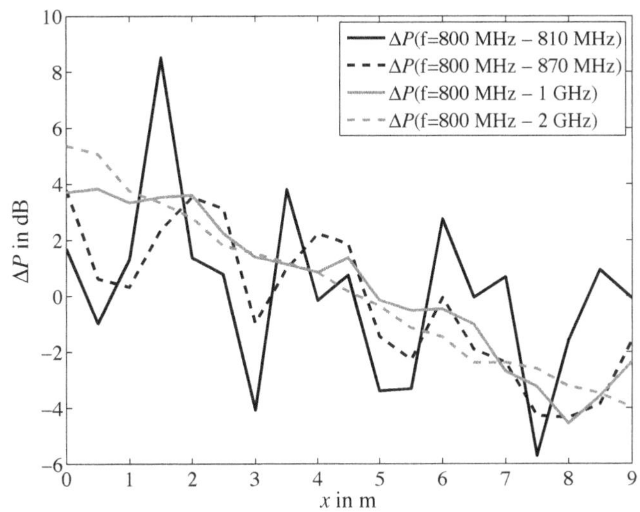

Fig. 8. Measured level difference at the output ports of a 9 m long distributed antenna system with four equally spaced antenna elements.

Figure 8 shows the measurement results carried out in a hall at the authors institute. As expected from the simulations location accuracy is increased significantly with arising bandwidth. Within this measurement campaign the required localization fidelity of 3 m is achieved at a system bandwidth of 70 MHz. Comparing measurement and simulation results stronger frequency selective fading effects occurred in the simulations at a signal bandwidth of 10 MHz. A reasonable explanation is that within the simulations much stronger multipath effects occurs due to walls with metallic surfaces.

5.3 Appraisal of results

Measurement and simulation results showed that localization with a distributed antenna system based on simple level detection is possible provided that the signal bandwidth is sufficiently large. On the other hand RMS level detection averages the available signal power over a certain period of time. Thus the system also offers the possibility localize narrowband signal sources with Frequency Hopping Spread Spectrum (FHSS). Further improvement can be expected by applying statistical tracking filters for parameter estimation (Kalman, 1960). It is intended to evaluate the performance of the system also with time domain signals in the future.

6 Conclusions

A localization concept for a wireless indoor positioning system has been presented. The algorithm is based on information from simple power measurements of received signals. Thus positioning with communication transceivers is possible without the need for too much additional hardware. This makes the implementation of the system very cost efficient. Both numerical system simulations and practical measurements in an exemplary indoor environment show that location accuracy of 3 m can be reached for signals with a bandwidth of at least 70 MHz.

References

Bill, R., Cap, C., Kofahl, M., and Mundt, T.: Indoor and Outdoor Positioning in Mobile Environments – a Review and some Investigations on WLAN-Positioning, Geographic Information Sciences, 10, 91–98, 2004.

Buddendick, H. and Eibert, T. F.: Acceleration of Ray-Based Radar Cross Section Predictions Using Monostatic-Bistatic Equivalence, IEEE Transactions on Antennas and Propagation, 58, 531–539, 2010.

Delogne, P. P. and Deryck, L.: Underground Use of a Coaxial Cable with Leaky Sections, IEEE Transactions on Antennas and Propagation, 28, 875–883, 1980.

Engelbrecht, J., Collmann, R., Birkel, U., and Weber, M.: Methodical leaky feeder design for indoor positioning considering multipath environments, in: Proc. IEEE Radio and Wireless Symposium (RWS), pp. 164–167, New Orleans, LA, 2010.

Farr, E. G. and Baum, C. E.: Time Domain Characterization of Antennas with TEM Feeds, http://www.farr-research.com/Papers/ssn426.pdf, 1998.

Gibson, P. J.: The Vivaldi Aerial, in: Proc. IEEE 9th European Microwave Conference, pp. 101–105, Brighton, UK, 1979.

Gu, Y., Lo, A., and Niemegeers, I.: A Survey of Indoor Positioning Systems for Wireless Personal Networks, IEEE Communications Surveys & Tutorials, 11, 30–32, 2009.

Hagan, M. T., Demuth, H. B., and Beale, M. H.: Neural Network Design, PWS Publishing, Boston, MA, 1996.

Kalman, R. E.: A New Approach to Linear Filtering and Prediction Problems, Transactions of the ASME–Journal of Basic Engineering, 82, 35–45, 1960.

Kunisch, J.: Implications of Lorentz Reciprocity for Ultra-Wideband Antennas, in: Proc. IEEE International Conference on Ultra-Wideband (ICUWB), pp. 214–219, Singapore, 2007.

Lange, J.: Interdigitated Stripline Quadrature Hybrid, IEEE Transactions on Microwave Theory and Techniques, 17, 1150–1151, 1969.

Ling, H., Chou, R. C., and Lee, S. W.: Shooting and bouncing rays: Calculating the RCS of an arbitrarily shaped cavity, IEEE Transactions on Antennas and Propagation, 37, 194–205, 1989.

Motley, A. J. and Palmer, D. A.: Reduced long-range signal reception with leaky feeders, Electronics Lett., 19, 714–715, 1983.

Ott, A. T. and Eibert, T. F.: A 433 MHz–22 GHz reconfigurable dielectric loaded biconical antenna, in: Proc. IEEE 4th European Conference on Antennas and Propagation, vol. 4, pp. 1721–1724, 2010.

Saha, S., Chaudhuri, K., Sanghi, D., and Bhagwat, P.: Location determination of a mobile device using IEEE 802.11b access point signals, in: Proc. IEEE Wireless Communications and Networking (WCNC), vol. 3, pp. 1987–1992, New Orleans, LA, USA, 2003.

Saleh, A. A. M., Rustako, Jr., A. J., and Roman, R. S.: Distributed Antennas for Indoor Radio Communications, IEEE Transactions on Communications, 35, 1245–1251, 1987.

Simpson, G. R.: A Generalized n-Port Cascade Connection, in: International Microwave Symposium Digest (IEEE MTT-S), pp. 507–509, Los Angeles, CA, 1981.

Walter, C. H.: Traveling Wave Antennas, McGraw-Hill, New York, NY, 1965.

Weber, M., Birkel, U., and Collmann, R.: Indoor RF Fingerprinting using leaky feeder cable considering environmental changes, in: Proceedings of the 6th International Conference on Mobile Technology, Application & Systems, vol. 18, New York, NY, USA, 2009.

Wilkinson, E.: An N-Way Hybrid Power Divider, IRE Transactions on Microwave Theory and Techniques, 8, 116–118, 1960.

28

Modeling of electrical transport in Zinc Oxide varistors

K. Bavelis, E. Gjonaj, and T. Weiland

Institut für Theorie Elektromagnetischer Felder, Technische Universität Darmstadt, Schloßgartenstr. 8, 64289 Darmstadt, Germany

Correspondence to: K. Bavelis (bavelis@temf.tu-darmstadt.de)

Abstract. The electrical transport in zinc oxide (ZnO) varistors is analyzed using microstructural material modeling. The fully three dimensional current distribution is computed by means of a nonlinear equivalent circuit model representing the assembly of current carrying grains and grain boundaries of the material. The investigation focuses on the phenomenon of current filamentation due to inhomogeneities of the varistor microstructure. Numerical results highlight the importance of 3-D percolation effects in the modeling of varistor currents as well as that of the grain bulk resistivity which so far has been neglected in previous studies.

1 Introduction

ZnO varistors are ceramic semiconductor materials that are characterized by a highly nonlinear electrical response. Their resistance drops remarkably with increasing voltage. Thus, a varistor based device behaving as an insulator under normal operating conditions, will draw large currents from the circuit as soon as an overvoltage event occurs (Meshkatoddini, 2011). ZnO varistors are, furthermore, capable of absorbing high amounts of energy (Gupta, 1990). For this reasons, they are most often applied as surge arresters for protection against temporary electrical overstresses, such as lightning and switching surges in power distribution lines, transformers and switches (Standler, 2002). Further applications of ZnO varistors include passive voltage stabilization for low power electronics (Eda, 1989).

The properties of ZnO varistors are a result of their crystalline microstructure. Varistor ceramics are composed of a large number of ZnO grains of different shape, size and orientation. Small amounts of doping additives such as Bi_2O_3 or Mn_3O_4 are usually present (Eda, 1984; Gupta, 1990). The nonlinear electrical response is attributed to the potential barriers forming at the boundaries between ZnO grains (Blatter and Greuter, 1986). These boundaries represent, in fact, Schottky-like microjunctions with an electrical response corresponding roughly to that of a back-to-back pair of Zener diodes. The grain boundary resistances determine varistor currents for voltages in the nonlinear breakdown voltage region of the electrical characteristics as well as in the linear pre-breakdown region. The resistance of the varistor at higher voltages becomes again Ohmic as it is essentially determined by grain bulk resistivity (Vojta et al., 1996).

Although this basic operation principle has been known for some time, the characterization of ZnO varistors for engineering applications remains a major challenge. Several numerical models have been proposed to describe electrical transport in varistors (Bartkowiak and Mahan, 1995; Vojta et al., 1996; Qingheng et al., 2002; Hofstaetter and Supancic, 2013). In these models, the varistor microstructure is first represented geometrically by a space filling mesh. The mesh cells correspond to the ZnO grains whereas cell interfaces (mesh edges in 2-D or mesh faces in 3-D) to the grain boundaries. Almost all of the models so far presented in the literature operate in 2-D planar geometry. They are based either on a Cartesian mesh representation (Vojta et al., 1996; Wen and Clarke, 1993) or on a Voronoi-type tessellation of the varistor sample (Bartkowiak et al., 1996a; Zhao et al., 2007). Given a microstructure representation, electrical transport can be modeled by means of an equivalent electrical circuit with voltage nodes and current branches associated with the grain centers and grain boundaries, respectively. The grain boundary resistances are taken into account by introducing diode-like elements with specified electrical characteristics (Vojta et al., 1996) in the current branches. The macroscopic electrical response is, then, obtained by the solution of the resulting circuit equations. It should be noted that varistor microstructure (grain sizes and shapes) as well

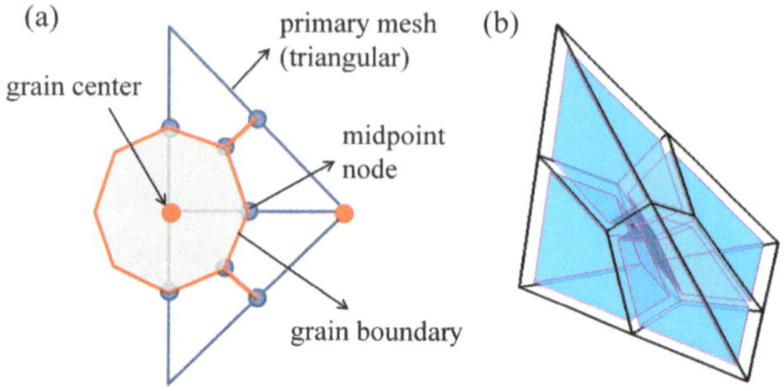

Figure 1. **(a)** Basic procedure for the construction of a polygonal grain tessellation on the basis of a simplicial mesh. **(b)** Example barycentric decomposition of a tetrahedron.

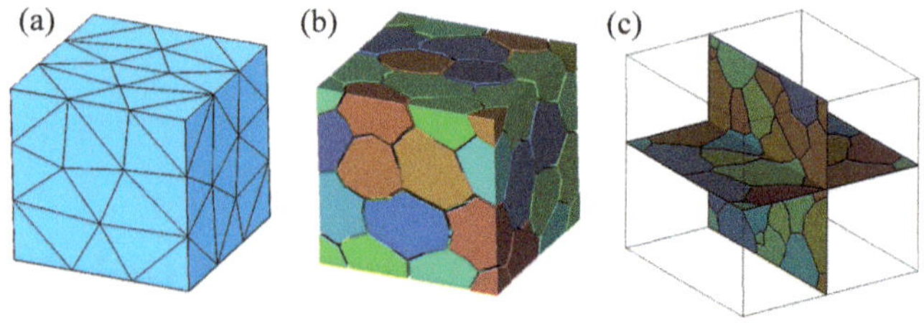

Figure 2. Grain structure representation of a cubic sample: **(a)** Primary mesh. **(b)** 3-D polygonal tessellation. **(c)** Cross-sectional cuts depicting the resulting grain structure.

as grain boundary resistances are subject to stochastic distributions which makes the modeling task in the general case very difficult.

In this work, we focus on two particular aspects of varistor modeling. The first concerns the influence of the 3-D geometry on modeling accuracy. Since percolation thresholds for 2-D and 3-D systems are in general very different, the commonly used 2-D models fail to reproduce properly the current distribution in the material and thus compute the varistor characteristic accurately. We perform a comparison between 2-D and 3-D varistor models using the approach of (Vojta et al., 1996) based on a Cartesian mesh representation. The second aspect is related to the effect of grain bulk resistivity. This effect involving both grain geometry and the topology of the grain network has been thoroughly neglected in previous studies. Finally, we present a simulation example featuring the current filamentation phenomenon using a 3-D modeling approach for a realistic varistor sample.

2 Varistor modeling

ZnO varistors are basically composed of polyhedral grains of various shapes and sizes, typically, varying in the range of 10–100 μm. The electrical current distribution in the material is strongly affected by its microstructure including grain shape, topology of the grain network as well as type and distribution of crystalline defects. The first modeling task consists, thus, in the construction of an appropriate geometrical model of the varistor microstructure.

2.1 Geometrical representation

The simplest geometrical model consists of a 2-D Cartesian mesh of square grains. This approach was used, e.g., in (Wen and Clarke, 1993; Vojta et al., 1996). Because of the simple geometry, it is not possible to take into account local grain size variations. Also, the resulting topology of the grain network is far from realistic. This model, however, allows for investigating current flow percolation effects in a systematic manner. For this purpose, we have implemented a 3-D Cartesian mesh model which is used in Sect. 3.1 to compare macroscopic varistor responses obtained by 2-D and 3-D simulations, respectively.

For a more general geometrical description, an unstructured mesh representation is needed. Unlike the commonly used 2-D tessellations (Bartkowiak et al., 1996a; Zhao et al., 2007; Hu et al., 2010), however, we aim at a fully 3-D polyhedral grain model. The following 3-D tessellation procedure is proposed. We start with a simplicial tetrahedral mesh on the varistor sample. Such a mesh can be easily constructed using, e.g., a conventional FEM tool. The only constrain imposed on this 'primary' mesh is that the average edge length should coincide with the sought grain size of the material.

A polyhedral grain structure is constructed on the basis of the primary mesh by applying a midpoint subdivision procedure (Li et al., 1995). According to this procedure, new mesh nodes are introduced at the midpoints of every edge, face and cell of the primary mesh, respectively. Mesh cells of polyhedral shape (representing the ZnO grains) are, then, constructed by connecting these nodes as exemplarily shown in Fig. 1a. By construction, each node of the primary mesh corresponds to one cell of the polyhedral mesh. These nodes can be considered as the generating germs of the resulting grain structure. The grain boundaries are represented by the generally non-planar interfaces between the neighboring polyhedral cells. For illustration, the midpoint subdivision of a tetrahedron into 4 hexahedrons is depicted in Fig. 1b.

Note that, in general, grain cell convexity is not guaranteed. Also, grains with sharp angle corners and very small edges may occur. It is, however, not the purpose of this construction to generate a high mesh quality in the FEM sense. This procedure allows in the first place for a more realistic representation of the varistor microstructure than that of a simple Cartesian mesh model. Furthermore, grain size, grain number and (to some extent) grain shapes can be easily controlled. Tessellations for arbitrary material samples of, e.g., cubic (cf. Fig. 2) or cylindrical shape can be generated. Due to the robustness of modern mesh generators it is, furthermore, possible to construct large varistor models with several hundreds of thousands of grains.

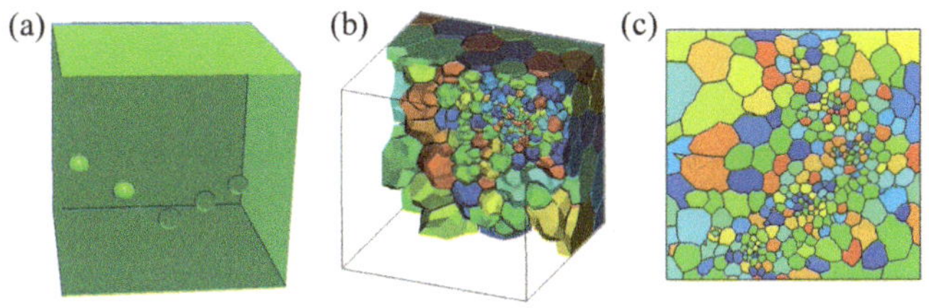

Figure 3. Modeling of a nonuniform grain size distribution: **(a)** Geometrical description of the defective pattern by small spheres located within the volume of a cubic varistor sample. 3-D cut **(b)** and cross-sectional view **(c)** of the resulting grain structure.

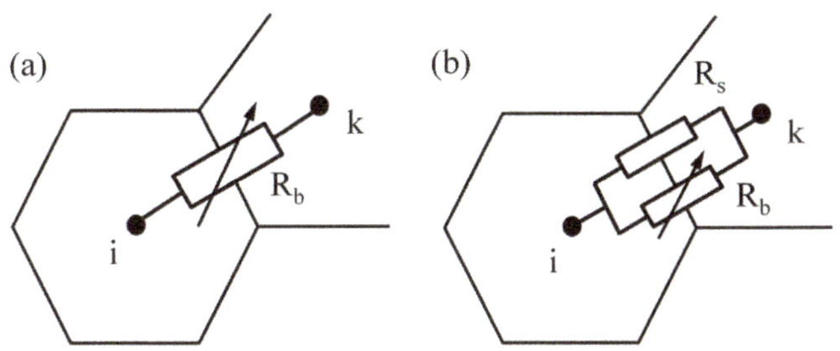

Figure 4. (a) Equivalent electrical circuit representation of a grain boundary according to Eq. (1). **(b)** Grain boundary equivalent circuit corresponding to the model of (Vojta et al., 1996).

Another strength of this approach is the ability to model microstructure inhomogeneities. This is important for the investigation of crystalline defects or inclusions and their impact on the varistor characteristic. Figure 3 demonstrates the model of a varistor with a non-uniform grain size distribution generated by this approach. The defective pattern is initially prescribed by means of a few small spheres properly located within the sample volume. The result is an nonuniform primary mesh and, therefore, a non-uniform polygonal grain tessellation with grain size distribution closely following the assumed defect distribution.

2.2 Grain boundary conductivity models

The $J - V$ characteristic of a single grain boundary in the breakdown region can be described by the empirical relation

$$J(V) = kV^{\alpha}, \tag{1}$$

where k is a parameter, α is the dynamic conductance, J is the current density and V the voltage drop across the boundary. In terms of an equivalent electrical circuit the contribution of a single grain boundary can be incorporated by introducing a nonlinear resistor between adjacent grains sharing this boundary (see Fig. 4a). This approach, however, does not count for the grain bulk resistivity. Furthermore, the reverse leakage diode currents which are responsible for the pre-breakdown varistor response are neglected.

A more general approach was proposed by (Vojta et al., 1996) and then later used by several authors (Zhao et al., 2007). The grain boundary characteristic is assumed to

$$J(V) = \frac{\sigma_s}{d}V + \frac{\sigma_g}{2d}V\{1 + \tan h[s(V - V_B]\}, \tag{2}$$

where V_B is the barrier voltage, σ_s and σ_g are Ohmic conductivities, d is the distance between the centers of the two grains sharing the boundary and s is the nonlinearity varistor parameter. A closer inspection of Eq. (2) reveals that this is equivalent to two resistors connected in parallel across the boundary (see Fig. 4b). This configuration is able to reproduce qualitatively the varistor response in the full voltage range. Note that σ_g is interpreted as the conductivity of the grain bulk material since it determines the characteristics at high voltages. Referring to Fig. 4b, however, it is not possible to separate the contributions of grain bulk and grain boundary at lower voltages. Thus, neither the shape nor the size of the single ZnO grains can be explicitly accounted for in this approach. The physical understanding that grain currents enter through its boundaries only, implies that bulk and boundary resistors should be connected in series.

2.3 Distributed grain conductance model

A more accurate model is obtained by employing a conductance matrix to describe current flow within the ZnO grains. In order to illustrate the idea we consider a single hexagonal grain as shown in Fig. 5a. Neglecting leakage resistances, the model of (Vojta et al., 1996) reduces to a star network of resistors. Applying an arbitrary voltage V_1 at one of the boundaries while keeping the rest at a fixed potential V_0, results in an equal current distribution among the remaining boundaries. This is true independently from the number of boundaries or the actual grain shape. As can be easily verified by simulations, however, most of the current should flow through the boundaries closest to the excited one. For the varistor problem, where the macroscopic characteristic depends strongly on the current flow pattern within the material, this approach may lead to inaccurate results.

We propose the elementary equivalent circuit shown in Fig. 5b. The grain bulk resistance is described by the conductance matrix **G** of a N-port network, where N is the number of grain boundaries. The grain boundary resistances are connected in series at each port of the network. They consist of a leakage resistor connected in parallel to a nonlinear one obeying the $J - V$ characteristic (Eq. 1). The conductance matrix can be determined numerically by 3-D FEM simulations for arbitrarily shaped grains. The elementary equivalent circuits constructed by this principle are, then, assembled into a global equivalent circuit which can be solved numerically for the current distribution within the varistor. The drawback of this approach is that a large number of FEM simulations may be needed to obtain conductance matrices for each individual grain within the material.

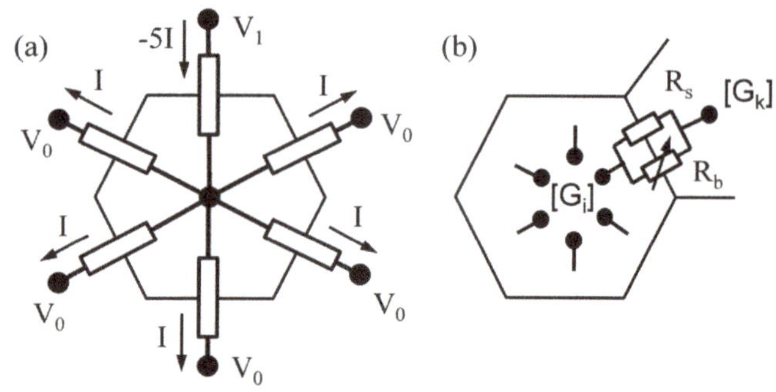

Figure 5. **(a)** Grain boundary resistors for a hexagonal grain and current distribution resulting from the model of (Vojta et al., 1996). **(b)** Equivalent circuit representation in the distributed grain conductance approach.

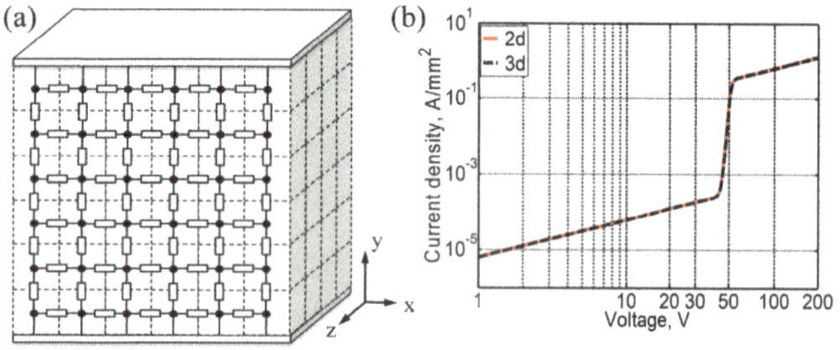

Figure 6. **(a)** Cartesian mesh grain structure and corresponding equivalent circuit. **(b)** Varistor characteristics resulting form 2-D and 3-D simulations, respectively, when a uniform distribution of the microjunction properties is assumed.

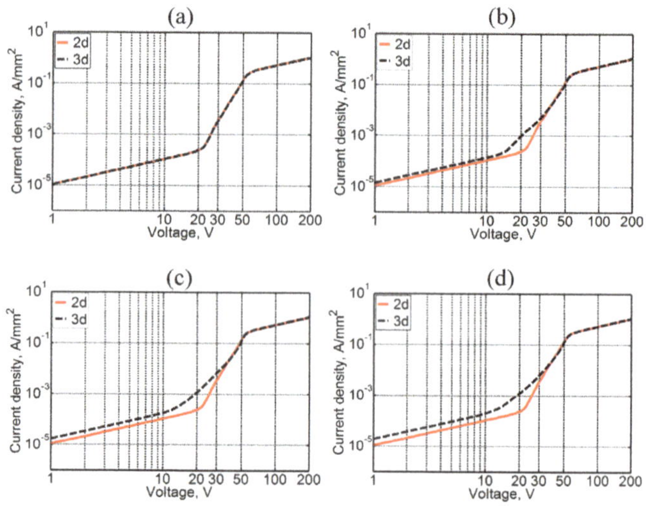

Figure 7. Varistor characteristics obtained when 20 % of the microjuctions are assumed to be inactive. A sequence of 3-D models is considered which is obtained by adding grain layers in the *z* direction: **(a)** 1 layer. **(b)** 3 layers. **(c)** 10 layers. **(d)** 15 layers.

3 Results

3.1 Influence of 3-D geometry

We consider the simplest varistor model using a Cartesian grain structure with microjunction characteristics given by Eq. (2). The 3-D model consists of a $15 \times 15 \times 15$ mesh of cubic grains with a side length of $10\mu m$. The 2-D geometry case is emulated by employing a mesh with $15 \times 15 \times 1$ grains of the same size. The situation is illustrated in Fig. 6a. For consistency, the global varistor characteristic computed with the two models for a uniform distribution of microjunction properties is shown in Fig. 6b. In this case, all grain boundaries are described by the same parameters, $\sigma_s = 10^{-3}\,\text{S/m}$, $\sigma_g = 1\,\text{S/m}$, $s = 8$ and $V_B = 3.2\,\text{V}$. The current flow between top and bottom electrodes is uniform, thus, 2-D and 3-D models yield the same varistor characteristic.

The influence of 3-D geometry becomes obvious when inhomogeneity is introduced into the model. We adopt the approach of (Vojta et al., 1996) and (Bartkowiak et al., 1996a) where a number of grain boundaries uniformly distributed within the sample volume are assumed to be electrically inactive. This is implemented by setting the diode parameters for these boundaries to $s = 0$ and $\sigma_s = 0\,\text{S/m}$. Figure 7 shows the results obtained when this procedure is applied for a fixed portion of inactive grain boundaries (20 %). A sequence of models is considered which is obtained by adding grain layers in the *z* direction to the 2-D model (single layer in the *z* direction) while keeping the grain size constant. The 3-D varistor characteristic is shifted toward lower voltages as the number of layers is increased. The lower switching voltage in the 3-D case compared to the 2-D one can be explained by the higher probability for 3-D currents to find a conductive path to the bottom electrode. By further increasing the number of layers (see Fig. 7c) this effect saturates and the varistor characteristics does not change any more. At this point, the percolation threshold for the current paths in 3-D geometry has been reached. In general, this phenomenon depends on grain size distribution and on the topology of the grain network. This simple example, however, already demonstrates the necessity of 3-D modelling in varistor simulations.

3.2 Influence of grain bulk conductivity

The same experiment is repeated for the distributed grain conductance model described in Sect. 2.3. The nominal grain boundary parameters are assumed to $k = 4.055 \times 10^{-18}\,\text{S/m}^2$, $\alpha = 45$ and $\sigma_s = 10^{-3}\,\text{S/m}$. Again, the same shifting of the switching voltage towards lower values is observed, when 20 % of the boundaries are assumed to be inactive. When a sufficient number of 3-D layers has been added the varistor characteristic converges to its final shape (see Fig. 8). Compared to the results obtained by the Vojta approach (Fig. 7), this characteristic appears much smoother in the transition region between breakdown and upturn voltages. This behavior is due to the influence of grain bulk

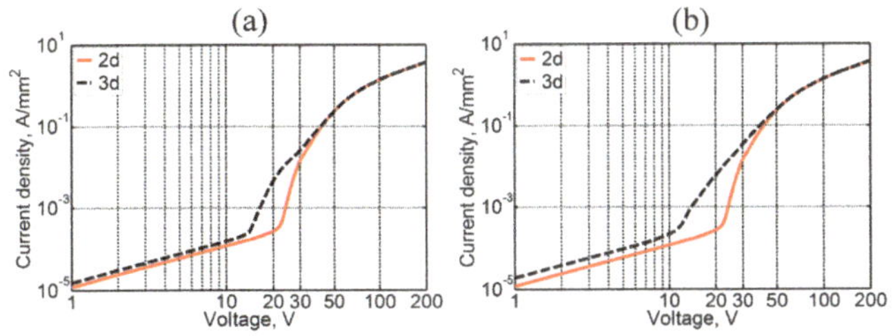

Figure 8. 3-D Varistor characteristics resulting from the distributed grain conductance model when 20 % of the microjuctions are assumed to be inactive. The 3-D models consist of **(a)** 3 grain layers and **(b)** 10 grain layers, respectively.

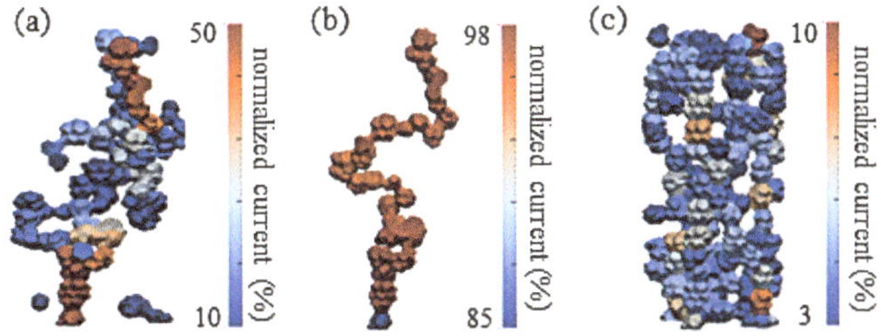

Figure 10. Current distribution within a cylindrical sample (see Fig. 9) containing 10 % inactive boundaries for different applied voltages: **(a)** Leakage region. **(b)** Breakdown voltage. **(c)** Upturn voltage.

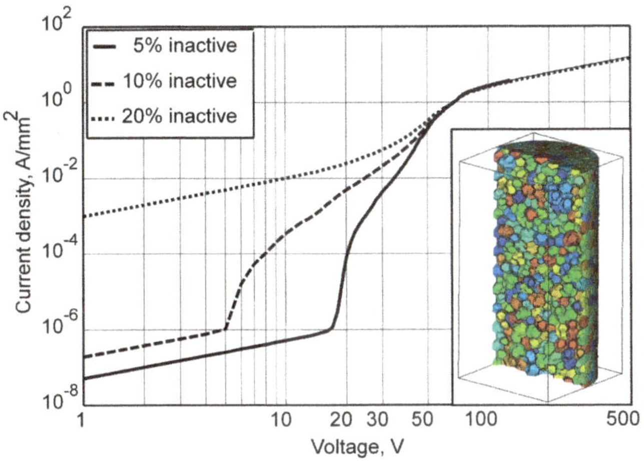

Figure 9. Grain structure representation of a cylindrical varistor sample and the computed characteristics for 5, 10 and 20 % inactive boundaries, respectively.

conductivity which is properly taken into account in the distributed conductance model.

3.3 Current filamentation and varistor characteristics

In the following, a more realistic varistor simulation is discussed. A cylindrical sample of radius $50\mu m$ and height $200\mu m$ with an average grain size of $10\mu m$ is considered. The microstructure geometry is constructed by applying the tessellation procedure described in Sect.2.1. The resulting 3-D grain model consists of a total of 1737 ZnO grains (see the inset of Fig. 9). Furthermore, the model contains 10 081 grain boundaries corresponding to the number of nonlinear resistors employed in the equivalent electrical circuit model.

To model the grain boundary microjunctions we apply the Vojta approach (Eq. 2) with nominal parameters, $\sigma_g = 1\,S/m$, $\sigma_s = 10^{-6}\,S/m$, $V_B = 3.2V$ and $s = 8$, respectively. Inhomogeneity is introduced into the model by assuming a certain number of inactive microjuctions randomly distributed among the available grain boundaries. Figure 10 shows the current distribution within a varistor containing 10 % inactive boundaries for three different voltages applied at the top and bottom electrodes. Depicted is the current flowing through each grain normalized to the total varistor current. For visualization purposes, only grains carrying a current above a certain threshold are shown. An inhomogeneous current distribution is observed already at low voltages (see Fig. 10a). In this case, a single grain can carry up to 50 % of the total varistor current. In the breakdown region, this phenomenon becomes substantially stronger. As shown in Fig. 10b, nearly the total varistor current flows along a single narrow path of width no larger than twice the grain size. As the voltage is increased above the upturn voltage, the current distribution becomes again more homogeneous (see Fig. 10c). The varistor response in this region is essentially Ohmic, thus, as expected the effect of inactive boundaries in the current flow distribution is less pronounced.

The influence of inhomogeneous current flow on the overall varistor performance is illustrated in Fig. 9. Three cases with 5, 10 and 20 % inactive boundaries, respectively, are considered. The $J-V$ characteristic for 5 % inactive boundaries corresponds essentially to the single grain boundary response (Eq. 2). Increasing the number of inactive boundaries causes the switching woltage to shift towards lower values while the effective dynamic conductance of the material is decreased. The transition from fully nonlinear to Ohmic varistor response is, however, not trivial. As already observed in (Bartkowiak et al., 1996b) using 2-D simulations, the $J-V$ characteristic of a varistor containing inactive grain boundaries manifests a double-knee pattern featuring a second inflection in the current-voltage curve in the region between switching and upturn voltages, respectively.

4 Conclusions

The simulation of electrical transport in ZnO varistors requires 3-D modeling. This was demonstrated by a direct comparison of the electrical varistor characteristics computed by 2-D and 3-D simulations. The simulations show a systematic shift of the switching voltage toward lower values in the 3-D case. This can be explained by the intrinsically inhomogeneous current flow pattern through the grain microstructure which implies that the percolation behavior of the current paths in 3-D space is different from the 2-D one. Furthermore, we propose a novel equivalent circuit approach employing a distributed grain conductance model. This approach allows to properly take into account the influence of grain bulk resistivity on the macroscopic varistor response for arbitrarily shaped ZnO grains. A procedure for the fully 3-D geometrical representation of varistor microstructures based on a particular 3-D tessellation method is introduced. Using this approach, simulations for a realistic varistor sample were performed. They show the experimentally observed current filamentation phenomenon resulting from microstructural inhomogeneities which strongly influence the electrical performance of ZnO varistors.

Acknowledgements. The authors thank the German Research Foundation (DFG) for its financial support.

Edited by: U. van Rienen

References

Bartkowiak, M. and Mahan, G. D.: Nonlinear currents in Voronoi networks, Phys. Rev. B, 51, 10825–10832, 1995.

Bartkowiak, M., Mahan, G. D., Modine, F. A., Alim, M. A., Lauf, R., and McMillan A.: Voronoi network model of ZnO varistors with different types of grain boundaries, J. Appl. Phys., 80, 6516–6522, 1996a.

Bartkowiak, M., Mahan, G. D., Modine, F. A. and Alim, M. A.: Influence of ohmic grain boundaries in ZnO varistors, J. Appl. Phys., 79, 273–281, 1996b.

Blatter G. and Greuter F.: Carrier transport through grain boundaries in semiconductors, Phys. Rev. B, 33, 3952–3966, 1986.

Eda K.: Destruction mechanism of ZnO varistors due to high currents, J. Appl. Phys., 56, 2948–2955, 1984.

Eda, K.: Zinc Oxide Varistors, IEEE Electr. Insul. M., 5, 28–41, 1989.

Gupta, T. K.: Application of Zinc Oxide Varistors, J. Am. Ceram. Soc., 73, 1817–1840, 1990.

Hofstätter, M. and Supancic, P.: 3-D Netzwerksimulationen von Varistoren mit verschiedenen Korngrößenverteilungen, BHM Berg- und Hüttenmännische Monatshefte, 158, 206–210, 2013.

Hu, J., Long, W., He, J. and Liu, J.: Simulation of Surge Repsonses of ZnO Varistors by Voronoi Network with Actual Grain Boundary Model, CEIDP, Annual Report Conference on Electrical Insulation and Dielectric Phenomena, West Lafayette, USA, 2010.

Li, T. S., McKeag, R. M., and Armstrong, C. G.: Hexahedral meshing using midpoint subdivision and integer programming, Comput. Methods Appl. M., 124, 171–193, 1995.

Meshkatoddini, M. R.: Metal Oxide ZnO-Based Varistor Ceramics, in: Advances in Ceramics – Electric and Magnetic Ceramics, Bioceramics, Ceramics and Environment, edited by: Sikalidis, C., InTech publishing, Chapter 14, doi:10.5772/23601, 2011.

Qingheng, C., Jinliang, H., Kexiong, T., Shuiming, C., Minyu, Y., and Jianxin, T.: Influence of grain size on distribution of temperature and thermal stress in ZnO varistor ceramics, Sci. China Ser. E, 45, 337–347, 2002.

Standler, R. B.: Protection of Electronic Circuits From Overvoltages, Dover Publications, Mineola New York, 2002.

Vojta, A., Wen, Q., and Clarke, D. R.: Influence of microstructural disorder on the current transport behavior of varistor ceramics, Comp. Mater. Sci., 6, 51–62, 1996.

Wen, Q. and Clarke, D. R.: Modeling the electrical characteristics of Polycrystalline Varistors using individual grain boundary properties, Grain boundaries and Interfacial Phenomena in electronic Ceramics, 41, 217–230, 1993.

Zhao, G., Joshi, R. P., Lakdawala, V. K., and Hjalmarson, H. P.: Electro-thermal Simulation Studies for Pulsed Voltage Induced Energy Absorption and Potential Failure in Microstructured ZnO Varistors, IEEE T. Dielect. El. In., 14, 1007–1015, 2007.

A mixed finite-element formulation for the modal analysis of electromagnetic waveguides featuring improved low-frequency resolution of transmission line modes

R. Baltes[1], **J. Al Ahmar**[2], **O. Farle**[1], **and R. Dyczij-Edlinger**[1]

[1]Chair of Electromagnetic Theory, Dept. of Physics and Mechatronics, Saarland University, Saarbrücken, Germany
[2]Chair of Microintegration and Reliability, Dept. of Physics and Mechatronics, Saarland University, Saarbrücken, Germany

Correspondence to: R. Baltes (r.baltes@lte.uni-saarland.de)

Abstract. This paper presents an improved finite-element formulation for axially uniform electromagnetic waveguides. It allows for both dielectric and conduction losses and covers the entire range from optics down to the static limit. Propagation coefficients of small magnitude, particularly those of transmission line modes in the low-frequency regime, are computed much more accurately than with previous approaches.

1 Introduction

Since electromagnetic waveguides constitute generalized transmission line systems, they are of fundamental importance for the design of electromagnetic devices. Because of complex geometries and inhomogeneous materials, analytical solutions for the modal field patterns and propagation coefficients are often difficult to obtain or even unavailable. Numerical field simulation methods provide a powerful remedy. In particular the finite-element (FE) method stands out for its high convergence rates and great flexibility in modeling geometry and materials.

Early approaches suffered from the occurrence of spurious modes, but with the advent of $\mathcal{H}(curl)$ conforming basis functions these problems were overcome. A great variety of FE formulations have been proposed: Some are in terms of the electric (Lee et al., 1991) or magnetic field intensity (Valor and Zapata, 1995), while others employ a magnetic vector potential and a electric scalar potential (Bardi and Biro, 1991; Polstyanko and Lee, 1995). They all provide accurate results for high frequencies, but very few can handle the low-frequency (LF) (Vardapetyan and Demkowicz, 2003) or even the static case (Lee et al., 2003; Farle et al., 2004). Moreover, they incorporate electric conductivity σ by means of an equivalent imaginary part ε'' of the electric permittivity (Lee, 1994), which breaks down when the angular frequency ω tends to zero, due to

$$\varepsilon'' = \frac{\sigma}{\omega}. \tag{1}$$

All methods above lead to generalized algebraic eigenvalue problems for the square of the propagation coefficient. As will be detailed in Sect. 4.2, this strongly amplifies numerical round-off error in propagation coefficients of small magnitude, e.g. waveguide modes close to cut-off and, more importantly, transmission line (TEM or quasi-TEM) modes in the LF regime.

To overcome these limitations, we propose in Sect. 2 the prototype of a mixed-field formulation in terms of the electric field intensity $\boldsymbol{E}$ and the magnetic flux density $\boldsymbol{B}$. Its distinguishing feature is that the sought eigenvalue is the propagation coefficient itself rather than its square. The suggested approach incorporates conduction losses quite naturally and minimizes round-off errors in propagation coefficients of small magnitude. In Sect. 3 we establish stability in the LF regime, by imposing suitable constraints.

The ability of a single formulation to predict the propagation characteristics of inhomogeneous transmission lines accurately from statics up to microwave frequencies is of great practical importance for mixed-signal analysis, particularly in the context of model-order reduction (Polstyanko et al., 1997; Bertazzi et al., 2002; Schultschik et al., 2008).

2 Formulation

We consider an axially uniform waveguide Ω pointing in the $\hat{z}$ direction. Its boundary Γ, with unit outward normal vector $\hat{\boldsymbol{n}}$, is assumed to consist of perfect electric walls Γ_E and perfect magnetic walls Γ_H. Let Ω be connected and Γ_E nonempty. We denote the wavenumber, speed of light, and characteristic impedance, respectively, of free space by k_0, c_0, and η_0. The relative magnetic permeability μ_r, the relative electric permittivity ε_r, and the electric conductivity σ are assumed to be scalar-valued. Thanks to uniformity along the waveguide axis, the modal fields are given by waves propagating in $\pm\hat{z}$ direction. For the negative $\hat{z}$ direction, we have the decomposition (Pozar, 2005, p. 93)

$$\boldsymbol{E} = \left(\boldsymbol{E}_t + E_z\hat{z}\right)\exp(\gamma z), \tag{2}$$
$$\boldsymbol{B} = \left(\boldsymbol{B}_t + B_z\hat{z}\right)\exp(\gamma z), \tag{3}$$

where subscripts t and z denote transversal and longitudinal components, respectively, and γ is the propagation coefficient. Thus the differential operators ∂_z and ∇ simplify to

$$\partial_z = \gamma, \tag{4}$$
$$\nabla = \nabla_t + \partial_z\hat{z} = \nabla_t + \gamma\hat{z}. \tag{5}$$

By applying Eqs. (2)–(5) to Faraday's law and Ampere's law,

$$\nabla \times \boldsymbol{E} = -\mathrm{j}k_0c_0\boldsymbol{B}, \tag{6}$$
$$\nabla \times \mu_r^{-1}c_0\boldsymbol{B} = (\mathrm{j}k_0\varepsilon_r + \sigma\eta_0)\boldsymbol{E}, \tag{7}$$

we arrive at a homogeneous boundary value problem, given by the set of partial differential equations

$$\hat{z} \times (\gamma\boldsymbol{E}_t - \nabla_t E_z) = -\mathrm{j}k_0c_0\boldsymbol{B}_t, \tag{8}$$
$$\hat{z} \cdot (\nabla_t \times \boldsymbol{E}_t) = -\mathrm{j}k_0c_0B_z, \tag{9}$$
$$\hat{z} \times \left(\gamma\mu_r^{-1}c_0\boldsymbol{B}_t - \nabla_t\mu_r^{-1}c_0B_z\right) = (\mathrm{j}k_0\varepsilon_r + \sigma\eta_0)\,\boldsymbol{E}_t, \tag{10}$$
$$\hat{z} \cdot (\nabla_t \times \mu_r^{-1}c_0\boldsymbol{B}_t) = (\mathrm{j}k_0\varepsilon_r + \sigma\eta_0)\,E_z, \tag{11}$$

subject to the boundary conditions

$$\left.\begin{aligned} E_z &= 0 \\ \hat{\boldsymbol{n}} \times \boldsymbol{E}_t &= \boldsymbol{0} \\ \hat{\boldsymbol{n}} \cdot \boldsymbol{B}_t &= 0 \end{aligned}\right\} \text{ on } \Gamma_E, \tag{12}$$

$$\left.\begin{aligned} \mu_r^{-1}B_z &= 0 \\ \hat{\boldsymbol{n}} \times \mu_r^{-1}\boldsymbol{B}_t &= \boldsymbol{0} \\ \hat{\boldsymbol{n}} \cdot \varepsilon_r\boldsymbol{E}_t &= 0 \end{aligned}\right\} \text{ on } \Gamma_H. \tag{13}$$

2.1 Finite-element representation

Let $\mathcal{H}^1$, $\mathcal{H}^{\mathrm{curl}}$, $\mathcal{H}^{\mathrm{div}}$, and $\mathcal{H}^0$ denote the Sobolev spaces with respect to the operator in superscript (Boffi et al., 2013, pp. 4). We define the subspaces

$$\mathcal{H}_E^1 := \left\{\lambda \in \mathcal{H}^1(\Omega) \mid \lambda = 0 \text{ on } \Gamma_E\right\}, \tag{14}$$
$$\mathcal{H}_E^{\mathrm{curl}} := \left\{\boldsymbol{u} \in \mathcal{H}^{\mathrm{curl}}(\Omega) \mid \hat{\boldsymbol{n}} \times \boldsymbol{u} = \boldsymbol{0} \text{ on } \Gamma_E\right\}, \tag{15}$$
$$\mathcal{H}_E^{\mathrm{div}} := \left\{\boldsymbol{u} \in \mathcal{H}^{\mathrm{div}}(\Omega) \mid \hat{\boldsymbol{n}} \cdot \boldsymbol{u} = 0 \text{ on } \Gamma_E\right\}, \tag{16}$$

and denote the corresponding finite-element spaces by $\mathcal{V}_E^1 \subset \mathcal{H}_E^1$, $\mathcal{V}_E^{\mathrm{curl}} \subset \mathcal{H}_E^{\mathrm{curl}}$, $\mathcal{V}_E^{\mathrm{div}} \subset \mathcal{H}_E^{\mathrm{div}}$, and $\mathcal{V}^0 \subset \mathcal{H}^0$. The transversal and axial fields $\boldsymbol{E}_t, E_z, \boldsymbol{B}_t, B_z$ are discretized by basis functions of lowest order (Zhu and Cangellaris, 2006, pp. 19), for their respective function spaces. Let N_N denote the number of free nodes, N_E the number of free edges, and N_F the number of faces in the mesh. Then

$$E_z = \sum_{k=1}^{N_N} x_{Ez,k}\lambda_k \qquad \text{with } \lambda_k \in \mathcal{V}_E^1, \tag{17}$$
$$\boldsymbol{E}_t = \sum_{k=1}^{N_E} x_{Et,k}\boldsymbol{w}_k \qquad \text{with } \boldsymbol{w}_k \in \mathcal{V}_E^{\mathrm{curl}}, \tag{18}$$
$$c_0B_z = \sum_{k=1}^{N_F} x_{Bz,k}W_k \qquad \text{with } W_k \in \mathcal{V}^0, \tag{19}$$
$$c_0\boldsymbol{B}_t = \sum_{k=1}^{N_E} x_{Bt,k}\boldsymbol{v}_k \qquad \text{with } \boldsymbol{v}_k \in \mathcal{V}_E^{\mathrm{div}}. \tag{20}$$

Note that Eqs. (19) and (20) have been scaled by c_0, for numerical stability. The FE representations of the differential operators grad, curl, and div are given by the incidence matrices $\mathbf{G} \in \mathbb{R}^{N_E \times N_N}$, $\mathbf{R} \in \mathbb{R}^{N_F \times N_E}$, and $\mathbf{D} \in \mathbb{R}^{N_F \times N_E}$. Also, in two dimensions, we have (Zhu and Cangellaris, 2006, pp. 26)

$$\mathbf{D} = \mathbf{R}. \tag{21}$$

Since Eqs. (8) and (9) are in terms of fields represented on the primal mesh, they are discretized in strong form. In contrast, Eqs. (10) and (11) are based on the fields $\boldsymbol{H} = \mu^{-1}\boldsymbol{B}$, $\boldsymbol{D} = \varepsilon\boldsymbol{E}$, and $\boldsymbol{J} = \sigma\boldsymbol{E}$. These reside on the dual complex and are thus considered in weak form. The resulting generalized algebraic eigenvalue problem reads

$$(\mathbf{A} + \mathrm{j}k_0\mathbf{C})\boldsymbol{x} = -\gamma\mathbf{B}\boldsymbol{x}. \tag{22}$$

At this point it can be seen that the eigenvalue is the propagation coefficient γ itself rather than its square. The vector and the matrices in Eq. (22) are given by

$$\boldsymbol{x} = \begin{bmatrix} \boldsymbol{x}_{Bt} \\ \boldsymbol{x}_{Bz} \\ \boldsymbol{x}_{Et} \\ \boldsymbol{x}_{Ez} \end{bmatrix}, \quad \mathbf{A} = \begin{bmatrix} \mathbf{0} & \mathbf{0} & \mathbf{0} & -\mathbf{G} \\ \mathbf{0} & \mathbf{0} & -\mathbf{R} & \mathbf{0} \\ \mathbf{0} & -\mathbf{R}^T\mathbf{T}_{Bz} & -\mathbf{T}_{\sigma_t} & \mathbf{0} \\ \mathbf{G}^T\mathbf{T}_{Bt} & \mathbf{0} & \mathbf{0} & \mathbf{T}_{\sigma z} \end{bmatrix},$$
$$\mathbf{B} = \begin{bmatrix} \mathbf{0} & \mathbf{0} & \mathbf{I} & \mathbf{0} \\ \mathbf{0} & \mathbf{0} & \mathbf{0} & \mathbf{0} \\ -\mathbf{T}_{Bt} & \mathbf{0} & \mathbf{0} & \mathbf{0} \\ \mathbf{0} & \mathbf{0} & \mathbf{0} & \mathbf{0} \end{bmatrix}, \quad \mathbf{C} = \begin{bmatrix} \mathbf{I} & \mathbf{0} & \mathbf{0} & \mathbf{0} \\ \mathbf{0} & \mathbf{I} & \mathbf{0} & \mathbf{0} \\ \mathbf{0} & \mathbf{0} & -\mathbf{T}_{Et} & \mathbf{0} \\ \mathbf{0} & \mathbf{0} & \mathbf{0} & \mathbf{T}_{Ez} \end{bmatrix}, \tag{23}$$

where $\mathbf{I}$ denotes the identity matrix. The other submatrices are defined as

$$T_{Bt,ij} = \int_\Omega \boldsymbol{v}_i \cdot \mu_r^{-1} \boldsymbol{v}_j \,\mathrm{d}\Omega \qquad \text{with } \boldsymbol{v}_i, \boldsymbol{v}_j \in \mathcal{V}_E^{\text{div}}, \tag{24}$$

$$T_{Bz,ij} = \int_\Omega W_i \mu_r^{-1} W_j \,\mathrm{d}\Omega \qquad \text{with } W_i, W_j \in \mathcal{V}^0, \tag{25}$$

$$T_{Et,ij} = \int_\Omega \boldsymbol{w}_i \cdot \varepsilon_r \boldsymbol{w}_j \,\mathrm{d}\Omega \qquad \text{with } \boldsymbol{w}_i, \boldsymbol{w}_j \in \mathcal{V}_E^{\text{curl}}, \tag{26}$$

$$T_{Ez,ij} = \int_\Omega \lambda_i \varepsilon_r \lambda_j \,\mathrm{d}\Omega \qquad \text{with } \lambda_i, \lambda_j \in \mathcal{V}_E^1, \tag{27}$$

$$T_{\sigma z,ij} = \eta_0 \int_\Omega \lambda_i \sigma \lambda_j \,\mathrm{d}\Omega \qquad \text{with } \lambda_i, \lambda_j \in \mathcal{V}_E^1, \tag{28}$$

$$T_{\sigma t,ij} = \eta_0 \int_\Omega \boldsymbol{w}_i \cdot \sigma \boldsymbol{w}_j \,\mathrm{d}\Omega \qquad \text{with } \boldsymbol{w}_i, \boldsymbol{w}_j \in \mathcal{V}_E^{\text{curl}}. \tag{29}$$

In absence of magnetic (dielectric) losses, we have $\mu_r \in \mathbb{R}^+$ ($\varepsilon_r \in \mathbb{R}^+$) everywhere, and the corresponding matrices $\mathbf{T}_{Bt}$ and $\mathbf{T}_{Bz}$ ($\mathbf{T}_{Et}$ and $\mathbf{T}_{Ez}$) are positive definite. In contrast, $\mathbf{T}_{\sigma t}$ and $\mathbf{T}_{\sigma z}$ are positive semi-definite, because $\sigma > 0$ holds in the conductive region only.

3 Low-frequency behavior

For $k_0 = 0$, the eigenvalue problem of Eq. (22) reduces to

$$\mathbf{A}\boldsymbol{x} = -\gamma \mathbf{B}\boldsymbol{x}. \tag{30}$$

The matrix $\mathbf{A} + \gamma\mathbf{B}$ turns out to be singular, for any value of γ. We will analyze this effect and propose a suitable regularization.

3.1 Analysis of instability

We decompose the waveguide cross-section Ω into the lossless region Ω_{ll}, with $\sigma = 0$, the lossy subdomain Ω_l, with $\sigma > 0$, and their common interface Γ_0:

$$\Omega = \Omega_l \cup \Omega_{ll} \cup \Gamma_0. \tag{31}$$

In view of Eq. (23), the eigenvalue problem for the static case, Eq. (30), is equivalent to

$$\gamma \boldsymbol{x}_{Et} = \mathbf{G}\boldsymbol{x}_{Ez}, \tag{32}$$

$$-\mathbf{R}\boldsymbol{x}_{Et} = \mathbf{0}, \tag{33}$$

$$\gamma \mathbf{T}_{Bt}\boldsymbol{x}_{Bt} + \mathbf{T}_{\sigma t}\boldsymbol{x}_{Et} = -\mathbf{R}^T\mathbf{T}_{Bz}\boldsymbol{x}_{Bz}, \tag{34}$$

$$\mathbf{G}^T\mathbf{T}_{Bt}\boldsymbol{x}_{Bt} + \mathbf{T}_{\sigma z}\boldsymbol{x}_{Ez} = \mathbf{0}. \tag{35}$$

Consider the vectors

$$\begin{bmatrix} \boldsymbol{x}_{Et,\gamma} \\ \boldsymbol{x}_{Ez,\gamma} \end{bmatrix} = \begin{bmatrix} \mathbf{G} \\ \gamma \mathbf{I} \end{bmatrix} \boldsymbol{x}_N, \tag{36}$$

$$\begin{bmatrix} \boldsymbol{x}_{Bt,\gamma} \\ \boldsymbol{x}_{Bz,\gamma} \end{bmatrix} = \begin{bmatrix} -\mathbf{T}_{Bt}^{-1}\mathbf{R}^T\mathbf{T}_{Bz} \\ \gamma \mathbf{I} \end{bmatrix} \boldsymbol{x}_F, \tag{37}$$

with $\boldsymbol{x}_F \in \mathbb{R}^{N_F}$ arbitrary and any $\boldsymbol{x}_N \in \mathbb{R}^{N_N}$ satisfying

$$x_{N,i} = \begin{cases} 0 & \text{if node } i \in \Omega_l \cup \Gamma_0, \\ \text{arbitrary} & \text{if node } i \in \Omega_{ll}. \end{cases} \tag{38}$$

Plugging Eqs. (36) and (37) into Eqs. (32)–(35) shows that these are satisfied for any arbitrary value of γ. The vectors

$$\boldsymbol{u}_\gamma = \begin{bmatrix} \boldsymbol{x}_{Bt,\gamma} \\ \boldsymbol{x}_{Bz,\gamma} \\ \boldsymbol{x}_{Et,\gamma} \\ \boldsymbol{x}_{Ez,\gamma} \end{bmatrix} \tag{39}$$

form the eigenspace corresponding to γ. To regularize the eigenvalue problem and eliminate the non-physical eigenvectors, we first analyze the vectors $\boldsymbol{u}_\gamma$. Let $\boldsymbol{E}_\gamma$ and $\boldsymbol{B}_\gamma$ denote the fields represented by Eqs. (36) and (37), respectively. In view of Eq. (38), $\boldsymbol{E}_\gamma$ is nonzero in the lossless region only and, by Eq. (4), it is a gradient field. Similarly, it can be shown that the magnetic field strength $\boldsymbol{H}$ corresponding to $\boldsymbol{B}_\gamma$ is a gradient, represented on the dual mesh. Plugging $\boldsymbol{E}_\gamma$ and $\boldsymbol{B}_\gamma$ into Eqs. (6) and (7) and taking the divergence yields

$$-\mathrm{j}k_0 c_0 \nabla \cdot \boldsymbol{B}_\gamma = 0, \tag{40}$$

$$\mathrm{j}k_0 c_0^{-1} \nabla \cdot \varepsilon_r \boldsymbol{E}_\gamma = 0. \tag{41}$$

It can be seen that, for $k_0 = 0$, these fields are no longer forced to be source-free.

3.2 Stabilization

To stabilize the method, we impose the conditions

$$\nabla \cdot \boldsymbol{B} = 0 \qquad \text{in } \Omega, \tag{42}$$

$$\nabla \cdot (\varepsilon_0 \varepsilon_r \boldsymbol{E}) = 0 \qquad \text{in } \Omega_{ll}, \tag{43}$$

similarly to (Polstyanko et al., 1997; Hiptmair et al., 2008), by introducing Lagrange multipliers $\boldsymbol{p}_F$ and $\boldsymbol{p}_N$. The discrete representations of Eqs. (42) and (43) read

$$\mathbf{D}\boldsymbol{x}_{Bt} + \gamma \boldsymbol{x}_{Bz} = \mathbf{R}\boldsymbol{x}_{Bt} + \gamma \boldsymbol{x}_{Bz} = \mathbf{0}, \tag{44}$$

$$\tilde{\mathbf{I}}_N \mathbf{G}^T \mathbf{T}_{Et} \boldsymbol{x}_{Et} - \gamma \tilde{\mathbf{I}}_N \mathbf{T}_{Ez} \boldsymbol{x}_{Ez} = \mathbf{0}, \tag{45}$$

where the matrix $\tilde{\mathbf{I}}_N$ selects the nodes belonging to the lossless regions Ω_{ll}.

Care must be taken if the structure is bounded by electric walls only ($\Gamma_H = \emptyset$). In this case, the transpose of the curl matrix $\mathbf{R}^T$, i.e. the gradient operator on the dual mesh, has a nontrivial nullspace of dimension one. It consists of constant fields and is represented by a coefficient vector $\boldsymbol{x}_c$ with

$$x_{c,i} = \operatorname{sign}(\det(\mathbf{J}_i)), \tag{46}$$

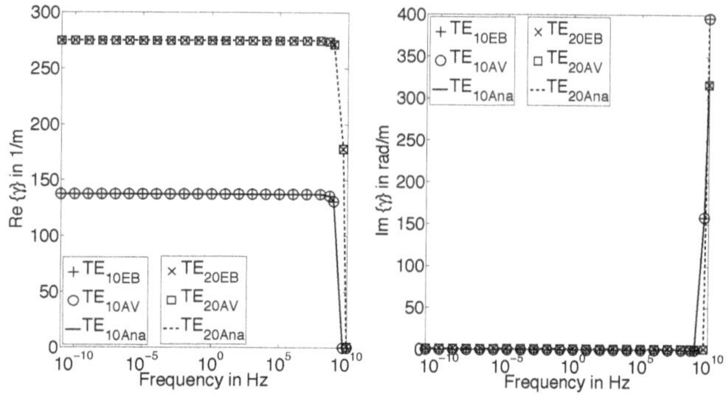

(a) Attenuation coefficient. (b) Phase coefficient.

Figure 1. Lossless rectangular waveguide: propagation coefficients of TE_{10} and TE_{20} modes versus frequency.

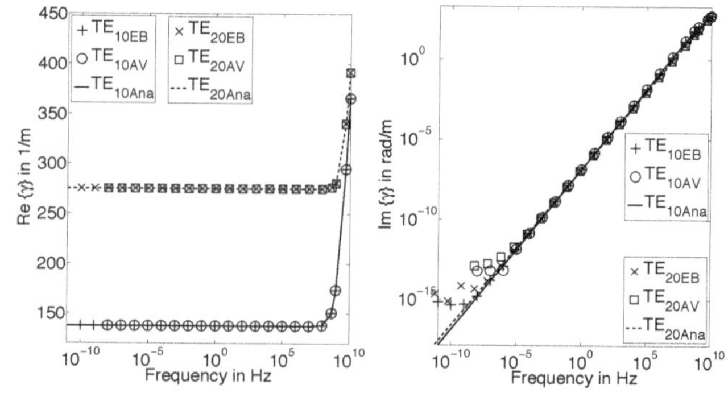

(a) Attenuation coefficient. (b) Phase coefficient.

Figure 2. Lossy RWG with $\sigma = 5$ S/m: propagation coefficients of TE_{10} and TE_{20} modes versus frequency.

where $\mathbf{J}_i$ is the Jacobian matrix of face i. In view of Eq. (23), the corresponding kernel vector of $\boldsymbol{A}$ reads

$$\boldsymbol{u}_c = \begin{bmatrix} \mathbf{0} \\ \mathbf{T}_{Bz}^{-1}\boldsymbol{x}_c \\ \mathbf{0} \\ \mathbf{0} \end{bmatrix}. \tag{47}$$

It is easy to verify that the fields of $\boldsymbol{u}_c$ satisfy Eqs. (32)–(35) and (45). Moreover, Eq. (44) implies

$$\gamma(\boldsymbol{u}_c) = 0. \tag{48}$$

Thus there exists a zero eigenvalue corresponding to a constant magnetic field strength $\boldsymbol{H}$ pointing in $\hat{z}$ direction. To exclude the eigenpair $(\gamma = 0, \boldsymbol{x} = \boldsymbol{u}_c)$, we impose the orthogonality constraint

$$\boldsymbol{x}_c^T \boldsymbol{x}_{Bz} = 0, \tag{49}$$

by introducing a Lagrange multiplier p_c. Thus, the stabilized eigenvalue problem for the case $\Gamma_H = \emptyset$ becomes

$$(\hat{\mathbf{A}} + \mathrm{j}k_0\hat{\mathbf{C}})\hat{\boldsymbol{x}} = -\gamma\overline{\mathbf{B}}\hat{\boldsymbol{x}}, \tag{50}$$

with

$$\hat{\mathbf{A}} = \begin{bmatrix} \mathbf{0} & \mathbf{0} & \mathbf{0} & -\mathbf{G} & \mathbf{0} & \mathbf{R}^T & \mathbf{0} \\ \mathbf{0} & \mathbf{0} & -\mathbf{R} & \mathbf{0} & \mathbf{0} & \mathbf{0} & \boldsymbol{x}_c \\ \mathbf{0} & -\mathbf{R}^T\mathbf{T}_{Bz} & -\mathbf{T}_{\sigma_t} & \mathbf{0} & \mathbf{T}_{Et}\mathbf{G}\tilde{\mathbf{I}}_N^T & \mathbf{0} & \mathbf{0} \\ \mathbf{G}^T\mathbf{T}_{Bt} & \mathbf{0} & \mathbf{0} & \mathbf{T}_{\sigma z} & \mathbf{0} & \mathbf{0} & \mathbf{0} \\ \mathbf{0} & \mathbf{0} & \tilde{\mathbf{I}}_N\mathbf{G}^T\mathbf{T}_{Et} & \mathbf{0} & \mathbf{0} & \mathbf{0} & \mathbf{0} \\ \mathbf{0} & \mathbf{R} & \mathbf{0} & \mathbf{0} & \mathbf{0} & \mathbf{0} & \mathbf{0} \\ \mathbf{0} & \boldsymbol{x}_c^T & \mathbf{0} & \mathbf{0} & \mathbf{0} & \mathbf{0} & 0 \end{bmatrix}$$

$$\overline{\mathbf{B}} = \begin{bmatrix} \mathbf{0} & \mathbf{0} & \mathbf{I} & \mathbf{0} & \mathbf{0} & \mathbf{0} & \mathbf{0} \\ \mathbf{0} & \mathbf{0} & \mathbf{0} & \mathbf{0} & \mathbf{0} & \mathbf{I} & \mathbf{0} \\ -\mathbf{T}_{Bt} & \mathbf{0} & \mathbf{0} & \mathbf{0} & \mathbf{0} & \mathbf{0} & \mathbf{0} \\ \mathbf{0} & \mathbf{0} & \mathbf{0} & \mathbf{0} & -\mathbf{T}_{Ez}\tilde{\mathbf{I}}_N^T & \mathbf{0} & \mathbf{0} \\ \mathbf{0} & \mathbf{0} & \mathbf{0} & -\tilde{\mathbf{I}}_N\mathbf{T}_{Ez} & \mathbf{0} & \mathbf{0} & \mathbf{0} \\ \mathbf{0} & \mathbf{I} & \mathbf{0} & \mathbf{0} & \mathbf{0} & \mathbf{0} & \mathbf{0} \\ \mathbf{0} & \mathbf{0} & \mathbf{0} & \mathbf{0} & \mathbf{0} & \mathbf{0} & 0 \end{bmatrix},$$

$$\hat{\mathbf{C}} = \begin{bmatrix} \mathbf{I} & \mathbf{0} & \mathbf{0} & \mathbf{0} & \mathbf{0} & \mathbf{0} & \mathbf{0} \\ \mathbf{0} & \mathbf{I} & \mathbf{0} & \mathbf{0} & \mathbf{0} & \mathbf{0} & \mathbf{0} \\ \mathbf{0} & \mathbf{0} & -\mathbf{T}_{Et} & \mathbf{0} & \mathbf{0} & \mathbf{0} & \mathbf{0} \\ \mathbf{0} & \mathbf{0} & \mathbf{0} & \mathbf{T}_{Ez} & \mathbf{0} & \mathbf{0} & \mathbf{0} \\ \mathbf{0} & \mathbf{0} & \mathbf{0} & \mathbf{0} & \mathbf{0} & \mathbf{0} & \mathbf{0} \\ \mathbf{0} & \mathbf{0} & \mathbf{0} & \mathbf{0} & \mathbf{0} & \mathbf{0} & \mathbf{0} \\ \mathbf{0} & \mathbf{0} & \mathbf{0} & \mathbf{0} & \mathbf{0} & \mathbf{0} & 0 \end{bmatrix}, \quad \hat{\boldsymbol{x}} = \begin{bmatrix} \boldsymbol{x}_{Bt} \\ \boldsymbol{x}_{Bz} \\ \boldsymbol{x}_{Et} \\ \boldsymbol{x}_{Ez} \\ \boldsymbol{p}_N \\ \boldsymbol{p}_F \\ p_c \end{bmatrix}. \tag{51}$$

All eigenvectors of Eq. (50) that exhibit non-vanishing Lagrange multipliers, i.e. non-zero $\boldsymbol{p}_N$, $\boldsymbol{p}_F$ or p_c, are known to span a subspace without physical meaning. This means the relevant subvectors do not solve the original eigenvalue problem Eq. (22). Rather than to purge such unwanted eigenvectors from the results, the authors prefer to shift the associated eigenvalues to infinity, so that they do not pollute the spectrum. This is easily accomplished, by setting the columns of $\overline{\mathbf{B}}$ that belong to Lagrange multipliers to zero. The modified matrix $\hat{\mathbf{B}}$ reads

$$\hat{\mathbf{B}} = \begin{bmatrix} \mathbf{0} & \mathbf{0} & \mathbf{I} & \mathbf{0} & \mathbf{0} & \mathbf{0} & \mathbf{0} \\ \mathbf{0} & \mathbf{0} & \mathbf{0} & \mathbf{0} & \mathbf{0} & \mathbf{0} & \mathbf{0} \\ -\mathbf{T}_{Bt} & \mathbf{0} & \mathbf{0} & \mathbf{0} & \mathbf{0} & \mathbf{0} & \mathbf{0} \\ \mathbf{0} & \mathbf{0} & \mathbf{0} & \mathbf{0} & \mathbf{0} & \mathbf{0} & \mathbf{0} \\ \mathbf{0} & \mathbf{0} & \mathbf{0} & -\tilde{\mathbf{I}}_N\mathbf{T}_{Ez} & \mathbf{0} & \mathbf{0} & \mathbf{0} \\ \mathbf{0} & \mathbf{I} & \mathbf{0} & \mathbf{0} & \mathbf{0} & \mathbf{0} & \mathbf{0} \\ \mathbf{0} & \mathbf{0} & \mathbf{0} & \mathbf{0} & \mathbf{0} & \mathbf{0} & 0 \end{bmatrix}, \tag{52}$$

and the final form of the eigenvalue problem is given by

$$(\hat{\mathbf{A}} + \mathrm{j}k_0\hat{\mathbf{C}})\hat{\boldsymbol{x}} = -\gamma\hat{\mathbf{B}}\hat{\boldsymbol{x}}. \tag{53}$$

4 Numerical examples

4.1 Rectangular waveguide

We consider a rectangular waveguide (RWG) of dimensions 22.86 mm×11.43 mm with perfectly conducting walls and homogenous material properties ($\epsilon_r = 4$, $\mu_r = 1$) in both the

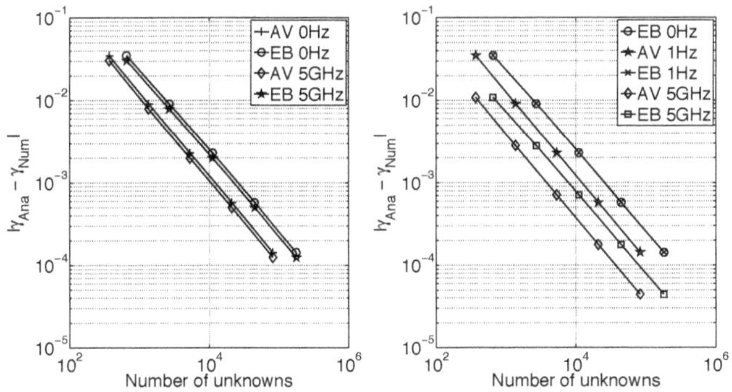

(a) Lossless case: $\sigma = 0$ S/m. (b) Lossy case: $\sigma = 5$ S/m.

Figure 3. Convergence analysis for the TE_{10} mode.

lossless case ($\sigma = 0$) and in presence of Ohmic losses ($\sigma = 5$ S/m). Our main goal is to demonstrate the correct function of the present approach and compare it to the $\boldsymbol{A}-V$ potential formulation of (Farle et al., 2004), a state-of-the-art method that solves for γ^2 and models conductivity via complex permittivity, Eq. (1). According to (Pozar, 2005, p. 108), the analytical solution for the propagation coefficient γ_{mn} is given by

$$\gamma_{mn} = \sqrt{\left(\frac{m\pi}{a}\right)^2 + \left(\frac{n\pi}{b}\right)^2 - \varepsilon_r\mu_r k_0^2 + \mathrm{j}\mu_r\sigma k_0\eta_0}, \qquad (54)$$

where a and b denote the width and height of the waveguide, respectively. Figures 1 and 2 present the dispersion curves of the TE_{10} and TE_{20} modes in the frequency range $[10^{-10}, 10^{10}]$ Hz, and Fig. 3b gives the error in propagation coefficient as a function of the number of unknowns.

In the lossless case, the solutions of both the $\boldsymbol{E}-\boldsymbol{B}$ and $\boldsymbol{A}-V$ methods are in very good agreement with the theory, over the whole frequency range; see Fig. 1. Figure 3a demonstrates that the convergence rates of both formulations are the same, for $f = 0$ Hz and $f = 5$ GHz, respectively.

In the lossy case (σ=5 S/m), it can be seen from Fig. 2 that the $\boldsymbol{E}-\boldsymbol{B}$ approach works over the entire frequency range whereas the $\boldsymbol{A}-V$ scheme fails to converge below 10^{-8} Hz, due to breakdown of Eq. (1). Above this threshold, the results of both methods agree very well with analytical results. The numerical noise visible in Fig. 2b for phase coefficients of very small magnitude, $\mathrm{Im}\,\gamma < 10^{-12}$ rad/m, is insignificant because, as seen in Fig. 2a, the corresponding attenuation coefficients are more than 14 orders of magnitude larger, $\mathrm{Re}\,\gamma > 120$. Figure 3b indicates that, within their respective range of validity, both numerical methods exhibit the same rate of convergence, independently of frequency. In case of the $\boldsymbol{E}-\boldsymbol{B}$ scheme, this also holds in the static case, $f = 0$ Hz.

4.2 Shielded microstrip line

The purpose of this example is to demonstrate the advantage of formulating the eigenvalue problem in terms of γ rather than γ^2, with respect to round-off error in propagation coefficients of small magnitude. Again, we compare the $\boldsymbol{E}-\boldsymbol{B}$ and $\boldsymbol{A}-V$ methods, by means of the lossless shielded microstrip line shown on the inset of Fig. 4. In contrast to the RWG of Sect. 4.1, the present structure supports a quasi-TEM mode, the phase coefficient of which is known to depend linearly on frequency in the LF regime. Figure 4 gives results for both numerical methods. While the $\boldsymbol{E}-\boldsymbol{B}$ data exhibit the expected behavior, the phase coefficients produced by the $\boldsymbol{A}-V$ method stagnate for frequencies below 10^4 Hz. This may come as a surprise, because the latter formulation was designed to be low-frequency stable (Farle et al., 2004).

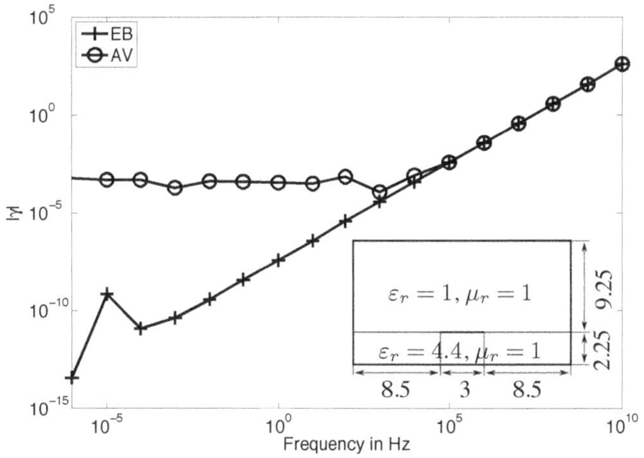

Figure 4. Shielded microstrip line: LF behavior of propagation coefficient of quasi-TEM mode. Inset shows cross-section of structure. Dimensions are in mm.

To clarify the situation, we present in Table 1 a comparison for not only the quasi-TEM wave but also the first two box modes. It can be seen that only the quasi-TEM mode is affected; the results for box modes are correct, even at 0 Hz. This behavior results from the fact that the static limit of $|\gamma|$ is zero for the quasi-TEM mode but non-zero for all others. At sufficiently low frequency values, $|\gamma_{\mathrm{qTEM}}|$ becomes so much smaller than all other $|\gamma|$ values that the eigenvalue solver is unable to resolve γ_{qTEM} to sufficient accuracy, due to numerical noise. Note that the gap in eigenvalue $|\lambda_1/\lambda_2|$ is

$$\left|\frac{\lambda_1}{\lambda_2}\right| = \begin{cases} \left|\frac{\gamma_{\mathrm{qTEM}}}{\gamma_{\mathrm{Box\,1}}}\right| & \text{for } \boldsymbol{E}-\boldsymbol{B} \text{ method,} \\ \left|\frac{\gamma_{\mathrm{qTEM}}}{\gamma_{\mathrm{Box\,1}}}\right|^2 & \text{for } \boldsymbol{A}-V \text{ method,} \end{cases} \qquad (55)$$

because the $\boldsymbol{E}-\boldsymbol{B}$ formulation solves for γ and the $\boldsymbol{A}-V$ scheme for γ^2. Figure 5 presents the eigenvalue ratio as a function of frequency. For both formulations, the eigenvalue solver produces significant round-off error for $|\lambda_1/\lambda_2| < 10^{-12}\ldots10^{-9}$. However, the frequency which this happens at is 10^4 Hz for the $\boldsymbol{A}-V$ method, whereas the proposed $\boldsymbol{E}-\boldsymbol{B}$ scheme produces accurate results down to 10^{-3} Hz, thanks to improved eigenvalue ratio in Eq. (55).

Table 1. Comparison of first three modes of microstrip line.

Mode	Method	0 Hz	10 kHz	100 kHz
quasi-TEM	$\boldsymbol{E}-\boldsymbol{B}$	$1.6653e-16$	$3.7160e-04j$	$3.7162e-03j$
	$\boldsymbol{A}-V$	$4.7821e-04j$	$8.6431e-04$	$3.7630e-03j$
Box 1	$\boldsymbol{E}-\boldsymbol{B}$	$1.5699e+02$	$1.5699e+02$	$1.5699e+02$
	$\boldsymbol{A}-V$	$1.5699e+02$	$1.5699e+02$	$1.5699e+02$
Box 2	$\boldsymbol{E}-\boldsymbol{B}$	$2.6987e+02$	$2.6987e+02$	$2.6987e+02$
	$\boldsymbol{A}-V$	$2.6987e+02$	$2.6987e+02$	$2.6987e+02$

Table 2. Computational data.

Example	Lossy RWG		Microstrip	
Method	$\boldsymbol{E}-\boldsymbol{B}$	$\boldsymbol{A}-V$	$\boldsymbol{E}-\boldsymbol{B}$	$\boldsymbol{A}-V$
Unknowns	11074	3969	13377	4416
Non-zeros $\hat{\mathbf{A}}+\hat{\mathbf{C}}$	73707	60167	107496	70602
Non-zeros $\hat{\mathbf{B}}$	19844	34375	29328	39386
Runtime in sec	1.52	0.27	2.96	0.24
Iterations	85	59	82	55

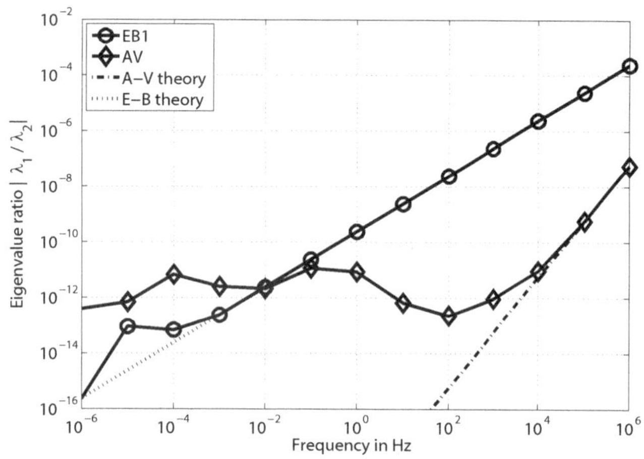

Figure 5. Eigenvalue ratio versus frequency.

5 Critical discussion and conclusions

We have presented a mixed finite-element formulation for electromagnetic waveguides. Stability in the LF regime is achieved by imposing source-free conditions for the electric and magnetic flux densities, with the help of Lagrange multipliers.

Since the sought eigenvalue is the propagation coefficient itself rather than its square, round-off errors in propagation coefficients of small magnitude are significantly reduced. This is of great practical importance for the mixed-signal analysis of transversally inhomogeneous transmission line structures, such as microstrips or coplanar waveguides: For the example of Sect. 4.2, the LF range is extended by 7 orders of magnitude, compared to γ^2 approaches. Another feature of the proposed method is that Ohmic losses enter the formulation directly rather than in the form of equivalent complex permittivity.

The advantages of the new method do not come for free, though. Table 2 shows that the number of variables is three times as large as in the $\boldsymbol{A}-V$ case. Nevertheless, the non-zeros in $\hat{\mathbf{A}}+\hat{\mathbf{C}}$ just increase by 50 %, and those in $\hat{\mathbf{B}}$ actually decrease. Iteration counts of the `eigs` solver of MATLAB-R2013a are 50 % higher for the $\boldsymbol{E}-\boldsymbol{B}$ approach. As a result, solution times for the lossy RWG are six times as long. For the microstrip, that ratio is twelve because, in the lossless case, the system matrices of the $\boldsymbol{A}-V$ method become real-valued, whereas those of the $\boldsymbol{E}-\boldsymbol{B}$ scheme remain complex.

Acknowledgements. This work was supported in part by grant DY 112/1-1 of the Deutsche Forschungsgemeinschaft (DFG).

Edited by: U. van Rienen

References

Bardi, I. and Biro, O.: An efficient finite-element formulation without spurious modes for anisotropic waveguides, IEEE T. Microw. Theory, 39, 1133–1139, 1991.

Bertazzi, F., Peverini, O. A., Goano, M., Ghione, G., Orta, R., and Tascone, R.: A fast reduced-order model for the full-wave FEM analysis of lossy inhomogeneous anisotropic waveguides, IEEE T. Microw. Theory, 50, 2108–2114, 2002.

Boffi, D., Brezzi, F., and Fortin, M.: Mixed Finite Element Methods and Applications, Spinger, Berlin, 2013.

Farle, O., Hill, V., and Dyczij-Edlinger, R.: Finite-Element Waveguide Solvers Revisited, IEEE T. Magn., 40, 1468–1471, 2004.

Hiptmair, R., Krämer, F., and Ostrowski, J.: A Robust Maxwell Formulation for All Frequencies, IEEE T. Magn., 44, 682–685, 2008.

Lee, J.-F.: Finite Element Analysis of Lossy Dielectric Waveguides, IEEE T. Microw. Theory, 42, 1025–1031, 1994.

Lee, J.-F., Sun, D. K., and Cendes, Z. J.: Full-Wave Analysis of Dielectric Waveguides Using Tangential Vector Finite Elements, IEEE T. Microw. Theory, 39, 1262–1271, 1991.

Lee, S.-C., Lee, J.-F., and Lee, R.: Hierarchical vector finite elements for analyzing waveguiding structures, IEEE T. Microw. Theory, 51, 1897–1905, 2003.

Polstyanko, S. V. and Lee, J.-F.: H_1(curl) Tangential Vector Finite Element Method for Modeling Anisotropic Optical Fibers, J. Lightwave Technol., 13, 2290–2295, 1995.

Polstyanko, S. V., Dyczij-Edlinger, R., and Lee, J.-F.: Fast Frequency Sweep Technique for the Efficient Analysis of Dielectric Waveguides, IEEE T. Microw. Theory, 45, 1118–1126, 1997.

Pozar, D. M.: Microwave Engineering, 3rd Edition, John Wiley & Sons, New York, 2005.

Schultschik, A., Farle, O., and Dyczij-Edlinger, R.: A Model Order Reduction Method for the Finite-Element Simulation of Inhomogeneous Waveguides, IEEE T. Magn., 44, 1394–1397, 2008.

Valor, L. and Zapata, J.: Efficient Finite Element Analysis of Waveguides with Lossy Inhomogeneous Anisotropic Materials Characterized by Arbitrary Permittivity and Permeability Tensors, IEEE T. Microw. Theory, 43, 2452–2459, 1995.

Vardapetyan, L. and Demkowicz, L.: Full-wave analysis of dielectric waveguides at a given frequency, Math. Comput., 72, 105–129, 2003.

Zhu, Y. and Cangellaris, A. C.: Multigrid Finite Element Methods for Electromagnetic Field Modeling, IEEE Press, New Jersey, 2006.

Analytical finite element matrix elements and global matrix assembly for hierarchical 3-D vector basis functions within the hybrid finite element boundary integral method

L. Li[1]**, K. Wang**[1]**, H. Li**[2]**, and T. F. Eibert**[1]

[1]Technische Universität München, Lehrstuhl für Hochfrequenztechnik, Arcisstrasse 21, 80333 Munich, Germany
[2]Politecnico di Torino, Electronic and Telecommunication, Corso Duca degli Abruzzi, 24, 10129 Torino, Italy

Correspondence to: L. Li (li.li@tum.de)

Abstract. A hybrid higher-order finite element boundary integral (FE-BI) technique is discussed where the higher-order FE matrix elements are computed by a fully analytical procedure and where the gobal matrix assembly is organized by a self-identifying procedure of the local to global transformation. This assembly procedure applys to both, the FE part as well as the BI part of the algorithm. The geometry is meshed into three-dimensional tetrahedra as finite elements and nearly orthogonal hierarchical basis functions are employed. The boundary conditions are implemented in a strong sense such that the boundary values of the volume basis functions are directly utilized within the BI, either for the tangential electric and magnetic fields or for the asssociated equivalent surface current densities by applying a cross product with the unit surface normals. The self-identified method for the global matrix assembly automatically discerns the global order of the basis functions for generating the matrix elements. Higher order basis functions do need more unknowns for each single FE, however, fewer FEs are needed to achieve the same satisfiable accuracy. This improvement provides a lot more flexibility for meshing and allows the mesh size to raise up to $\lambda/3$. The performance of the implemented system is evaluated in terms of computation time, accuracy and memory occupation, where excellent results with respect to precision and computation times of large scale simulations are found.

1 Introduction

The Finite Element Boundary Integral (FE-BI) method (Jin, 2002; Tzoulis and Eibert, 2005; Eibert and Hansen, 1997) is an efficient numerical technique for solving electromagnetic field problems. Traditional finite element methods rely on utilizing the local information of the FEs. The fixed local node order forces the local matrix elements to be transformed into global ones. Facing low order (LO) basis functions, the local-global transformation is easy as edge related elements only follow the edge directions. When it comes to higher order (HO) basis functions (Jin, 2002; Sun et al., 2001; Ismatullah and Eibert, 2009; Jorgensen et al., 2004; Graglia et al., 1997; Jorgensen et al., 2005), the basis functions are also related to faces, volumes or even more complicated structures. The local-global transformation procedure introduces then considerably more difficulties. In this paper, a self-identified hierarchical basis function method is illustrated. This method effectively overcomes the problem mentioned above and provides more feasibility within FE-BI. Without fixing the node order or the sequence order of the basis functions for the local FEs, the self-identified hierarchical basis function organization allows a simple assembly of the global equation system. Simultaneously, this method guarantees the compatibility between FE and BI (Ismatullah and Eibert, 2009) fluently. All arbitrarily shaped components are meshed into tetrahedra (Volakis et al., 1998) apart from perfect electric conductors (PEC) or perfect magnetic conductors (PMC) where $\boldsymbol{E}$ and $\boldsymbol{H}$ are forced to vanish inside the volume. As FE-BI solves for the field distribution inside the volume together with the corresponding equivalent surface currents (Ismatullah and Eibert, 2009; Rao et al., 1982), the self-identified hierarchical basis functions describe the distribution of fields within the tetrahedra. When observation points tend to the enclosed boundary surface, the boundary condition determines the continuity of $\boldsymbol{E}$ and $\boldsymbol{H}$ fields (Harrington, 1961; Bladel, 1964; Mautz and Harrington, 1978).

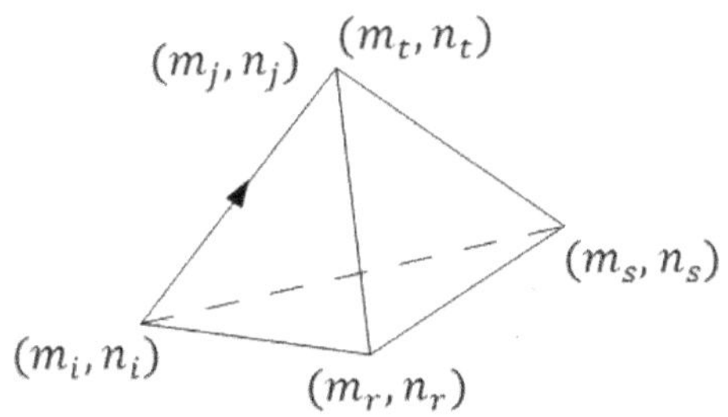

Figure 1. The definition of subscripts based on a single tetrahedron. Every vertex index can be used as a row index m or as a column index n.

So it is useful to guarantee that the basis functions for FE are the same as the basis functions for the BI.

Derived from the equivalence theorem, basis functions for BI are used to compose equivalent 2-D surface currents (Rao et al., 1982; Chew et al., 2001; Ylä-Oijala and Taskinen, 2003). The currents are relevant to the surface unit normal vector and the polarization of the fields, thus the tangential component of FE basis functions on the surface is perpendicular to the current basis functions. With respect to the corresponding sources (electric current $\boldsymbol{J}_s$ or magnetic current $\boldsymbol{M}_s$) and the surface normal unit vector, the corresponding subspaces ensure the compatibility between FE and BI.

The LO basis functions have shortcomings when the simulation accuracy and the large number of unknowns are considered. Precision and efficiency of LO are difficult to improve with increasing numbers of unknowns. The solution with LO basis functions demands the mesh size to be around $\lambda/20$ to $\lambda/8$. With coarser meshes, the elements may introduce inaccurate waveforms for field reconstruction.

The well-known mixed order basis functions are successful for electromagnetic field distributions and surface current reconstructions. *Rao–Wilton–Glisson (RWG)* (Rao et al., 1982) basis functions are inherited as LO. As *RWG* is very effective for BI, it is implemented as the first order of Rotational Subspace (0th order) in FE tetrahedra. The Nedelec HO basis functions also form the first order of Gradient Subspace (1st order), the second order Rotational Subspace (2nd order), an so on, where this paper is restricted to 2nd order. Apart from BI, LO and HO basis functions are also utilized within FE and they improve the accuracy for field computations. In FE-BI, 0th and 1st order basis functions for FE and BI are easy to match. Both of them are edge related and follow the same direction of the edge vector. The situations for 2nd order basis functions are more complicated. 2nd order basis functions are face related, so that to achieve compatibility between FE and BI, the basis functions of them have to maintain the same global node order. The tetrahedral FE basis functions defined in Table 1 have a format represented by subscripts $k = (i, j)$ and $k = (r, s, t)$ as illustrated in Fig. 1. As elements of matrices need row and column positions, the subscripts (m_i, m_j) and (m_r, m_s, m_t) are introduced for row basis functions and subscripts (n_i, n_j) and (n_r, n_s, n_t) are assigned to column basis functions. In this work, (m_i, m_j) and (n_i, n_j) contain the global order of local node numbers for edge related basis functions, whereas (m_r, m_s, m_t) and (n_r, n_s, n_t) represent the global order of nodes for face related basis functions, where $k = (i, j)$ and $k = (r, s, t)$ store the local node number of finite elements.

In FE analysis, the self-identified hierarchical basis functions are derived from the geometrical information of the tetrahedra. In the mesh file, the data structure for each tetrahedra contains six edge identities and four face identities. Each edge is constructed with two node numbers and the order of the two nodes determines the edge global direction. Every face has the identity of the corresponding outside boundary triangles, inside volume triangles or inside boundary triangles. The outside boundary triangles are described by three edges in certain directions. The first edge gives out the first two nodes of the outside boundary triangle, the last node can be found through another two edges. The order of these three nodes are inherited as global node order. The inside volume triangles and inside boundary triangles are constructed by three nodes directly and the node order is viewed as the global node order. Through the index of edge and face identities, the tetrahedral FEs can easily consult the corresponding edges and triangles. Thus the determined global node order can be set for the LO and HO basis functions. As shown in Fig. 1, (m_i, n_i) always represent the starting point of the edge, (m_j, n_j) represent the ending point and (m_r, n_r), (m_s, n_s), (m_t, n_t) represent the node order of the triangle. Practically, the local node numbers are arrayed in the unique global order and assigned to the corresponding subscripts. When generating the system matrices, the assigned subscripts are set to the corresponding positions into the list of basis functions. Thus elements of system matrices are automatically assigned to global edges and faces. If HO basis functions are implemented into FE, the matrix elements can be calculated analytically and precisely. Since these results are commonly not available in FE literature, it is a major contribution of this paper to present these analytical matrix elements up to 2nd order. As the order of basis functions is enlarged, the accuracy of the boundary integral (BI) should also be improved. As it turns out, the integration order for the testing surface integrals should be increased, the adaptive numbers of quadrature points in the singularity cancellation technique have to grow and larger maximum numbers for spherical harmonic expansion terms are needed within the Multilevel Fast Multipole Method (Eibert, 2005).

HO achieves satisfactory accuracy with larger mesh size and it provides a better solution for non uniform finite elements. Good radar cross section (RCS) results of PEC structures coated by dielectric materials are acquired. Since LO is inherited by the hierarchical bases, orthogonality or near-orthogonality of basis functions is usefull for HO. Based on the structure of tetrahedra, system matrices built from nearly orthogonal basis functions converge faster and the solution is more accurate, so that the flexibility of mesh size provided by HO gives a more feasible solution. Meanwhile, HO improves the accuracy and also reduces the number of finite elements.

FE-BI solutions for coated spheres and the Flamme aircraft referring to the self-identified hierarchical nearly orthogonal basis functions are explicitly illustrated. The material of the layered sphere is homogeneous, isotropic and lossy. A variety of FE-BI simulations up to 3 million unknowns based on self-identified hierarchical basis functions are presented. The accuracy of HO testing cases is good, and the simulation results based on HO basis functions are also compared with LO situations.

2 Finite element variational formulation

Consider the configuration of an arbitrary volume as shown in Fig. 2. A typical FE model consists of a finite volume V_a with possibly anisotropic and inhomogeneous materials inside. The materials are characterized by relative permittivity $\overleftrightarrow{\epsilon}_r(\boldsymbol{r})$ and relative permeability $\overleftrightarrow{\mu}_r(\boldsymbol{r})$. A_d is the assembled enclosed envelope and $\widehat{n}(\boldsymbol{r})$ is the normal vector pointing out of V_a.

Assuming a field expansion $\boldsymbol{E} = \sum u_n \boldsymbol{\alpha}_n$ and $\boldsymbol{H} = \sum i_n \boldsymbol{\alpha}_n$ as well as a suppressed time dependence $e^{j\omega t}$, a linear system of equations (Jin, 2002; Tzoulis and Eibert, 2005) is achieved as

$$([\mathbf{R}_{mn}]-k_0^2[\mathbf{S}_{mn}])[u_n]+jk_0Z_0[\mathbf{T}_{mn}][i_n]=-jk_0Z_0[\boldsymbol{w}_m]. \quad (1)$$

The system matrices $[\mathbf{R}_{mn}]$, $[\mathbf{S}_{mn}]$, $[\mathbf{T}_{mn}]$ and the right hand side vector $[\boldsymbol{w}_m]$ are defined as

$$[\mathbf{R}_{mn}]=\iiint_{V_a}\{\nabla\times\boldsymbol{a}_m(\boldsymbol{r})\cdot\overleftrightarrow{\mu}_r^{-1}(\boldsymbol{r})\cdot\nabla\times\boldsymbol{a}_n(\boldsymbol{r})\}dv, \quad (2)$$

$$[\mathbf{S}_{mn}]=\iiint_{V_a}\{\boldsymbol{a}_m(\boldsymbol{r})\cdot\overleftrightarrow{\epsilon}_r(\boldsymbol{r})\cdot\boldsymbol{a}_n(\boldsymbol{r})\}dv, \quad (3)$$

$$[\mathbf{T}_{mn}]=\oint\!\!\oint_{A_d}\boldsymbol{a}_m(\boldsymbol{r})\cdot(\boldsymbol{a}_n(\boldsymbol{r})\times\widehat{n}(\boldsymbol{r}))da, \quad (4)$$

$$[\boldsymbol{w}_m]=\iiint_{V_a}\{\boldsymbol{a}_m(\boldsymbol{r})\cdot\boldsymbol{J}_d(\boldsymbol{r})\}dv, \quad (5)$$

where $\boldsymbol{a}_m$ and $\boldsymbol{a}_n$ are field basis functions, $k_0=\omega\sqrt{\epsilon_0\mu_0}$ is the free space wave number, $Z_0=\sqrt{\mu_0/\epsilon_0}$ is the free space intrinsic impedance, and $\boldsymbol{J}_d$ is a volume current density source.

The hierarchical basis functions are based on the normalized barycentric (simplex) coordinates in FE tetrahedra $(\lambda_0, \lambda_1, \lambda_2, \lambda_3)$, where $\lambda_0+\lambda_1+\lambda_2+\lambda_3=1$. The functions are also related to the gradients $\nabla\lambda_i (i=0,1,2,3)$ and the volume of the tetrahedron V_T. Equation (2) is related to the curl of the basis functions. Equations (2) and (3) form the system matrices related to electric field unknowns $[u_n]$.

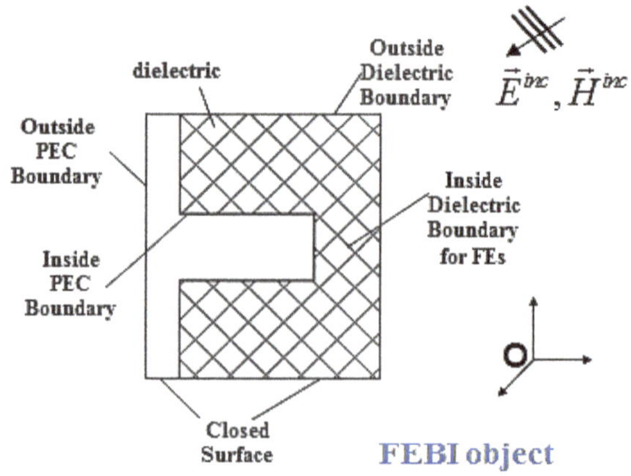

Figure 2. The general geometrical configuration for finite element–boundary integral method.

Equation (4) generates the matrices related to surface magnetic field unknowns $[i_n]$ where the integral region is the envelope of V_a. Equation (5) is related to the current excitation inside the volume. As $V_a = \sum V_e$ and $A_d = \sum A_e$, integration intervals of Eqs. (2), (3), (4) and (5) are the combination of finite elements. With simplex coordinates, the properties of basis functions allow for an analytical solution of the matrices in Eq. (1).

3 3-D HO self-identified basis functions

The general format of FE-BI basis functions for different orders is displayed in Table 1. $\boldsymbol{A}_i$ $(i=0,1,2,3)$ is the local face area vector related to surfaces of the tetrahedron pointing into the volume, V_T is the tetrahedral volume. Face vectors are independent in the tetrahedron and the surface tangential part of field basis functions compared with surface current basis functions are perpendicular. The field basis functions distribute inside the finite element volume. When tending to the tetrahedral boundary edges and faces, the corresponding tangential parts of edge basis functions for 0th and 1st order exist on the adjacent connected surfaces, they are zero on other edges or faces. The tangential components of face based functions for 2nd order only exist on the related face, but vanish on the other unrelated faces and on all edges.

The compatibility between FE and BI is easily achieved by utilizing self-identified hierarchical bases defined by global node order. The identified hierarchical basis functions $\boldsymbol{\alpha}_i$ are composed of rotational first order (0th order) represented by $\boldsymbol{a}_1$, gradient first order (1st order) represented by $\boldsymbol{b}_1$ and rotational second order (2nd order) related to $\boldsymbol{c}_2$ and $\boldsymbol{d}_2$. Properties of identified hierarchical basis functions are displayed in Table 1. (i, j) and (r, s, t) in Table 1 contain the local node numbers in the series defined by global order. Also, the curl of identified hierarchical basis functions is given.

4 Analytical calculation of FE system matrices

With self-identified hierarchical basis functions, analytical solutions for system matrices can be achieved through the properties of simplex coordinates in tetrahedra. The inner products of simplex coordinates (Lapidus and Pinder, 1982) are given as

$$L_2(\lambda_p;\lambda_q)=\iiint\limits_{V_T}\lambda_p\lambda_q dv=6V_T\frac{X!}{5!}$$

$$X=\begin{cases}2, p=q\\ 1, p\neq q\end{cases}, \{p,q\}=\{0,1,2,3\}, \tag{6}$$

$$L_3(\lambda_r;\lambda_p,\lambda_q)=\iiint\limits_{V_T}\lambda_r\lambda_p\lambda_q dv=6V_T\frac{X!}{6!}$$

$$X=\begin{cases}2, r=p,q\\ 1, r\neq p,q\end{cases}, \{r,p\neq q\}=\{0,1,2,3\}, \tag{7}$$

$$L_4(\lambda_r,\lambda_s;\lambda_p,\lambda_q)=\iiint\limits_{V_T}\lambda_r\lambda_s\lambda_p\lambda_q dv=6V_T\frac{X!Y!}{7!}$$

$$X=\begin{cases}2, r=p,q\\ 1, r\neq p,q\end{cases}, \{r,p\neq q\}=\{0,1,2,3\}$$

$$Y=\begin{cases}2, s=p,q\\ 1, s\neq p,q\end{cases}, \{s,p\neq q\}=\{0,1,2,3\}. \tag{8}$$

The dot and cross products between face vectors are defined as

$$I_{pq}=\boldsymbol{A}_p\cdot\boldsymbol{A}_q, \{p\neq q\}=\{0,1,2,3\}, \tag{9}$$

$$\boldsymbol{T}_{pq}=\boldsymbol{A}_p\times\boldsymbol{A}_q, \{p\neq q\}=\{0,1,2,3\}. \tag{10}$$

Combining identified basis functions and their properties (Table 1, Eqs. 6–10), the system matrix $[\mathbf{R}_{mn}]$ is calculated in Table 2 and $[\mathbf{S}_{mn}]$ is given in Table 3. From Eqs. (2) and (3), matrices $[\mathbf{R}_{mn}]$ and $[\mathbf{S}_{mn}]$ are symmetric, so that the elements at symmetric positions in Tables 2 and 3 are identical and it is explicit that the functions are the same along columns and rows.

System matrices $[\mathbf{T}_{mn}]$ depend on the surface boundaries of the volume, so that the dimension of the hierarchical space is reduced. In fact Eq. (4) can be written as

$$[\mathbf{T}_{mn}]=\oiint\limits_{A_d}(\boldsymbol{a}_m(\boldsymbol{r})\times\boldsymbol{a}_n(\boldsymbol{r}))\cdot\widehat{n}(\boldsymbol{r}))da, \tag{11}$$

and the hierarchical space turns out to be

$$\boldsymbol{a}_s=\lambda_i\nabla\lambda_j-\lambda_j\nabla\lambda_i=\frac{1}{2A_T}(\lambda_i\boldsymbol{l}_j-\lambda_j\boldsymbol{l}_i)$$
$$\{i,j\}=\{1,2,3\},\{i<j\}, \tag{12}$$

$$\boldsymbol{b}_s=\lambda_i\nabla\lambda_j+\lambda_j\nabla\lambda_i=\frac{1}{2A_T}(\lambda_i\boldsymbol{l}_j+\lambda_j\boldsymbol{l}_i)$$
$$\{i,j\}=\{1,2,3\},\{i<j\}, \tag{13}$$

$$\boldsymbol{c}_s=\lambda_r\lambda_s\nabla\lambda_t+\lambda_s\lambda_t\nabla\lambda_r-2\lambda_r\lambda_t\nabla\lambda_s$$
$$=\frac{1}{2A_T}(\lambda_r\lambda_s\boldsymbol{l}_t+\lambda_s\lambda_t\boldsymbol{l}_r-2\lambda_r\lambda_t\boldsymbol{l}_s)$$
$$\{r,s,t\}=\{1,2,3\},\{r<s<t\}, \tag{14}$$

$$\boldsymbol{d}_s=\lambda_r\lambda_s\nabla\lambda_t-\lambda_s\lambda_t\nabla\lambda_r$$
$$=\frac{1}{2A_T}(\lambda_r\lambda_s\boldsymbol{l}_t-\lambda_s\lambda_t\boldsymbol{l}_r)$$
$$\{r,s,t\}=\{1,2,3\},\{r<s<t\}. \tag{15}$$

With properties of surface triangle finite elements (Lapidus and Pinder, 1982), it is shown that

$$\iint\limits_{A_T}\lambda_1^{m_1}\lambda_2^{m_2}\lambda_3^{m_3}da=2A_T\frac{m_1!m_2!m_3!}{(m_1+m_2+m_3+2)!}, \tag{16}$$

$$\boldsymbol{l}_i\times\boldsymbol{l}_j=-2\boldsymbol{A}_T(-1)^{j-i},\{i<j\}, \tag{17}$$

$$\boldsymbol{A}_T=\frac{1}{2}\boldsymbol{t}_{12}\times\boldsymbol{t}_{13}. \tag{18}$$

Based on Eq. (11), $[\mathbf{T}_{mn}]$ is computed as

$$[\mathbf{T}_{mn}]=\iint\limits_{A_T}\{[\boldsymbol{a}_s,\boldsymbol{b}_s,\boldsymbol{c}_s,\boldsymbol{d}_s]^T\times[\boldsymbol{a}_s,\boldsymbol{b}_s,\boldsymbol{c}_s,\boldsymbol{d}_s]\cdot\widehat{n}(\boldsymbol{r})\}da$$
$$=\frac{1}{6}[\widehat{n_A}\cdot\widehat{n}(\boldsymbol{r})]\cdot\begin{bmatrix}0&1&1&-1/2&1/2&1/2&1/4&1/4\\-1&0&1&-1/2&1/2&-1/2&1/2&0\\-1&-1&0&1/2&1/2&-1/2&1/4&-1/4\\1/2&1/2&-1/2&0&0&0&-3/20&1/20\\-1/2&-1/2&-1/2&0&0&0&0&-1/10\\-1/2&1/2&1/2&0&0&0&3/20&1/20\\-1/4&-1/2&-1/4&3/20&0&-3/20&0&-1/10\\-1/4&0&1/4&-1/20&1/10&-1/20&1/10&0\end{bmatrix}. \tag{19}$$

From Eq. (5), $[\boldsymbol{w}_m]$ can be written as

$$[\boldsymbol{w}_m]=\iiint\limits_{V_T}\left\{\begin{bmatrix}\boldsymbol{a}_1\\\boldsymbol{b}_1\\\boldsymbol{c}_2\\\boldsymbol{d}_2\end{bmatrix}\cdot\boldsymbol{J}_d\right\}dv, \tag{20}$$

where the current source $\boldsymbol{J}_d$ can be used directly for the product with self-identified 3-D basis functions and vector $[\boldsymbol{w}_m]$ can be easily solved.

Thus, with analytical solutions for the hierarchical basis functions, the system matrices of FE integrals can be generated analytically and this avoids any numerical error accumulations and generates accurate matrices.

Table 1. Hierarchical basis functions and properties within the tetrahedron (Sun et al., 2001).

		FE Vector Basis Functions $\boldsymbol{\alpha}_i$		$\nabla \times \boldsymbol{\alpha}_i$	
R^1	$\boldsymbol{a}_1$	$\lambda_i \nabla\lambda_j - \lambda_j \nabla\lambda_i$	$\frac{\lambda_i \boldsymbol{A}_j - \lambda_j \boldsymbol{A}_i}{3V_T}$	$2\nabla\lambda_i \times \nabla\lambda_j$	$\frac{2}{9V_T^2}(\boldsymbol{A}_i \times \boldsymbol{A}_j)$
G^1	$\boldsymbol{b}_1$	$\lambda_i \nabla\lambda_j + \lambda_j \nabla\lambda_i$	$\frac{\lambda_i \boldsymbol{A}_j + \lambda_j \boldsymbol{A}_i}{3V_T}$	$\mathbf{0}$	$\mathbf{0}$
R^2	$\boldsymbol{c}_2$	$\lambda_r\lambda_s\nabla\lambda_t + \lambda_s\lambda_t\nabla\lambda_r - 2\lambda_r\lambda_t\nabla\lambda_s$	$\frac{\lambda_r\lambda_s\boldsymbol{A}_t + \lambda_s\lambda_t\boldsymbol{A}_r - 2\lambda_r\lambda_t\boldsymbol{A}_s}{3V_T}$	$3\lambda_r\nabla\lambda_s \times \nabla\lambda_t - 3\lambda_t\nabla\lambda_r \times \nabla\lambda_s$	$\frac{1}{3V_T^2}(\lambda_r\boldsymbol{A}_s \times \boldsymbol{A}_t - \lambda_t\boldsymbol{A}_r \times \boldsymbol{A}_s)$
	$\boldsymbol{d}_2$	$\lambda_r\lambda_s\nabla\lambda_t - \lambda_s\lambda_t\nabla\lambda_r$	$\frac{\lambda_r\lambda_s\boldsymbol{A}_t - \lambda_s\lambda_t\boldsymbol{A}_r}{3V_T}$	$\lambda_r\nabla\lambda_s \times \nabla\lambda_t + 2\lambda_s\nabla\lambda_r \times \nabla\lambda_t + \lambda_t\nabla\lambda_r \times \nabla\lambda_s$	$\frac{1}{9V_T^2}(\lambda_r\boldsymbol{A}_s \times \boldsymbol{A}_t + 2\lambda_s\boldsymbol{A}_r \times \boldsymbol{A}_t + \lambda_t\boldsymbol{A}_r \times \boldsymbol{A}_s)$

Table 2. Analytical solution for $[\mathbf{R}_{mn}]$ matrix elements.

	$\nabla\times\boldsymbol{a}_{1n}$	$\nabla\times\boldsymbol{b}_{1n}$	$\nabla\times\boldsymbol{c}_{2n}$	$\nabla\times\boldsymbol{d}_{2n}$
$\nabla\times\boldsymbol{a}_{1m}$	$\frac{4\overleftrightarrow{\mu}_r^{-1}(\boldsymbol{r})}{81V_T^3}\boldsymbol{T}_{ij}\cdot\boldsymbol{T}_{ij}$	0	$\frac{4\overleftrightarrow{\mu}_r^{-1}(\boldsymbol{r})}{81V_T^3}\{\boldsymbol{T}_{ij}\cdot\boldsymbol{T}_{st} - \boldsymbol{T}_{ij}\cdot\boldsymbol{T}_{rs}\}$	$\frac{\overleftrightarrow{\mu}_r^{-1}(\boldsymbol{r})}{162V_T^3}\{\boldsymbol{T}_{ij}\cdot\boldsymbol{T}_{st} + 2\boldsymbol{T}_{ij}\cdot\boldsymbol{T}_{rt} + \boldsymbol{T}_{ij}\cdot\boldsymbol{T}_{rs}\}$
$\nabla\times\boldsymbol{b}_{1m}$	0	0	0	0
$\nabla\times\boldsymbol{c}_{2m}$	$\frac{4\overleftrightarrow{\mu}_r^{-1}(\boldsymbol{r})}{81V_T^3}\{\boldsymbol{T}_{ij}\cdot\boldsymbol{T}_{st} - \boldsymbol{T}_{ij}\cdot\boldsymbol{T}_{rs}\}$	0	$\frac{\overleftrightarrow{\mu}_r^{-1}(\boldsymbol{r})}{9V_T^4}\{L_2(r;r)\boldsymbol{T}_{st}\cdot\boldsymbol{T}_{st} - L_2(r;t)\boldsymbol{T}_{st}\cdot\boldsymbol{T}_{rs} - L_2(t;r)\boldsymbol{T}_{rs}\cdot\boldsymbol{T}_{st} + L_2(t;t)\boldsymbol{T}_{rs}\cdot\boldsymbol{T}_{rs}\}$	$\frac{\overleftrightarrow{\mu}_r^{-1}(\boldsymbol{r})}{27V_T^4}\{L_2(r;r)\boldsymbol{T}_{st}\cdot\boldsymbol{T}_{st} + 2L_2(r;s)\boldsymbol{T}_{st}\cdot\boldsymbol{T}_{rt} + L_2(r;t)\boldsymbol{T}_{st}\cdot\boldsymbol{T}_{rs} - L_2(t;r)\boldsymbol{T}_{rs}\cdot\boldsymbol{T}_{st} - 2L_2(t;s)\boldsymbol{T}_{rs}\cdot\boldsymbol{T}_{rt} - L_2(t;t)\boldsymbol{T}_{rs}\cdot\boldsymbol{T}_{rs}\}$
$\nabla\times\boldsymbol{d}_{2m}$	$\frac{\overleftrightarrow{\mu}_r^{-1}(\boldsymbol{r})}{162V_T^3}\{\boldsymbol{T}_{ij}\cdot\boldsymbol{T}_{st} + 2\boldsymbol{T}_{ij}\cdot\boldsymbol{T}_{rt} + \boldsymbol{T}_{ij}\cdot\boldsymbol{T}_{rs}\}$	0	$\frac{\overleftrightarrow{\mu}_r^{-1}(\boldsymbol{r})}{27V_T^4}\{L_2(r;r)\boldsymbol{T}_{st}\cdot\boldsymbol{T}_{st} + 2L_2(r;s)\boldsymbol{T}_{st}\cdot\boldsymbol{T}_{rt} + L_2(r;t)\boldsymbol{T}_{st}\cdot\boldsymbol{T}_{rs} - L_2(t;r)\boldsymbol{T}_{rs}\cdot\boldsymbol{T}_{st} - 2L_2(t;s)\boldsymbol{T}_{rs}\cdot\boldsymbol{T}_{rt} - L_2(t;t)\boldsymbol{T}_{rs}\cdot\boldsymbol{T}_{rs}\}$	$\frac{\overleftrightarrow{\mu}_r^{-1}(\boldsymbol{r})}{81V_T^4}\{L_2(r;r)\boldsymbol{T}_{st}\cdot\boldsymbol{T}_{st} + 2L_2(r;s)\boldsymbol{T}_{st}\cdot\boldsymbol{T}_{rt} + L_2(r;t)\boldsymbol{T}_{st}\cdot\boldsymbol{T}_{rs} + 2L_2(s;r)\boldsymbol{T}_{rt}\cdot\boldsymbol{T}_{st} + 4L_2(s;s)\boldsymbol{T}_{rt}\cdot\boldsymbol{T}_{rt} + 2L_2(s;t)\boldsymbol{T}_{rt}\cdot\boldsymbol{T}_{rs} + L_2(t;r)\boldsymbol{T}_{rs}\cdot\boldsymbol{T}_{st} + 2L_2(t;s)\boldsymbol{T}_{rs}\cdot\boldsymbol{T}_{rt} + L_2(t;t)\boldsymbol{T}_{rs}\cdot\boldsymbol{T}_{rs}\}$

Table 3. Analytical solution for $[\mathbf{S}_{mn}]$ matrix elements.

	a_{1n}	b_{1n}	c_{2n}	d_{2n}
a_{1m}	$\frac{\overleftrightarrow{\epsilon}_r(\mathbf{r})}{9V_T^2}\{L_2(i;i)I_{jj} - L_2(i;j)I_{ji} - L_2(j;i)I_{ij} + L_2(j;j)I_{ii}\}$	$\frac{\overleftrightarrow{\epsilon}_r(\mathbf{r})}{9V_T^2}\{L_2(i;i)I_{jj} + L_2(i;j)I_{ji} - L_2(j;i)I_{ij} - L_2(j;j)I_{ii}\}$	$\frac{\overleftrightarrow{\epsilon}_r(\mathbf{r})}{9V_T^2}\{L_3(i;r,s)I_{jt} + L_3(i;s,t)I_{jr} - 2L_3(i;r,t)I_{js} - L_3(j;r,s)I_{it} - L_3(j;s,t)I_{ir} + 2L_3(j;r,t)I_{is}\}$	$\frac{\overleftrightarrow{\epsilon}_r(\mathbf{r})}{9V_T^2}\{L_3(i;r,s)I_{jt} - L_3(i;s,t)I_{jr} - L_3(j;r,s)I_{it} + L_3(j;s,t)I_{ir}\}$
b_{1m}	$\frac{\overleftrightarrow{\epsilon}_r(\mathbf{r})}{9V_T^2}\{L_2(i;i)I_{jj} + L_2(i;j)I_{ji} - L_2(j;i)I_{ij} - L_2(j;j)I_{ii}\}$	$\frac{\overleftrightarrow{\epsilon}_r(\mathbf{r})}{9V_T^2}\{L_2(i;i)I_{jj} + L_2(i;j)I_{ji} + L_2(j;i)I_{ij} + L_2(j;j)I_{ii}\}$	$\frac{\overleftrightarrow{\epsilon}_r(\mathbf{r})}{9V_T^2}\{L_3(i;r,s)I_{jt} + L_3(i;s,t)I_{jr} - 2L_3(i;r,t)I_{js} + L_3(j;r,s)I_{it} + L_3(j;s,t)I_{ir} - 2L_3(j;r,t)I_{is}\}$	$\frac{\overleftrightarrow{\epsilon}_r(\mathbf{r})}{9V_T^2}\{L_3(i;r,s)I_{jt} - L_3(i;s,t)I_{jr} + L_3(j;r,s)I_{it} - L_3(j;s,t)I_{ir}\}$
c_{2m}	$\frac{\overleftrightarrow{\epsilon}_r(\mathbf{r})}{9V_T^2}\{L_3(i;r,s)I_{jt} + L_3(i;s,t)I_{jr} - 2L_3(i;r,t)I_{js} - L_3(j;r,s)I_{it} - L_3(j;s,t)I_{ir} + 2L_3(j;r,t)I_{is}\}$	$\frac{\overleftrightarrow{\epsilon}_r(\mathbf{r})}{9V_T^2}\{L_3(i;r,s)I_{jt} + L_3(i;s,t)I_{jr} - 2L_3(i;r,t)I_{js} + L_3(j;r,s)I_{it} + L_3(j;s,t)I_{ir} - 2L_3(j;r,t)I_{is}\}$	$\frac{\overleftrightarrow{\epsilon}_r(\mathbf{r})}{9V_T^2}\{L_4(r,s;r,s)I_{tt} + L_4(r,s;s,t)I_{tr} - 2L_4(r,s;r,t)I_{ts} + L_4(s,t;r,s)I_{rt} + L_4(s,t;s,t)I_{rr} - 2L_4(s,t;r,t)I_{rs} - 2L_4(r,t;r,s)I_{st} - 2L_4(r,t;s,t)I_{sr} + 4L_4(r,t;r,t)I_{ss}\}$	$\frac{\overleftrightarrow{\epsilon}_r(\mathbf{r})}{9V_T^2}\{L_4(r,s;r,s)I_{tt} - L_4(r,s;s,t)I_{tr} + L_4(s,t;r,s)I_{rt} - L_4(s,t;s,t)I_{rr} - 2L_4(r,t;r,s)I_{st} + 2L_4(r,t;s,t)I_{sr}\}$
d_{2m}	$\frac{\overleftrightarrow{\epsilon}_r(\mathbf{r})}{9V_T^2}\{L_3(i;r,s)I_{jt} - L_3(i;s,t)I_{jr} - L_3(j;r,s)I_{it} + L_3(j;s,t)I_{ir}\}$	$\frac{\overleftrightarrow{\epsilon}_r(\mathbf{r})}{9V_T^2}\{L_3(i;r,s)I_{jt} - L_3(i;s,t)I_{jr} + L_3(j;r,s)I_{it} - L_3(j;s,t)I_{ir}\}$	$\frac{\overleftrightarrow{\epsilon}_r(\mathbf{r})}{9V_T^2}\{L_4(r,s;r,s)I_{tt} - L_4(r,s;s,t)I_{tr} + L_4(s,t;r,s)I_{rt} - L_4(s,t;s,t)I_{rr} - 2L_4(r,t;r,s)I_{st} + 2L_4(r,t;s,t)I_{sr}\}$	$\frac{\overleftrightarrow{\epsilon}_r(\mathbf{r})}{9V_T^2}\{L_4(r,s;r,s)I_{tt} - L_4(r,s;s,t)I_{tr} - L_4(s,t;r,s)I_{rt} + L_4(s,t;s,t)I_{rr}\}$

5 Surface integral formulation

With Huygen's principle (Ismatullah and Eibert, 2009; Harrington, 1961; Chew et al., 2001), the radiation sources can be replaced by equivalent electric surface currents $\boldsymbol{J}_s$ and magnetic surface currents $\boldsymbol{M}_s$ on the volume enclosed boundary A_d for evaluation of free space radiation. The numerical solution for the surface currents are based on the well known EFIE, MFIE and CFIE (Ismatullah and Eibert, 2009; Rao et al., 1982; Ylä-Oijala and Taskinen, 2003), given as

$$EFIE: \frac{1}{2}\boldsymbol{M}_s + \widehat{n} \times [\Lambda(\boldsymbol{M}_s) + Z_0 K(\boldsymbol{J}_s)] = -\widehat{n} \times \boldsymbol{E}^{\mathrm{inc}}, \tag{21}$$

$$MFIE: \frac{1}{2}\boldsymbol{J}_s + \widehat{n} \times \left[\Lambda(\boldsymbol{J}_s) - \frac{1}{Z_0}K(\boldsymbol{M}_s)\right] = \widehat{n} \times \boldsymbol{H}^{\mathrm{inc}}, \tag{22}$$

$$CFIE: \quad \alpha EFIE + Z_0(1-\alpha)\widehat{n} \times MFIE, \tag{23}$$

where the vector operators Λ and K are calculated as

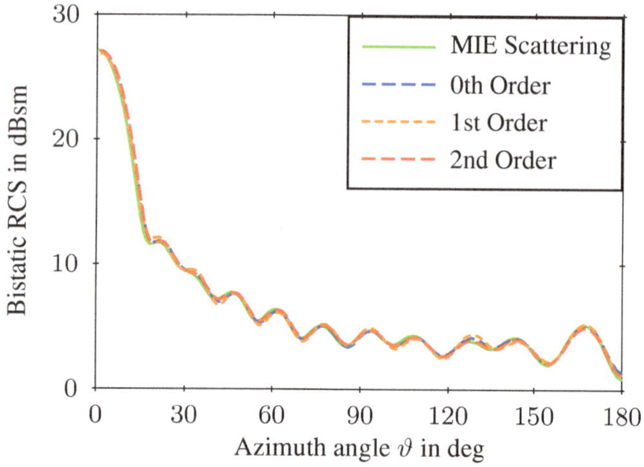

Figure 3. Bistatic RCS of Coated Sphere I @ 550 MHz on xz cut half plane ($\varphi = 0°$).

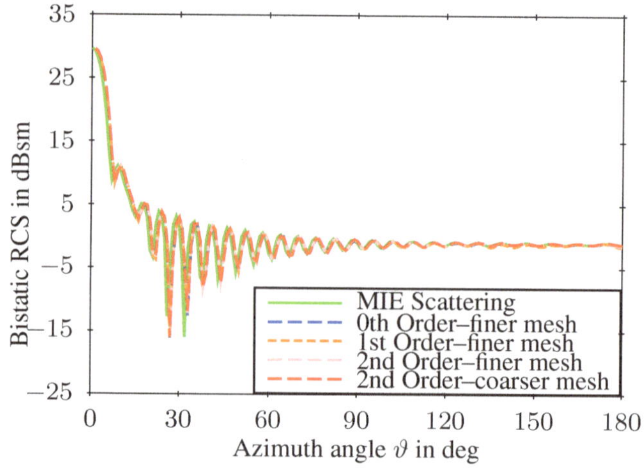

Figure 4. Bistatic RCS of coated sphere II @ 3 GHz on xz cut half plane ($\varphi = 0°$).

$$\Lambda(\boldsymbol{X}) = -\oiint_{A_d} (\nabla G_0 \times \boldsymbol{X}) da, \tag{24}$$

$$K(\boldsymbol{X}) = -jk_0 \left[\frac{1}{k_0^2} \nabla \oiint_{A_d} G_0 \nabla' \cdot \boldsymbol{X} da + \oiint_{A_d} \boldsymbol{X} G_0 da \right]. \tag{25}$$

$\boldsymbol{E}^{\text{inc}}$ and $\boldsymbol{H}^{\text{inc}}$ are incident electric and magnetic fields, G_0 is the scalar Green's function in free space and α is a number between zero and one. $\boldsymbol{J}_s$ and $\boldsymbol{M}_s$ are given as

$$\boldsymbol{J}_s = \widehat{n} \times \boldsymbol{H}, \quad \boldsymbol{M}_s = -\widehat{n} \times \boldsymbol{E}. \tag{26}$$

Thus the electric and magnetic currents can be written as

$$\boldsymbol{J}_s = \sum i_n \widehat{n} \times \boldsymbol{a}_n, \quad \boldsymbol{M}_s = -\sum u_n \widehat{n} \times \boldsymbol{a}_n. \tag{27}$$

With $\boldsymbol{f}_n = \widehat{n} \times \boldsymbol{a}_n$ (Ylä-Oijala and Taskinen, 2003), $\boldsymbol{J}_s$ and $\boldsymbol{M}_s$ are expanded as

$$\boldsymbol{J}_s = \sum_{n=1}^{N_I} i_n \boldsymbol{f}_n, \quad \boldsymbol{M}_s = -\sum_{n=1}^{N_M} u_n \boldsymbol{f}_n, \tag{28}$$

where $\boldsymbol{f}_n$ is the 2-D surface hierarchical basis function for BI, i_n and u_n in Eq. (27) are coefficients with respect to the surface elements and the total number of BI unknowns is $N = N_I + N_M$.

The discretized solution for MoM can be achieved through Galerkin's process and fast solutions as given by (Ismatullah and Eibert, 2009, 2008; Notaros, 2002; Eibert, 2005; Notaros, 2008) can be utilized. The definition of HO $\boldsymbol{f}_n$ is in 2-D derived from 3-D HO $\boldsymbol{a}_n$, thus the system matrices from MoM and FE will be compatible based on the same geometrical structure information of the object. The system matrices from BI are described in Ismatullah and Eibert (2009).

6 Linear algebraic equation system

To solve the electric field $[u]$ and the magnetic field $[i]$, the subsystem from FE in form of Eq. (1) and the subsystem generated by BI (Ismatullah and Eibert, 2009) must be combined as a complete system. The BI subsystem based on *EFIE* may introduce resonances into the final system, so it is necessary to utilize *CFIE* with similarly satisfiable accuracy. As a result, the subsystems can be regarded as

$$\left[M_u^{\text{FE}}\right][u] + \left[M_i^{\text{FE}}\right][i] = \left[V^{\text{FE}}\right], \tag{29}$$

$$\left[M_u^{\text{BI}}\right][u] + \left[M_i^{\text{BI}}\right][i] = \left[V^{\text{BI}}\right], \tag{30}$$

where $M_{u,i}^{\text{FE,BI}}$ is the sub-matrix derived from FE-BI for corresponding u and i. M_u^{FE} is square, symmetrical and sparse, M_i^{FE} is rectangular and sparse, M_u^{BI} is rectangular and fully staffed, M_i^{BI} is square, symmetrical and fully occupied, $V^{\text{FE,BI}}$ are excitation vectors for FE-BI. Thus, the complete combined system is written as

$$\begin{bmatrix} M_u^{\text{FE}} & M_i^{\text{FE}} \\ M_u^{\text{BI}} & M_i^{\text{BI}} \end{bmatrix} \begin{bmatrix} u \\ i \end{bmatrix} = \begin{bmatrix} V^{\text{FE}} \\ V^{\text{BI}} \end{bmatrix}. \tag{31}$$

The complete system solves the electric and magnetic fields simultaneously, thus equivalent surface electric and magnetic currents can be achieved.

7 Numerical results

To testify the accuracy of the analytical matrix elements and the global matrix assembly in FE-BI, several numerical simulation results are shown in this section. A cogent demonstration is to utilize a coated sphere testing case, where a PEC sphere is enclosed by a layer of dielectric material. The analytical RCS is well known as MIE Scattering (Balanis, 1989). Good matching of RCS between analytical solution

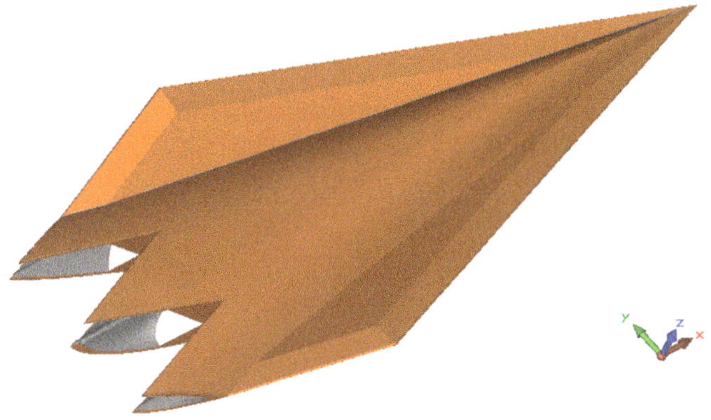

Figure 5. The geometry of a Flamme airplane.

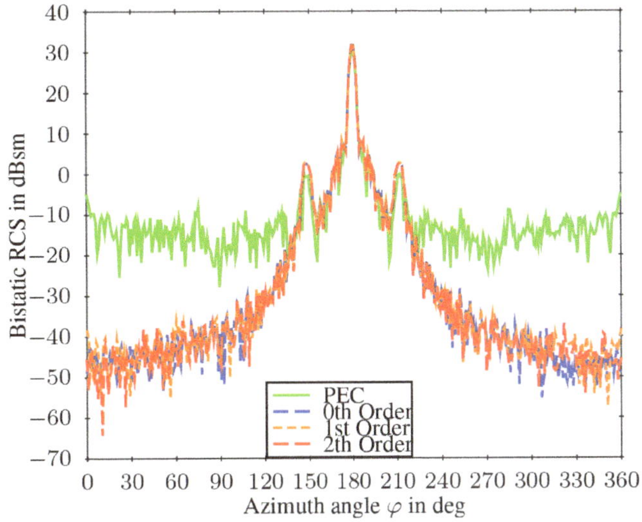

Figure 6. Bistatic RCS of Flamme @ 2.5 GHz on xy cut plane ($\vartheta = 90°$).

and numerical method verifies the efficacy of FE-BI. As a more complicated testing case, a second sphere is displayed. With higher frequency and finer mesh, more unknowns are handled. Moreover, an example of FE-BI application in very large scale simulations is shown through the RCS of the Flamme aircraft. As 0th order of FE-BI has been verified in many published articles (Tzoulis and Eibert, 2005; Eibert and Hansen, 1997; Eibert, 2007), it can be utilized as a reference for HO FE-BI. Efficiency of FE-BI based on different orders of self-identified basis functions are presented. The sphere simulations were performed on a PC with Intel(R) Core(TM)2 Quad CPU Q9550 @ 2.83 GHz processor, installed memory (RAM) 16.0 GB and 64-bit operating system. The simulation of the Flamme aircraft was operated on a Server with Intel(R) Xeon(R) CPU E5630 @ 2.53 GHz (2 processors), installed memory (RAM) 96.0 GB and 64-bit operating system. All simulations were computed on one core.

7.1 Coated sphere I

For the coated sphere, the RCS from 0th, 1st and 2nd order of self-identified basis functions for FE-BI are compared with MIE scattering as shown in Fig. 3. The coated sphere consists of a PEC core with radius 0.9 m and a layer of dielectrics with thickness 0.1 m. The dielectric layer properties are given by $\epsilon_r = 3 - 0.1j$ and $\mu_r = 1.0$. The incident wave is 550 MHz and propagating towards $+z$ direction and the electric field is 100 V/m along x direction ($E_x = 100$ V/m).

For 0th order, the mesh size was set to 0.04 m in Hypermesh software (HyperWorks, 2012), and the mean edge length is 4.65 cm, with minimum edge length 2.24 cm and maximum edge length 8.11 cm. The total unknowns are 154 822. The running time was 1314.8 s. For 1st order, the same mesh as for 0th order was utilized. The total unknowns are 309 644. The running time was 4763.4 s. For 2nd order, the mesh size is set 0.1 m, and the mean edge length is 10.93 cm, with minimum edge length 6.48 cm and maximum edge length 18.08 cm. The total unknowns are 70 448. The running time was 419.6 s.

The mesh size of LO is around $\lambda/8$ and the mesh size of HO is enlarged up to around $\lambda/3$. HO with coarser mesh and fewer unknowns achieves also accurate result as LO with finer mesh.

7.2 Coated sphere II

For the second coated sphere testing case, numerical RCS from different orders are also compared with MIE scattering. The second coated sphere contains a PEC sphere core with radius 0.5 m, and the PEC core is enclosed with a dielectric layer with thickness 0.0025 m. The properties of the dielectric layer are presented with $\epsilon_r = 2.5 - 0.5j$ and $\mu_r = 1.0$. The incident wave is 3 GHz and propagating toward $+z$ direction and the electric field is 100 V/m along x direction ($E_x = 100$ V/m). The results for the RCS are shown in Fig. 4.

For 0th order, the mesh size was set to 0.01 m, and the mean edge length is 0.858 cm, with minimum edge length 0.250 cm and maximum edge length 1.631 cm. The total unknowns are 411 339. The running time was 3525.6 s. For 1st order, the same mesh as 0th order is utilized. The total unknowns are 822 678. The running time was 4271.3 s. For 2nd order, there are two testing cases. One uses the same mesh as 0th and 1st order. The total unknowns are 1 973 604. The running time was 7725.5 s. Another simulation utilizes mesh size 0.03 in Hypermesh, and the mean edge length is 2.554 cm. The minimum edge is 0.250 cm and maximum edge length 4.316 cm. The total unknowns are 202 522. The running time was 1357.6 s.

The numerical RCS results are compared with MIE. The mesh size of LO is around $\lambda/8$, while HO with finer mesh size is also working well though a lot more unknowns are

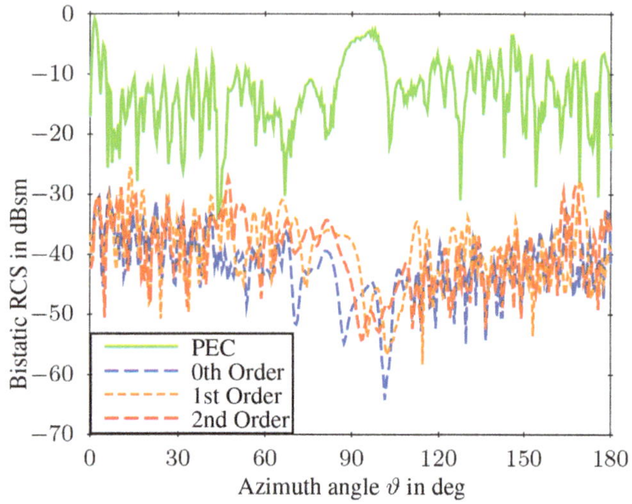

Figure 7. Bistatic RCS of Flamme @ 2.5 GHz on xz cut half plane ($\varphi = 0°$).

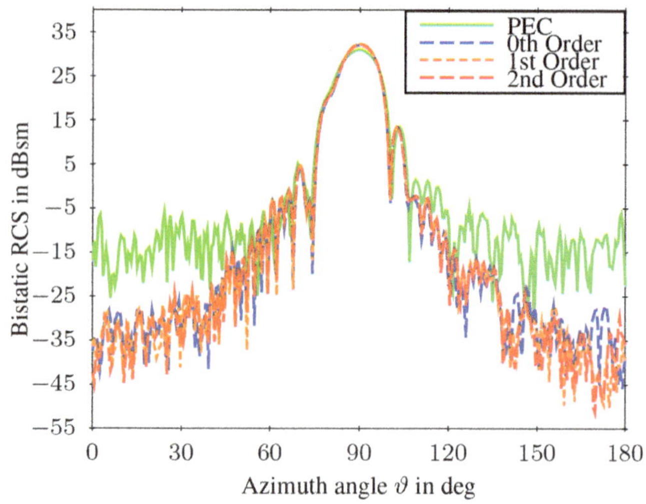

Figure 9. Bistatic RCS of Flamme @ 2.5 GHz on xz cut half plane ($\varphi = 180°$).

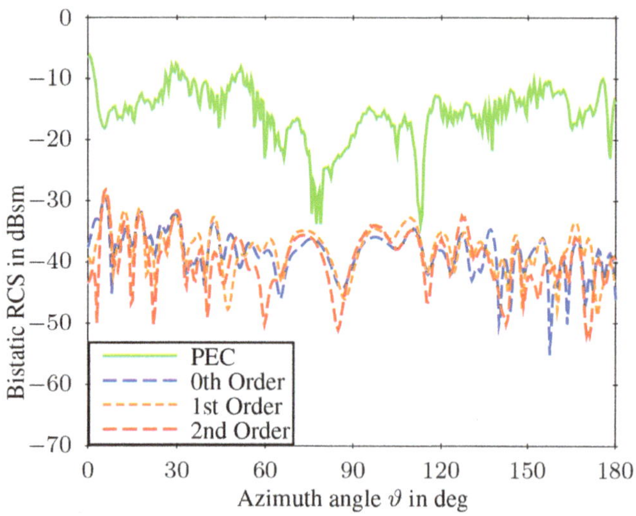

Figure 8. Bistatic RCS of Flamme @ 2.5 GHz on yz cut half plane ($\varphi = 90°$).

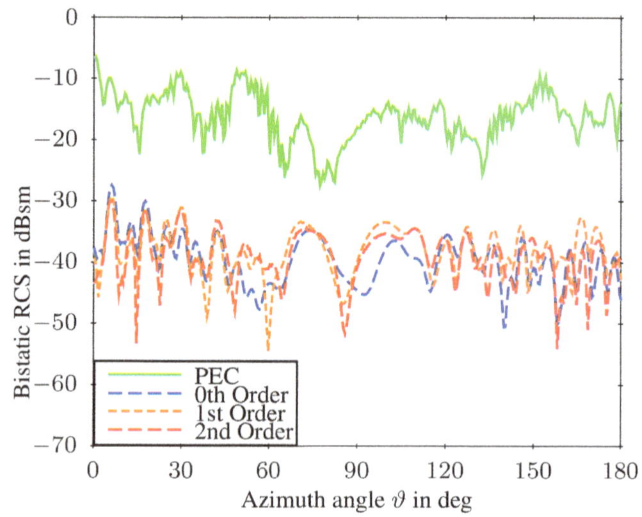

Figure 10. Bistatic RCS of Flamme @ 2.5 GHz on yz cut half plane ($\varphi = 270°$).

settled. When the mesh size of HO increases up to around $\lambda/4$, HO with coarser mesh for FE-BI maintains good precision as the results from LO with finer mesh as shown in Fig. 4.

7.3 Flamme

The Flamme case is an application of FE-BI for very large scale simulation. The Flamme is located in the xy plane, with nose heading along the $+x$ axis, as shown in Fig. 5. The Flamme is enclosed by a layer of lossy dielectric material with thickness of approximately 1 cm. The permittivity of the dielectric material is $\epsilon_r = 1.21 - 10j$ and the permeability is $\mu_r = 1$. The simulation frequency is 2.5 GHz. The incident plane wave propagates towards $-x$ direction, with electric field ($E_z = 100\,\text{V/m}$). To visualize absorbing effects of the lossy dielectric material, a PEC Flamme simulated with BI with 0th order of basis functions is utilized for comparison.

The RCS of PEC and layered Flamme in different cut planes are shown in Figs. 6–10. The PEC Flamme is simulated through BI with 0th order of self-identified basis functions, the layered Flamme is simulated through FE-BI with 0th, 1st and 2nd order of self-identified basis functions. As the efficacy of 0th order with finer mesh has been verified, here it is used as a reference. The RCS comparison shows that most of input power goes over the Flamme. In scattered directions, the scattered power is evidently absorbed by the dielectric material. Figure 11 shows the equivalent surface electric current of the covered Flamme.

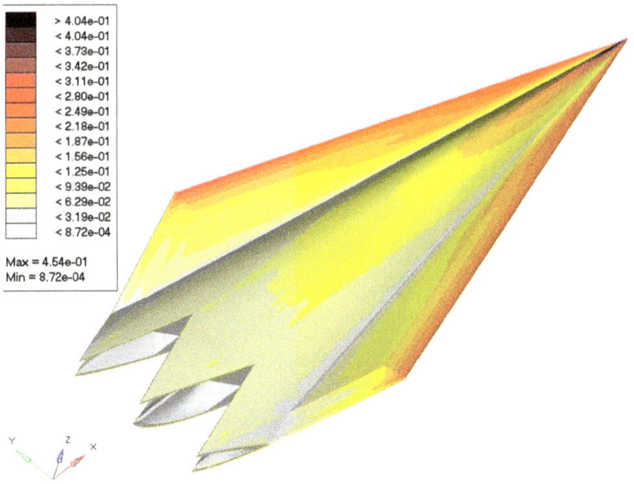

Figure 11. Electric surface current density $|\boldsymbol{J}|$ in A/m distribution of a covered *Flamme* airplane with plane wave incidence @ 2.5 GHz.

For PEC, the mesh size of the PEC Flamme was set to 0.01 m, and the mean edge length is 1.011 cm, with minimum edge length 0.100 cm and maximum edge length 2.559 cm. The total unknowns are 692 952. The number of BI electric currents is 692 952 and the number of BI magnetic currents is 0. The number of levels for MLFMM is 8 and the peak memory consumption was 4083.746 MBytes. The running time was 75 529.3 s. For 0th order, the mesh size of the layered Flamme was set to 0.01 m, and the mean edge length is 1.083 cm, with minimum edge length 0.044 cm and maximum edge length 2.671 cm. The total unknowns are 2 081 547. The number of BI electric currents is 690 960 and the number of BI magnetic currents is 602 284. The number of levels for MLFMM is 8 and the peak memory consumption was 11 384.43 MBytes. The run time was 23 511.6 s. For 1st order, the mesh size of layered Flamme was set to 0.02 m, the mean edge length is 1.770 cm, with minimum edge length 0.065 cm and maximum edge length 3.933 cm. The total unknowns are 1 252 430. The number of BI electric currents is 427 144 and the number of BI magnetic currents is 360 976. The number of levels for MLFMM is 7 and the peak memory consumption was 8825.734 MBytes. The running time was 12 141.3 s. For 2nd order, the mesh size was set to 0.02 m, and the mean edge length is 1.770 cm, with minimum edge length 0.065 cm and maximum edge length 3.933 cm. The total unknowns are 2 941 242. The number of BI electric currents is 712 284 and the number of BI magnetic currents is 602 004. The number of levels for MLFMM is 8 and the peak memory consumption was 18 536.68 MByte. The running time was 47 871.3 s.

8 Conclusion

Self-identified hierarchical 3-D vector basis functions were generated for the hybrid finite element (FE) – boundary integral (BI) technique, where analytical solutions for the FE matrix elements have been presented up to 2nd order. Self-identified basis functions provide feasibility for FE and effectively maintain compatibility with BI. Going from 1st to 2nd order, FE-BI allows for a mesh size increase from $\lambda/8$ up to $\lambda/3$. From coated sphere testing cases, good accuracy was found and the Flamme simulations displayed that FE-BI based on self-identified basis functions can be applied for very large scale simulations.

Edited by: U. van Rienen

References

Balanis, C. A.: Advanced Engineering Electromagnetics, John Wiley & Sons, New York, 1989.

Bladel, J. V.: Electromagnetic Fields, McGraw-Hill, New York, 1964.

Chew, W. C., Jin, J. M., and Michielssen, E.: Fast and Efficient Algorithms in Computational Electromagnetics, Artech House, Boston, MA, 2001.

Eibert, T. F.: A Diagonalized Multilevel Fast Multipole Method With Spherical Harmonics Expansion of the k-Space Integrals, IEEE Trans. Antennas Propag., 53, 814–817, 2005.

Eibert, T. F.: Some scattering results computed by surface-integral-equation and hybrid finite-element-Boundary-integral techniques, accelerated by the multilevel fast multipole method, IEEE Antennas Propag. Mag., 49, 61–69, 2007.

Eibert, T. F. and Hansen, V.: 3-D FEM/BEM-Hybrid Approach Based on a General Formulation of Huygens' Principle for Planar Layered Media, IEEE Trans. Microw. Theory Tech., 45, 1105–1112, 1997.

Graglia, R., Wilton, D., and Peterson, A.: Higher order interpolatory vector bases for computational electromagnetics, IEEE Trans. Antennas Propag., 3, 329–342, 1997.

Harrington, R. F.: Time Harmonic Electromagnetic Fields, McGraw-Hill, New York, 1961.

HyperWorks, http://www.altairhyperworks.com/, last access: 27 June 2014, 2012.

Ismatullah and Eibert, T. F.: Adaptive Singularity Cancellation for Efficient Treatment of Near-Singular and Near-Hypersingular Integrals in Surface Integral Equation Formulations, IEEE Trans. Antennas Propag., 56, 274–278, 2008.

Ismatullah and Eibert, T. F.: Surface Integral Equation Solutions by Hierarchical Vector Basis Functions and Spherical Harmonics Based Multilevel Fast Multipole Method, IEEE Trans. Antennas Propag., 57, 2084–2093, 2009.

Jin, J.: The Finite Element Method in Electromagnetics, John Wiley & Sons, Inc, 111 River Street, Hoboken, 2002.

Jorgensen, E., Volakis, J. L., Meincke, P., and Breinbjerg, O.: Higher order hierarchical Legendre basis functions for electromagnetic modeling, IEEE Trans. Antennas Propag., 52, 2985–2995, 2004.

Jorgensen, E., Kim, O., Meincke, P., and Breinbjerg, O.: Higher order hierarchical Legendre basis functions in integral equation formulations applied to complex electromagnetic problems, Proc. IEEE Int. Symp. Antennas and Propag. Society., 3A, 64–67, 2005.

Lapidus, L. and Pinder, G. F.: Numerical Solutions of Partial Differential Equations in Science and Engineering, John Wiley & Sons, New York, 1982.

Mautz, J. and Harrington, R.: H-field, E-field, and combined-field solutions for conducting bodies of revolution, Archiv für Elektronik und Übertragungstechnik (Electron. Commun.), 32, 157–164, 1978.

Notaros, B. M.: A highly effective preconditioner for solving the finite element-boundary integral matrix equation of 3-D scattering, IEEE Trans. Antennas Propag., 50, 1212–1222, 2002.

Notaros, B. M.: Higher order frequency-domain computational electromagnetics, IEEE Trans. Antennas Propag., 56, 2251–2276, 2008.

Rao, S., Wilton, D., and Glisson, A.: Electromagnetic scattering by surfaces of arbitrary shape, IEEE Trans. Antennas Propag., 30, 409–418, 1982.

Sun, D. K., Lee, J. F., and Cendes, Z.: Construction of nearly orthogonal Nedelec bases for rapid convergence with multilevel preconditioned solvers, SIAM J. Sci. Comput., 23, 1053–1076, 2001.

Tzoulis, A. and Eibert, T. F.: A Hybrid FEBI-MLFMM-UTD Method for Numerical Solutions of Electromagnetic Problems Including Arbitrarily Shaped and Electrically Large Objects, IEEE Trans. Antennas Propag., 53, 3358–3366, 2005.

Volakis, J. L., Chatterjee, A., and Kempel, L. C.: Finite Element Method for Electromagnetics, IEEE, New York, 1998.

Ylä-Oijala, P. and Taskinen, M.: Calculation of CFIE Impedance Matrix Elements With RWG and $n \times$RWG Functions, IEEE Trans. Antennas Propag., 51, 1837–1846, 2003.

31

Local microwave heating of sand molds as a means to overcome design limitations in sand mold casting

O. Wiedenmann[1], R. Ramakrishnan[2], P. Saal[2], E. Kılıç[1], U. Siart[1], T. F. Eibert[1], and W. Volk[2]

[1]Technische Universität München, Lehrstuhl für Hochfrequenztechnik, Arcisstr. 21, 80333 Munich, Germany
[2]Technische Universität München, Lehrstuhl für Umformtechnik und Gießereiwesen, Walther-Meißner-Straße 4, 85747 Garching, Germany

Correspondence to: O. Wiedenmann (oliver.wiedenmann@tum.de)

Abstract. Microwave induced selective heating outmatches conventional mold heating by convective heat transfer in means of energy efficiency and cycle time. Moreover, it provides the opportunity of a local manipulation of solidification and cooling processes within the sand casting mold. In this paper, we investigate the suitability of different highly absorbing materials to indirectly heat up the mostly microwave transparent sand mold. The temperature-dependent permittivity of the involved materials is determined by resonator experiments and subsequently used to simulate the electromagnetic field and the thermodynamic response of the sand mold prior to a metal casting process. Experimental results are presented and compared with the outcome of the coupled electromagnetic-thermodynamic simulations and the influence of local microwave heating on the solidification and cooling of the cast is studied.

1 Introduction

Applying microwaves during the production and processing of industrial goods is a widely accepted and popular technology. Especially in the food and ceramic industries drying and heating by microwaves outmatches conventional heating methods in many areas (Meredith, 1998). With regards to the metal casting industry microwave technology is merely used for drying and firing of ceramic mold materials (Robak, 2005). Furthermore, Chandrasekaran et al. (2011) investigated the heating of different metals by microwave energy. This paper focuses on the generation of a custom temperature field inside a sand mold by selective microwave heating.

In die casting processes temperature control by means of heating and cooling channels is commonly used and necessary to assure a controlled solidification front and thus the requested mechanical properties of the cast part. In sand mold casting, however, the options for a temperature control are limited. Only cooling elements, so called chills, or exothermic feeders are used frequently. Whereas convective heating of the mold is not reasonable, as the heat input can not be locally defined, microwave heating offers a possibility to generate an inhomogeneous temperature distribution. In this work, a custom temperature field in sand molds is achieved by positioning highly absorbing materials, so called susceptors, within the mold. Hence, we combine microwave transparent bulk materials with susceptor materials in a single sand mold. This rigged sand mold is subjected to a microwave field thus generating a locally heated sand mold prior to the casting process.

The first step is the selection of a suitable susceptor material, which fulfills all requirements for an application in a sand casting process. All involved materials are dielectrically characterized by a resonator technique in a way that allows to determine the permittivity of the material samples depending on the temperature. This knowledge is used to develop a simulation model to predict the electromagnetic and thermodynamic behavior of the experimental setup. Due to the strong temperature dependency of the permittivity of the susceptor material, electrical and thermal phenomena are coupled to each other, so that electromagnetic and thermodynamic simulations are no longer independent. Instead, a simulation model has to be derived that takes into account the coupling between the electromagnetic and the thermodynamic simulations. Finally, the results of the numerical methods and the casting experiments are compared and the effect of the local heating on the solidification and cooling process is investigated.

2 Selection of susceptor materials

The microwave experiments for the selection of susceptor materials are carried out in a laboratory microwave furnace MKE – 0.8/2.45 – 0.6/5.8 manufactured by Linn High Therm GmbH. Microwaves are applied to the cylindrical chamber with a frequency of 2.45 GHz which has proven to be suitable for most needs (Clark, 2005). The inner chamber diameter is 330 mm with a depth of 260 mm. The packing space for the microwave furnace is specified by the manufacturer with a volume of $100 \times 100 \times 100\,mm^3$ in the center of the cylinder.

The key to accomplish selective sand mold heating by microwave energy is the right choice of materials. The microwave energy is used to heat the susceptors, which consist of a material that shows a high microwave absorption rate. For the investigations of appropriate susceptor materials a simple two-part sand block is used, hereinafter called susceptor carrier. The bulk of the susceptor carrier, where no heating is desired, should be as microwave transparent as possible in order to convey the microwave field to the susceptors which lie within the sand. Commonly used molding sands on quartz basis meet this requirement, since these materials do not heat up significantly in a microwave field. For our experiments we manufactured the susceptor carriers from laser croning sand Resital® QLS 50 CG (Hüttenes-Albertus Chemische Werke GmbH) by selective laser sintering (SLS) in an EOSINT S700 (EOS GmbH) laser sintering machine. The susceptor carrier consists of a bottom block with a pocket for the susceptor material and a top block for covering the susceptor. Each bottom and top block has a size of 80 mm × 80 mm × 50 mm. In the bottom part of a susceptor carrier one susceptor specimen can be positioned. Figure 1 schematically shows an example of the susceptor carrier with a susceptor specimen.

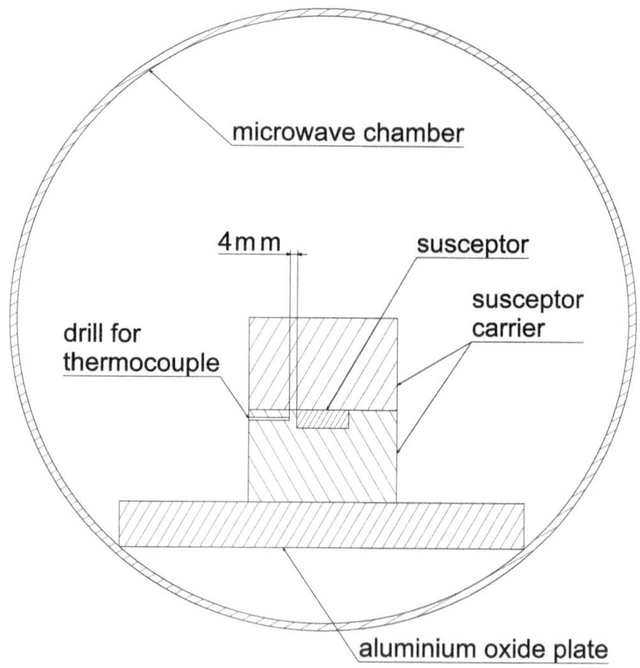

Figure 1. Scheme of the microwave chamber with susceptor carrier and susceptor positioned inside. For temperature measurements a thermocouple can be inserted when the magnetron is turned off.

Inside the cylindrical microwave chamber a supporting surface out of aluminum oxide is positioned, which can be assumed as microwave transparent. On top of the plate the susceptor carrier including the susceptor is positioned. The presence of a metallic thermocouple inside a microwave field has a great influence on the field distribution inside the microwave chamber. Arcing and microwave interference with temperature measurement are common problems (Grellinger and Janney, 1993; Chandrasekaran et al., 2011). Therefore, thermocouples have to be inserted shortly after the microwave is turned off. The prepared susceptor carrier with one susceptor is positioned in the designated packing space of the microwave furnace. After a microwave irradiation the temperature is measured in a distance of 4 mm from the susceptor surface. This is the designated distance – and minimum mold wall thickness – between the susceptor and the melt in the later casting process.

As the desired microwave heat treatment of a sand mold takes place before casting, the mold stability required for a satisfying casting result has to be kept up during the irradiation. Binders in molding sands lose their stability by mechanical or thermal stress. With rising susceptor temperatures the desired effect on the casting process will be more significant. However, high susceptor temperatures increase the probability of erosion within the mold due to binder degradation. Exothermic reactions excited in susceptor materials by the microwave process can also reduce the stability of the susceptor itself or surrounding molding sand. Therefore, high reactivity of the susceptor materials has to be avoided generally, regardless of whether the susceptors are in direct contact to the casting cavity, thus shape the cast part, or surrounded by molding material.

Further demands on susceptor materials are an easy fabrication of variably shaped susceptors and that they do not represent a danger for health during fabrication, irradiation or casting. Based on a literature review the following materials are investigated as possible susceptor materials. (Rhee, 2002; Lukasz, 2005; Lühken, 2005; Bott, 1991; Gerdes et al., 1998):

- Silicon carbide (SiC)
- Tungsten carbide (WC)
- Copper oxide (CuO)

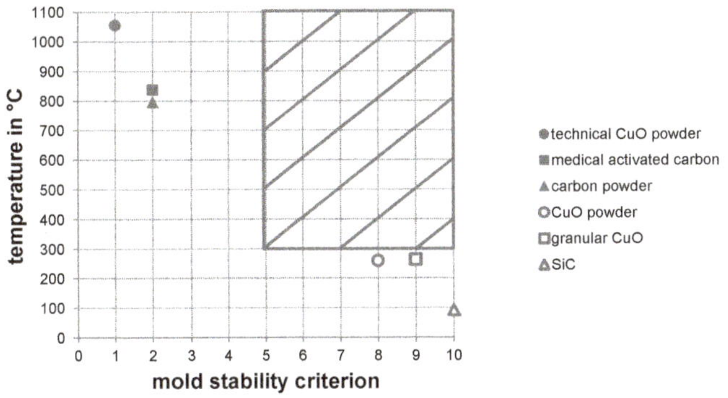

Figure 2. Maximum temperatures in 4 mm distance to the susceptor (pure materials) after an irradiation time of 300 s plotted over the mold stability.

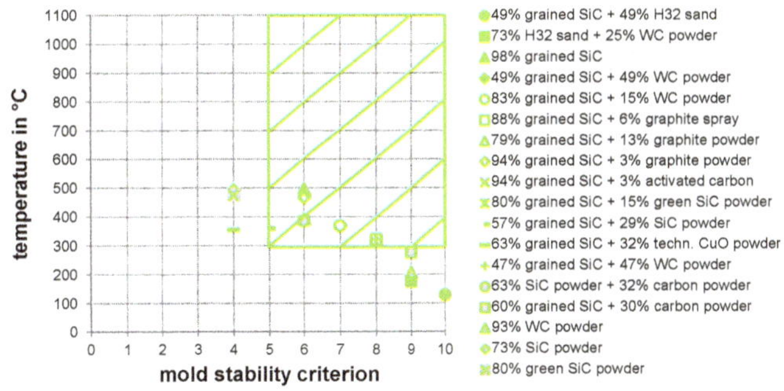

Figure 3. Maximum temperatures in 4 mm distance to the susceptor (compressed and bound mixtures, weight balance equals binder amount) after an irradiation time of 300 s plotted over the mold stability.

– Graphite modifications (C)

These materials are investigated in different modifications according to the reached temperature during a defined microwave irradiation and their reactivity, which goes along with mold stability.

First the mentioned susceptor materials are used in powder form with different grain sizes. The powders are filled in the susceptor carrier pocket and irradiated for 300 s. For the first experiments a reduced susceptor volume is used. The pocket volume filled with susceptor powder is 1000 mm^3 for graphite and 2000 mm^3 for other materials, preventing expected reactions of graphite.

The results of the experiments are summed up in Fig. 2. It shows the maximum temperatures measured in a distance of 4 mm to the susceptor materials after an irradiation time of 300 s over a mold stability criterion. This criterion was defined by visually evaluating (pairwise comparison) the damage inflicted on the sand mold in the susceptor's vicinity. Low mold stability criterion values indicate severe binder degradation and therefore strong mold disruption. To affect the solidification and cooling behavior of the cast metal significantly, an increase in temperature of 300 °C is defined as the desired lower limit. The target area, defined by the desired minimal temperature rise of 300 °C and the acceptable mold stability criterion of 5 or higher, is highlighted.

The powder materials used in the stage 1 experiments (Fig. 2) did either not achieve the desired temperature rise of 300 °C or heated up so much, that the stability of susceptor and surrounding molding sand was destroyed and thus a defined temperature measurement in a distance of 4 mm was not possible. The temperatures reached in these susceptors amount to approximately 1000 °C. One reason for these high temperatures is an overlay of microwave heating and exothermal reaction. In the case of graphite, microwave irradiation induces combustion of graphite powder with surrounding air, which destroys the mold stability. In order to combine the good heating conditions reached with some materials and the required inert characteristics of others, the initial materials extended by tungsten carbide are mixed in different portions and bound with varied amounts of INOTEC® binder and compressed using a hand press. Silicon carbide and quartz sand (H32) were used as carrier materials, as they show low reactivity and good wettability characteristics for the INOTEC® binder. Using these mixtures, the susceptor volume could be increased to 8000 mm^3 to induce a higher amount of energy without losing mold stability. Figure 3 shows the results of the corresponding experiments summarized. Considerable improvements could be reached in terms of mold stability. However silicon carbide (SiC) also showed an unexpected high maximum temperature which we attribute to the compression of the susceptor. Having one of the lowest temperatures without compression, the SiC susceptor reached the highest temperature in the stage 2 experiments. Based on this result, SiC bound with INOTEC® and compressed was selected as the most suitable material for susceptor applications.

3 Material characterization by resonator experiments

For the electromagnetic simulation, the dielectric properties of all involved materials have to be determined. Therefore, a resonator technique is employed which is based on a cylindrical cavity resonator as illustrated in Fig. 4. A sample of the material to be analyzed can be inserted into the center of the resonator. By measuring the resonance frequency and the quality factor of the detuned resonator, real and imaginary part of the permittivity of the sample material can be computed by an inversion algorithm. This approach also takes into account the effects of lossy walls and of the filling hole by imposing inhomogeneous surface impedance boundary conditions (Kılıç et al., 2013).

The method starts by converting the vector wave equation for the magnetic field into a variational form. The finite conductivity of the resonator walls is accounted for by enforcing an impedance boundary condition along the metallic surface. The filling hole, through which dielectric material

Figure 4. Cylindrical cavity resonator with an opening for the insertion of material samples.

samples can be plunged into the cavity, is modeled as a circular hollow waveguide. Therefore, the surface impedance in the cross section of the filling hole can be approximated by the scalar wave impedance of evanescent waveguide modes.

The variational expression for the magnetic field can be expanded by the natural modes of the lossless empty cavity, which form a complete set. This reduces the problem to a generalized eigenvalue problem for the unknown material parameters in matrix form. In contrast to the direct problem, in which the wave number is determined from the permittivity, the problem at hand is the other way around. Hence, it is an inverse problem, where the known quantity is the complex wave number which is determined by the measured resonance frequency and the quality factor. The inverse problem is solved by a generalized Schur decomposition. This procedure provides an accurate way to obtain the permittivity of a material sample.

The described configuration allows heating up the material sample prior to measurement, so that the temperature dependency of the permittivity can be accounted for. Thus, the temperature of the heated sample was recorded with an infrared camera during the cooling process, while the scattering parameters were measured in parallel with a vector network analyzer. The setup for the measurement of temperature and electrical scattering parameters is illustrated in Fig. 5.

Since the usage of an infrared camera only returns the temperature on the material surface, the samples have been heated up a second time, but this time the core temperature has been measured as well by inserting a thermocouple into a tiny drilled hole. This offers the possibility to assign the surface temperature measured during the resonator experiment to the corresponding core temperature. Measuring the temperature with the thermocouple during the network analyzer measurements is not feasible, because the metallic thermocouple disturbs the fields and, thus, renders the measured electric data useless.

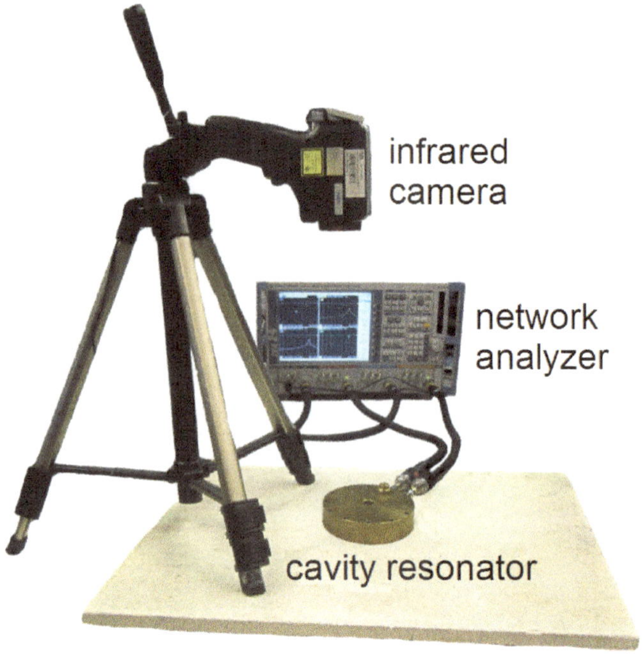

Figure 5. Measurement setup for the simultaneous determination of temperature and electrical scattering parameters comprising a network analyzer and an infrared camera.

The results for the complex relative permittivity $\varepsilon_r' - j\varepsilon_r''$ of silicon carbide are illustrated in Fig. 6. It can be discerned that there is a distinctive dependency on the temperature, which has to be taken into account in the numerical simulation. A rational function is used to interpolate between the measurement points, thereby creating a functional dependency between temperature and permittivity, which is required as input to simulate the electric field distribution.

4 Numerical simulation of electric field and heat distribution

In order to compute the expected temperature distribution inside the sand mold after microwave heating, an electromagnetic full-wave simulation is necessary. The calculated electric field serves as input for a subsequent thermodynamic simulation which computes the heat flow between the susceptors and the sand mold. For the computation of the electromagnetic field distribution, the software package COMSOL Multiphysics is employed to solve Maxwell's

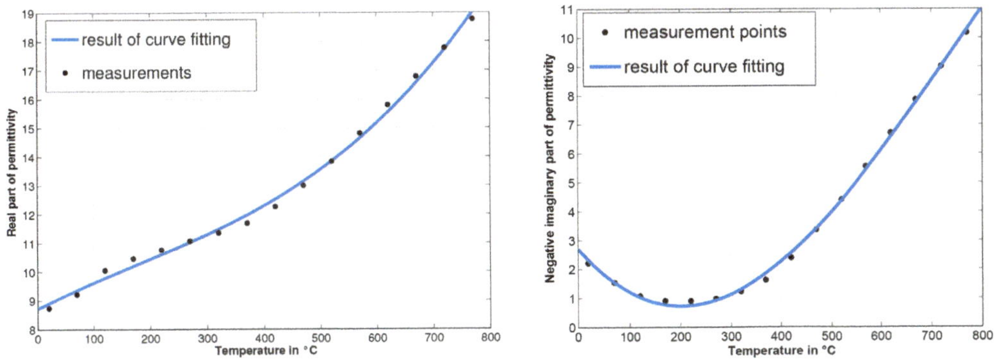

Figure 6. Results for the temperature-dependent, complex permittivity of the susceptor material silicon carbide obtained through resonator experiments.

equations (Maxwell, 1873), which are given by

$$\nabla \times \boldsymbol{H} = \boldsymbol{J} + \frac{\partial \boldsymbol{D}}{\partial t} \qquad \text{(Ampère-Maxwell law)} \tag{1}$$

$$\nabla \times \boldsymbol{E} = -\frac{\partial \boldsymbol{B}}{\partial t} \qquad \text{(Faraday's law)} \tag{2}$$

$$\nabla \cdot \boldsymbol{B} = 0 \qquad \text{(Gauss's law for magn. fields)} \tag{3}$$

$$\nabla \cdot \boldsymbol{D} = \rho_e, \qquad \text{(Gauss's law)} \tag{4}$$

where $\boldsymbol{E}$ is the electric field, $\boldsymbol{H}$ the magnetic field, $\boldsymbol{D}$ the electric displacement current, $\boldsymbol{B}$ the magnetic flux density, ρ_e the electric charge density and $\boldsymbol{J}$ the electric current density. This set of equations is solved numerically by the finite element method (Jin, 2002). The resulting electromagnetic loss distribution acts as thermal heat source in a subsequent thermodynamic simulation. The solution of the heat conduction equation

$$\rho_m c \frac{\partial T}{\partial t} = \nabla (k \nabla T) + q \tag{5}$$

provides the temperature distribution after a specified time t. Here, ρ_m denotes the mass density, T the temperature, k the thermal conductivity, c the specific heat capacity and q the absorbed power in a unit volume. The absorbed power can be written as

$$q = 2\pi f \varepsilon_0 \varepsilon_r'' |E_{\text{eff}}|^2 \tag{6}$$

and depends on the frequency f, the permittivity of free space ε_0, the imaginary part of the relative permittivity ε_r'' and the square of the effective value of the electric field E_{eff}.

In order to take into account the temperature dependency of the permittivity, electromagnetic and thermodynamic simulation have to be coupled. This means that after every time step of the transient thermal simulation, it has to be verified that the material parameters have not changed significantly due to the propagating heat. Each time the material parameters have altered to a certain degree, a new electromagnetic simulation has to be performed with updated material parameters according to the actual temperature distribution. The refreshed heat sources are then used for the next time steps of the transient thermal simulation, until the electromagnetic problem part has to be recomputed again.

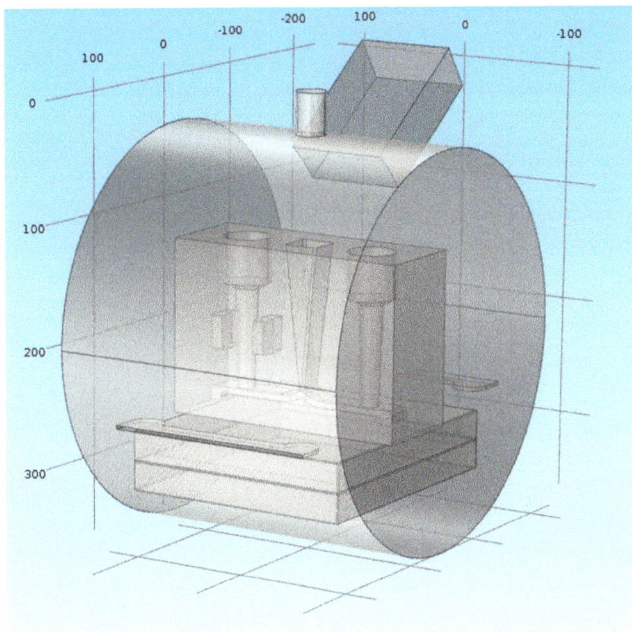

Figure 7. Simulation setup of a sand mold with susceptors inside the microwave chamber.

This simulation setup is used to compute the expected temperature distribution inside a sand mold after a certain amount of time. In Fig. 7, a model of the experimental setup is shown which includes a sand mold with two susceptors inside a microwave chamber. The temperature distribution after 210 s of heating is illustrated in Fig. 8. It should be denoted that the sand mold absorbs almost no electromagnetic energy, whereas the susceptors heat up significantly. Moreover, it can be discerned that the heat is propagating well inside the sand mold, so that a potential influence on the solidification process of the cast metal is expected. The different temperature

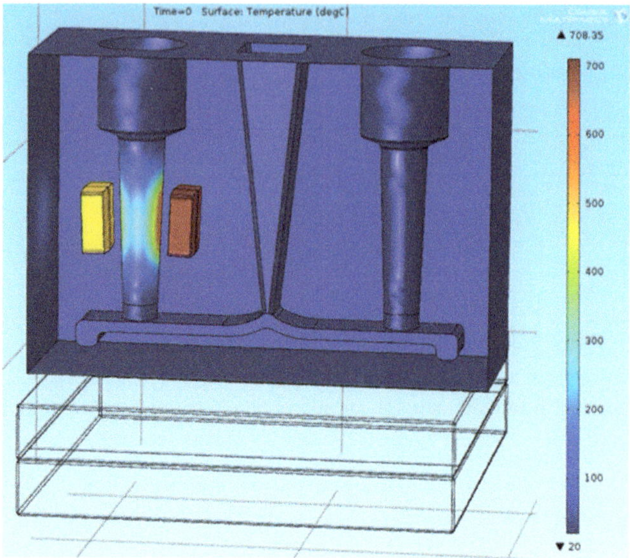

Figure 8. Temperature distribution inside sand mold after heating.

of the two susceptors is due to the inhomogeneous electric field. This issue can be addressed in future applications by embedding a mode stirring component (Wiedenmann et al., 2012).

5 Experimental results

5.1 Comparison with simulation results

To validate the simulation setup, numerical results are compared to results obtained from experimental measurements. For this purpose, the setup illustrated in Fig. 1 is adapted. Instead of a thermocouple, an infrared camera is used to measure the susceptor temperature during the heating phase. Here a SiC susceptor is embedded in a sand block which serves as susceptor carrier. Line of sight for the infrared camera is achieved by a cylindrical hole in the top part of the susceptor carrier. Microwave heating is applied for 90 s and the temperature is recorded using an infrared camera during heating and the subsequent cooling process. The measurement results are compared with the numerical results obtained from a simulation model with COMSOL Multiphysics in Fig. 9. Since the thermal conductivity of SiC and sand could not be measured, reference values obtained from literature were used instead. Silicon carbide shows rather high thermal conductivity, so that a possible error influences the results only negligibly. The considerably lower thermal conductivity of the sand effectively limits the heat transport away from the susceptor. Therefore, small variations in this parameter have a strong influence on the temperature characteristics. Whereas the thermal conductivity of the considered sand should be around $0.8\,\mathrm{W\,m^{-1}\,K^{-1}}$, a better agreement with the measurement curve can be observed if the thermal conductivity is decreased by 25 %. If the thermal conductivity was increased instead, the heat flow in the sand would be improved so that the surface temperature of the susceptor would be lowered.

For a closer matching of experimental results and numerical simulation, a more accurate knowledge of the thermophysical parameters such as thermal conductivity and specific heat capacity would be required.

5.2 Casting experiments

The described process of heating sand molds selectively by microwave energy gives us the possibility to generate a temperature distribution inside the mold prior to casting. Hereby we are able to influence the solidification characteristics of the aluminum alloy AlSi12 for example. For the casting experiments we manufactured molds from H32 and SLS sand with compressed SiC bound with INOTEC® as susceptor material. Figure 10 shows an opened sand mold with the susceptors inside.

This mold is closed and placed into the microwave furnace for heating (see also Fig. 7). After that the aluminum alloy is cast into the mold's inlet at a temperature of approx. 750 °C and fills the specimen cavities from bottom to top. This way turbulent filling of the mold can be avoided. By casting into a series of these molds we could produce specimens which cooled naturally on the one side and were heated due to the localized susceptors on the other side of the mold. The cast specimens were cut in the center and analyzed with respect to their dendrite arm spacing (DAS). During solidification the primary aluminum phase forms tree-shaped arms, the so called dendrites. The DAS is defined as the distance between two dendrite arms and is a common metallographic parameter. Short solidification times result in small DAS and vice versa. Hereby we found, that specimens which cooled slower – since they were heated by the susceptors – had a larger DAS than naturally cooled ones. The results for different mold materials and heating times are shown in Fig. 11.

Although there is a relatively large deviation among the measurements of the DAS we could observe a clear trend that specimens which solidified and cooled near a susceptor had a considerable higher DAS. This effect could be shown with both mold materials and also with different heating times. Here longer heating times lead to a larger difference in DAS which is a consequence of the stronger heating in the susceptor area. These results correspond with the well-known link between solidification time and dendrite arm spacing and show that the solidification time of the aluminum alloy AlSi12 can be influenced by localized microwave heating. It should be noticed that DAS in this context is a measure for solidification time only. The connection of DAS and mechanical properties is not of interest in this case as the scope of the presented work is to overcome design limitations by locally manipulating solidification times.

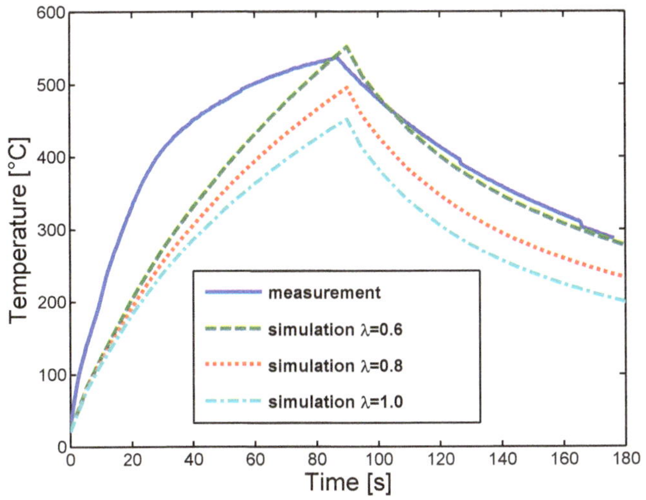

Figure 9. Comparison between experimental and simulation results for the setup shown in Fig. 1. The influence of the thermal conductivity of the sand is illustrated by considering the effect of an increase/decrease of the thermal conductivity by 25 % on the temperature on the susceptor surface.

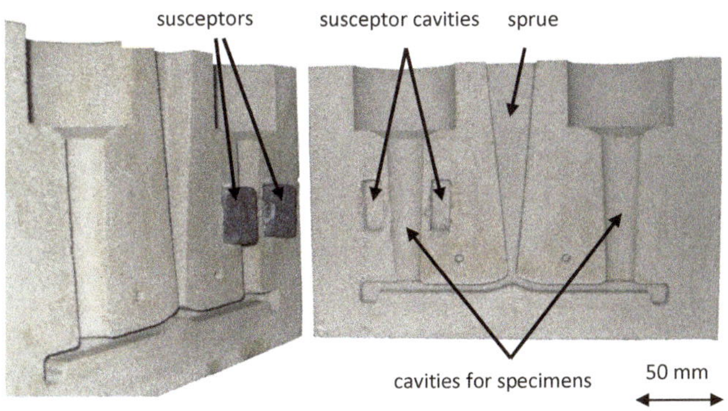

Figure 10. Opened sand mold with two inserted susceptors (black) for casting of two round specimens.

6 Conclusions and outlook

In this paper we presented a method for selectively heating a sand mold prior to a casting process. By using such a selectively heated sand mold it is possible to influence the thermal properties during the casting process and, hence, achieve desirable casing properties. The temperature field inside the sand mold is generated by a microwave process. The sand mold – which is mostly microwave transparent – is locally equipped with susceptors, which strongly absorb the microwave energy and function as a heat source. Based on a selection of microwave absorbing materials we found compressed and chemically bound silicon carbide to be the most suitable material with respect to heating performance and mold stability. The susceptors were dielectrically characterized using a resonator technique and the temperature dependency was taken into account by parallel temperature measurements with an infrared camera. A coupled electro-thermal simulation model was created so that the temperature distribution inside a sand mold can be predicted and optimized. Comparisons between microwave heating experiments and numerical results showed fairly good agreement and first casting experiments confirmed the effect of local heating on the cooling process.

In future works, the electro-thermal simulation shall be directly coupled to a casting simulation, so that the whole process chain can be modeled and optimized.

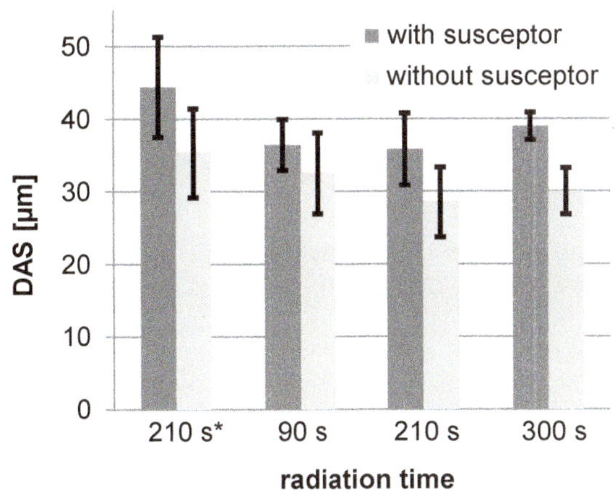

Figure 11. Dendrite Arm Spacing of specimens cast with (dark grey) and without (light grey) the heating influence of a susceptor for different radiation times. The column marked with * represents the only available experiment series with SLS molds. The three other displayed series were performed in H32 molds.

Acknowledgements. The authors thank "Deutsche Forschungsgemeinschaft" (DFG) for partially funding this work under grant EI 352/9-2 and HO 2165/38-2.

Edited by: U. van Rienen

References

Bott, P.-M.: Untersuchungen zur Mikrowellenabsorption und zum Reduktionsverhalten oxidischer Systeme, Ph.D. thesis, Universität Tübingen, 1991.

Chandrasekaran, S., Basak, T., and Ramanathan, S.: Experimental and theoretical investigation on microwave melting of metals, J. Mater. Process. Tech., 211, 482–487, 2011.

Clark, D. E. and Folz, D. C.: Microwave Solutions for Ceramic Engineers, The American Ceramic Society, Westerville, 2005.

Gerdes, T., Willert-Porada, M., and Roediger, K.: Mikrowellensintern elektrisch leitfähiger Werkstoffe, Mikrowelleneinsatz in den Materialwissenschaften, der chemischen Verfahrenstechnik und in der Festkörperchemie, Shaker Verlag, Aachen, 1998.

Grellinger, D. J. and Janney, M. A.: Temperature measurement in a 2.45 GHz microwave furnace, Ceramic Transactions Microwaves: theory and applications in materials processing, Am. Ceram. Soc. Bull., 36, 529–538, Westerville, 1993.

Jin, J.: The Finite Element Method in Electromagnetics, John Wiley & Sons, New York, NY, 2002.

Kılıç, E., Siart, U., Wiedenmann, O., Faz, U., Ramakrishnan, R., Saal, P., and Eibert, T. F.: Cavity resonator measurement of dielectric materials accounting for wall losses and a filling hole, IEEE T. Instrum. Meas., 62, 401–407, doi:10.1109/TIM.2012.2215072, 2013.

Lühken, A.: Ultraschall und Mikrowellenstrahlung im Chemieunterricht – Entwicklung und Erprobung einfacher Experimente zum nichtklassischen Energieeintrag, Ph.D. thesis, Johann W. Goethe-Universität Frankfurt a. Main, 2005.

Lukasz, R.: Mikrowellenunterstützte Wärme- und Stoffübertragung beim Trocknen und Entbindern technischer Keramik, Ph.D. thesis, TU Bergakademie Freiberg, 2005.

Maxwell, J. C.: A Treatise on Electricity and Magnetism, Oxford University Press, London, UK, 1873.

Meredith, R.: Engineers Handbook of Industrial Microwave Heating, The Institution of Electrical Engineers, London, England, 1998.

Rhee, S.: Mikrowellenprozesstechnik für keramische Werkstoffe der Mikrosystemtechnik, Ph.D. thesis, Forschungszentrum Karlsruhe GmbH, 2002.

Robak, L.: Mikrowellenunterstützte Wärme- und Stoffübertragung beim Trocknen und Entbindern technischer Keramik, Ph.D. thesis, Technische Universität Bergakademie Freiberg, Germany, 2005.

Wiedenmann, O., Ramakrishnan, R., Kılıç, E., Saal, P., Siart, U., and Eibert, T. F.: A multi-physics model for microwave heating in metal casting applications embedding a mode stirrer, Microwave Conference (GeMiC), 2012 The 7th German, Ilmenau, Germany, 12–14 March 2012, 1–4, 2012.

Comparison of matching layers for automotive radome design

F. Fitzek[1], **R. H. Rasshofer**[1], **and E. M. Biebl**[2]

[1]BMW Group Research and Technology, Munich, Germany
[2]Technische Universität München – Fachgebiet Höchstfrequenztechnik, Munich, Germany

Abstract. Hidden integration of 79 GHz sensors behind plastic and painted fascia represents a challenging task since both electromagnetic and car body design constraints have to be met. This paper compares different possibilities for low-cost integration of radar sensors. Based on a model for stratified media, a study of the most important parameters such as bandwidth, angle and tolerances is shown. Our results suggest that for plastic fascia, the requirements of future radar sensors can be met with low-cost matching. Even with metallic paints, the requirements imposed by modern 79 GHz radar sensors can be met under certain conditions.

1 Introduction

Automotive radar systems, first introduced on the customer market in the late 1990s, show a steady trend from comfort (Advanced Driver Assistance, short ADAS) to safety oriented vehicle functions (Active Safety, short AS). Radar-based systems tend to address scenarios of growing complexity, targeting urban or city traffic rather than highway scenarios, which were the main application in recent years. Resolution in range, azimuth angle and target velocity is the key feature to deal with demanding scenarios of high complexity and high dynamics (Rasshofer, 2007). In the European Union, a new frequency band from 77 to 81 GHz – the 79 GHz band – has been allocated to enable future ADAS and AS systems replacing the previously used 24 GHz ultrawide band radar sensors. To start the development of 79 GHz band technology, the European funded project "Radar on Chip for Cars" (short RoCC) started in fall 2008 (RoCC, 2009).

Integration of 79 GHz sensors into modern cars represents a challenging task since electromagnetic and car body design constraints have to be met simultaneously. This problem is even more evident when more than one sensor has to be integrated, because hidden integration of the sensors behind a plastic bumper is often required. As is clear from both theoretical investigations and from previous experience with 76 GHz radar systems, operation of these sensors behind painted, plastic fascia will lead to performance degradation if no optimization of the stratified radome structures is performed. In the millimetre wave regime, the electromagnetic field propagates quasi-optically, which results in high reflection and loss factors for obstacles or media in the line-of-sight. This is the case if the radar sensor is to be integrated in the bumper or behind other fascia of the car.

The performance degradation is due to the reflections between the bumper and the sensor. The bumper consists generally of a substrate and some layers of paint. This paper shows a detailed analysis of important electromagnetic radome properties like bandwidth, angle and tolerances for substrate only and painted substrates. Additionally, the cost of integration should be at a minimum, hence this paper considers low-cost matching possibilities for hidden integration of radar sensors.

2 Mathematical model

The electromagnetic theory of the fields in stratified media can be derived from the plane wave assumption and the boundary conditions of electromagnetic waves (time-harmonic convention $e^{j\omega t}$). In Kurt Altenburg (1953) this theory is derived. It can be found today in many books for optics, electromagnetics or microwave theory (Hecht, 2005). In the case of perpendicular incidence of a TE-wave (derivation for TM-waves accordingly) forward- and backward waves are defined, which summarize all waves in the same direction (important difference to other approaches). The complete derivation can be found in Hecht (2005) and a summary of it in Fitzek and Rasshofer (2009). For each layer i with the material parameters n_i as the index of refraction ($k_i = \omega/c_0 n_i$ respectively), d_i as the thickness, and Z_i as the wave impedance of the semi-infinite layer the following charateristic matrix can be derived.

Correspondence to: F. Fitzek
(frerk.fitzek@bmw.de)

$$[\mathbf{M}^i] = \begin{bmatrix} m^i_{11} & m^i_{12} \\ m^i_{21} & m^i_{22} \end{bmatrix} \tag{1}$$

$$= \begin{bmatrix} \cos(k_i d_i) & j Z_i \sin(k_i d_i) \\ j \frac{1}{Z_i} \sin(k_i d_i) & \cos(k_i d_i) \end{bmatrix} \tag{2}$$

The amplitude transmission- and reflection factors of the layer i embedded between semi-infinite layers $i-1$ and $i+1$ are defined as

$$r = \frac{Z_{i+1} m^i_{11} + m^i_{12} - Z_{i+1} Z_{i-1} m^i_{21} - Z_{i-1} m^i_{22}}{Z_{i+1} m^i_{11} + m^i_{12} + Z_{i+1} Z_{i-1} m^i_{21} + Z_{i-1} m^i_{22}} \tag{3}$$

$$t = \frac{2 Z_{i+1}}{Z_{i+1} m^i_{11} + m^i_{12} + Z_{i+1} Z_{i-1} m^i_{21} + Z_{i-1} m^i_{22}} \tag{4}$$

With these equations, the transmission and reflection factor can be derived using the loads and the characteristic matrix M^i. For a multilayer media, the characteristic media matrices have to be multiplied in the right order.

3 Theory and practice of matching

The theory of matching is well known from transmission line theory. The aim is always to get a high transmission next to a low reflection (power matching). The losses in the material can't be compensated, which results in a decreased transmission. For semi-infinite layers, the methods shown here can be used as well. Nevertheless, for mm-wave applications the constraints of matching should be carefully considered (e.g. Fitzek and Rasshofer, 2009).

For a bumper consistent of one layer of substrate and several layers of paint, there exists different options for matching. One is to optimize the thickness of the substrate. For a homogeneous material this results in a thickness of multiple of half-wavelengths. For stratified media, the solution can be found with minimization of the reflection coefficient through the mathematical model. This matching has only one degree of freedom – the decreased thickness of the substrate material. Another possibility is an additional matching layer which is placed directly onto the bumper. This material can have a desired thickness and relative permittivity, hence has two degrees of freedom. One could also consider artificial structures or even metamaterials as an additional matching layer, but as mentioned above, low-cost matching is the focus in this paper, hence these options are not considered.

The third matching option is a combination of these two matching methods. Therefore, a structure (e.g. rib profile) is assembled into the subtrate material of the bumper. Because the overall thickness stays constant, this matching method has only two degrees of freedom (depth and relative permittivity of the integrated structure). Figure 1 depicts the different possibilities. For all matching methods the mathematical model can be used to calculate the optimal values for the desired matching. Therefore, an optimization algorithm based on a direct search method is used. Two different simulations

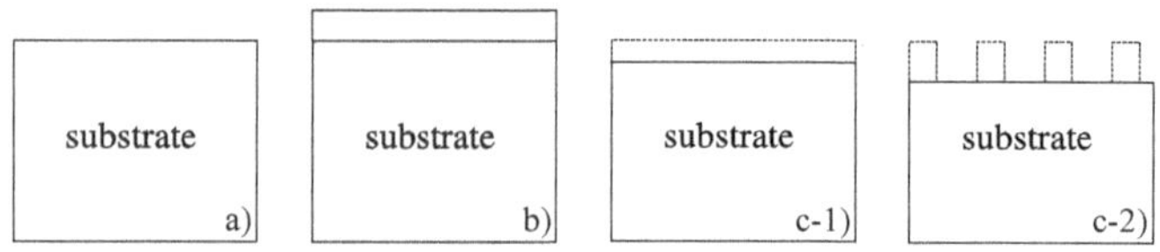

Fig. 1. **(a)** substrate material of bumper (no matching), **(b)** substrate with additional matching layer, **(c-1)** substrate with integrated matching layer (thickness optimization), **(c-2)** substrate with integrated matching layer (rib profile).

can be performed where the thickness and relative permittivity for the matching layer are optimized. This approach neglects the roughness of the material and assumes a non-magnetic, plane material.

For the additional matching layer, the bumper materials (substrate or paint) are *not* changed. On the other hand, for an integrated matching layer, the thickness of the substrate material is changed according to the thickness of this layer. This results in a decreased thickness for the substrate material of the bumper. With this integrated matching the integrated structure and the thickness optimization can be modeled. For the latter the relative permittivity is fixed to one, whereas it is a degree of freedom for the integrated structure.

3.1 Integrated matching structure

The integrated matching structure is a structured dielectric (e.g. rib profile) and can be any structure which fullfills the homogeneity limit of $p < \lambda/4$ with p as the cell size in the direction of propagation, in order to avoid parasitic scattering effects. For easy integration the rib profile is one possibility. There exists several methods to calculate the relative permittivity of such structures (Biber, 2003). Here a method based on transmission line models is used in which the resulting relative permittivity is described as the mean of the relative characteristic impedances ($Z' = \sqrt{\frac{\mu_r}{\varepsilon_r}}$ with $\mu_r = 1$) of free space and the substrate depending on the ratio of ribs and bars. Hence, a relative permittivity between 1 and $\varepsilon_{r,s}$ of the substrate can be reached.

$$\varepsilon_r = \left(\frac{t_{\mathrm{bar}} Z'_{\mathrm{substrate}} + t_{\mathrm{rib}} Z'_0}{t_{\mathrm{bar}} + t_{\mathrm{rib}}} \right)^{-2} \tag{5}$$

If the thickness of the ribs (t_{rib}) are predefined due to manufacturing constraints, one can easily calculate the thickness of the bars (t_{bar}) to reach the desired value for ε_r as calculated by the optimization algorithm.

Figure 2 shows the dependency of the relative permittivity versus the ratio of bars to ribs with constant rib thickness. It can be seen that the variation in relative permittivity for an increased bar thickness (resulting in higher ε_r values) is lower, than for smaller values of the bar thickness. This will be an important design possibility later on.

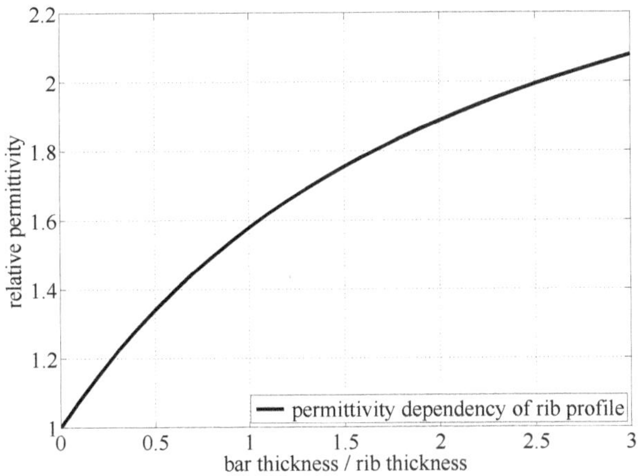

Fig. 2. Permittivity dependency of rib profile.

4 Comparison of matching layers

For the comparison of these different approaches the optimization algorithm is visualized by graphs with all possible degrees of freedom. This means that for all possible thickness of matching layer, all possible relative permittivities ε_r are shown. Obviously this can not be done for really all possibilities, so discrete values are chosen in steps of 0.1 for the relative permittivity.

This is done for one frequency $f = 78.5$ GHz and for the considered bandwidth $B = 76 - 81$ GHz. For the bandwidth the maximum reflection (meaning the worst case) within the band is considered for the evaluation. The best solution of the bandwidth analysis is taken for an angle variation in the range of ± 40 deg. With the more exact solution (ε_r stepsize 0.01) of the algorithm a tolerance analysis is shown with variations of $\sigma = \pm 0.1$ mm. For the additional matching layer the variation affects only the thickness of this layer. For the thickness optimization (integrated matching layer) the variation only affects the thickness of the subtrate material of the bumper. The variation of the integrated structure (integrated matching layer) affects on the one hand the thickness of the substrate material and on the other hand in our case the thickness of the bars.

4.1 Additional matching layer for substrate

The results for an additional matching layer for the subtrate material of the bumper (no paint considered yet) are depicted in Fig. 3a and b. In these figures the first (+), best (diamond) and last (x) graph are always highlighted to see the influence of increasing relative permittivity. The substrate material has a relative permittivity in the range of $\varepsilon_{r,s} = 2.86 - 0.06i$ and a thickness of about $d_s = 3$ mm. Considering low-cost materials the relative permittivity for the matching layer is varied between $\varepsilon_r = 1$ (initial condition) and $\varepsilon_r = 3$. It can be seen that for a relative permittivity of $\varepsilon_r = 2.6$ and a thickness of $d = 0.43$ mm the reflection can be decreased by about 45 dB

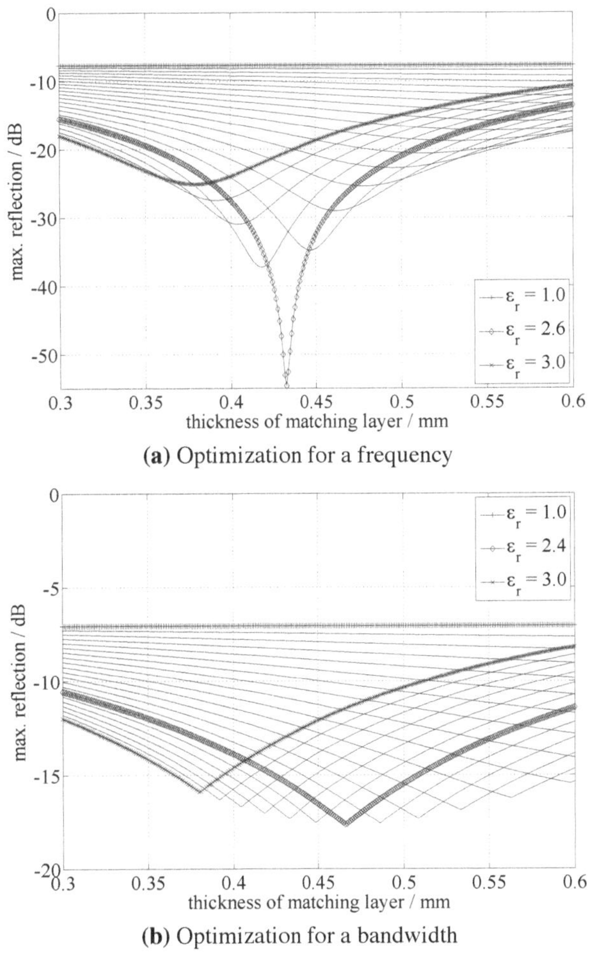

Fig. 3. Frequency and bandwidth analysis.

for one frequency ($f = 78.5$ GHz). Considering the maximum reflection within the band ($B = 76 - 81$ GHz) the improvement is about 10 dB, resulting in a maximum reflection of -17 dB. This is achieved for a relative permittivity of $\varepsilon_r = 2.4$ and a thickness of $d = 0.47$ mm.

The angular dependency is depicted in Fig. 4a. Here the optimal values for the relative permittivity and thickness combination from the bandwidth analysis are used to show the behaviour. The combination mentioned before also gives the best result for the angular dependency. Hence, the maximum reflection coefficient within an angle of ± 40 deg is still below -10 dB. In Fig. 4b the reflection is shown over frequency. The maximum reflection coefficient of about -17 dB can be seen at the edges of the band (76/81 GHz) as calculated before. Next to that the tolerance analysis for a variation of $\sigma = \pm 0.1$ mm is shown. The maximum reflection is below -12 dB.

4.2 Integrated matching layer for substrate

The results for an integrated matching layer for the subtrate material of the bumper (again no paint considered) are

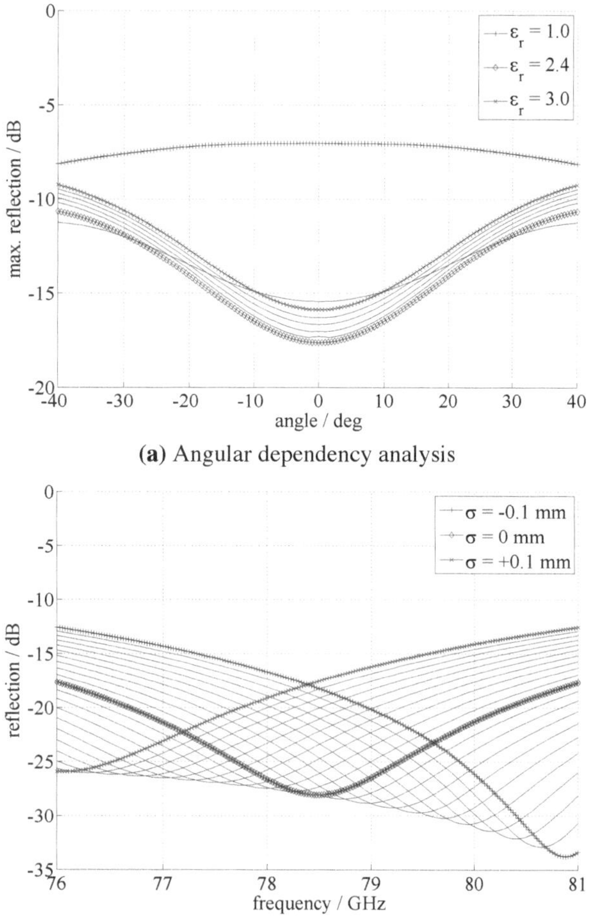

(a) Angular dependency analysis

(b) Tolerance analysis

Fig. 4. Angular and tolerance analysis.

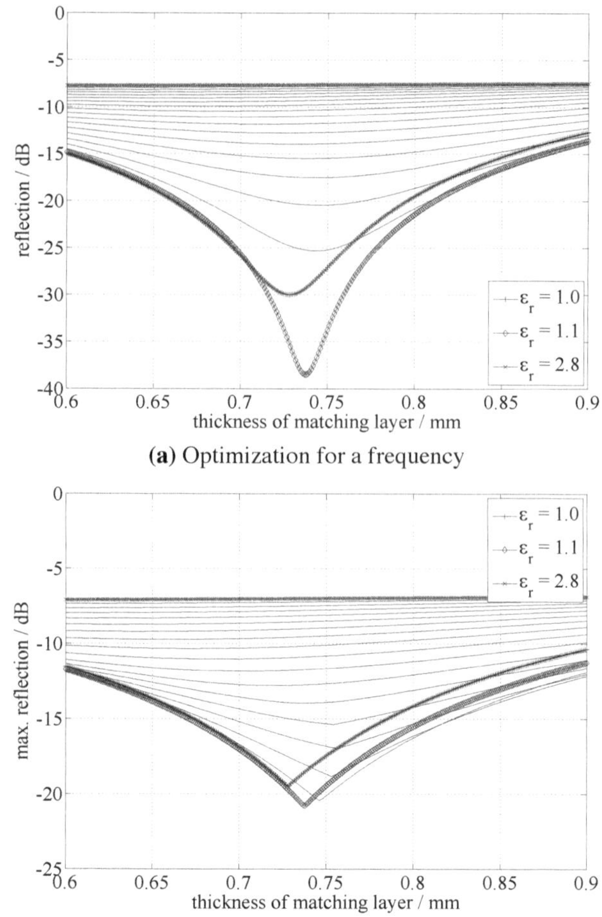

(a) Optimization for a frequency

(b) Optimization for a bandwidth

Fig. 5. Frequency and bandwidth analysis.

depicted in Fig. 5a and b. This is done for the same material like before. The only difference is the above mentioned reduction of the substrate material. In the figures $\varepsilon_r = 1$ shows the results for a thickness optimization, because only the thickness of the matching layer is varied. This results in a decreased thickness of the subtrate material of the bumper. The initial condition is represented by the relative permittivity of $\varepsilon_r = 2.8$, which is nearly the same as the substrate material itself. The optimal solution for one frequency and for the bandwidth is $\varepsilon_r = 1.1$ with a thickness of d =0.74 mm. The maximum reflection within the band is about -21 dB, 4 dB lower than that of the additional matching layer. The angular dependency is depicted in Fig. 6. The solution of the bandwidth analysis shows no improvement concerning the maximum reflection within the angular range. Nevertheless for a smaller range (e.g. ± 30 deg) the maximum reflection would be below -10 dB. To cover the complete angular range one ends up with another solution, $\varepsilon_r = 1.4$, which reduces the maximum reflection below -10 dB, but increases the reflection for 0 deg by about 2 dB. Concerning only the angular dependency this would be the optimal solution. This shows

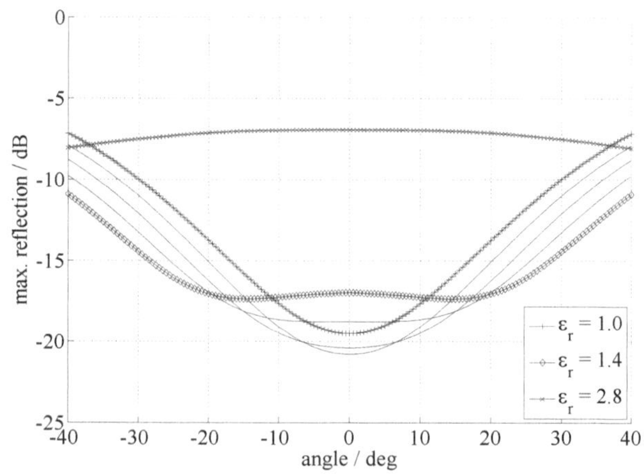

Fig. 6. Angular analysis.

that already in this stage the meaning of "optimal" is not distinct. In the Sect. 4.3 this will be discussed. Nevertheless we have now two possible "optimal" solutions. The results of the tolerance analysis are shown in Fig. 7a for $\varepsilon_r = 1.4$ and

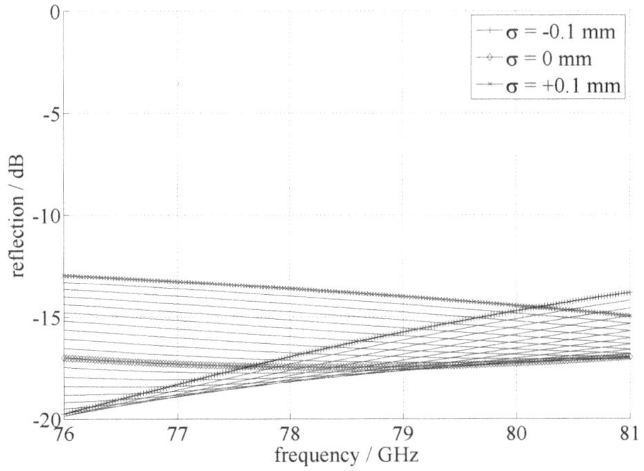

(**a**) Tolerance analysis for a rel. permittivity of 1.4

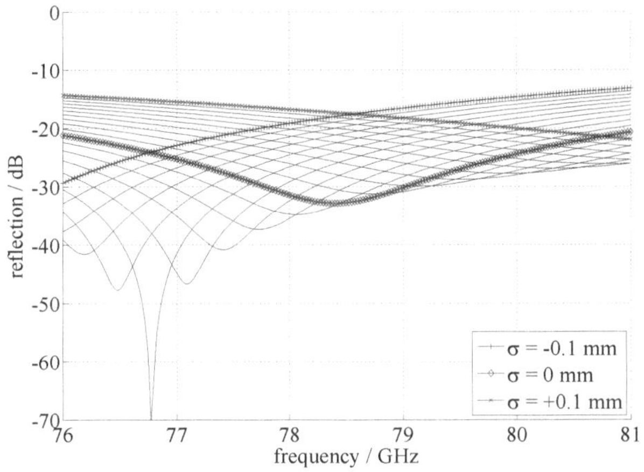

(**b**) Tolerance analysis for a rel. permittivity of 1.1

Fig. 7. Tolerance analysis.

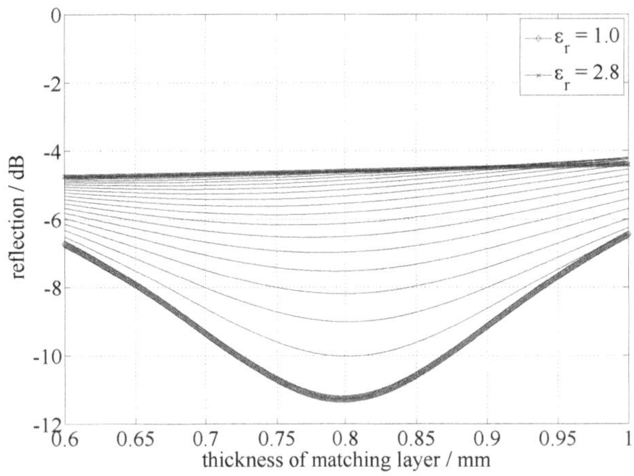

(**a**) Optimization for a frequency

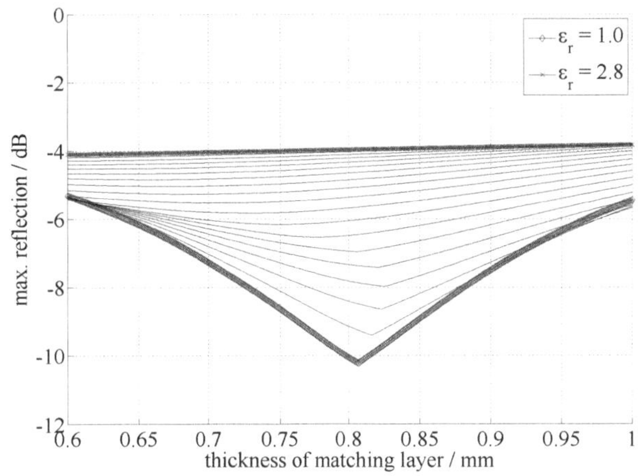

(**b**) Optimization for a bandwidth

Fig. 8. Frequency and bandwidth analysis.

7b for $\varepsilon_r = 1.1$. Here the before mentioned coherence between a higher permittivity and a smaller variation in relative permittivity can be seen.

4.3 Discussion of substrate materials

The analysis of the reflections coming from the bumper shows that one optimal solution can not be defined without knowing the sensor requirements. Nevertheless it is possible to define a cost function from sensor requirements which clearly defines the optimal solution. Together with the detailed analysis shown above the optimal parameters can hence be calculated. Additionally, the evaluation shows that the integrated matching layer is superior over the additional matching layer concerning maximum reflection within the bandwidth. Because this is a strong requirement for future radar sensors, the evaluation is focused on integrated matching layers in the following section.

4.4 Integrated matching layer for substrate and paint

The results for an integrated matching layer for the subtrate material of the bumper with metallic paint (permittivity of metallic paint $\varepsilon_r > 4$ – condition depends on the substrate's permittivity) are depicted in Fig. 9a and 9b. The substrate material is the same as before but with a slightly decreased thickness of $d_s = 2.7$ mm. Again, the integrated matching layer results in a reduction of the substrate material. In the figures $\varepsilon_r = 1$ is coexistent for a thickness optimization and $\varepsilon_r = 2.8$ represents the intial condition. The figures show that the reflections are increased for the substrate material with metallic paint. The optimal solution for one frequency and for the bandwidth is $\varepsilon_r = 1$ with a thickness of $d = 0.81$ mm[1]. The maximum reflection within the band is about -10 dB. The angular dependency is depicted in Fig. 9a for the best solutions of the bandwidth analysis. The maximum reflection within an angular range of ± 40 deg is about -4.5 dB. This can be optimized with a relative permittivity of $\varepsilon_r = 1.6$ below -6 dB, but with higher reflections within the band.

[1] For paints with $\varepsilon_r < 4$ the optimal solution differs.

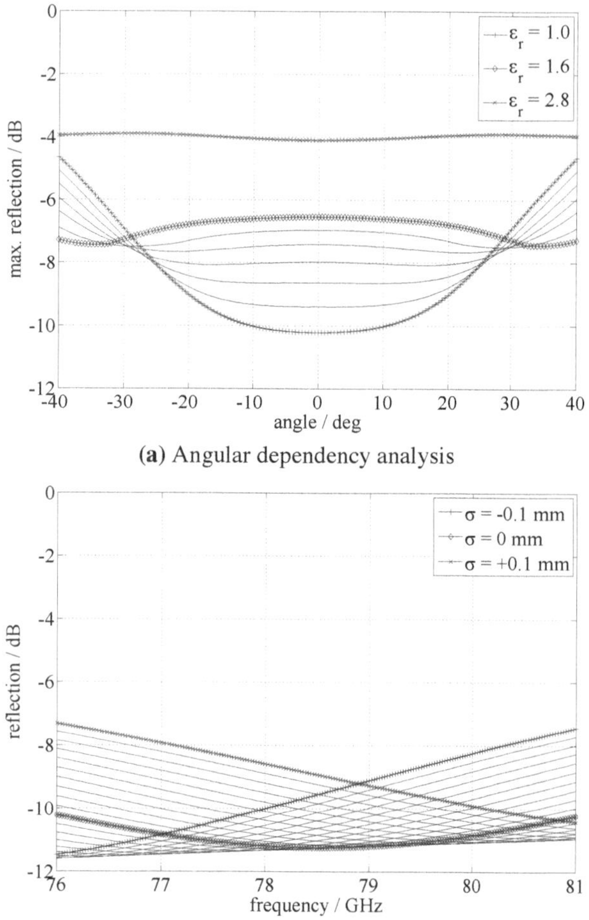

(**a**) Angular dependency analysis

(**b**) Tolerance analysis

Fig. 9. Angular and tolerance analysis.

Nevertheless for a reduced angular range of ± 10 deg the maximum reflection within the angular range and the bandwidth are below -10 dB. The tolerance analysis shows a maxmimum reflection below -7 dB. For $\sigma = 0$ mm the result of the bandwidth analysis can be checked with a maximum reflection within the band below -10 dB.

4.5 Discussion of painted materials

The evaluation of the different parameters show that it is possible to reduce the maximum reflection within the bandwidth below -10 dB even with metallic paints. Metallic paints have the highest permittivities of all paints, hence considering a worst case. It is clearly shown that the angular dependency and the tolerances play an important role for the design of an optimal matching layer. Again an "optimal" solution only exists for specific sensor requirements.

5 Conclusions

This paper compares different possibilities for low-cost integration of radar sensors for the 79 GHz band. The assumptions of the model described are a plane wave incident on a non-magnetic, plane material without roughness. It is shown for plastic fascia, that the bandwidth requirements $B = 76 - 81$ GHz, the angular requirements ± 40 deg and tolerance influences $\sigma = \pm 0.1$ mm can be met coexistently with low-cost matching. This can be realized using a rib profile (integrated matching layer).

For painted fascia the bandwidth requirement can be met with a maximum reflection of below -10 dB. The thickness optimization (integrated matching layer) delivers the best solution considering the mentioned realization possibilities. Because the angular dependency and the tolerance analysis are based on this evaluation, they suffer from the requirement of 5 GHz bandwidth. Nevertheless the maximum reflection can be reduced below -10 dB under certain conditions. The optimisation of the angular dependency and the tolerance influence is considered for future work. An overall optimal solution can be found knowing the detailed requirements of future radar sensors.

Acknowledgements. The authors wish to thank the German Federal Ministry of Education and Research (BMBF) for the funding of the RoCC joint project.

References

Alitalo, P., Luukkonen, O., Vehmas, J., and Tretyakov, S. A.: Impedance-Matched Microwave Lens, 7, 187–191, 2008.

Biber, S., Richter, J., Martius, S., and Schmidt, L.-P.: Design of artificial dielectrics for anti-reflection-coatings, in: European Microwave Conference, 33rd, October 2003, pp. 1115–1118, 2003.

Fitzek, F. and Rasshofer, R. H.: Automotive Radome Design – Reflection Reduction in stratified media, Antennas and Wireless Propagation Letters, IEEE, 8, 2009.

Hecht, E.: Optik, Oldenbourg Wissenschaftsverlag, 2005.

Jones, E. and Cohn, S.: Surface matching of dielectric lenses, in: Proc. IRE International Convention Record, vol. 2, pp. 46–53, 1954.

Kurt Altenburg, S. K.: Wellenausbreitung in geschichteten Medien bei senkrechtem Einfall [...], Annalen der Physik, 448, 1–43, 1953.

Rasshofer, R. H.: Functional Requirements of Future Automotive Radar Systems, in: Proc. of the European Radar Conference, Munich, Germany, 2007.

RoCC press release, http://www.eenova.de/news-presse/rocc/, 2009.

A reconfigurable software defined ultra-wideband impulse radio transceiver

M. D. Blech[1], **A. T. Ott**[2], **P. Neumeier**[3], **M. Möller**[1], **and T. F. Eibert**[2]

[1]Institut für Hochfrequenztechnik, Universität Stuttgart, 70550 Stuttgart, Germany
[2]Lehrstuhl für Hochfrequenztechnik, Technische Universität München, 80290 Munich, Germany
[3]Abteilung Leistungsoptimierte Systeme, Fraunhofer-Institut für Integrierte Schaltungen IIS, 90411 Nürnberg, Germany

Abstract. An ultra-wideband (UWB) software defined radio (SDR) implementation is presented. The developed impulse radio (IR) transceiver employs first order bandpass (BP) sampling at a conversion frequency which is four times the channel bandwidth. The subsampling architecture directly provides the RF signal avoiding any non-ideal mixer stages and reduces the requirements of digital signal processing implemented in a field programmable gate array (FPGA). The transmitter consists basically of a multi-Nyquist digital to analog converter (DAC), whereas the implemented matched filter (MF) receiver prototype employs a standard digitizing oscilloscope. This design can be adaptively reconfigured in terms of modulation, data rate, and channel equalization. The reconfigurable design is used for an extensive performance analysis of the quadrature phase shift keying (QPSK) modulation scheme investigating the influence of different antennas, amplifiers, narrowband interferers as well as different equalizer lengths. Even for distances up to 7 m in a multipath environment robust communication was achieved.

1 Introduction

The approval of UWB technology in the US and in Europe has caused enormous research activities on broadband wireless communication schemes (Reed, 2005). Up to now mainly solutions based on orthogonal frequency division multiplexing (OFDM) (Batra et al., 2004) or direct sequence spread spectrum (DSSS) (Fisher et al., 2004) are commercially available. IR devices are still a niche product (Time Domain Corporation, 2006).

An SDR concept offers the best opportunities to combine the advantages of impulse based UWB communication with the flexibility required by modern transceiver designs. Recently, some investigations dealing with UWB SDR concepts have been published (Chen and Brodersen, 2007; Keller et al., 2007) mainly focusing on the receiver implementation and using analog pulse generators on the transmitter side.

The flexibility of an UWB SDR transceiver implementation must be gained by a huge computational effort necessary for the wideband digital signal processing of a bandwidth in excess of 500 MHz as required by the regulation. In ordinary SDR designs the signal on an intermediate frequency (IF) is digitized whereas up- and downconversion to the radio frequency (RF) band afford additional analog mixer stages (see Fig. 1 top) causing signal distortions due to their non-ideal transfer characteristic. Employing bandpass sampling (Vaughan et al., 1991) a subsampling direct conversion transceiver can be implemented avoiding any analog circuitry except for bandpass filters and optional amplifiers (see Fig. 1 bottom). First order sampling on the one hand relaxes the requirements of the DAC stage, but on the other hand limits the usable frequency range (Vaughan et al., 1991). Second order sampling offers more advantages in terms of frequency agility, but in turn requires more complex digital signal processing (Coulson et al., 1994). The application of either of these concepts is difficult in narrowband communications because of noise aliasing, but in UWB systems the performance degradation due to noise folding is negligible (Chen and Brodersen, 2007).

Thus, a novel subsampling UWB direct conversion SDR IR transceiver is presented. This reconfigurable prototype is designed for laboratory use for the analysis of various UWB IR communication schemes. Its transmitter basically consists of a field programmable gate array (FPGA) (Avnet, Inc., 2006) and a multi-Nyquist DAC (Maxim Integrated Products, Inc., 2006) enabling adaptive predistortion to equalize the channel impulse response, whereas a commercial digitizing oscilloscope is used on the receiver side of the transceiver prototype. Signal processing of the receiver is carried out offline, but with full applicability in real-time on adequate hardware. The trend of semiconductor technology towards

Correspondence to: M. D. Blech
(mail@ihf.uni-stuttgart.de)

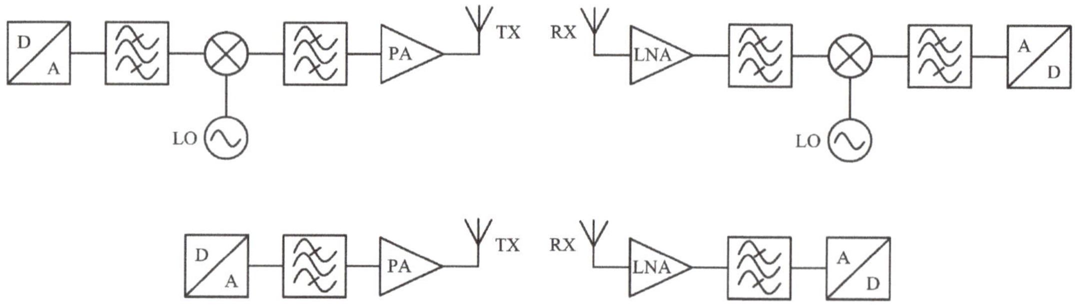

Fig. 1. Block diagram of a conventional SDR transceiver (top) and a subsampling architecture (bottom).

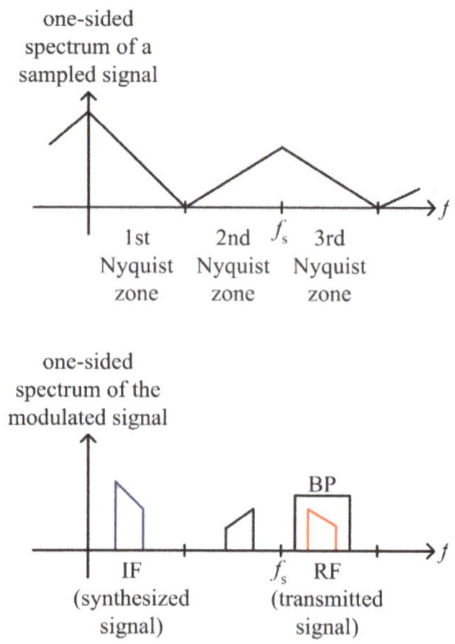

Fig. 2. Spectrum of a time discrete signal (top) and spectrum of a band-limited modulated signal (bottom).

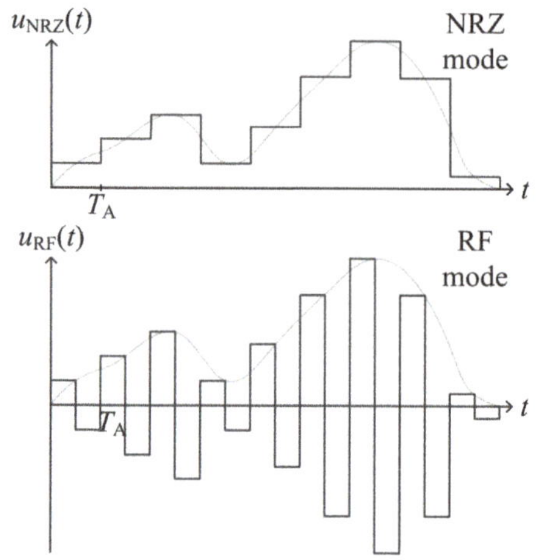

Fig. 3. DAC output signals (black) of a digitized analog waveform (gray) in different modes.

higher frequencies and lower power consumption will make UWB SDR concepts applicable in consumer products in the near future.

In this paper the transceiver architecture of the developed prototype is described in Sect. 2. In Sect. 3 the performance of the SDR employing pulsed QPSK modulation is analyzed by numerical simulations and by measurements. Finally the work is concluded in Sect. 4.

2 SDR transceiver

The developed transceiver uses the subsampling architecture depicted in Fig. 1 (bottom). It uses first order sampling and is designed for the RF frequency range of 2–3 GHz, which is outside the approved UWB frequency band, but adequate to investigate the basic properties of an UWB communication system in a laboratory environment.

The sampling frequency f_s of both the DAC in the transmitter and the ADC in the receiver is 2 GHz, enabling the transmission of impulses exhibiting a bandwidth of up to 1 GHz. The employed QPSK modulation scheme uses Gaussian impulses as described in (Blech et al., 2008) with a -10 dB bandwidth of 500 MHz and a pulse repetition frequency (PRF) f_{sym}=0.5...62.5 MHz.

2.1 Transmitter architecture

Each time discrete signal has a periodic spectrum as shown in Fig. 2 (top). Synthesizing a signal on an intermediate frequency (IF) in the first Nyquist zone does also generate multiple attenuated replicas in the higher Nyquist zones (see Fig. 2 bottom). Using an anti-aliasing filter exhibiting a BP instead of a lowpass (LP) characteristic directly provides the desired RF signal.

The employed DAC offers different modes, namely the non return to zero (NRZ), the return to zero (RZ), and the so called RF mode. Each of them is dedicated to applications in different Nyquist zones. The according output signals of an exemplary waveform in all aforementioned modes are depicted in Fig. 3.

In the NRZ mode, which is usually used for LP signals, each step of the sampled waveform is described by a rectangular impulse

$$h_{\mathrm{NRZ}}(t) = \mathrm{rect}\left(\frac{t}{T_\mathrm{s}}\right) \circ\!\!-\!\!\bullet\, T_\mathrm{s}\,\mathrm{sinc}(\pi f T_\mathrm{s}) = H_{\mathrm{NRZ}}(f)\,, \tag{1}$$

where T_s denotes the sampling interval, $h_{\mathrm{NRZ}}(t)$ is the time domain waveform, and $H_{\mathrm{NRZ}}(f)$ stands for the frequency domain representation of one impulse.
The same holds for the bipolar RF impulse

$$h_{\mathrm{RF}}(t) = \mathrm{rect}\left(\frac{t+\frac{T_\mathrm{s}}{4}}{\frac{T_\mathrm{s}}{2}}\right) - \mathrm{rect}\left(\frac{t-\frac{T_\mathrm{s}}{4}}{\frac{T_\mathrm{s}}{2}}\right)$$
$$\circ\!\!-\!\!\bullet\, \mathrm{j}T_\mathrm{s}\frac{1-\cos(\pi f T_\mathrm{s})}{\pi f T_\mathrm{s}} = H_{\mathrm{RF}}(f)\,, \tag{2}$$

where the impulse is bisected, has no direct current (DC) component and exhibits higher spectral amplitudes in the 2nd and 3rd Nyquist zone compared to the NRZ mode (Maxim Integrated Products, Inc., 2006).
The pulse train to be transmitted is composed of analog impulses $g(t)$ (Blech et al., 2008), which are sampled at the sampling interval T_s yielding the signal

$$g_\mathrm{s}(t) = g(t)\sum_{n=-\infty}^{\infty}\delta(t-nT_\mathrm{s}) = \sum_{n=-\infty}^{\infty} g(nT_\mathrm{s})\delta(t-nT_\mathrm{s})\,. \tag{3}$$

In order to obtain the spectrum of the DAC output signal $g_{\mathrm{DAC}}(t)$, the sampled pulse train must be convolved with the corresponding impulse waveform of the chosen DAC mode

$$g_{\mathrm{DAC}}(t) = h(t) * g_\mathrm{s}(t)\,. \tag{4}$$

This yields the output spectra

$$G_{\mathrm{s,NRZ}}(f) = H_{\mathrm{NRZ}}(f)\frac{1}{T_\mathrm{s}}\sum_{k=-\infty}^{\infty} G_\mathrm{s}\left(f-\frac{k}{T_\mathrm{s}}\right), \tag{5}$$

$$G_{\mathrm{s,RF}}(f) = H_{\mathrm{RF}}(f)\frac{1}{T_\mathrm{s}}\sum_{k=-\infty}^{\infty} G_\mathrm{s}\left(f-\frac{k}{T_\mathrm{s}}\right). \tag{6}$$

In Fig. 4 the measured output spectra of the NRZ and RF mode are depicted. The signal is measured directly at the output of the DAC board synthesizing a Gaussian shaped impulse train with full scale amplitude. The RF mode provides the highest output amplitude in the third Nyquist zone and is used in this work. There the output signal needs further amplification to exploit the maximum permitted power spectral density (PSD) of −41.3 dBm/MHz.

The PSD of a modulated pulse train analytically results in

$$\mathrm{PSD}(f) = \frac{\sigma_\mathrm{b}^2}{T_{\mathrm{sym}}}\frac{|G_{\mathrm{s,RF}}(f)|^2}{Z_\mathrm{c}} + \frac{\mu_\mathrm{b}^2}{T_{\mathrm{sym}}^2}\sum_{m=-\infty}^{\infty}\frac{|G_{\mathrm{s,RF}}(f)|^2}{Z_\mathrm{c}}\,\delta\left(f-\frac{m}{T_{\mathrm{sym}}}\right) \tag{7}$$

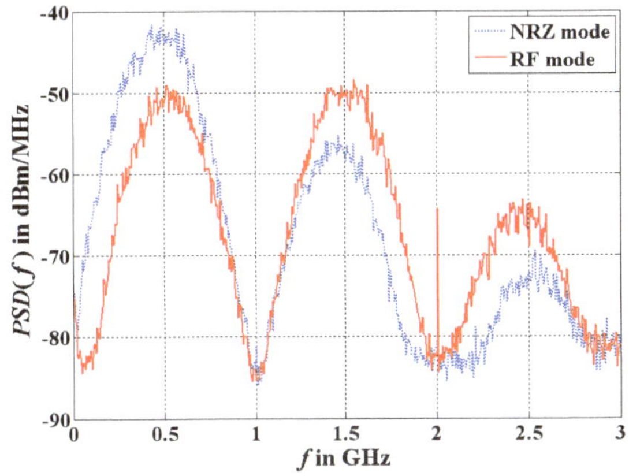

Fig. 4. Measured PSD of the QPSK output signal in different DAC modes (f_s=2 GHz, f_{sym}=62.5 MHz, f_c=500 MHz, B=500 MHz).

for an equidistant sequence of symbols b spaced by T_{sym} (Proakis, 2001). In Eq. (7) Z_C stands for the characteristic impedance, in this case 50 Ω, whereas μ_b and σ_b^2 denote the expected value and the variance of the transmitted symbols, respectively. As a pseudo random binary sequence (PRBS) generator is used μ_b=0 and the second part of the sum vanishes.

In order to filter the desired RF signal in the third Nyquist zone a substrate integrated waveguide (SIW) filter (Nam et al., 2005) is used to suppress spectral components in the first and second Nyquist zone and to avoid aliasing. This filter exhibits a highpass (HP) characteristic with an almost constant group delay and with a cut-off frequency of 2.15 GHz as well as a rejection of more than 40 dB below 2 GHz. Due to the limited rise time of the output signal of the DAC and the LP characteristic of the output network on the DAC board, spectral components in the Nyquist zones four or higher do actually not exist and thus need not to be filtered explicitly. However, amplification is necessary.

The very high sampling frequency of 2 GHz requires an immense parallelization of the transmitter design in the FPGA implementation. The 12 bit DAC already includes a 4:1 multiplexer splitting up the data bus in four branches (A-D), each consisting of 12 low voltage differential signaling (LVDS) lines. But as the Virtex-5 FPGA from Xilinx (Avnet, Inc., 2006) must meet some timing constraints, another 2:1 multiplexer consisting of the internal output serializer deserializer (OSERDES) blocks is necessary reducing the FPGA clock to $f_\mathrm{s}/8 = 250$ MHz. Each QPSK waveform is stored in two random access memory (RAM) clusters (Fig. 5 left), whose data is initialized via a universal serial bus (USB) interface by a host PC. The programmed waveform of arbitrary shape and user-defined length N is directly linked to the symbol clock frequency f_{sym}. The according clock signals and the composition of a waveform are plotted in Fig. 5 (right). As shown in the diagram, Gray-coding is used. Though using only 11 of the 12 differential data lines of each data bus,

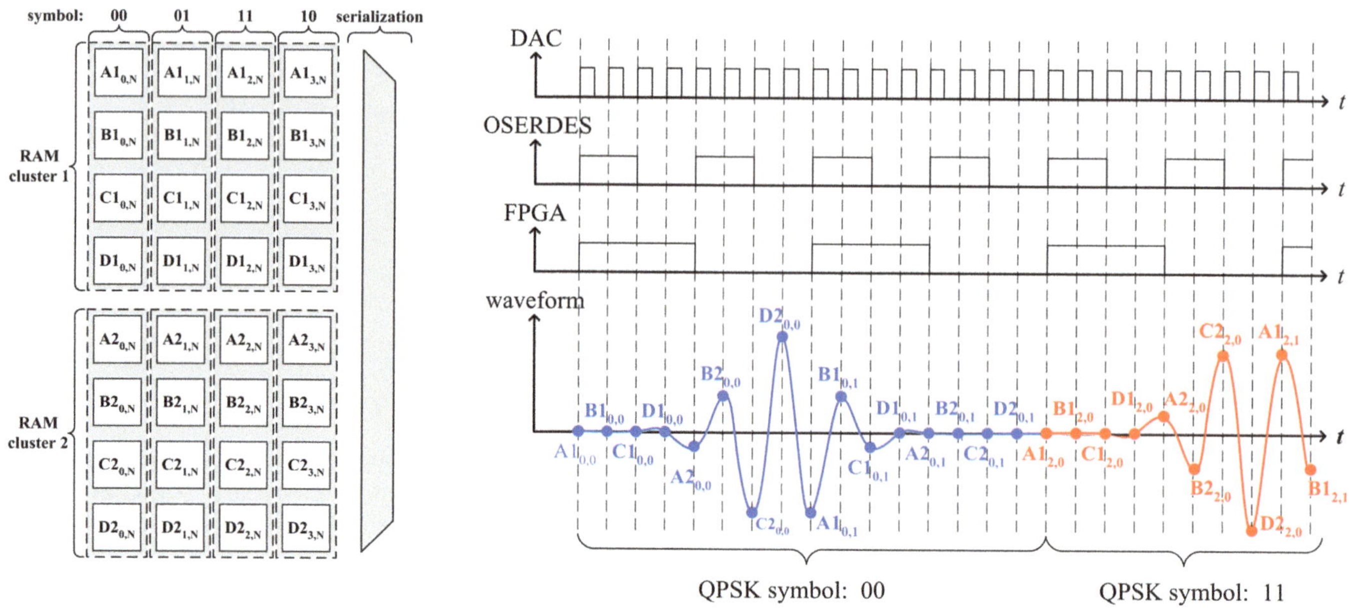

Fig. 5. Memory organization of the transmit symbols (left) and according transmit waveform (right).

still 88 lines between FPGA and DAC and an update rate of 22 Gbit/s are required.

2.2 Receiver design

A correlation receiver is implemented in MATLAB for offline use as well as for real-time reception in Handel-C for an FPGA design. The receiver uses the digitized waveform

$$v_{\mathrm{s,RX}}(t) = v_{\mathrm{RX}}(t) \sum_{n=-\infty}^{\infty} \delta(t - nT_{\mathrm{s}}) \quad (8)$$

obtained after sampling the analog receive signal $v_{\mathrm{RX}}(t)$. The according spectrum

$$V_{\mathrm{s,RX}}(f) = V_{\mathrm{RX}}(f) * \frac{1}{T_{\mathrm{s}}} \sum_{k=-\infty}^{\infty} \delta\left(f - \frac{k}{T_{\mathrm{s}}}\right) \quad (9)$$

is periodic and a simple LP filter provides the desired replica in the first Nyquist zone. In the receiver prototype the receive signal is acquired by a real-time oscilloscope DSO81004A from Agilent running at 2 GS/s. The RF frontend only consists of a BP filter (Mini-Circuits, 2008) and an optional low noise amplifier (LNA) with a gain of 48 dB and a noise figure of 1.6 dB.

A block diagram of the MF receiver is given in Fig. 6. It demodulates the acquired receive signal $v_{\mathrm{RX}}(t)$ offline, but it meets the requirements for a real-time application on e.g. an FPGA running at $f_{\mathrm{s}}/4$. All filters are of finite impulse response (FIR) type, which is advantageous for parallelization.

The equalizer impulse response $h_{\mathrm{equ}}(t)$ is estimated using a preamble of up to 100 symbols and a time domain deconvolution technique (Preis, 1977) to determine the channel impulse response. In the QPSK modulation scheme four matched filters ($h_{\mathrm{MF,00}}, \ldots, h_{\mathrm{MF,11}}$) and a subsequent maximum likelihood (ML) decision provide the digital receive symbols $\hat{b}$.

A portation for an FPGA based receiver has been carried out already, but due to the lack of an appropriate wideband ADC all results presented in this paper have been computed offline employing the aforementioned oscilloscope for the signal acquisition and the equivalent MATLAB implementation of the receiver.

3 Performance analysis

A complete model of the transceiver has been implemented in MATLAB in order to compare the simulated and measured system performance. In this paper exemplary results of a line-of-sight (LOS) QPSK IR communication in a distance ranging from 1 to 7 m are presented. In this investigation a symbol frequency $f_{\mathrm{sym}} = 62.5$ MHz, a center frequency $f_{\mathrm{c}} = 2.5$ GHz, and a bandwidth $B = 500$ MHz are used.

3.1 Simulations

In the simulations the transmit waveform is generated taking into account the jitter affecting the transmit signal. However, the jitter of the commercial real-time oscilloscope on the receiver side is negligible. In UWB applications the impulse response of an antenna is an important issue. Thus the normalized impulse response (Farr and Baum, 1998) of the various investigated UWB antennas is computed using CST Microwave Studio. From the transfer functions $H_{\mathrm{N,TX}}$ and

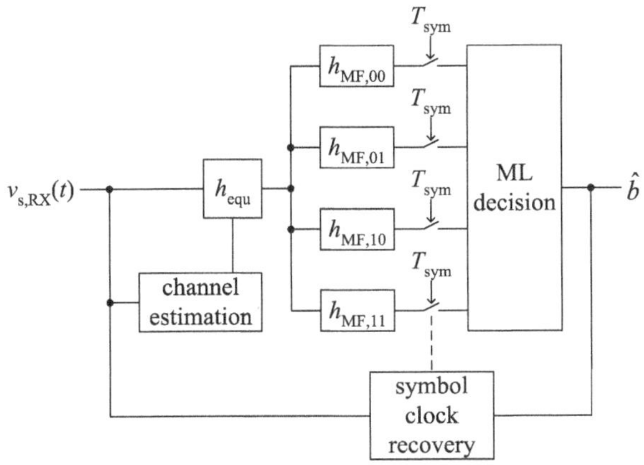

Fig. 6. Block diagram of the implemented matched filter receiver.

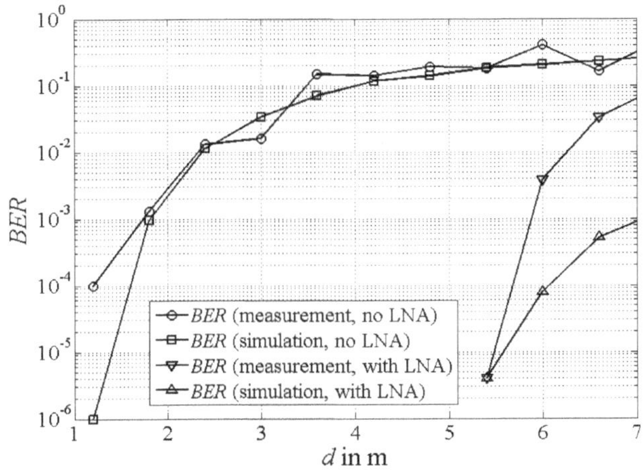

Fig. 7. Influence of the LNA on the simulated and measured BER for a LOS scenario employing directional vivaldi antennas.

$H_{N,RX}$ of the transmit and receive antenna, respectively, the modified "new" normalized transfer functions

$$H_{N,new,TX}(\omega) = j\omega \frac{1}{c_0}\sqrt{\frac{\omega}{\pi}} H_{N,TX}(\omega) \tag{10}$$

and

$$H_{N,new,RX}(\omega) = \frac{1}{c_0}\sqrt{\frac{\omega}{\pi}} H_{N,RX}(\omega) \tag{11}$$

are derived according to Ott et al. (2009). Regarding the entire radio link, this enables the extraction of the channel transfer function H_{CH} from the antenna transfer characteristic defined in (Farr and Baum, 1998).

Thus the transmission parameter between transmitter and receiver can now be written as

$$S_{21}(\omega) = H_{N,new,TX}(\omega)\, H_{CH}(\omega)\, H_{N,new,RX}(\omega)\ . \tag{12}$$

Knowing the characteristics of the transmit and receive antennas the channel impulse response is synthesized using the

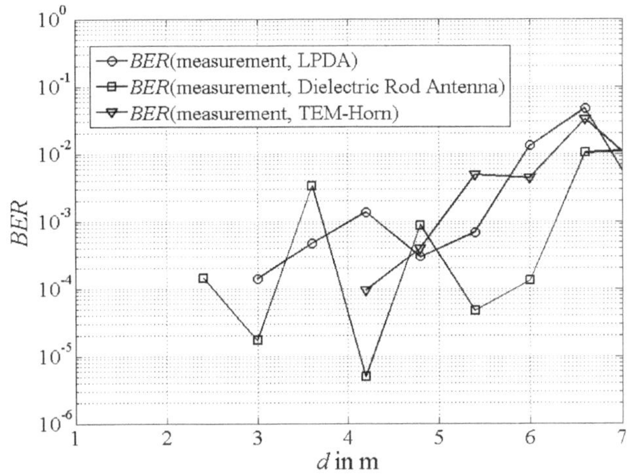

Fig. 8. Simulated and measured BER for a LOS scenario using different types of directional antennas.

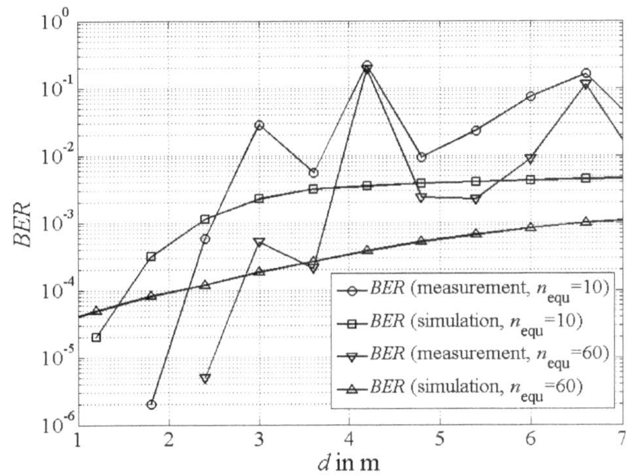

Fig. 9. Simulated and measured BER for a LOS scenario using omnidirectional biconical antennas and different equalizer lengths.

empirical UWB channel model presented in (Ghassemzadeh et al., 2005). For this purpose the parameters representing the environment at the authors' institute have been determined by a measurement campaign (Ott et al., 2009).

The receive voltage $v_{RX}(t)$ is obtained by convolving all impulse responses of all individual components of the transfer chain. The demodulation is accomplished by the MF receiver implementation described in Sect. 2.2.

3.2 Measurements

Measurements in one exemplary test site have been carried out using the SDR IR transceiver described in Sect. 2. The receive signal is acquired by a digitizing oscilloscope DSO81004A from Agilent, which can record up to 1 MSymbols. For the demodulation and the evaluation of the BER the same receiver program is used as in the simulation.

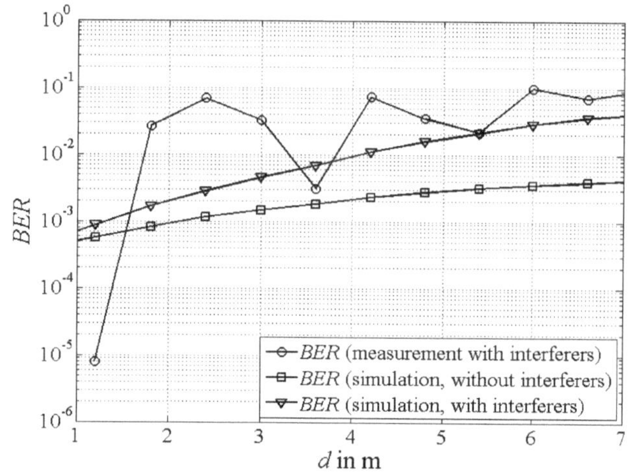

Fig. 10. Comparison of the measured and simulated BER for a LOS scenario with and without interferers using omnidirectional biconical antennas.

In Fig. 7 the measured BER achieved with two directive Vivaldi antennas pointing straight from the transmitter to the receiver is presented. The measurements are compared to simulation results obtained for an ideal LOS channel. It can be seen that the optional LNA can improve the BER significantly, whereas the imperfections of the channel are negligible.

Furthermore different types of highly directive antennas are used to measure the BER. The antennas can be divided in two classes: On the one hand a logarithmic periodic dipole antenna (LPDA) which suffers from dispersion and on the other hand antennas for impulse radio applications like the TEM horn or the dielectric rod antenna presented by (Blech and Eibert, 2007), which all exhibit an excellent transient behavior, are considered. As depicted in Fig. 8 a low BER is achieved with all types of antennas which means that the influence of the antennas can be compensated by the equalizer of the correlation receiver.

Figure 9 shows a LOS scenario of two omnidirectional biconical antennas which are more sensitive to multipath propagation. In this comparison the LNA has always been used and the simulation results showing the average BER have been computed using the channel model of (Ghassemzadeh et al., 2005). The number of taps n_{equ} of the FIR equalizer h_{equ} is varied. As depicted in Fig. 9 a longer filter can slightly improve the BER as inter symbol interference (ISI) can be compensated in addition.

Comparing the measured BER in Fig. 10 in an commercial building with an ideal simulation, strong deviations can be seen. By considering the influence of the Wi-Fi network in the system simulation a much better correspondence to the measurement is achieved. Thus narrowband interferers have an impact on the signal quality but they do not completely corrupt the UWB signal.

4 Conclusions

A prototype of a flexible subsampling UWB software defined IR transceiver has been presented. It can be reconfigured and used for the analysis of the performance of various IR modulation schemes such as e.g. QPSK, which is investigated in this paper. The design uses first order BP sampling avoiding any non-ideal mixer stages. As numerical system simulations and practical measurements of the BER in an exemplary environment show, the presented IR concept is a real alternative to OFDM systems offering the speed and radio link quality demanded by modern wireless UWB communications.

References

Avnet, Inc.: Xilinx Virtex-5 LX50 Development Kit, http://www.avnet.com, 2006.

Batra, A., and 78 co-authors: IEEE Multi-band OFDM Physical Layer Proposal for IEEE 802.15 Task Group 3a, 2004.

Blech, M. D. and Eibert, T. F. : A Dipole Excited Ultrawideband Dielectric Rod Antenna With Reflector, IEEE Transactions on Antennas and Propagation, 55, 1948–1954, 2007.

Blech, M. D., Geier, D., and Eibert, T. F.: Concept of an UWB Impulse Radio B-/QPSK Transmitter Based on Standard Logic Components, in: Proc. IEEE ICUWB 2008, pp. 225–228, Hannover, Germany, 2008.

Coulson, A. J., Vaughan, R. G., and Poletti, M. A.: Frequency-Shifting Using Bandpass Sampling, IEEE Transactions on Signal Processing, 42, 1556–1559, 1994.

Chen, S.-W. M. and Brodersen, R. W.: A Subsampling Radio Architecture for Ultrawideband Communications, IEEE Transactions on Signal Processing, 55, 5018–5031, 2007.

Farr, E. G. and Baum, C. E.: Time Domain Characterization of Antennas with TEM Feeds, http://www.farr-research.com/Papers/ssn426.pdf, 1998.

Fisher, R., Kohno, R., McLaughlin, M., and Welborn, M.: DS-UWB Physical Layer Submission to 802.15 Task Group 3a, 2004.

Ghassemzadeh, S. S., Greenstein, L. J., Sveinsson, T., Kavcic, A., and Tarokh, V.: UWB Delay Profile Models for Residential and Commercial Indoor Environments, IEEE Transactions on Vehicular Technology, 54, 1235–1243, 2005.

Keller, C. M., Burkhart, J. M., and Phuong, T. T.: Ultra-Wideband Direct Sampling Receiver, in: Proc. IEEE ICUWB 2007, pp. 387–392, Singapore, Singapore, 2007.

Maxim Integrated Products, Inc.: 12-Bit, 2.3Gsps, Multi-Nyquist DAC with Selectable Frequency Response, http://www.maxim-ic.com, 2006.

Mini-Circuits: Coaxial Bandpass Filter VBFZ-2575+, http://www.mini-circuits.com, 2008.

Nam, H., Yun, T.-S., Kim, K.-B., Yoon, K.-C., and Lee, J.-C.: Ku-Band Transition Between Microstrip and Substrate Integrated Waveguide (SIW), in: Proc. IEEE APMC 2005, vol. 1, Suzhou, China, 2005.

Ott, A. T., Blech, M. D., Kraemer, M., and Eibert, T. F.: Enhanced Parameter Determination of an Empirical UWB Channel Model for European Building Standards, in: Proc. GeMiC 2009, Munich, Germany, 2009.

Preis, D.: Least-Squares Time-Domain Deconvolution for Transversal-Filter Equalisers, IEEE Electronics Lett., 13, 356–357, 1977.

Proakis, J. G.: Digital Communications, McGraw-Hill, New York, NY, 4th edn., 2001.

Reed, J. H.: An Introduction to Ultra Wideband Communication Systems, Prentice Hall, Upper Saddle River, NJ, 2005.

Time Domain Corporation: PulsOn P210 Evaluation Kit, http://www.timedomain.com, 2006.

Vaughan, R. G., Scott, N. L., and White, D. R.: The Theory of Bandpass Sampling, IEEE Transactions on Signal Processing, 39, 1973–1984, 1991.

Modeling of field singularities at dielectric edges using grid based methods

C. Classen[1], **E. Gjonaj**[2], **U. Römer**[2], **R. Schuhmann**[1], **and T. Weiland**[2]
[1]University of Paderborn, Fachgebiet Theoretische Elektrotechnik, Germany
[2]TU Darmstadt, Institut Theorie Elektromagnetischer Felder, Germany

Abstract. Electric field singularities at sharp metallic edges or at a dielectric contact line can be described analytically by asymptotic expressions. The a priori known form of the field distribution in the vicinity of these edges can be used to construct numerical methods with improved accuracy. This contribution focuses on a modified Finite Integration Technique and on a Discontinuous Galerkin Method with singular approximation functions. Both methods are able to handle field singularities at perfectly electric conducting as well as at dielectric edges. The numerical accuracy of these methods is investigated in a number of simulation examples including static and dynamic field problems.

1 Introduction

One of the fundamental principles of electrodynamics states that the electromagnetic field energy within a finite domain is finite. This remains valid even when the electromagnetic field within the domain becomes singular as may be the case at perfectly conducting edges and at dielectric contact lines. Electromagnetic field singularities are restricted, according to the finite energy principle, to be no stronger than $\rho^{-1+\chi}$, where $\chi > 0$ and ρ is the distance from the edge. Asymptotic expressions for the singular fields at a perfectly conducting edge have been early presented in the literature (Meixner, 1972; Hurd, 1976). More than a decade later (Olyslager, 1994), singular field solutions for the general case of a contact edge between a number of dielectric and/or magnetic materials were found.

The idea of using the a priori known asymptotic behavior of singular fields in numerical simulations to improve numerical accuracy appears natural. This approach has been proposed by several authors in the context of different discretization methods. In Mur (1981) singular correction terms are incorporated into the Finite Difference Time Domain (FDTD) method for the time domain modeling of high frequency problems, see also Beilenhoff and Heinrich (1993), Przybyszewski and Mrozowski (1998). In the context of Finite Element Methods (FEM), several approaches based on the use of specialized scalar and vector basis functions incorporating the singular field behavior have been proposed (Webb, 1988; Graglia and Lombardi, 2004). Recent developments include the Extended Finite Element Method (XFEM) (Moës et al., 1999) and the Partition of Unity Finite Element Method (PUFEM) (Melenk and Babuška, 1996).

The main difficulty with most of the discretization methods using singularity correction techniques is the increased numerical and implementation complexity. This is, e.g., the case for the FEM where specialized singular functions, fullfilling global continuity conditions, must be used in the approximation. The continuity condition imposes an important constraint which limits the flexibility and, as indicated by numerical results (Chahine et al., 2006), the accuracy of the method. The application of this approach for geometrically complicated problems (e.g. for sharp edges interfacing at several dielectric and perfectly conducting domains) is difficult. Furthermore, from the numerical point of view, it may be advantageous to apply a standard discretization method which includes a simple to implement (probably less accurate but numerically more efficient) technique for singular field correction.

In this contribution, two new numerical approaches to handle field singularities are proposed. The first one consists in a modification of the Finite Integration Technique (FIT) (Weiland, 1977) for right angle corner edges in Cartesian grid problems. The second approach uses a flexible singular basis approximation on unstructured grids based on the Discontinuous Galerkin (DG) method. The numerical accuracy of these approaches is investigated for a number of singular field problems with different types of field singularities.

Correspondence to: C. Classen
(classen@tet.upb.de)

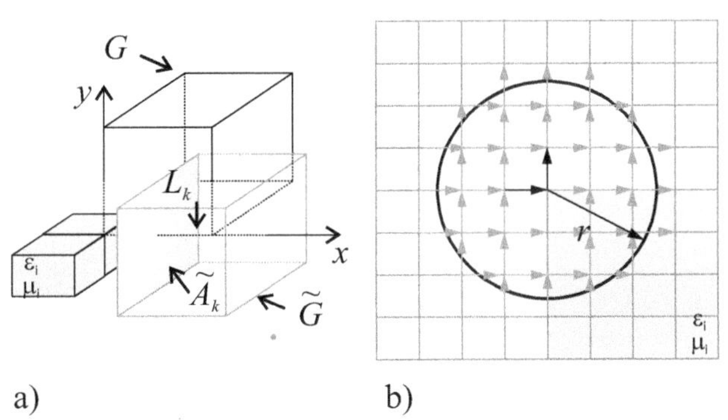

Fig. 1. (a) FIT Scheme: The grid flux $\widehat{\widehat{d}}_i$ is defined on the dual Grid $\widetilde{G}$, the grid voltage $\widehat{e}_i$ corresponds to the primary grid G. **(b)** Correction area: All grid edges within a radius r are corrected. The corrected edges are marked with an arrow.

2 Singular field solutions

In the direct vicinity of edges of perfectly conducting – or dielectric wedges, the asymptotic form of the electromagnetic field solution is given by

$$E_\rho \approx a_0(\varphi)\rho^{t-1};\ E_\varphi \approx b_0(\varphi)\rho^{t-1};\ E_z \approx c_0(\varphi)\rho^{t-1} \quad (1)$$

$$H_\rho \approx \alpha_0(\varphi)\rho^{t-1};\ H_\varphi \approx \beta_0(\varphi)\rho^{t-1};\ H_z \approx \gamma_0(\varphi)\rho^{t-1}. \quad (2)$$

The singularity index t obeys the edge condition and has to be in the range $t \in (0,1)$. It is possible to determine t exactly. A simple analytical procedure to compute t in the general case of a contact edge interfacing to a number of dielectric, magnetic and perfectly conducting materials can be found in (Olyslager, 1994).

3 FIT with singularity correction

The numerical field unknowns in the FIT formulation for electrostatics problems are given by the electric grid voltages $\widehat{\mathbf{e}}$ and dielectric grid fluxes $\widehat{\widehat{\mathbf{d}}}$ defined by

$$\widehat{e} = \int_L \boldsymbol{E}\, ds \quad \text{and} \quad \widehat{\widehat{d}} = \int\int_{\widetilde{A}} \boldsymbol{D}\, dA, \quad (3)$$

respectively. At the discrete grid level, the relation between these quantities needs to be approximated by the matrix equation

$$\widehat{\widehat{\mathbf{d}}} = \mathbf{M}_{\epsilon;\text{FIT}}\widehat{\mathbf{e}}, \quad (4)$$

where $\mathbf{M}_{\epsilon;\text{FIT}}$ is the permittivity matrix of the FIT formulation. In the case of staggered Cartesian grids (see Fig. 1a), a typical approximation for $\mathbf{M}_{\epsilon;\text{FIT}}$ consists in a diagonal matrix with entries

$$\left(M_{\epsilon;\text{FIT}}\right)_k := \frac{\Delta\widetilde{y}\Delta\widetilde{z}}{\Delta x}\epsilon, \quad (5)$$

corresponding to the one-to-one relation between the k-th voltage $\widehat{\mathbf{e}}$ and the k-th flux $\widehat{\widehat{\mathbf{d}}}$ on the grid. In Eq. (5), $\Delta\widetilde{y}$ and $\Delta\widetilde{z}$ are the dual grid lengths in y- and z-directions (face $\widetilde{A}$), respectively, and Δx is the primary grid length in x-direction. In the case of singular field problems, it has been early realized that approximation (5) leads to an extremely slow numerical convergence. The overall numerical accuracy, even at grid points far away from sharp metallic or dielectric edges is drastically reduced compared to regular field problems. A new variant of a singular field correction approach has recently been proposed in the context of FIT (Classen et al., 2010). Hereby, the idea is to derive a modified permittivity matrix by using the singular field behavior (1) in both integral expressions (3). Thus, instead of using the purely geometrical approximation (5), the modified coefficients in the material matrix (for each dual pair of edges and faces in the grid) become

$$\epsilon_{a_k} = \frac{\widehat{\widehat{d}}_{a_k}}{\widehat{e}_{a_k}} = \epsilon\Delta\widetilde{z}\frac{\int_{\Delta\widetilde{y}} E_x(x=\frac{\Delta x}{2}, y)dy}{\int_{\Delta x} E_x(x, y=0)dx} = \left(M_{\epsilon;\text{FIT}}\right)_k K_k, \quad (6)$$

where the correction factor K_k involves a numerical integration of the asymptotic expressions (1). Note that in this approximation, the analytically known singularity index, t, is used depending on geometry and on the electrical properties of the materials adjacent to the edge. Furthermore, this correction can be applied also at grid points not lying on the singular edge but which are sufficiently close to be influenced by the field singularity. The approach taken in this work is to apply the correction (6) within a small sphere surrounding the singular point (see Fig. 1b). The radius r of this sphere represents a free parameter of the formulation. The permittivity matrix $\mathbf{M}_\epsilon$ with singularity correction can be inserted as usual in the discrete FIT equations, e.g., for electrostatics as

$$\widetilde{\mathbf{S}}_{\text{FIT}}\mathbf{M}_{\epsilon;\text{FIT}}(\widetilde{\mathbf{S}}_{\text{FIT}})^T\mathbf{v} = \mathbf{q}, \quad (7)$$

where $\widetilde{\mathbf{S}}$ is the source operator, $\mathbf{q}$ are the volume charges and $\mathbf{v}$ the nodal potentials. Note, however, that the same approach can be used to derive a corrected permeability matrix for magnetic field and high frequency problems.

4 DG with singular basis functions

The local discontinuous Galerkin method is chosen in this work because of its flexibility in the choice of approximation spaces. Following the derivation by Arnold et al. (2002), in the electrostatic case, the potential V and the electric field $\boldsymbol{E}$ are approximated within every element K of the mesh by local approximation spaces $P(K)$ and $\Sigma(K)$, respectively. The standard choice is $P(K) = \mathcal{P}_p$; the space of polynomials of degree at most p, and $\Sigma(K) = (\mathcal{P}_p)^2$. Now, let $V_s(\rho,\varphi) = \rho^t\Psi(\varphi)$ denote a function corresponding to the asymptotic electric potential solution at the singular point. Although being finite at every point, V_s is referred to as singular function due to its low regularity. The approximation

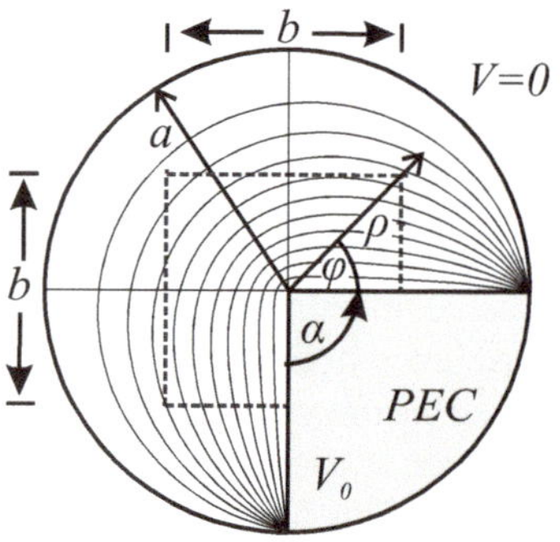

Fig. 2. Simulation setup for a 2-D electrostatics problem with PEC edge singularity. The exact potential solution for this problem is given by the series (13).

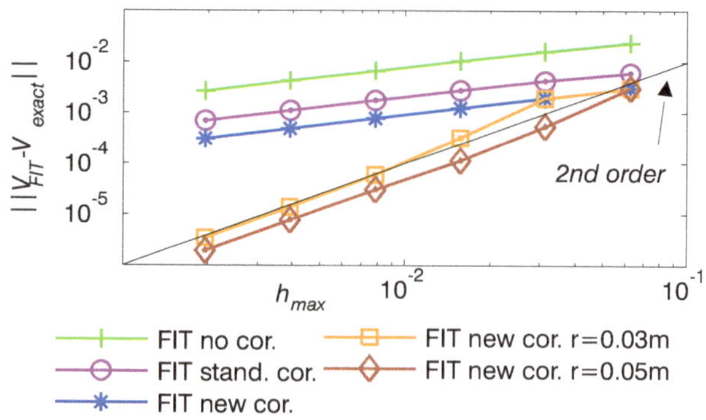

Fig. 3. Maximum electrical potential error for the PEC edge problem using FIT.

spaces for all elements lying completely or partially inside a region of radius r around the singular point can be easily enriched with additional basis functions of the type V_s. Let $< V_s >$ denote the space of scalar multiples of V_s, then for these elements the modified local approximation spaces are given by

$$\tilde{P}(K) = \mathcal{P}_p(K) + < V_s\Big|_K >, \tag{8}$$

$$\tilde{\Sigma}(K) = (\mathcal{P}_p(K) + < \partial_x V_s\Big|_K >) \times (\mathcal{P}_p(K) + < \partial_y V_s\Big|_K >). \tag{9}$$

Note that since the DG approximation is inherently discontinuous, the singular basis functions can be simply defined according to the asymptotic field expressions (1). There is no need to adapt these functions to a particular mesh in order to fulfill the continuity condition as is the case for the FEM. This flexibility represents also the main advantage of the method for the solution of singular field problems.

Omitting details on the involved finite element spaces V_h and Σ_h, the numerical fluxes are defined as

$$\hat{v}_K = \{v_h\}, \tag{10}$$

$$\hat{\boldsymbol{\tau}}_K = \{\boldsymbol{\tau}_h\} - C_{11}[\![v_h]\!]. \tag{11}$$

In Eqs. (10), (11), $v_h \in V_h$ and $\boldsymbol{\tau}_h \in \Sigma_h$ approximate scalar and vector quantities, respectively. Furthermore, $\{\cdot\}$ denotes the mean value and $[\![\cdot]\!]$ the jump whereas C_{11} is a suitably chosen stabilization parameter, see also Castillo et al. (2000).

For the numerical integration of terms involving singularities, higher order (or even adaptive) quadrature rules should be used. Using this approximation procedure, the discrete DG equations can be obtained as a matrix equation which is formally similar to Eq. (7):

$$(\mathbf{S}_{\mathrm{DG}}\mathbf{M}_{\epsilon;\mathrm{DG}}(\mathbf{S}_{\mathrm{DG}})^T + \mathbf{C}_{\mathrm{DG}})\mathbf{v} = \mathbf{q}, \tag{12}$$

where $\mathbf{S}_{\mathrm{DG}}$, $\mathbf{M}_{\epsilon;\mathrm{DG}}$ and $\mathbf{v}$ are the DG counterparts of the source, the permittivity matrix and the potential, respectively, and $\mathbf{C}_{\mathrm{DG}}$ is a stabilization matrix.

5 Numerical results

The FIT edge correction approach and the DG with singular basis functions are validated for a number of simulation setups.

5.1 PEC corner singularity

First tests are performed using a 2-D electrostatics problem containing a sharp perfectly conducting edge. The parameters in Fig. 2 are chosen as: $a = 1$ m, $b = 0.5$ m, $V_0 = 1$ V, $\alpha = \pi/2$ and the boundary values at the surrounding box with edge length b are imprinted using the potential:

$$V(\rho,\varphi) = V_0 + \textstyle\sum_{n=0}^{\infty} E_n \rho^{\frac{\pi(2n+1)}{\beta}} \sin\left(\frac{\pi(2n+1)}{\beta}\varphi\right). \tag{13}$$

Equation (13) represents the exact solution of the problem with $\beta = 2\pi - \alpha$, and E_n given by

$$E_n = \frac{-4V_0}{a^{\frac{\pi(2n+1)}{\beta}}\pi(2n+1)}.$$

The singularity index t and the functions given in Eqs. (1) and (2), can be derived using the general formalism by Olyslager or – in this special case – directly by the derivative of Eq. (13). The problem is discretized with rectangles for the FIT and triangles for the DG method, respectively.

The FIT results for different values of the parameter r are plotted in Fig. 3. The maximum norm, comparing numerical results to analytic results at the grid points, is plotted over the mesh step size $h_{\max}$. The green curve (+) shows the results of standard FIT without edge correction while the purple curve (○) represents the results using a standard edge correction scheme as given in Beilenhoff and Heinrich (1993). Obviously the total error is decreased, although the order of convergence is not improved. The blue curve ($*$) represents the new edge correction approach which is only applied to the grid edges directly located at the singular point (as it is also done in the standard correction scheme). In this case, the results can be further improved. It can be clearly observed in these cases that no second order convergence can be achieved

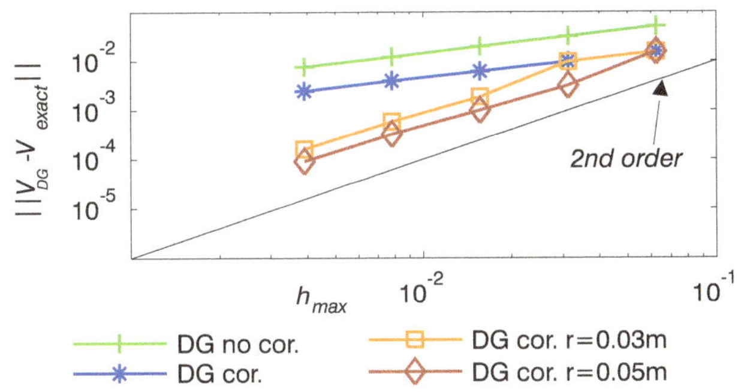

Fig. 4. Maximum electrical potential error for the PEC edge problem using DG.

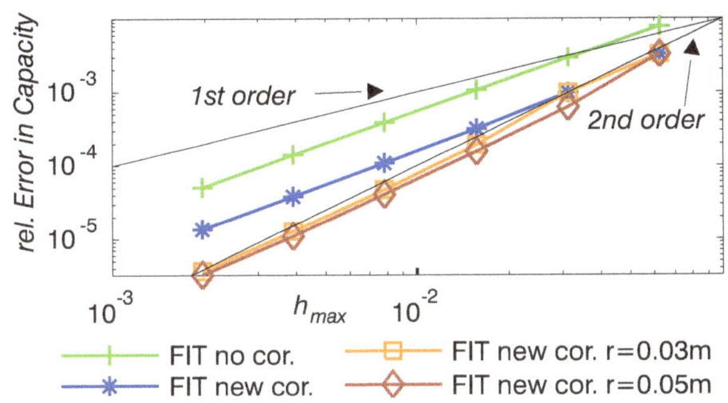

Fig. 5. Numerical error (electrical capacity) for the dielectric edge problem using FIT.

as it would be the case for a regular field problem. On the other hand, correcting all grid edges lying at least partially inside a fixed area with radius r (see Fig. 1b), second order convergence can be restored. As observed in Fig. 3, in order for this to occur, the edge length has to become smaller than the parameter r.

The results of the DG method with singular basis functions and first order polynomials are depicted in Fig. 4. Also here, a fixed area of enrichment with singular basis functions is needed to obtain a higher order of convergence. Using singular basis functions only in elements having a common point with the singularity (blue curve) ($*$), does not improve the order of convergence, see also Laborde et al. (2005).

5.2 Dielectric edge singularity

The setup consists in a dielectric loaded capacitor (see Fig. 1b). The total extension of the domain is 0.5 m in y- and x-direction. At the top and bottom of the domain Neumann boundary conditions are imposed; on the left and right Dirichlet boundary conditions with the fixed potentials $\pm 1\,V$ are applied. The dielectric inset has the permittivity $\epsilon_r = 10$. In this case the singularity is given by $V_s(\rho,\varphi) = \rho^t \Psi(\varphi)$, with $t \approx 0.73$.

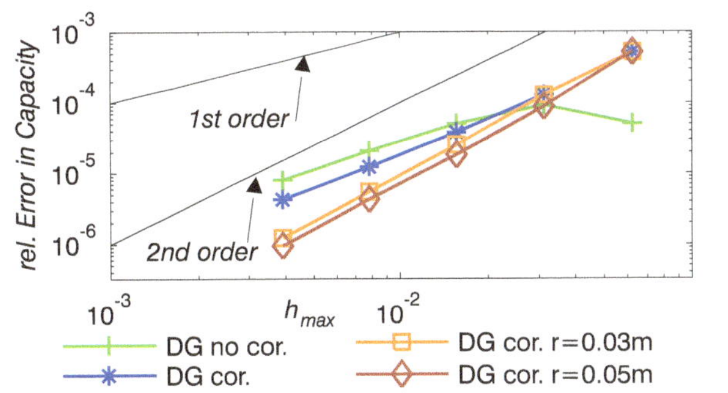

Fig. 6. Numerical error (electrical capacity) for the dielectric edge problem using DG.

It should be mentioned that both methods necessitate a slight modification when different materials surrounding the edge are present. The angular part of the singular function $\Psi(\varphi)$ can be written in the form

$$\Psi(\varphi) = C\sin(t\varphi) + \cos(t\varphi), \qquad (14)$$

where the constant C, depending on the permittivities, the angular material distribution and the singularity index is different in each material. Therefore the singularity correction and the singular approximation functions have to be adapted to each material, refer to Olyslager (1994) for details.

We compare the numerically computed electrical capacity to the one calculated using a very fine mesh and a higher order FEM simulation. The FIT results are presented in Fig. 5; the DG results for first order polynomials in Fig. 6. In both cases, the convergence behavior concerning mesh refinement is identical to the previous example. These results indicate that the described corrections are well suited for treating dielectric type singularities just as in the PEC case.

5.3 Dielectric contact singularity

A more complicated setup concerns singularities arising at the contact line between three or more dielectric materials. To avoid additional effects from a non-conformal geometry discretization in a rectangular FIT grid, this setup is investigated only for the DG method on a triangular grid. The case of three dielectric wedges, with relative permittivities $\epsilon_1 = 1$, $\epsilon_2 = 80$ and $\epsilon_3 = 4$ is depicted in Fig. 7. For illustration purposes, also an exemplary triangular grid as well as the region used for DG singular basis enrichment are shown. The singularity is given by $V_s(\rho,\varphi) = \rho^t \Psi(\varphi)$, with $t \approx 0.61$, corresponding to a contact angle $\alpha = 120°$. The boundary conditions are identical to the previous example and $a = 0.5$ m. The reference result is obtained by a 2-D FEM calculation using a heavily refined mesh in the vicinity of the contact line. Also in this case, numerical results indicate that the error in the electric potential converges at second order when the DG approach with singular basis functions (within a fixed region) and first order polynomials is used (see Fig. 8).

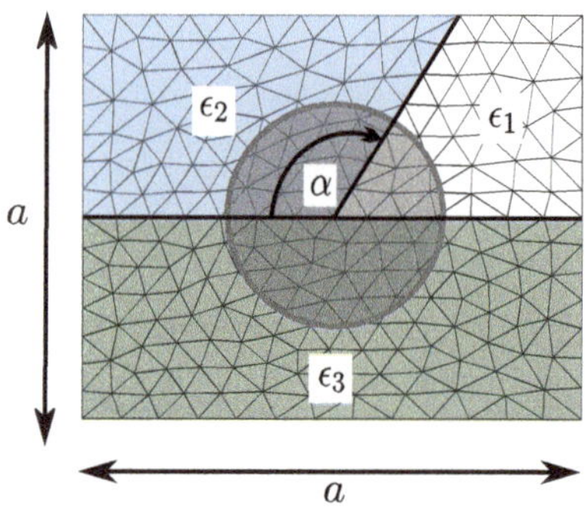

Fig. 7. Domain triangulation and radius of influence for the DG method with singular basis functions.

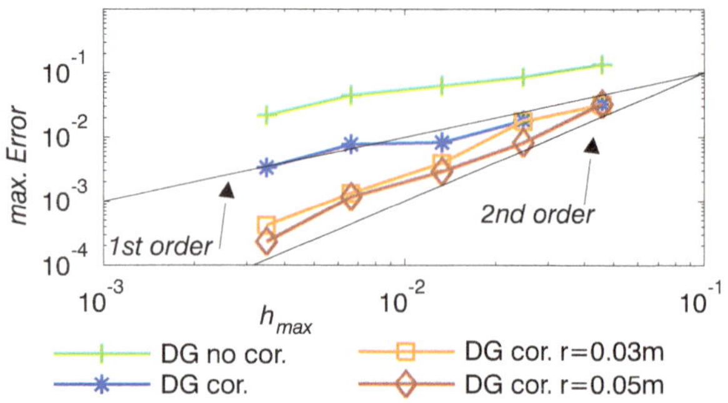

Fig. 8. Numerical error for the triple dielectric edge problem using DG.

5.4 3-D eigenvalue problem with reentrant corner

The last example is the solution of the curl-curl eigenvalue problem for a cavity with reentrant corner. The structure is a hollow, PEC bounded rectangular box (see Fig. 9) ($x = 0..1$ m, $y = 0..0.7$ m, $z = 0..0.8$ m) in which a PEC rectangular box is inserted ($x = 0..0.5$ m, $y = 0..0.35$ m, $z = 0..0.8$ m). In this setup only the first with the singular point corresponding grid edges are corrected. The implementation of an influence region for three-dimensional problems is part of current work. However, in this case both material matrices ($\mathbf{M}_{\mu;\mathrm{FIT}}, \mathbf{M}_{\epsilon;\mathrm{FIT}}$) have to be corrected to solve the curl curl eigenvalue equation:

$$\mathbf{M}_{\epsilon;\mathrm{FIT}}^{-1}(\mathbf{C}_{\mathrm{FIT}})^T\mathbf{M}_{\mu;\mathrm{FIT}}^{-1}\mathbf{C}_{\mathrm{FIT}}\widehat{\mathbf{e}} = \omega^2\widehat{\mathbf{e}}, \tag{15}$$

where $\mathbf{C}_{\mathrm{FIT}}$ is the discrete curl matrix of FIT. The convergence of the first resonance frequency with respect to the grid size is shown in Fig. 10. As observed in the previous examples the standard correction technique (◯) and the new edge correction ($*$) improve the accuracy, meanwhile the order of convergence remains unchanged.

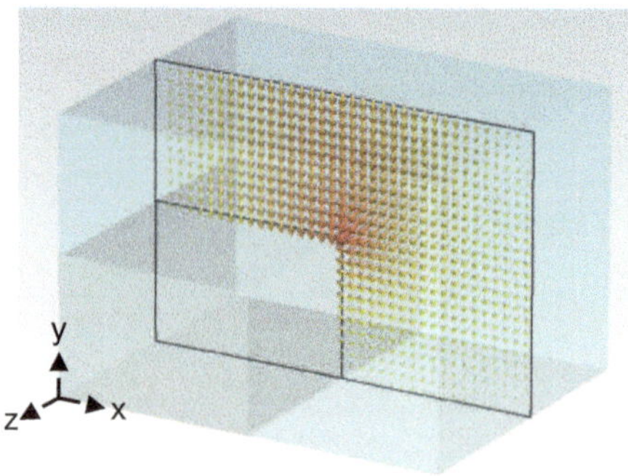

Fig. 9. Arrowplot of the electric field distribution for the eigenvalue problem.

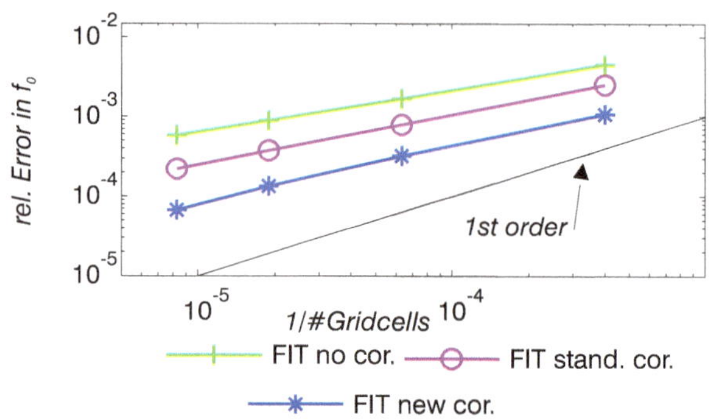

Fig. 10. Convergence of the resonance frequency for the singular eigenvalue problem using FIT.

6 Conclusions

Two numerical approaches for the simulation of singular field problems have been presented. The first consists in a special modification of the material matrices of FIT, whereas the second represents a DG method with singular basis functions. Both methods lead to an immense improvement in numerical accuracy in the case of PEC as well as for dielectric edge singularities. Numerical simulations for a number of setups with typical electromagnetic field singularities show, in particular, that the optimal convergence order for regular field problems can be fully recovered when the respective singular corrections are applied within a small but finite region of influence surrounding the singularity.

Acknowledgements. One author (Christoph Classen) wishes to acknowledge the assistance and support of the Deutsche Forschungsgemeinschaft: GRK 1464.

References

Arnold, D. N., Brezzi, F., Cockburn, B., and Marini, L. D.: Unified Analysis of Discontinuous Galerkin Methods for Elliptic Problems, SIAM Journal on Numerical Analysis, 39, 1749–1779, 2002.

Beilenhoff, K. and Heinrich, W.: Treatment of field singularities in the finite-difference approximation, Microwave Symposium Digest, 1993., IEEE MTT-S International, 2, 979–982, 1993.

Castillo, P., Cockburn, B., Perugia, I., and Schötzau, D.: An A Priori Error Analysis of the Local Discontinuous Galerkin Method for Elliptic Problems, SIAM Journal on Numerical Analysis, 38, 1676–1706, 2000.

Chahine, E., Laborde, P., and Renard, Y.: Crack tip enrichment in the XFEM method using a cut-off function, Int. J. Numer. Meth. Engng., 75, 629–646, 2006.

Classen, C., Bandlow, B., Schuhmann, R., and Tsukerman, I.: FIT & FLAME for Sharp Edges in Electrostatics, in: EMTS 2010 – International Symposium on Electromagnetic Theory, Berlin, Germany, 2010.

Graglia, R. D. and Lombardi, G.: Singular Higher Order Complete Vector Bases for Finite Methods, Antennas and Propagation, IEEE Transactions on, 52, 1672–1685, 2004.

Hurd, R. A.: The Edge Condition in Electromagnetics, Antennas and Propagation, IEEE Transactions on, 24, 70–73, 1976.

Laborde, P., Pommier, J., Renard, Y., and Salan, M.: High-order extended finite element method for cracked domains, Int. J. Numer. Meth. Engng., 64, 354–381, 2005.

Meixner, J.: The behavior of electromagnetic fields at edges, Antennas and Propagation, IEEE Transactions on, 20, 442–446, 1972.

Melenk, J. M. and Babuška, I.: The partition of unity finite element method: Basic theory and applications, Comput. Methods Appl. Mech. Engrg., 139, 289–314, 1996.

Moës, N., Dolbow, J., and Belytschko, T.: A finite element method for crack growth without remeshing, Int. J. Numer. Meth. Engng., 46, 130–150, 1999.

Mur, G.: The Modeling of Singularities in the Finite-Difference Approximation of the Time-Domain Electromagnetic-Field Equations, Microwave Theory and Techniques, IEEE Transactions on, 29, 1073–1077, 1981.

Olyslager, F.: The behavior of electromagnetic fields at edges in bi-isotropic and bi-anisotropic materials, Antennas and Propagation, IEEE Transactions on, 42, 1392–1397, 1994.

Przybyszewski, P. and Mrozowski, M.: A conductive wedge in Yee's mesh, Microwave and Guided Wave Letters, IEEE, 8, 66–68, 1998.

Webb, J. P.: Finite element analysis of dispersion in waveguides with sharp metal edges, Microwave Theory Tech., IEEE Transactions on, 36, 1819–1824, 1988.

Weiland, T.: A discretization method for the solution of Maxwell's equations for six-component Fields, Electronics and Communications AEÜ, Devices and Fields, 31, 116–120, 1977.

Impact of system parameter selection on radar sensor performance in automotive applications

H.-L. Blöecher[1], M. Andres[2], C. Fischer[1], A. Sailer[2], M. Goppelt[2], and J. Dickmann[1]

[1]Daimler AG, Group Research and Advanced Engineering, 89081 Ulm, Germany
[2]University Ulm, Institute of Microwave Techniques, 89081 Ulm, Germany

Correspondence to: H.-L. Blöcher (hans-ludwig.bloecher@daimler.com)

Abstract. The paper deals with the investigation of relevant boundary conditions to be considered in order to operate 77/79 GHz narrow and ultra wide band automotive radar sensors in the automotive platform and the automotive environment.

1 Introduction

Millimeterwave radar is a vital component of present and future automotive environmental sensing systems. Information given by these systems is essential for the realization of automotive safety- and driver assistance functions. Automotive radar sensors operating in the 77/79 GHz band will, due to frequency regulation constraints, offer the only remaining option to operate at UWB bandwidth (>500 MHz). Moreover, due to the decreasing sensor dimensions with increasing frequency, W-band radar will offer increased angular resolution using antennas with moderate aperture dimensions.

The frequency range 77 GHz to 81 GHz (79 GHz), has been defined by the European Commission in 2004 as the frequency allocation for automotive short range radar systems. The initial intention and objective of this 79 GHz project is to establish and speed up the worldwide harmonised frequency allocation for vehicular radars in that frequency range. The 79 GHz automotive frequency band, once available worldwide, will promote and accelerate the transition from the currently used 24 GHz band, which is an intermediate regulatory solution, to the 79 GHz band.

As a consequence of the 24/79 GHz frequency allocation for UWB automotive radar in Europe by 2005, industry had to show honest efforts to develop and provide a cost efficient technology platform at the 77/79 GHz band. Thus, the German BMBF public funded projects KOKON (2004–2007) and RoCC (2008–2011) were established. Within the BMBF public funded project RoCC – "radar-on-chip for cars", the necessary cost efficient SiGe based technology for the realization of W-band radar is developed. Furthermore, relevant technical constraints to be considered for operation of these sensors are investigated and the consequences with respect to sensor specifications are derived. Within the paper, some very essential constraints are discussed:

- Covered/buried placement and packaging of W band radar within the automotive platform, which leads to the need to design car body parts as radomes.
- Necessary operational bandwidth, derived from relevant applications and used for the realization of adequate range and angular resolution.
- Car to car interference and possible countermeasures, in particular with respect to realization of safety functions and proposed market penetration rates of automotive radar.

Finally, the potential of DBF/DoA approaches to efficiently contribute to the solution of these problems will be discussed.

2 Selection of application driven radar parameters

Due to diverse urban scenarios, radar sensors have to cope with different target types, like pedestrians, motorbikes, cars and trucks. Hence, urban scenarios require highly sophisticated radar sensor performance. High range and angular resolution are essential to detect two closely neighbouring targets with unequal radar cross sections and the same velocity. For example, a pedestrian waiting in between two parked cars and intending to cross the street, requires high range and

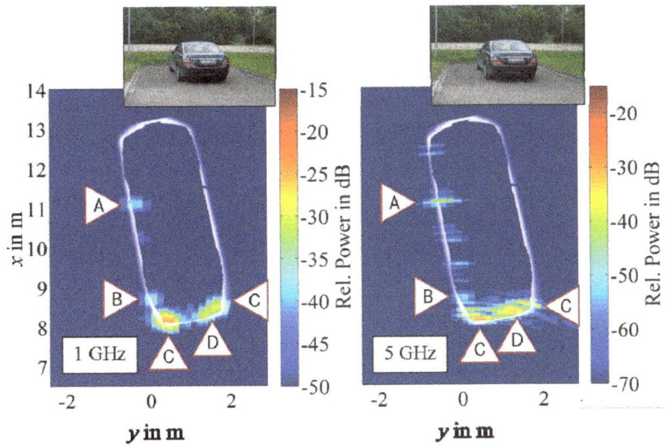

Fig. 1. Measurement result of the contour determination of a car using a bandwidth of 1 GHz (left) and 5 GHz (right) at a carrier frequency of 77 GHz and an angular resolution of 1°.

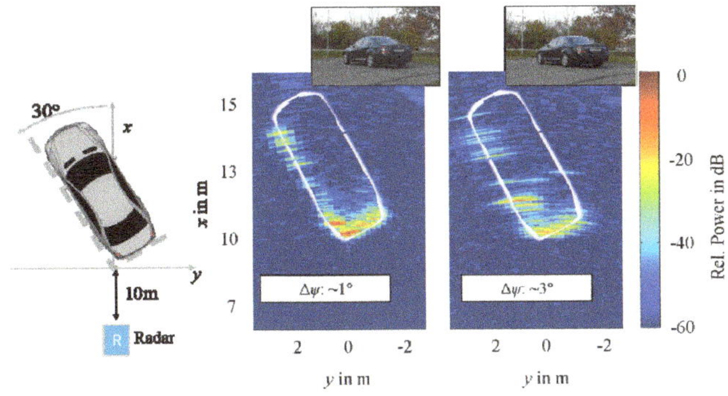

Fig. 2. Measurement result of the contour determination of a car using a bandwidth of 6 GHz and an angular resolution of ~1° (left) and ~3° at a carrier frequency of 77 GHz.

angular resolution of the radar sensor. To estimate the time to collision, the length, width and orientation of the car and the exact position of the pedestrian have to be determined. Having preliminary knowledge of the main locations of the scattering centers enables the evaluation of the situation to be more precise. Strong reflection regions of a car are located at the rearview mirror, transition between running surface and asphalt, light units and the license plate (Andres, 2011).

To extract the contour of a car, the required bandwidth and angular resolution have to be determined. In Fig. 1, a 77 GHz FMCW radar frontend used in (Feil, 2010) with a high directive antenna was attached to a mechanical scanning unit. The radar sensor was equipped with a double folded reflect array antenna with a half power beam width (HPBW) of approximately 1° (two way pattern). Therefore, the angular stepping size of the mechanical scanning unit was adjusted to 1°. The radar sensor was placed behind the car at the origin of the coordinate system at a distance of 8 m. In Fig. 1 left a bandwidth of 1 GHz and in Fig. 1 right a bandwidth of 5 GHz was adjusted to the FMCW radar. Comparing these two images the increase of the bandwidth enables a more precise extraction of the contour of the vehicle. However, the main scattering centers like the left rearview mirror (A), the transition between running surface of the left wheel and asphalt (B), the rear light units (C) and the license plate (D) could be extracted even at the bandwidth of 1 GHz.

The angular resolution of the radar sensor has an impact on the contour determination of the target as well. To analyze this influence two different antenna types were used to scan the same measurement scenario. In Fig. 2 left and right the car was located at a distance of 10 m with a gear angle of 30°. In Fig. 2 left, the previous reflected fold array antenna with a HPBW of 1° was applied on the radar sensor and in Fig. 2 right a lens horn antenna with a HPBW of approximately 3° (two way) was attached. For both measurements a bandwidth of 6 GHz was adjusted. The measurement results show that by reducing the angular resolution capability from 1° to 3°, the uncertainty of the width of the car increases and hence, the determination of orientation is more difficult. Furthermore, the interference between two neighbouring scattering centers increases. This effect can cause a destructive interference and hence a loss of the detectable scattering center.

In conclusion, using a bandwidth larger than 1 GHz, and an angular resolution better than 1°, enables estimation of the contour of a vehicle. This leads to an improved detection of the target, resulting in higher accuracy for the post processed tracking.

3 Lateral resolution and flexible antenna pattern adaption

The number of people living in cities and also the number of cars in the streets is expected to increase significantly within the next years (IAE, 2008). Due to this development the cars and also the sensors used inside the cars need to be adapted to this environment with growing complexity. The requirements for the sensors used in an inner city environment differ largely from those used on highways or motor-ways where the number of cars is small. Also close encounters of similar sensors occur seldom and only for a short time. The inner city scenario is the exact opposite: One encounters many cars and therefore has the necessity to differentiate between these with high reliability. These cars may also be equipped with similar sensors. Therefore interference free coexistence must also be assured.

The challenge to be faced in the inner city scenario is the reliability of the measurements. Resolution of the measurements in range was already covered but also the lateral resolution and accuracy is becoming more and more important. In the environment faced on highways movement is mostly in the range domain. In scenarios with turning cars, tracking of lateral movement also comes into play. In radar sensors lateral accuracy is directly correlated with angular accuracy

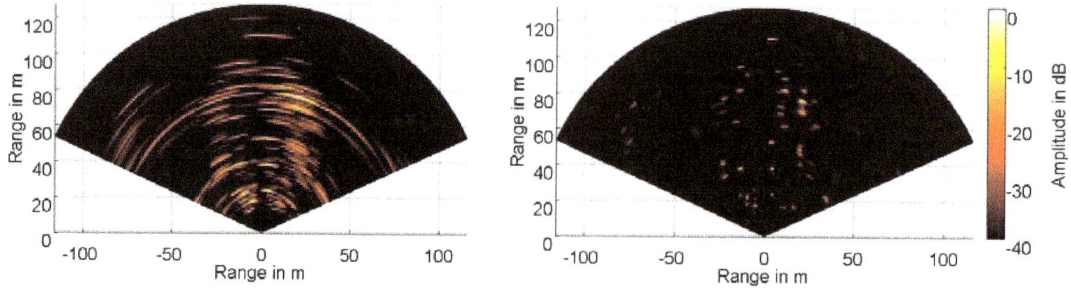

Fig. 3. Radar plot of a scenario processed with traditional beamforming (left) and beamforming using linear prediction with 5 coefficients (right). Scale is the same in both plots.

which depends on the aperture of the antenna. The aperture is proportional to the antenna's size and its operating frequency. As space is very limited in today's cars, the size of the sensor and therefore the antenna size is limited. To increase the resolution of a sensor one can either move to a higher frequency, which is limited by currently available technology, or use Digital Beamforming in combination with super resolution techniques (Feil, 2007). These techniques trade antenna space for computational complexity.

Figure 3 gives an example for the potential of super resolution techniques. In this case the method of linear prediction with five coefficients was applied as described in detail e.g. in Mayer (2008). One can see that with traditional beamforming many scatterers are blurred. By applying additional signal processing this blur can be reduced significantly and contours become more clearly visible. Another possible advantage of this high angular resolution is that interferers may be suppressed with higher selectivity. The necessary steps for this adaptive beamshaping will be the subject of further research.

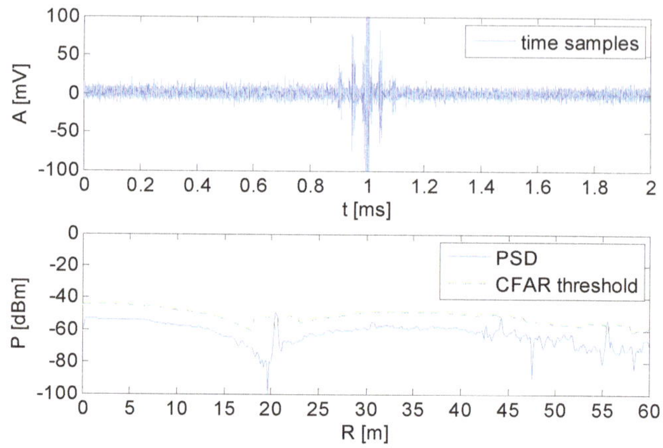

Fig. 4. FMCW radar signal with interference from a CW interferer. Upper diagram: the time samples taken by the analog-to-digital converter. Lower diagram: the signal in the frequency domain (power spectral density) along with the CFAR threshold.

4 Sensor – sensor interference and interference mitigation

Daimler's activities in the RoCC project with respect to sensor interference and interference mitigation aim at understanding interference mechanisms as well as testing and evaluating countermeasures (interference mitigation methods). For this purpose a radar simulator was created, analytic calculations were performed, and measurements with different radar frontends were conducted in the laboratory and with a vehicle.

For example, a vehicle with a radar frontend using FMCW modulation was moved towards a stationary frontend emitting a CW signal. The effect of the interfering CW signal at 77.5 GHz intersecting the FMCW frequency ramp (with a rise time of 2 ms from 77 GHz to 78 GHz) can be seen in Fig. 4 in the time domain (upper diagram) and in the frequency domain (lower diagram). The interference creates a wideband signal at the IF port of the mixer which reduces the sensitivity of the radar (Goppelt, 2011).

One possible way to reduce the effect of the interference is to detect and weight (multiply) the affected samples with the following weighting function:

$$w(i) = \left(1 - 0.5 \cdot \left(1 - \cos\left(\frac{2 \cdot \pi \cdot (i - i_1)}{i_2 - i_1}\right)\right)\right)^4, \quad (1)$$

with

$$i_1 \leq i \leq i_2. \quad (2)$$

It can be seen in Fig. 5 that this countermeasure significantly reduces the interference, resulting in a greatly improved sensitivity. Furthermore, it is easy to implement, since it consists purely of digitally processing the time samples without the need to modify the frontend.

5 Covered platform integration of 77/79 GHz automotive radar sensors

Within the car, short range radar (SRR) sensors are typically mounted invisibly behind painted bumpers or other layered

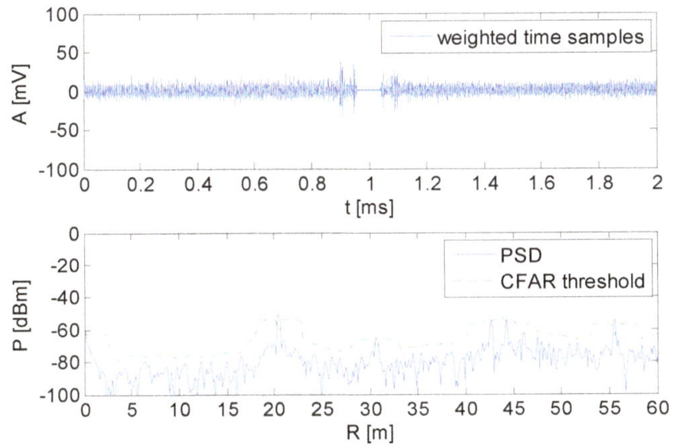

Fig. 5. The FMCW radar signal after applying the countermeasure (weighting of the time samples).

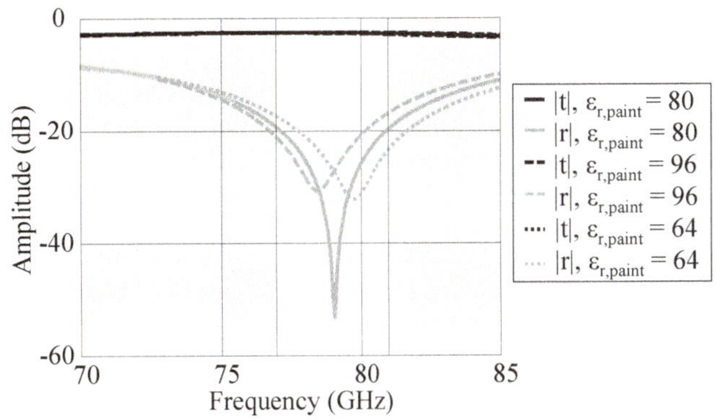

Fig. 6. Simulation of transmission $|t|$ and reflection $|r|$ through a bumper with metallic paint. Variation of the permittivity of the paint $\varepsilon_{r,\mathrm{paint}}$ shifts the reflection minimum.

components. Any cover of an antenna of a radar sensor has to be carefully designed, in order to avoid performance degradation due to transmission losses, reflections, and edge effects. This is a crucial issue concerning sensor systems working in the frequency range from 76 GHz to 81 GHz even more than for lower operating frequencies. Bumpers and other kinds of components mounted on a vehicle's front or rear-end have to be considered as radome structures. For the desired frequency range, their thickness is commonly in the order of not more than a couple of wavelengths.

The OEMs among the project partners, Daimler and BMW, have to consider a wide range of sensor packaging aspects in RoCC. Special attention has to be turned on the impact of the necessary frequency bandwidth around 79 GHz. This parameter is influenced by the electromagnetic characteristics of bulk material and painting (permittivity and loss tangent), manufacturing tolerances, multiple paintings (especially repair paintings) and covering of radomes with water, snow, ice, dust or salt, etc.

In Bloecher (2009a, b), first results concerning sensor integration behind painted bumpers and particular LRR radomes are depicted. Extensive studies of manufacturing tolerances and the composition of substrate materials as well as paint were carried out by Daimler within the project RoCC. As an exemplary result, Fig. 6 shows the transmission $|t|$ and the reflection $|r|$ through a bumper with metallic paint. The simulation was carried out with a multi-layer model implemented in Matlab. For the ideal case of properly chosen substrate thickness $d_{\mathrm{sub}} = 3.44$ mm, the reflection $|r|$ is found to be better than -20 dB for the whole SRR frequency band of 77 to 81 GHz. Introducing a variation of the equivalent permittivity of the paint would shift the desired minimum of $|r|$ to lower or higher frequencies. Nevertheless, the transmission $|t|$ stays approximately constant.

A radome that is robust against variation of the material parameters and thickness of substrate and paint has to provide sufficient transmission bandwidth. As a consequence, variations of dedicated scale would not seriously affect the radar's performance. Considering manufacturing tolerances, varying material combinations and required bandwidth, thickness optimization of the substrate material was found to be a good compromise between RF performance, costs and effort. Nevertheless, this compromise is strongly dependent on specific sensor requirements and the specific car model's needs. In the case where thickness matching does not provide sufficient RF performance, one could apply broad band design methods including frequency selective surface structures or inductive structures on the back side of the bumper as shown in (Pfeiffer, 2009). In conclusion, meeting the desired RF performance always means balancing effort in RF engineering with keeping dimensional accuracy and material composition.

6 Conclusions

Automotive radar is an essential component of present and future automotive environmental sensing systems used for the realization of automotive active safety and comfort functions. Due to regulatory reasons as well as the ability to use the same technology platform for long-, mid- and short range automotive radar, the 77/79 GHz frequency band is the future choice. The development of a low cost and robust SiGe based technology platform has been done within the BMBF public funded projects KOKON and RoCC. Beneath the availability of an adequate technology platform, further technical and operational requirements have to be fulfilled to operate the sensor within the automotive platform and the environment. Within the project's methods, solutions with respect to platform integration, sensor to sensor interference and application driven sensor parameter selection and have been developed and characterized successfully. These results are to be transferred into series development of the radar sensors. Furthermore, DBF/DoA/adaptive beamforming algorithms could potentially be used to obtain improved angular resolution with limited physical antenna dimensions. The flexibility of the adaptive antenna pattern will be used for suppression of interfering signals to keep and improve, for example, the radar sensitivity.

Acknowledgements. The authors wish to thank the German Federal Ministry of Education and Research (BMBF) for the funding of the RoCC project. The authors also wish to thank Stacey Rukezo (currently with Daimler AG) for proofreading the manuscript.

References

Andres, M., Feil, P., Menzel, W., Bloecher, H.-L., and Dickmann, J.: Analysis of automobile scattering center locations by SAR measurements, Proc. of Radar Conference 2011, Kansas City, USA, Mai, 2011.

Bloecher, H.-L., Sailer, A., Rollmann, G., and Dickmann, J.: 79 GHz UWB automotive short range radar – Spectrum allocation and technology trends, Adv. Radio Sci., 7, 61–65, 2009, http://www.adv-radio-sci.net/7/61/2009/.

Bloecher, H.-L., Sailer, A., Andres, M., Goppelt, M., and Dickmann, J.: Trends in Development of SiGe based 76–81 GHz Automotive Radar, IRS 2009 – International Radar Symposium, Hamburg, 2009.

Feil, P., Mayer, W., Menzel, W., and Guehl, R.: Various Cross Range Resolution Techniques Applied to an Eight Channel 77 GHz Sensor, EuRAD 2007, Munich, Germany, 59–62, 2007.

Feil, P., Mayer, W., Kraus, T., Migliaccio, C., and Menzel, W.: Industrial and Security Applications of Imaging Millimeter Wave Sensors, Technisches Messen, July–August, 2010.

Goppelt, M., Blöcher, H.-L., and Menzel, W.: Analytical investigation of mutual interference between automotive FMCW radar sensors, Proc. of GEMIC 2011, Darmstadt, Germany, 2011.

International Energy Agency (IEA): World Energy Outlook 2008, 2008.

Mayer, W.: Abbildender Radarsensor mit sendeseitig geschalteter Gruppenantenne, Dissertation, Ulm, 2008.

Pfeiffer, F. and Biebl, E. M.: Inductive Compensation of High-Permittivity Coatings on Automobile Long-Range Radar Radomes, IEEE T. Microw. Theory., 57, 2627–2632, November, 2009.

Eigenmode analysis of the electromagnetic field scattered by an elliptic cone

M. Kijowski and L. Klinkenbusch

Institute of Electrical and Information Engineering, Christian-Albrechts-Universität zu Kiel, Germany

Abstract. The vector spherical-multipole analysis is applied to determine the scattering of a plane electromagnetic wave by a perfectly electrically conducting (PEC) semi-infinite elliptic cone. From the eigenfunction expansion of the total field in the space outside the elliptic cone, the scattered far field is obtained as a multipole expansion of the free-space type by a single integration over the induced surface currents. As for the evaluation of the free-space-type expansion it is necessary to apply suitable series transformation techniques, a sufficient number of eigenfunctions has to be considered. The eigenvalues of the underlying two-parametric eigenvalue problem with two coupled Lamé equations belong to the Dirichlet- or the Neumann condition and can be arranged as so-called eigenvalue curves. It has been observed that the eigenvalues are in two different domains: In the first one Dirichlet- and Neumann eigenvalues are either nearly coinciding, while in the second one they are strictly separated. The eigenfunctions of the first (coinciding) type look very similar to free-space modes and do not contribute to the scattered field. This observation allows to significantly improve the determination of diffraction coefficients.

1 Introduction

Electromagnetic scattering by a PEC elliptic cone is of practical importance for several reasons. First, the geometry includes a tip, and the related tip-diffraction coefficient could be used to improve asymptotically valid methods like GTD and UTD in describing scattering by complex electrically large systems. Second, as the geometry can be treated mostly analytically using sphero-conal coordinates and the vector spherical-multipole expansion, the obtained results (e.g., for a finite elliptic cone with the degenerations sector and circular cone) can serve as a reference for numerical computations. Scattering by circular and (less often) by elliptic cones have been treated by many authors in the literature. Summaries on the different approaches can be found in Bowman et al. (1987); Klinkenbusch (2007). This paper starts with a description of sphero-conal coordinates which are used to describe the elliptic cone geometry. For an incident plane electromagnetic wave the exact total field outside and the corresponding surface current on the elliptic cone are determined by a spherical-multipole (eigenfunction) expansion based on suitable solutions of the scalar homogeneous Helmholtz equation. The far field is then found as a free-space type spherical-multipole expansion from an integration over the surface current. In sphero-conal coordinates the solution of the scalar Helmholtz equation leads to a two-parametric eigenvalue problem with two coupled Lamé differential equations, that is, the differential equations of the periodic and of the non-periodic Lamé functions. The two separation parameters can be arranged on so-called eigenvalue curves on which the Dirichlet- and Neumann eigenvalues are discretely distributed. It turns out that the numerical determination of some of these eigenvalues is difficult, however, as will be shown exactly these eigenvalues do not significantly contribute to the scattered far field and can be automatically sorted out. Finally, this procedure leads to an improved accuracy in determining the scattering coefficient magnitudes and phases.

Correspondence to: L. Klinkenbusch
(lbk@tf.uni-kiel.de)

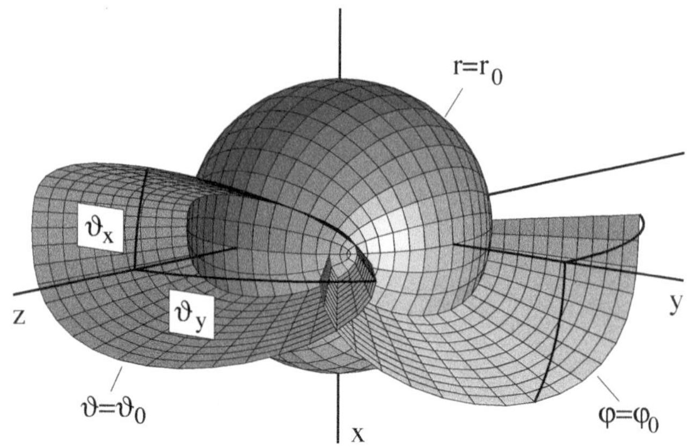

Fig. 1. Sphero-conal coordinate surfaces.

2 Sphero-conal coordinates

Sphero-conal coordinates r, ϑ, φ are related to Cartesian coordinates by (Boersma and Jansen, 1990)

$$x = r\sin\vartheta\cos\varphi$$
$$y = r\sqrt{1-k^2\cos^2\vartheta}\,\sin\varphi \qquad (1)$$
$$z = r\cos\vartheta\sqrt{1-k'^2\sin^2\varphi}$$

The ranges of the values are $0 \le r < \infty$, $0 \le \vartheta \le \pi$, $0 \le \varphi \le 2\pi$ and the ellipticity parameters k and k' satisfy

$$0 \le k,\, k' \le 1; \quad k^2 + k'^2 = 1. \qquad (2)$$

The coordinate surfaces are shown in Fig. 1. The elliptic cone is identical to the coordinate surface $\vartheta = \vartheta_0$ and can be characterized by the half opening angles $\vartheta_x = \vartheta_0$ and $\vartheta_y = \arccos(k\cos\vartheta_0)$. The normalized metric scaling coefficients are given by

$$s_\vartheta = \frac{1}{r}\left|\frac{\partial \boldsymbol{r}}{\partial\vartheta}\right| = \sqrt{\frac{k^2\sin^2\vartheta + k'^2\cos^2\varphi}{1-k^2\cos^2\vartheta}} \qquad (3)$$

$$s_\varphi = \frac{1}{r}\left|\frac{\partial \boldsymbol{r}}{\partial\varphi}\right| = \sqrt{\frac{k^2\sin^2\vartheta + k'^2\cos^2\varphi}{1-k'^2\sin^2\varphi}} \qquad (4)$$

Note that the elliptic cone includes several interesting degenerations: For $k = 1$ the elliptic cone turns into a circular cone while $\vartheta_0 = \pi$ describes a plane angular sector with half-opening angle $\arccos(k)$.

3 Eigenfunction expansion of the total field

In a linear, homogeneous, and isotropic domain outside the elliptic cone the total electromagnetic field can be derived as a multipole (eigenfunction) expansion in sphero-conal coordinates which is based on the corresponding solution of the scalar homogeneous Helmholtz equation

$$\Delta\Psi(\boldsymbol{r}) + \kappa^2\Psi(\boldsymbol{r}) = 0 \qquad (5)$$

where $\kappa = \omega\sqrt{\mu\varepsilon}$ denotes the wave number. A first separation ansatz

$$\Psi(r,\vartheta,\varphi) = R(r)Y(\vartheta,\varphi) \qquad (6)$$

with a first separation constant $\nu(\nu+1)$ leads to the differential equation of the spherical Bessel functions with solutions $z_\nu(\kappa r)$ and to the eigenvalue equation of surface spherical harmonics which are referred to as Lamé products in case of sphero-conal coordinates

$$(\boldsymbol{r}\times\nabla)^2 Y_\nu(\vartheta,\varphi) + \nu(\nu+1)Y_\nu(\vartheta,\varphi) = 0 \qquad (7)$$

For the given problem it is necessary to have solutions which are (at least) 2π – periodic in φ and fulfill canonical boundary conditions at $\vartheta = \vartheta_0$:

$$\text{Dirichlet condition:} \quad Y_\sigma(\vartheta,\varphi)|_{\vartheta=\vartheta_0} = 0 \qquad (8)$$

$$\text{Neumann condition:} \quad \left.\frac{\partial Y_\tau(\vartheta,\varphi)}{\partial\vartheta}\right|_{\vartheta=\vartheta_0} = 0 \qquad (9)$$

A second separation ansatz of the form

$$Y_\nu(\vartheta,\varphi) = \Theta_\nu(\vartheta)\Phi_\nu(\varphi) \qquad (10)$$

with a second separation constant λ yields two ordinary differential equations:

$$\sqrt{1-k'^2\sin^2\varphi}\,\frac{d}{d\varphi}\left(\sqrt{1-k'^2\sin^2\varphi}\,\frac{d\Phi_\nu}{d\varphi}\right) + \left[\lambda - \nu(\nu+1)k'^2\sin^2\varphi\right]\Phi_\nu = 0 \qquad (11)$$

$$\sqrt{1-k^2\cos^2\vartheta}\,\frac{d}{d\vartheta}\left(\sqrt{1-k^2\cos^2\vartheta}\,\frac{d\Theta_\nu}{d\vartheta}\right) + \left[\nu(\nu+1)(1-k^2\cos^2\vartheta) - \lambda\right]\Theta_\nu = 0 \qquad (12)$$

Equation (11) is referred to as the differential equation of the periodic Lamé functions $\Phi_\nu(\varphi)$ while Eq. (12) represents the differential equation of the non-periodic Lamé functions $\Theta_\nu(\vartheta)$. The periodic Lamé functions can be described as infinite Fourier series and the non-periodic Lamé functions are expanded into infinite series using associated Legendre functions of the 1st kind (Boersma and Jansen, 1990). In case of solutions in the free unbounded space the eigenvalues are integers ($\nu = n = 1, 2, 3, ..$), the series become finite, and consequently the solutions turn into periodic and non-periodic Lamé polynomials $\Phi_{nm}(\varphi)$ and $\Theta_{nm}(\vartheta)$, respectively.

For a given value of ν and of the parameter k^2 the corresponding second separation constant λ can be numerically determined. The resulting (ν,λ)-pairs lie on characteristic eigenvalue curves sorted by numbers $m = 0, 1, 2, \ldots$ Any arbitrary pair of eigenvalues (ν,λ) lying on the eigenvalue curves

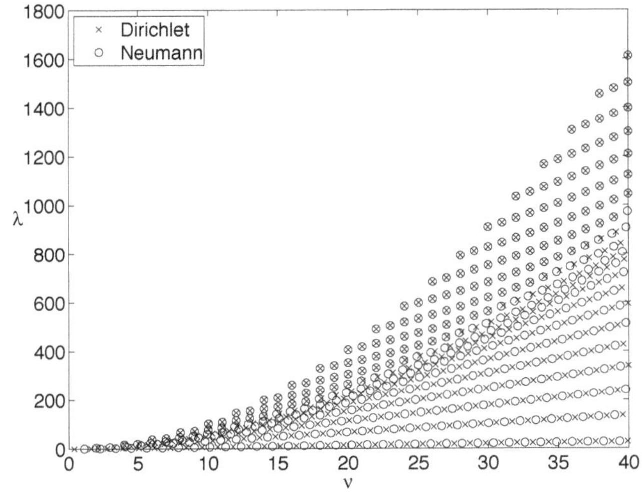

Fig. 2. Eigenvalue curves $\lambda(\nu)$ for $k^2 = 0.5$ with Dirichlet eigenvalues ($\times$) and Neumann eigenvalues ($\circ$).

leads to a valid solution of the eigenvalue equation of the Lamé products Eq. (7). Additional Dirichlet- and Neumann boundary conditions imposed upon the non-periodic Lamé functions Θ_ν at ϑ_0 result in a discrete spectrum of eigenvalue pairs (ν_i, λ_i) $(i = 1,2,3,..)$ lying on the eigenvalue curves. Figure 2 exemplarily shows the eigenvalue curves with discrete Dirichlet- and Neumann eigenvalues. Due to the Sturm-Liouville properties of the Lamé differential equations the discrete Dirichlet- and Neumann eigenvalues strictly must alternate on the eigenvalue curves (Boersma and Jansen, 1990).

Outside the PEC elliptic cone the total electromagnetic field can be expressed in the form of a spherical-multipole (eigenfunction) expansion (Stratton, 1941; Blume and Klinkenbusch, 1999)

$$\boldsymbol{E}^{\text{tot}}(\boldsymbol{r}) = \sum_\sigma a_\sigma \boldsymbol{N}_\sigma(\boldsymbol{r}) + \frac{Z}{j}\sum_\tau b_\tau \boldsymbol{M}_\tau(\boldsymbol{r}) \tag{13}$$

$$\boldsymbol{H}^{\text{tot}}(\boldsymbol{r}) = \frac{j}{Z}\sum_\sigma a_\sigma \boldsymbol{M}_\sigma(\boldsymbol{r}) + \sum_\tau b_\tau \boldsymbol{N}_\tau(\boldsymbol{r}) \tag{14}$$

where the expansion functions which are referred to as the vector spherical-multipole functions can be derived from the elementary solutions of the scalar homogeneous Helmholtz equation $\Psi_\nu(\boldsymbol{r})$ by

$$\begin{aligned}\boldsymbol{M}_\nu(\boldsymbol{r}) &= (\boldsymbol{r}\times\nabla)\Psi_\nu(\boldsymbol{r}) \\ &= z_\nu(\kappa r)\boldsymbol{m}_\nu(\vartheta,\varphi)\end{aligned} \tag{15}$$

$$\begin{aligned}\boldsymbol{N}_\nu(\boldsymbol{r}) &= \frac{1}{\kappa}[\nabla\times(\boldsymbol{r}\times\nabla)]\Psi_\nu(\boldsymbol{r}) \\ &= -\frac{z_\nu(\kappa r)}{\kappa r}n(n+1)Y_\nu(\vartheta,\varphi)\hat{r} \\ &\quad -\frac{1}{\kappa r}\frac{d}{dr}[r z_\nu(\kappa r)]\boldsymbol{n}_\nu(\vartheta,\varphi)\end{aligned} \tag{16}$$

with $\hat{r} = \boldsymbol{r}/r$ denoting the unit vector and $\kappa = \omega\sqrt{\varepsilon_0\mu_0}$ being the wave number in the free space. The transverse vector functions are defined as

$$\boldsymbol{m}_\nu(\vartheta,\varphi) = -\frac{1}{s_\varphi}\frac{\partial Y_\nu(\vartheta,\varphi)}{\partial\varphi}\hat{\vartheta} + \frac{1}{s_\vartheta}\frac{\partial Y_\nu(\vartheta,\varphi)}{\partial\vartheta}\hat{\varphi} \tag{17}$$

$$\boldsymbol{n}_\nu(\vartheta,\varphi) = \frac{1}{s_\vartheta}\frac{\partial Y_\nu(\vartheta,\varphi)}{\partial\varphi}\hat{\vartheta} + \frac{1}{s_\varphi}\frac{\partial Y_\nu(\vartheta,\varphi)}{\partial\vartheta}\hat{\varphi} \tag{18}$$

and the electric and magnetic multipole amplitudes are given by a_σ and b_τ, respectively. Note that the indices σ and τ symbolize the Dirichlet- and Neumann conditions as defined in Eqs. (8) and (9) to ensure the vanishing of the tangential electric field on the surface of the PEC elliptic cone. The incident plane wave is realized by locating a Hertzian dipole at infinity and multiplying the resulting field by an appropriate factor (Blume and Klinkenbusch, 1999). For a plane wave with amplitude E_0 incident from $(\theta^{\text{inc}}, \phi^{\text{inc}})$ and electrically polarized in the direction $\hat{C}$, the multipole amplitudes of the total field are found as

$$\alpha_\sigma = 4\pi E_0 \frac{e^{j(\sigma+1)\frac{\pi}{2}}}{\sigma(\sigma+1)}[\boldsymbol{n}_\sigma\cdot\hat{C}] \tag{19}$$

$$\beta_\tau = 4\pi\frac{E_0}{Z}\frac{e^{j(\tau+1)\frac{\pi}{2}}}{\tau(\tau+1)}[\boldsymbol{m}_\tau\cdot\hat{C}] \tag{20}$$

where $Z = \sqrt{\mu_0/\varepsilon_0}$ is the intrinsic impedance of the free space.

4 Spherical-multipole expansion of the scattered field

The scattered field is determined from the surface current $\boldsymbol{J}_S = -\hat{\vartheta}\times\boldsymbol{H}^{\text{tot}}|_{\vartheta_0}$ on the cone's surface by

$$\boldsymbol{E}(\boldsymbol{r}) = \int_V \Gamma_0(\boldsymbol{r},\boldsymbol{r}')\cdot\boldsymbol{J}_s(\boldsymbol{r}')dv' \tag{21}$$

where the dyadic Green's function of the free space in bilinear form is deduced as

$$\Gamma_0(\boldsymbol{r},\boldsymbol{r}') = j\kappa\sum_{n,m}\frac{\boldsymbol{N}^{II}_{n,m}(\boldsymbol{r})\boldsymbol{N}^{I}_{n,m}(\boldsymbol{r}')}{n(n+1)} + j\kappa\sum_{n,m}\frac{\boldsymbol{M}^{II}_{n,m}(\boldsymbol{r})\boldsymbol{M}^{I}_{n,m}(\boldsymbol{r}')}{n(n+1)} \tag{22}$$

At a time-factor $e^{j\omega t}$ the upper indices I and II stand for the use of spherical Bessel functions of the first kind ($z_n = j_n$) and of spherical Hankel functions of the second kind ($z_n = h_n^{(2)}$), respectively. It has been shown (Klinkenbusch, 2006) that the scattered electric far field can be written in form of a multipole expansion

$$\boldsymbol{E}^{sc}(\boldsymbol{r}) = \frac{e^{-j\kappa r}}{\kappa r}\left[-\sum_{n,m}a^{sc}_{n,m}j^n\boldsymbol{n}_{n,m} + \sum_{n,m}\frac{Z}{j}b^{sc}_{n,m}j^{n+1}\boldsymbol{m}_{n,m}\right] \tag{23}$$

with the multipole amplitudes

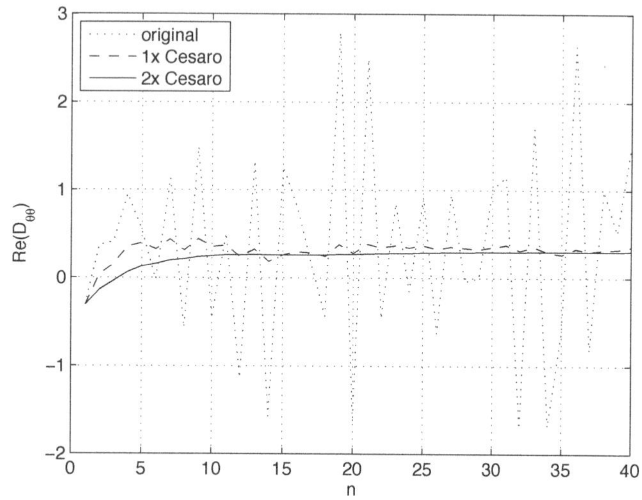

Fig. 3. Partial sum sequence of the real part of the scattering coefficient $D_{\theta\theta}$ for the maximum order of $n_{\max}=40$. Dotted curve is original partial sum sequence, dashed curve is single Cesàro transformed sequence and solid curve is double Cesàro transformed sequence.

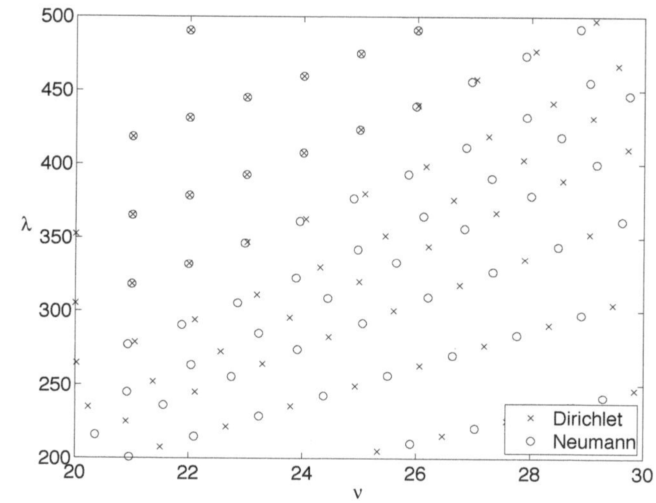

Fig. 4. Detailed view of the eigenvalue curves of Lamé functions for $k^2=0.5$ with Dirichlet- ($\times$) and Neumann ($\circ$) eigenvalues.

$$
\begin{aligned}
a^{sc}_{n,m} &= -j\Theta_{n,m}(\vartheta_0) \\
&\left\{\sqrt{1-k^2\cos^2\vartheta_0}\sum_\sigma \alpha_\sigma \left.\frac{d\Theta_\sigma}{d\vartheta}\right|_{\vartheta_0}\right. \\
&\int_0^\infty \frac{j_n(\kappa r)}{r} j_\sigma(\kappa r)\kappa r dr \int_0^{2\pi} \frac{\Phi_{n,m}(\varphi)\Phi_\sigma(\varphi)}{\sqrt{1-k'^2\sin^2\varphi}} d\varphi \\
&-\sum_\tau \frac{Z}{j}\beta_\tau\Theta_\tau(\vartheta_0) \\
&\int_0^\infty \frac{j_n(\kappa r)}{r}\frac{[rj_\tau(\kappa r)]'}{\kappa r}\kappa r dr \int_0^{2\pi} \Phi_{n,m}(\varphi)\frac{d\Phi_\tau(\varphi)}{d\varphi} d\varphi \\
&+\sum_\tau \frac{Z}{j}\beta_\tau\Theta_\tau(\vartheta_0)\frac{\tau(\tau+1)}{n(n+1)} \\
&\left.\int_0^\infty \frac{[rj_n(\kappa r)]'}{\kappa r}\frac{j_\tau(\kappa r)}{r}\kappa r dr \int_0^{2\pi} \frac{d\Phi_{n,m}(\varphi)}{d\varphi}\Phi_\tau(\varphi) d\varphi\right\}
\end{aligned} \tag{24}
$$

$$
\begin{aligned}
\frac{Z}{j} b^{sc}_{n,m} &= j \left.\frac{d\Theta_{n,m}}{d\vartheta}\right|_{\vartheta_0} \sqrt{1-k^2\cos^2\vartheta_0}\sum_\tau \frac{Z}{j}\beta_\tau\Theta_\tau(\vartheta_0) \\
&\frac{\tau(\tau+1)}{n(n+1)}\int_0^\infty j_n(\kappa r)\frac{j_\tau(\kappa r)}{r}\kappa r dr \int_0^{2\pi} \frac{\Phi_{n,m}(\varphi)\Phi_\tau(\varphi)}{\sqrt{1-k'^2\sin^2\varphi}} d\varphi .
\end{aligned} \tag{25}
$$

Finally, the scattered far field can be written as a function of the incident field by means of a scattering matrix as

$$
\begin{pmatrix} E^{sc}_\theta(\theta,\phi) \\ E^{sc}_\phi(\theta,\phi) \end{pmatrix} = \frac{e^{-j\kappa r}}{\kappa r}\begin{pmatrix} D_{\theta\theta} & D_{\theta\phi} \\ D_{\phi\theta} & D_{\phi\phi} \end{pmatrix}\begin{pmatrix} E^{\mathrm{inc}}_\theta(\theta',\phi') \\ E^{\mathrm{inc}}_\phi(\theta',\phi') \end{pmatrix} \tag{26}
$$

While the series in Eqs. (24) and (25) converge and yield stable multipole amplitudes of the scattered field, the resulting series in Eq. (23) do not converge. In order to obtain a meaningful limiting value it is necessary to apply a suitable sequence transformation. In contrast to nonlinear techniques (like the Shanks transform) linear sequence transformations always yield consistent results. For the linear Cesàro transform the transformed partial sum sequence s'_n is obtained from the original partial sum sequence s_n by

$$
s'_n = \frac{s_0+s_1+s_2+\ldots+s_n}{n+1}, \quad n=0,1,2,.. \tag{27}
$$

The sequence transformation can be repeatedly applied to enforce faster convergence of the resulting partial sum sequence. Figure 3 shows the double transformed partial sum sequence of the scattering coefficent $D_{\theta\theta}$ as a function of n. Clearly, a higher order of the original series and hence a higher number of eigenvalues is desired to obtain more accurate results. In order to increase the maximum number of available eigenvalues and eigenfunctions it is necessary to investigate the relevance of eigenvalues and eigenmodes which will be sketched in the following section.

5 Eigenmode analysis

In Fig. 2 it has been shown that the discrete (ν,λ)-pairs arranged in an eigenvalue-curve scheme which can approximately be divided into an upper region where Dirichlet- and

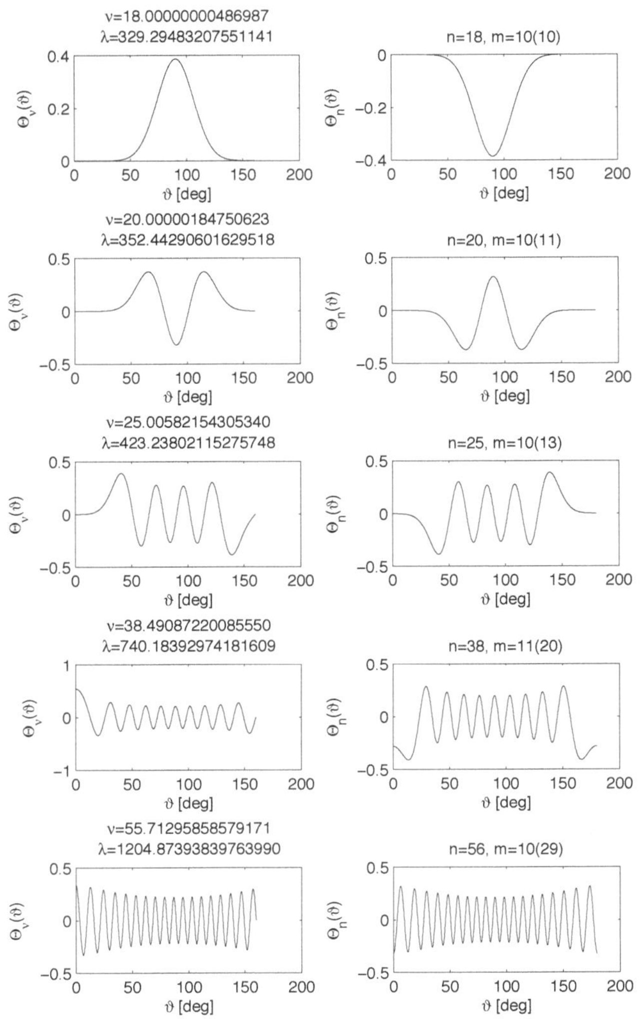

Fig. 5. Sequence of non-periodic Lamé functions $\Theta_\nu(\vartheta)$ to the Dirichlet boundary condition at $\vartheta_0 = 160°$, $k^2 = 0.5$ (left column) and (corresponding) non-periodic Lamé polynomials $\Theta_n(\vartheta)$ (right column), each as function of the arguments.

Neumann eigenvalues nearly coincide and into a lower region where Dirichlet- and Neumann eigenvalues are strictly separated. Figure 4 shows a closer view of this phenomenon revealing that the coinciding eigenvalues all are very near to integral values (integers) of ν. Since the numerical computation of these coinciding eigenvalues turns out to imply some numerical difficulties limiting the maximum number of computable eigenvalues we will now investigate that case in more detail.

The left column in Fig. 5 shows a sequence of plots of non-periodic Lamé functions each satisfying the Dirichlet boundary condition at $\vartheta_0 = 160°$ as a function of the argument ϑ. In the right column we see plots of the non-periodic Lamé polynomials (with integral eigenvalues $\nu = n$) at (n, λ)-pairs on the same eigenvalue curve nearest by those ones of the corresponding non-periodic Lamé functions.

We observe that for nearly integral eigenvalues of the non-periodic Lamé functions not only their values at ϑ_0 are vanishing but also their derivatives. Moreover, these eigenfunctions look very similar to the corresponding non-periodic Lamé polynomials. At non-integral eigenvalues only the values of the non-periodic Lamé functions vanish but not their derivatives, and their curves are different from the corresponding non-periodic Lamé polynomials, at least in the vicinity of the boundaries $\vartheta = 0$ and $\vartheta = \pi$. This general behavior is typical and can be observed for any other eigenvalue as well.

Due to numerical reasons, the computation of the nearly integral eigenvalues and -functions turns out to be difficult. However, as we can deduce from the representations of the multipole amplitudes Eqs. (24) and (25) the modes belonging to these eigenvalues do not significantly contribute to the scattered far field. Each part of Eqs. (24) and (25) has a factor of one of the following forms

$$\Theta_\tau(\vartheta_0)$$
$$\left.\frac{d\Theta_\sigma(\vartheta)}{d\vartheta}\right|_{\vartheta=\vartheta_0}$$

Clearly, if both function and derivative of a non-periodic Lamé function are small at ϑ_0, the corresponding scattering mode is also small compared to the other scattering modes. In other words, these eigenmodes of the PEC cone do not significantly lead to a surface current on the cone, or, the cone is nearly invisible for these eigenmodes. Consequently, they are very similar to free-space modes, which are characterized by integral eigenvalues.

Following this observation, these nearly-integral eigenvalues and eigenfunctions don't need to be exactly calculated, and the modified algorithm allows to calculate much more relevant eigenvalues and eigenfunctions to come to more accurate scattering coefficients.

6 Scattering coefficients

Figure 6 shows the amplitude and the phase of the electric far field scattered by a PEC semi-infinite elliptic cone illuminated by a plane wave electrically polarized in the xz plane and incident from $\theta^{\mathrm{inc}} = 105°$, $\phi^{\mathrm{inc}} = 0°$. The amplitude of the scattering coefficient $D_{\theta\theta}$ is shown for the maximum order $n_{\max} = 40$ including the integral-eigenvalue modes and $n_{\max} = 60$ excluding these non-contributing modes. The comparison between the phases shows marginal differences, however, the differences in amplitudes reveal the improvement of the results by considering more relevant eigenmodes.

Finally, Fig. 7 proves that the errors of amplitudes and phases of the scattering coefficient are actually marginal when all of the near-integer eigenvalues are neglected.

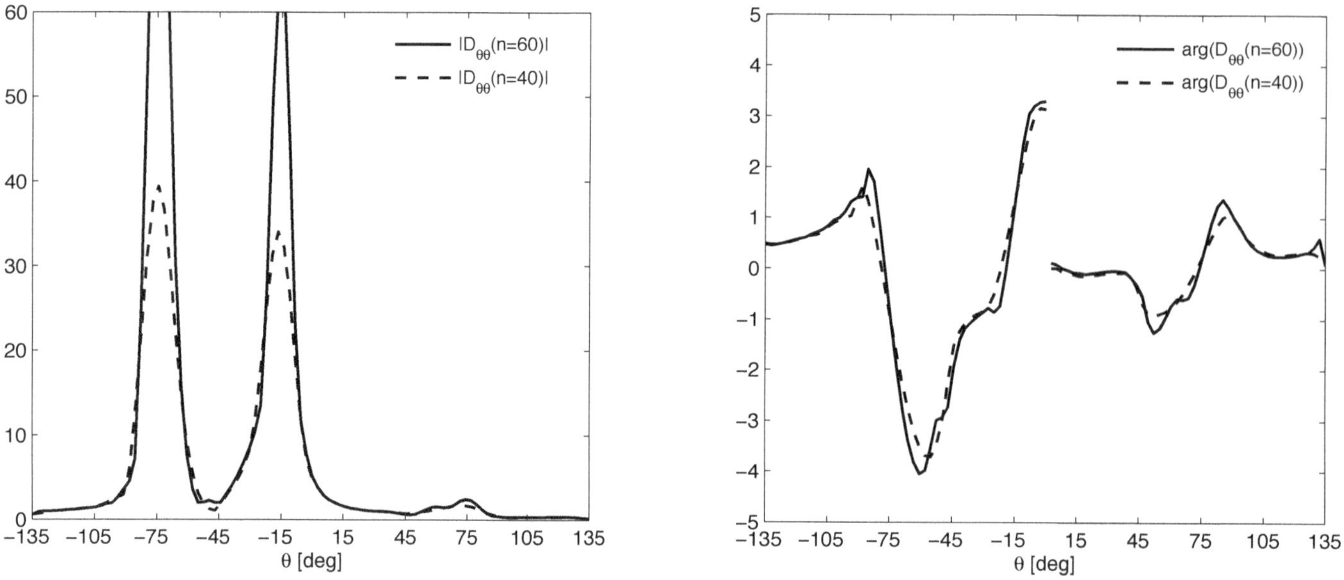

Fig. 6. Amplitude and phase of the scattering coefficient $D_{\theta\theta}$ in the xz plane of a PEC semi-infinite elliptic cone with the half opening angles $\alpha_x = 45°$, $\alpha_y = 60°$ for the order of the multipole expansion (23) $n_{\max} = 40$ (dashed line) and $n_{\max} = 60$ (solid line). The plane wave is incident from $\theta^{\text{inc}} = 105°$, $\phi^{\text{inc}} = 0°$.

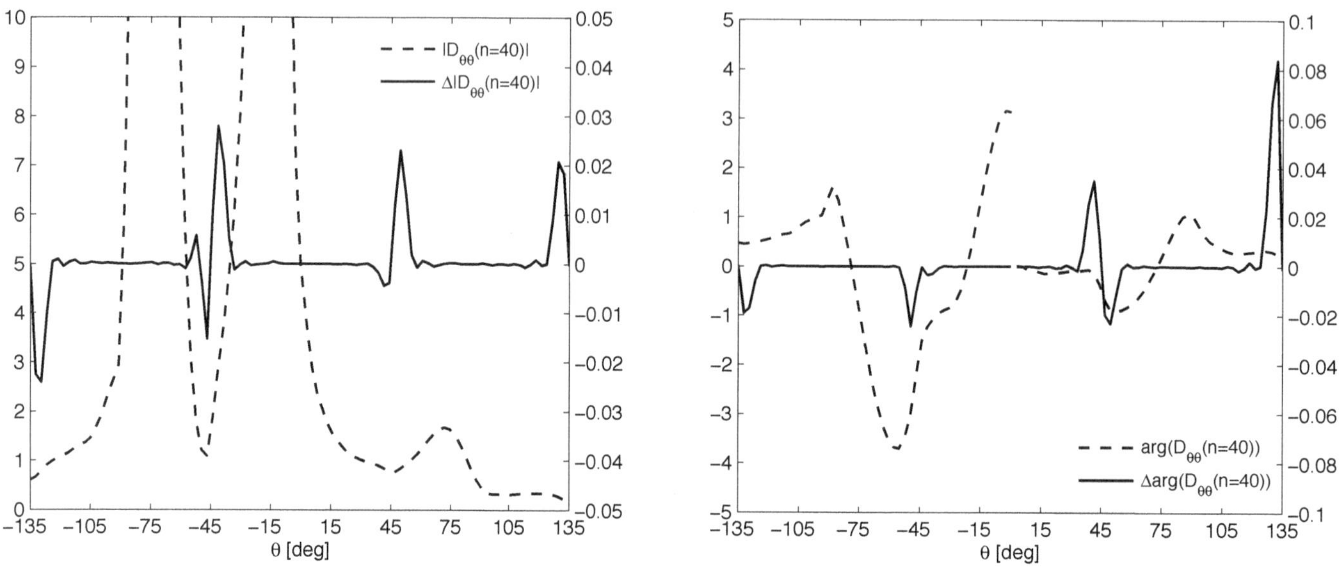

Fig. 7. Amplitude and phase of the scattering coefficient $D_{\theta\theta}$ (dashed line) in the xz plane of a PEC semi-infinite elliptic cone with the half opening angles $\alpha_x = 45°$, $\alpha_y = 60°$ and $n_{\max} = 40$. The plane wave is incident from $\theta^{\text{inc}} = 105°$, $\phi^{\text{inc}} = 0°$. The solid line using the right scale is the difference between the first amplitude (phase) resulting from all eigenvalues and the second amplitude (phase) resulting from all eigenvalues except nearly integer eigenvalues.

7 Conclusions

It has been found that the accuracy of computed scattering coefficients for a PEC elliptic cone can be greatly improved if computationally difficult but non-relevant modes of the scattered field are neglected. These non-scattered modes have nearly integral eigenvalues and are very similar to the free-space modes of the incident field. Further work will include an investigation into the nature of these non-scattered modes.

Acknowledgements. This work was supported by the Deutsche Forschungsgemeinschaft (DFG).

References

Blume, S. and Klinkenbusch, L.: Spherical-multipole analysis in electromagnetics, in: Frontiers in Electromagnetics, edited by: Werner, D. H. and Mittra, R., 553–606, IEEE Press, New York, 1999.

Boersma, J. and Jansen, J. K. M.: Electromagnetic field singularities at the tip of an elliptic cone, EUT-Report 90-WSK-01, Faculty of Mathematics and Computing Science, Eindhoven University of Technology, Eindhoven, 1990.

Bowman, J. J., Senior, T. B. A., and Uslenghi, P. L. E.: Electromagnetic and acoustic scattering by simple shapes (Revised Printing), Hemisphere Pub. Corp., New York, 1987.

Klinkenbusch, L.: Modal Analysis of the GTD Diffraction Coefficients for the Half-Plane's Edge, Electr. Eng., 88, 337–342, 2006.

Klinkenbusch, L.: Electromagnetic scattering by semi-infinite circular and elliptic cones, Radio Sci., 42, RS6S10, doi:10.1029/2007RS003649, 2007.

Magnus, W., Oberhettinger, F., and Soni, R. P.: Formulas and theorems for the special functions of mathematical physics, Springer-Verlag, New York, 1966.

Stratton, J. A.: Electromagnetic Theory, McGraw-Hill, New York, 1941.

Eigenvalue extraction from time domain computations

T. Banova[1,2], W. Ackermann[2], and T. Weiland[2]

[1]Technische Universität Darmstadt, Graduate School of Computational Engineering, Dolivostraße 15, 64293 Darmstadt, Germany
[2]Technische Universität Darmstadt, Institut für Theorie Elektromagnetischer Felder (TEMF), Schlossgartenstraße 8, 64289 Darmstadt, Germany

Correspondence to: T. Banova (banova@gsc.tu-darmstadt.de)

Abstract. In this paper we address a fast approach for an accurate eigenfrequency extraction, taken into consideration the evaluated electric field computations in time domain of a superconducting resonant structure. Upon excitation of the cavity, the electric field intensity is recorded at different detection probes inside the cavity. Thereafter, we perform Fourier analysis of the recorded signals and by means of fitting techniques with the theoretical cavity response model (in support of the applied excitation) we extract the requested eigenfrequencies by finding the optimal model parameters in least square sense. The major challenges posed by our work are: first, the ability of the approach to tackle the large scale eigenvalue problem and second, the capability to extract many, i.e. order of thousands, eigenfrequencies for the considered cavity. At this point, we demonstrate that the proposed approach is able to extract many eigenfrequencies of a closed resonator in a relatively short time. In addition to the need to ensure a high precision of the calculated eigenfrequencies, we compare them side by side with the reference data available from CEM3D eigenmode solver. Furthermore, the simulations have shown high accuracy of this technique and good agreement with the reference data. Finally, all of the results indicate that the suggested technique can be used for precise extraction of many eigenfrequencies based on time domain field computations.

1 Introduction

The field of quantum chaos encompasses the study of the manifestations of classical chaos in the properties of the corresponding quantum or more generally, wave-dynamical system (nuclei, atoms, quantum dots, and electromagnetic or acoustic resonators). Prototypes are billiards of arbitrary shape. In its interior a point-like particle moves freely and is reflected specularly at the boundaries. Depending on the shape its properties could exhibit chaotic dynamics.

Within this work, we investigate quantum billiards with its statistical eigenvalue properties, which reveal the periodic orbits in the quantum spectra and give the quantum chaotic scattering (Dembowski et al., 2005). Specifically, we simulate microwave resonator with chaotic characteristics, see Fig. 1, and we compute the eigenfrequencies that are needed for its level spacing analysis (Dembowski et al., 2002). Accordingly, the eigenfrequency level spacing analysis for determining the statistical properties requires many (in order of thousands) eigenfrequencies to be calculated and the accurate determination of the eigenfrequencies has a crucial significance. Moreover, considering that the problem is to compute a large number of eigenfrequencies, they can be often located in different ranges, i.e. left-most, right-most or interior portions of the spectrum could be sought.

In many scientific and engineering applications, as well as, in computational science, this results in solving one of the fundamental problems, the large scale eigenvalue problem. We list below just a few of the applications areas (Saad, 2011) where eigenvalue calculations arise: acceleration of charged particles, structural dynamics, electrical networks, Markov chain techniques, combustion processes, chemical reactions, macro-economics, control theory, etc. The above-mentioned realistic applications frequently challenge the limit of both computer hardware and numerical algorithms, as the involved matrices are of large scale.

The frequency domain methods (Jacobi-Davidson, Arnoldi, Lanczos, Krylov-Schur etc.) for eigenvalue calculation for some cavity structures might result in an extremely time consuming simulation, along with slow convergence and huge storage space. Particularly, structures with complicated geometry require a large number of grid points to achieve accurate simulation results. An additional challenge from our application is that the dimension of the desired eigen subspace is large, namely, one might possibly need thousands of eigen pairs located in a specified range, also referred to as a "window", of matrices with dimension in excess of several millions. Despite the fact that many algorithms for eigenvalue determination exist, accompanied by the numerical models that are becoming increasingly more sophisticated, not as many are specifically adapted for computing a large number of interior eigen pairs. They can usually calculate limited number of extreme or interior eigenfrequencies, and relatively few are designed for effectively reusing a large number of good initial presumptions, when they are available. Finally, computing a large number of interior eigenvalues remains one of the most difficult problems in computational linear algebra today.

In this context, our investigations comprise efficient, robust and accurate computations of many desired eigenfrequencies in a reasonable time, which also constitutes the main aim of our study. Within this work, we cover an approach for extraction of resonant frequencies given the output from time domain computations of closed resonators. The proposed approach uses the advantage that one single time domain simulation can provide the whole response of an electromagnetic system in a wide frequency band, whereas a frequency domain formulation uses one computation for each individual frequency. In addition, due to the fact that the time domain computations in the field of electromagnetics are already highly developed and considerably more efficient, as well as the fact that the transient solver contained in CST Microwave Studio (CST, 2012) uses a high degree of parallelization provided with modern graphics processing units (GPUs) feature the simulation can be dramatically accelerated. Therefore, we can easily and quickly get the time domain responses for a wide frequency band. In this way, significant reduction in computation time can be achieved and therefore, our high interest leads to time domain computations for electromagnetic problems.

The paper proceeds as follows. In Sect. 2 we briefly present the proposed approach for high precision eigenfrequency extraction given the available electric field computations. Here, we describe the fundamental modeling of the analyzed structure, followed by a description of the used signal postprocessing techniques. Section 3 investigates the simulation scenarios together with the obtained results. Lastly, conclusions are addressed in Sect. 4.

2 Time domain approach for eigenfrequency extraction

This section provides a brief overview of a precise time-domain (TD) approach for eigenfrequency extraction, which is applicable for different cavity structures. In a two-step process the modeling and simulation of a specific cavity structure is initially done, and afterward a postprocessing of the acquired time domain simulation results in MATLAB (MATLAB, 2011) is conducted. A descriptive sketch of the proposed TD approach is given in Algorithm 1. Additionally, the critical implementation points and details are covered, as well as, discussed within this section.

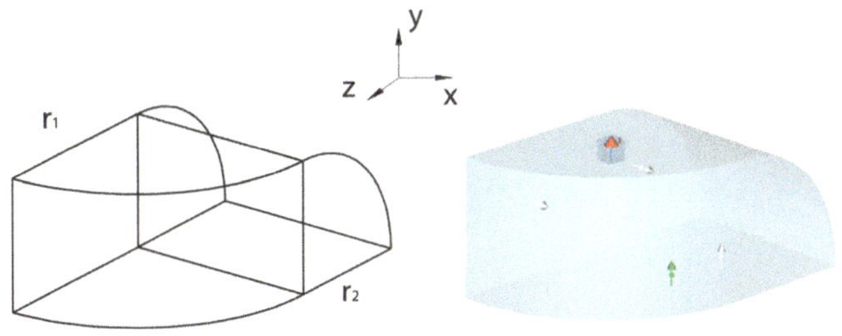

Fig. 1. Left: desymmetrized version of the three-dimensional generalized stadium billiard, consisted of two quarter cylinders with radii $r_1 = 200.0$ mm and $r_2 = 141.4$ mm. Right: CST setup of the billiard cavity with an exciting antenna and detection probes placed at various positions.

2.1 Field simulation in time domain

The proposed approach for eigenfrequency extraction (given in the following) is utilized within the project for determining the statistical properties of a cavity with chaotic characteristics. The requirements for chaotic characteristics are met by using a superconducting resonator with the shape of a desymmetrized three-dimensional stadium billiard (see Fig. 1-left). The billiard consists of two quarter cylinders with radii $r_1 = 200.0$ mm and $r_2 = 141.4$ mm, respectively, and is made of niobium, which becomes superconducting at temperatures below 9.2 K.

The cavity of interest is modeled in CST Microwave Studio (CST MWS) and a small tiny exciting antenna (as used in a physical model) is put properly in a way that the modes within a specific frequency range would be excited (see Fig. 1-right). Intentionally, the excitation signal applied at the antenna input is wide band signal, i.e. a Gaussian-modulated sinusoidal signal is chosen, which certainly covers the range of eigenfrequencies being sought. The field simulation with hexahedron discretization mesh in time domain is carried out with the transient solver from CST MWS and it detects and records the electric field intensity at specific field detection probes placed at various positions inside the cavity (see lines 1–6 in Algorithm 1). Later, the obtained time domain signals are used for further postprocessing in MATLAB, based on fitting techniques with a proposed model of the cavity response, as follows below.

2.2 Postprocessing of the CST signals

2.2.1 Limitations from finite simulation time

In view of the fact that we are dealing with superconducting walls, the response of a cavity could stay for a long time as

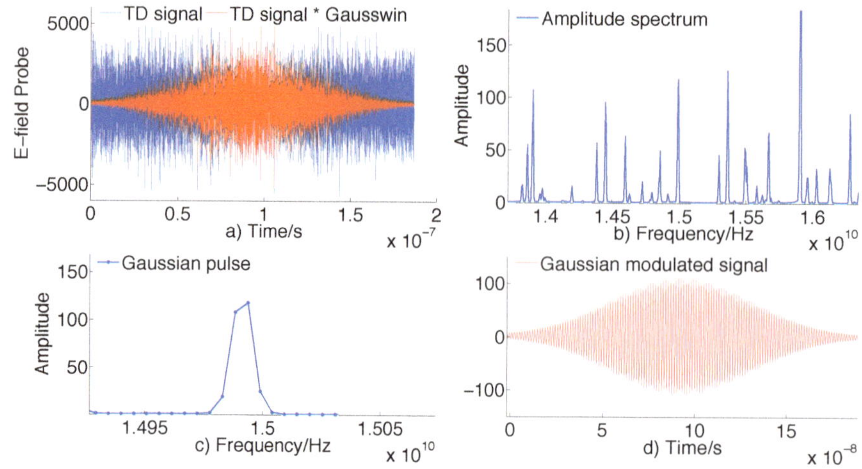

Fig. 2. Operating signals during the process of eigenfrequency extraction, listed in the respective sub-figures. **(a)** Acquired time domain signal and Gaussian windowed time domain signal. **(b)** Amplitude spectrum of the Gaussian windowed time domain signal. **(c)** Isolated Gaussian pulse within the eigenmode spectrum. **(d)** Gaussian modulated signal.

```
Input: Given a cavity with a specified probe P(x,y,z).
/* CST simulation                                              */
Data: run transient simulation with the predefined simulation
      time.
Data: export a TD_sig for the electric field intensity in
      P(x,y,z).
/* Post processing and frequency
   extraction in MATLAB                                        */
TD_sig = TD_sig * gausswin(size(TD_sig));
fft_data = fft(TD_sig);
amp_spec = abs(fft_data);
forall the Gaussian pulses in amp_spec do
    gauss_pulse = locate_gauss_pulse();
    /* ind_start=starting index
       ind_end=ending index                                    */
    GoF1 = cftool(gauss_pulse, gauss_model());
    /* GoF = Goodness of Fitting                               */
    if GoF1 ≻ 0 then
        gausswin_fft_data(ind_start : ind_end) =
        fft_data * gausswin(ind_end − ind_start);
        forall the ind ∉ gauss_pulse do
            gausswin_fft_data(ind) = 0;
        end
        gauss_mod_sig = real(ifft(gausswin_fft_data));
        GoF2 =
        cftool(gauss_mod_sig, gauss_mod_model());
        if GoF2 ≃ 100 then
            extract the eigenfrequency;
        end
    end
end
```

Algorithm 1: A sketch of the time domain approach for extracting eigenfrequencies.

the power losses in the walls are negligible. Theoretically, the response of an ideal-conducting cavity is Dirac impulse sequence in frequency domain, i.e. a summation of sinusoidal signals with the associated eigenfrequencies in time domain. Nevertheless, due to the limited simulated time interval the ideal Dirac impulses of the true spectrum are widened. Moreover, due to the finite conductivity and the inserted antenna, the amplitude spectrum of the signal does not contain Dirac delta pulses, but it has pulses of finite width.

An important issue coming from the limitation in time is the discontinuities at the edges of the measured time (Mandal and Asif, 2010). Given that sharp discontinuities have broad frequency spectra, these will lead to a higher side lobes level and each spectral line of the signal's frequency spectrum will be spread out in the same way. In other words, the spreading means that signal energy, which should be concentrated only at one eigenfrequency, leaks instead into other frequencies, the so called "spectral leakage". Consequently, the whole spectrum is distorted and some weak impulses, i.e., eigenfrequencies, can be masked by the resulting convolution with neighboring strong pulses. This leads to the idea of multiplying the original signal within the measurement time by Gaussian function (see Fig. 2a) that smoothly reduces the signal to zero at the end points of the measurement time: therefore avoiding discontinuities overall (see line 9 in Algorithm 1). As a result, in case of an ideal cavity, whose spectrum theoretically is constituted of Dirac impulses located at the eigenfrequencies, we would have Gaussian pulses instead (see Fig. 2b).

Furthermore, manifestation of a spectral distortion occurs as result of a reduced spectral resolution. Namely, this is an important issue, and the minimum separation needed between two frequency components must be determined, so that they can be resolved. In our approach, we choose the frequency resolution good enough to recover the sought frequency data. As a result, a large number of simulation time samples might be needed for a reliable characterization of the resonance frequency of the structure, given that better frequency resolution unescapably requires a longer simulation time. In spite of this, the modern GPUs feature a large number of processing cores and the simulation is speeded up significantly in comparison with a simulation running on single CPU.

2.2.2 Fast Fourier transformation

As soon as the limitations that come from finite simulation time or reduced frequency resolution are overcame, the step from the recorded time domain response to the frequency domain response is computed using the fast Fourier transformation (see lines 10–11 in Algorithm 1, see Fig. 2b).

The very classical approach in finding eigenfrequencies is to look for local maxima of the eigenmode spectrum. However, this way is not efficient when the interest is in precise determination of the resonant frequencies and has some drawbacks. Firstly, the spectrum is discrete with a certain resolution and the peak value may not be entirely located on a sample point. The eigenfrequency could be a value that is somewhere in the range between two samples given with the frequency resolution for certain discrete frequency values ("bins"), implying that the local maximum is not always the frequency that is sought. Secondly, a more serious focus was placed that the neighboring modes contribute a certain amount to the total response at the resonance of the mode being analyzed and affect slightly the resonant frequency. To deal with these problems, refined modal extraction methods based on signal processing techniques have been developed (see lines 12–31 in Algorithm 1).

2.2.3 Technique for locating a Gaussian pulse

For further analysis, local Gaussian pulses (see Fig. 2c) within the spectrum should be located properly (see line 13 in Algorithm 1). The location process is divided into several steps. Primarily, a local Gaussian pulse is found as a set of samples with a local maximum. Thereafter, supplementary check is conducted if some other samples might be added to the right-most/left-most side of the current Gaussian pulse. Namely, if the average amplitude of the succeeding right/left outer triple of samples is less than the average amplitude of the right-most/left-most triple of samples, the outer triple of samples is added appropriately to the right/left part of the Gaussian pulse. After that, the "empirical rule" for the current located Gaussian pulse is applied. For this purpose, the standard deviation for the pulse is estimated and four standard deviations are accounted for the resulting pulse. At the end, the distance from the both ends of the Gaussian pulse to its maximum is equally adjusted.

2.2.4 Fitting models

By invoking the acquired knowledge of the eigenmode spectrum of the cavity, the fitting model should be nonlinear in the parameters, i.e. we should use Gaussian nonlinear model to obtain the parameters. In this direction, the parametric fitting is involved as essential technique for precise determination of the eigenfrequencies and reducing the amount of data required for a given resolution. Relying on the above discussion, a custom Gaussian model is modeled within the MATLAB curve fitting tool (see line 16 in Algorithm 1), which suits to the specific curve fitting needs, as shown below:

$$y_i = a_i \epsilon^{\frac{-(f-f_i)^2}{2\sigma_i^2}}, \tag{1}$$

where i goes from unity to the number of the located local Gaussian pulses in the amplitude spectrum for the analyzed structure. With this model all of the local Gaussian pulses are fitted and in each i-th fitting the values for the found parameters a_i, f_i, and σ_i are saved. The parameters a_i and f_i are used as initial reasonable starting values of the parameters in the further fit, whereas the parameter σ_i is included to increase the numerical robustness. The values of the parameter f_i are good candidates for eigenfrequencies, since they give the position of the maximum value of each Gaussian pulse.

However, following this way we are restricted to a limited number of samples, same as the number of samples which constitute the local Gaussian pulse. The limitation in the samples causes that the coefficient, which represents the goodness of fitting, is very low. That means that the accuracy of the eventual eigenfrequency cannot be high. Additionally, with this approach only the amplitude information of the signal is used, which is not sufficient for precise extraction.

These disadvantages lead to extending our approach, in a sense where we can get also the phase information of the signal in a form that is suitable for implementation. Namely, after fitting a local Gaussian pulse in frequency domain the values of the parameters and the coefficients for the goodness of the fit are obtained. If the goodness of the fit for some Gaussian pulse is positive then we window this Gaussian pulse with Gaussian windowing function and we go back in time domain using Inverse Fast Fourier Transform (IFFT) (see lines 18–31 in Algorithm 1). Otherwise, we do not take into consideration this pulse, which is most probably noise. From signal processing is known that shifting in frequency domain means modulation in time domain. So, the Gaussian-modulated sinusoidal signal $y_i = a_i \sin(2\pi f_i t - \phi_i)\epsilon^{\frac{-(t-t_0)^2}{2\sigma_i^2}}$ is expected in time domain. The frequency of the modulation f_i is exactly the frequency that is sought. Consequently, we fit the resulting signal in time domain with a custom Gaussian modulated model of the cavity response and again by finding the optimal model parameters in the least squares sense we determine the "true" eigenfrequency.

The main advantage comparing the proposed approach with the classical approach for finding the peaks, is that within this approach the phase information of the signal is used and parametric fitting with all of the data available in the time domain representation is applied. Although, the fitting based only on the amplitude information of the signal results in poor fit, now using the phase information of the signal, very good value for the goodness of the fit can be reached. In this way, as shown in Sect. 3, we gain a very high accuracy in the eigenfrequency extraction from time domain computations.

2.2.5 Nonlinear parametric fitting

The fitting process uses the method of nonlinear least squares, presented in Nash and Sofer (1996) and Lewis et al. (2006), which minimizes the summed square of residuals, identified as error associated with the data, given by

$$S = \sum_{i=1}^{n} r_i^2 = \sum_{i=1}^{n} (y_i - \hat{y}_i)^2, \tag{2}$$

where the residual r_i for the i-th data point is defined as the difference between the observed response value y_i and the fitted response value $\hat{y}_i$ while n is the number of data points included in the fit. In case of nonlinear models the parameters cannot be estimated using simple matrix techniques and these models are particularly sensitive to the starting points. This leads to difficult fit and the starting points are adjusted. Therefore, at the beginning initial reasonable starting values for each parameter are provided. Then, the iterative approach follows some steps until the fit reaches the specified convergence criteria. At the end, the direction and magnitude of the adjustment of the parameters depends on the Levenberg-Marquardt fitting algorithm proposed by Levenberg (1944), Marquardt (1963), and Moré (1978).

3 Simulation results

The accuracy of the TD method for eigenfrequency extraction is tested for both analytically and non-analytically resolvable problems. Additionally, the proper functionality of the method has been checked during its implementation process. In the numerical tests, several resonators are considered. Namely, due to verification purposes, rectangular, cylindrical and spherical resonant structures are analyzed, whose exact solution can be analytically evaluated. Firstly, the eigenvalues of the above mentioned resonators are computed from their analytical expressions and following this way, a logarithmic relative error is calculated as

$$\text{relative error} = \log_{10} \frac{|f_{\text{analytical}} - f_{\text{numerical}}|}{f_{\text{analytical}}}, \tag{3}$$

by considering the first computed mode eigenfrequency, unless otherwise stated.

Besides the analytically resolvable resonators, the relative error is also measured for the chaotic billiard resonator, which has a clustered eigenvalue distribution (Dembowski et al., 2002). A clustered distribution containing too close eigenfrequencies might cause difficulties in the eigenfrequency extraction. As reference data, eigenvalues calculated with CEM3D eigenmode solver for different degrees of freedom (DOF) are computed. Here, CEM3D eigenmode solver is a parallel program, developed by (Ackermann and Weiland, 2012) for the accurate calculations of eigenfrequencies for a given structure. It implements a FEM formulation by means of higher order curvilinear elements.

In all simulations, we first modeled and meshed the related geometries in CST MWS with hexahedrons and saved the electric field intensity during the transient simulation. Afterward, the eigenvalue extraction is calculated in MATLAB.

In the simulation studies, we experienced that for the time domain field computations, a single personal computer is suited for problems with a moderate number of degrees of freedom (say, up to 10^6 DOF). To be precise, a computer with a single core Pentium 3 GHz processor and 6 GB main memory was used. The same PC configuration was also utilized for the TD method for eigenvalue extraction. On the other hand, a more powerful computer was necessarily used to enable the handling of meshes with hundreds of thousands up to several millions DOF. The larger-scale simulations were performed on a GPU computer, e.g. 2.00 GHz (quadcore) having 32 GB RAM memory and 4 nVIDIA Quadro GPUs.

3.1 Rectangular, cylindrical, and spherical cavity

As a first experiment, we consider a rectangular cavity with perfectly conducting walls, containing a perfect vacuum. The degeneracy is broken by making the side lengths different, i.e. rectangular resonator with dimensions $a = 20\,\text{cm}$, $b = 10e\,\text{cm}$, and $c = 10\pi$ cm. The resonance frequency of the rectangular microwave cavity for the TE_{011} mode (the mode with the lowest cutoff frequency for a rectangular waveguide where $c > b > a$) is found by imposing boundary conditions on the electromagnetic field expressions in which all of the field components vary sinusoidally at a single frequency (Pozar, 1998).

Along this line, the fundamental mode in a cylindrical cavity (Reitz and Milford, 1960) with radius $R = 20\,\text{cm}$ and length $L = 10\pi$ cm is calculated, too. For the analyzed cylindrical cavity, since we do not have large L fulfilling $L > 2.03R$, the TM_{010} mode constitutes the fundamental oscillation. The mode of interest has azimuthal symmetry and the electric field has no longitudinal variation ($\delta E/\delta z = 0$).

Lastly, the first TM_{101} mode of a spherical resonator with a radius $R = 1\,\text{m}$ is computed from analytical expressions given in (Gallagher and Gallagher, 1985) by employing a root finding algorithm of transcendental equations which is simply explained in (Pozar, 1998).

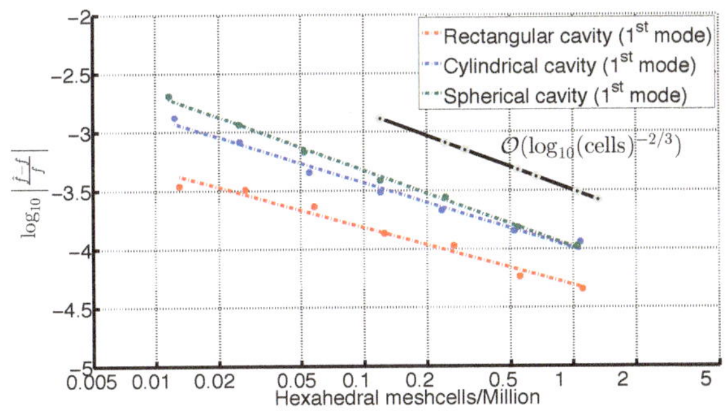

Fig. 3. Relative deviation of the numerically obtained values $\hat{f}$ for the lowest eigenfrequency to the analytical result f as a function of the degrees of freedom (DOF) for a rectangular, cylindrical, and spherical resonator.

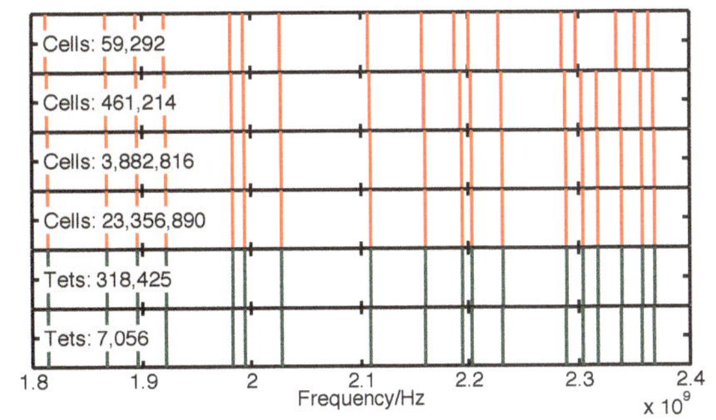

Fig. 4. Convergence study showing a comparison between the eigenfrequencies calculated with the proposed TD approach (red color) and the reference eigenfrequencies obtained with CEM3D eigenmode solver based on higher order curvilinear elements (green color). For the TD approach four different hexahedron discretization meshes are used. At the same time, the reference data are obtained using two different tetrahedral discretization meshes.

Specifically, for the time domain field simulations several different discretization meshes have been used and the convergence study based on the calculation of a relative error, given with Eq. (3), is shown in Fig. 3. As the number of discretization mesh cells increases, the difference between the analytical and the numerical solutions becomes smaller and absolute error in order of 10^{-4} is present. So, fast convergence is observable and it should be emphasized. As suggested by the convergence study, good accordance of the numerical with the analytical results is evident.

3.2 Billiard cavity

Using the previosly described algorithm, about 900 eigenfrequencies up to 7 GHz have been calculated for the billiard cavity structure. Unfortunately, for this shape of resonator an analytical solution for the electromagnetic problem is not available and in order to validate the obtained results, reference data from CEM3D eigenmode solver are used (see Fig. 4). The structures are modeled and meshed with curvilinear tetrahedrons and the corresponding meshes are imported to CEM3D eigenvalue solver in order to compute the requested eigenfrequencies for the analyzed structure.

In Fig. 4, a part of the results that are found with the TD approach compared to the reference data, is presented. On the abscissa are given the frequencies in an a priori selected frequency band, i.e. from 1.8 up to 2.4 GHz. The ordinate shows the eigenfrequencies obtained with the time domain approach using four different discretization meshes, and the reference data calculated using a field simulation in frequency domain with two different tetrahedral meshes. The total time for the transient simulations is 3.5×10^{-5} s, which results in frequency resolution of 30 kHz.

The results in Fig. 4 indicate that when fine mesh is used the number of eigenfrequencies found with the proposed TD approach is exactly the same as the reference data, i.e. no additional frequency is added or no missed. In addition, such check was conducted for all of the 900 calculated eigen-

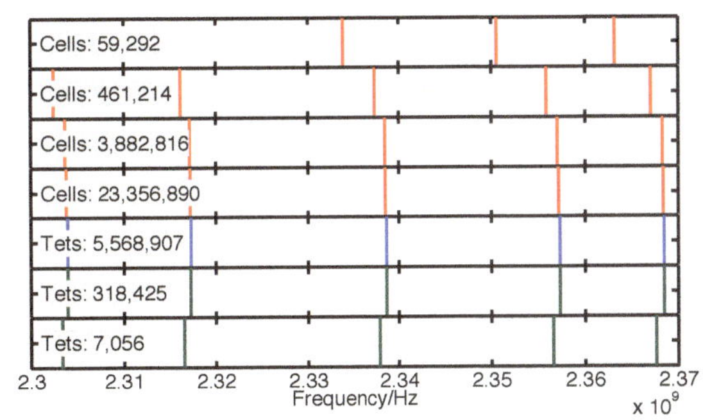

Fig. 5. Convergence study showing a comparison between the eigenfrequencies calculated with the proposed TD approach (red color) and the reference eigenfrequencies obtained with CEM3D eigenmode solver based on higher order curvilinear elements (green and blue color). For the TD approach four different hexahedron discretization meshes are used. At the same time, the reference data are obtained using three different tetrahedral discretization meshes.

frequencies. Concerning the accuracy of the obtained data, slightly shifting of the frequencies can be observed in case of coarse meshes. As the number of mesh cells increases, a good agreement with the reference data is clearly observed. For this purpose, a part of Fig. 4 is enlarged and shown in Fig. 5. According to Fig. 5, in case of fine meshes a good agreement of TD results with the reference data can be obtained. Furthermore, in Fig. 5 an additional row is added representing extremely accurate reference data for the eigenvalues in the range [2.30, 2.37] GHz. These results are obtained with almost 6 million of tetrahedral mesh cells and are used to calculate the logarithmic relative error for the modes: 2.36 and 2.37 GHz (see Fig. 6). From this figure a second order convergence error is observed. Consequently, the proposed

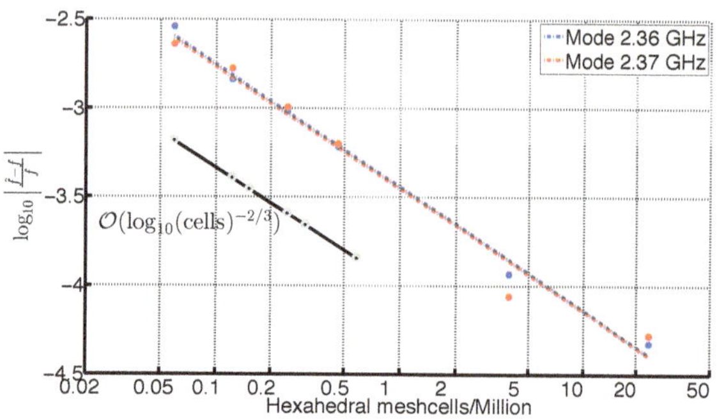

Fig. 6. Relative deviation of the numerically obtained values $\hat{f}$ to the reference results f as a function of the degrees of freedom (DOF) for a billiard resonator. Two eigenfrequencies are considered: 2.36 GHz and 2.37 GHz mode eigenfrequency.

approach can be used for precise extraction of eigenfrequencies from time domain computations.

4 Conclusions

In conclusion, we have presented an approach for eigenfrequency extraction when the time domain response for a superconducting closed resonator is available and when plenty of eigenfrequencies are required. The approach is relatively fast and very simple compared to others presented in literature. In addition, we have inspected the robustness and convergence of the proposed technique in time domain with the reference data determined by the CEM3D eigenmode solver with higher order curvilinear elements. Hereby, the conclusion shows that the proposed approach results in solutions which agree well with the reference data, gaining high accuracy and efficiency in eigenvalue extraction. Finally, all of the results indicate that the proposed TD technique can be used in different areas of applications, where a precise extraction of plenty of eigenfrequencies takes a crucial role.

Acknowledgements. The work of Todorka Banova is supported by the "Excellence Initiative" of the German Federal and State Governments and the Graduate School of Computational Engineering at Technische Universität Darmstadt.

References

Ackermann, W. and Weiland, T.: High Precision Cavity Simulations, Proceedings of the 11th International Computational Accelerator Physics Conference, 1–5, 2012.

CST AG, CST 2012, Darmstadt, Germany, MICROWAVE STUDIO, 2012.

Dembowski, C., Dietz, B., Gräf, H.D., Heine, A., Papenbrock, T., Richter, A., and Richter, C.: Experimental Test of a Trace Formula for a Chaotic Three Dimensional Microwave Cavity, Phys. Rev. Lett., 89, 064101, doi:10.1103/PhysRevLett.89.064101, 2002.

Dembowski, C., Dietz, B., Friedrich, T., Gräf, H.D., Harney, H.L., Heine, A., Miski-Oglu, M., and Richter, A.: Distribution of resonance strengths in microwave billiards of mixed and chaotic dynamics, Phys. Rev. E, 71, 046202, doi:10.1103/PhysRevE.71.046202, 2005.

Gallagher, S. and Gallagher, W. J.: The Spherical Resonator, IEEE Trans. on Nuclear Sci. Vol NS-32 5, 1985.

Mandal, M. and Asif, A.: Continuous and Discrete Time Signals and Systems, Posts & Telecom Press, 2010.

Marquardt, D.: An Algorithm for Least Squares Estimation of Nonlinear Parameters, SIAM J. Appl. Math, 11, 431–441, 1963.

MATLAB R2011b, The MathWorks Inc., Natick, MA, 2011.

Moré, J. J.: The Levenberg-Marquardt Algorithm: Implementation and Theory, Lecture Notes in Mathematics, v630, edited by: Watson, G., Springer, 1978.

Nash, S. G. and Sofer, A.: Linear and Nonlinear Programming, McGraw-Hill Companies, Inc, New York, ISBN 0-07-046065-5, 1996.

Levenberg, K.: A Method for the Solution of Certain Problems in Least Squares, Quart. Appl. Math, 2, 164–168, 1944.

Lewis, J. M., Lakshmivarahan, S., and Dhall, S.: Dynamic Data Assimilation: A Least Squares Approach (Encyclopedia of Mathematics and its Applications), Cambridge University Press, 2006.

Pozar, D. M.: Microwave Engineering, 2nd edition, John Wiley and Sons, Inc., New York, NY, 1998.

Reitz, J. R. and Milford, F. J.: Foundations of Electromagnetic Theory, Addison-Wesley, 1960.

Saad, Y.: Numerical Methods for Large Eigenvalue Problems, Revised Edition, Society for Industrial and Applied Mathematics, 2011.

The Method of Lines in the time domain

S. F. Helfert

FernUniversität in Hagen, Hagen, Germany

Correspondence to: S. F. Helfert (stefan.helfert@fernuni-hagen.de)

Abstract. The Method of Lines (MoL) is a semi-analytical numerical algorithm that has been used in the past to solve Maxwell's equations for waveguide problems. It is mainly used in the frequency domain. In this paper it is shown how the MoL can be used to solve initial value problems in the time domain. The required expressions are derived for one-dimensional structures, where the materials may be dispersive. The algorithm is verified with numerical results for homogeneous structures, and for the concatenation of standard dielectric and left handed materials.

1 Introduction

For developing waveguide structures numerical methods are widely used, because only in exceptional cases Maxwell's equations can be solved analytically. For studying e.g. the propagation of short pulses, time domain methods are best suited.

Recently there has been a huge interest in so called left handed materials (LHM see e.g. Eleftheriades and Balmain, 2005 Caloz and Itoh, 2006; other names found in the literature are e.g. double negative index material or negative index material).

In such materials the phase velocity is negative and the wave vector and the Poynting vector are anti-parallel. Due to the negative phase velocity it is not obvious that information is transmitted in forward direction. Therefore, time domain computations are performed to check if causality is violated.

As an example for such time domain methods we would like to mention the well known FDTD (finite difference time domain method see e.g. Taflove, 1995 Ziolkowski and Heyman, 2001, Ziolkowski, 2003). In this method all derivatives in time and space are approximated with finite differences.

An established technique in frequency domain is the Method of Lines (MoL) where besides the discretization also analytic expressions are used. Details can be found e.g. in the book articles Pregla and Pascher, 1989; Pregla, 1995 or the book Pregla, 2008.

Though the MoL is used in electromagnetics mainly as frequency domain method also time-domain computations have been made with this algorithm. In (Schiesser, 1991) the time dependency is treated with finite differences.

Another approach was followed by the group of Itoh (Nam et al., 1989a,b, 1988a,b, 1989c). A time dependency according to $\sin\omega t$ and $\cos\omega t$ was used and mainly the cut-off frequencies of waveguides were determined.

Recently, the MoL in frequency domain was combined with the discrete Fourier transformation to obtain time-dependent solutions (see Gerdes, 2007).

In the time domain MoL (TD-MoL) that is described in this paper, a different approach is followed. We will solve Maxwell's equation as initial value problem (as e.g. in the FDTD).

In the next section, we will derive suitable expressions. The basic principles are described in one dimension for dispersion free materials. Following, we show how material dispersion is taken into account. After that we present numerical results. To test the algorithm, we begin with a homogeneous structure. Once the correct implementation of the algorithm has been established, the concatenation of standard dielectric material with left materials is studied as further example. It should be mentioned that LHMs must be strongly dispersive (see e.g. Smith and Kroll, 2000). The paper ends with a short summary.

2 Theory

2.1 Dispersion free materials

In this section we will show the principles of the Method of Lines in the time domain. To develop suitable expressions, we start with Maxwell's equations:

$$\nabla \times \boldsymbol{E} = -\frac{\partial \boldsymbol{B}}{\partial t} \qquad \nabla \times \boldsymbol{H} = \frac{\partial \boldsymbol{D}}{\partial t} \tag{1}$$

For materials without dispersion (at least in the considered frequency domain) the relative permittivity resp. permeability are introduced, leading to the well known expressions:

$$\boldsymbol{D} = \varepsilon_0 \varepsilon_r \boldsymbol{E} \qquad \boldsymbol{B} = \mu_0 \mu_r \boldsymbol{H} \tag{2}$$

Next, the magnetic field is normalized according to

$$\widetilde{\boldsymbol{H}} = Z_0 \boldsymbol{H}$$

with the free space impedance Z_0 and the speed of light in vacuum c_0 determined as

$$\sqrt{\frac{\mu_0}{\varepsilon_0}} = Z_0 \qquad \frac{1}{\sqrt{\mu_0 \varepsilon_0}} = c_0$$

Then, Maxwell's Equations (1) can be writtten as

$$\nabla \times \boldsymbol{E} = -\frac{\mu_r}{c_0} \frac{\partial \widetilde{\boldsymbol{H}}}{\partial t} \qquad \nabla \times \widetilde{\boldsymbol{H}} = \frac{\varepsilon_r}{c_0} \frac{\partial \boldsymbol{E}}{\partial t} \tag{3}$$

For 3-D-considerations, we would use Eqs. (3) as starting point.

In the following, however, we restrict ourselves to one-dimensional problems. This has two reasons: (a) The basic principles of the algorithm are identical for 1-D–3-D problems and (b) obviously the numerical effort is lower for 1-D so that we start with this most simple case.

For one–dimensional problems with $\partial/\partial x = \partial/\partial y = 0$ we take the field components E_x and H_y and introduce the following abbreviations:

$$E = E_x \qquad H = \widetilde{H}_y$$

Then, Maxwell's equations (3) become:

$$\frac{\partial H}{\partial t} = -\frac{c_0}{\mu_r} \frac{\partial E}{\partial z} \qquad \frac{\partial E}{\partial t} = -\frac{c_0}{\varepsilon_r} \frac{\partial H}{\partial z} \tag{4}$$

As usual in the MoL, we discretize some of the derivatives with finite differences and use analytic expressions for the remaining ones. For the time domain algorithm, we discretize in z-direction and treat the time dependency analytically. The situation is shown in Fig. 1, where we see lines in t-direction, which may be seen as justification for the name "MoL" here.

As can be seen from Eq. (4) the electric and magnetic fields are coupled by first derivatives with respect to z. Therefore, as usual in finite difference methods, they are determined on different positions, i.e. shifted by half a discretization distance, as shown in Fig. 1. Since the computational window in z-direction must be finite, perfectly matched layers, as described e.g. in (Werner and Mittra, 1997) were introduced to avoid (resp. minimize) reflections at the boundaries.

For the further analysis the discretized fields are combined into a vector $[F]$ and the discretized derivatives into the operator matrix $\mathbf{Q}$. Then the expressions in Eq. (4)become:

$$\frac{\partial}{\partial t}[F] = -\mathbf{Q}[F] \qquad \text{with} \qquad [F] = \begin{bmatrix} [E] \\ [H] \end{bmatrix} \tag{5}$$

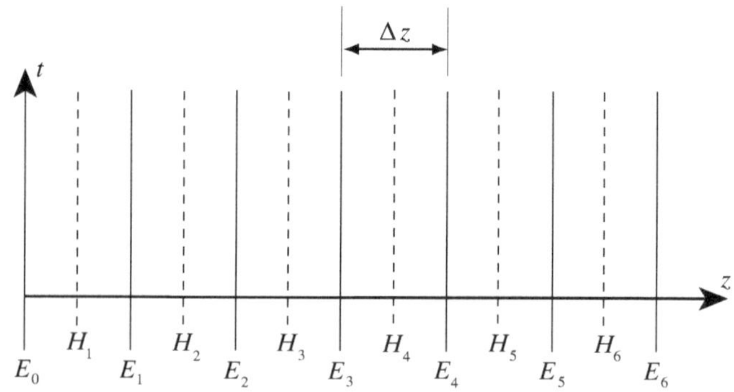

Fig. 1. Discretization in z-direction with finite differences

Note: the mathematical vectors containing the fields at different positions z were written in brackets to distinguish them from the physical vectors, which contain the spatial components of the fields and which were written in boldface.

With the exception that an analytic derivative with respect to t occurs, Equation (5) looks very similar to the discretized generalized transmission line equations (see e.g. Pregla, 2008 pp. 19–20) which are often used as starting point in the MoL in frequency domain.

The following steps, however, are different from the ones in frequency domain described e.g. in (Pregla, 2008). Particularly, we want to use Eq. (5) to solve an initial value problem with analytic expressions for $E_0(t)$ as input parameters. To obtain suitable expressions, we split the matrix $\mathbf{Q}$ into two parts as shown below:

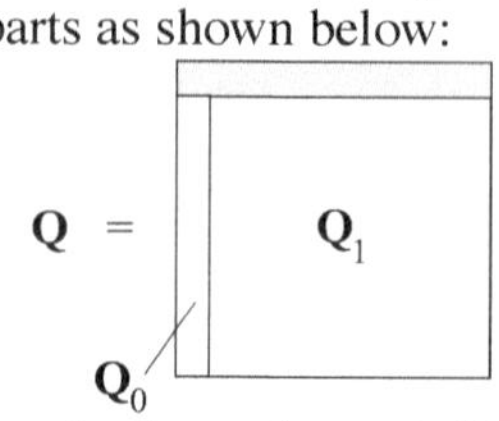

$$[F] = \begin{bmatrix} E_0 \\ [F_1] \end{bmatrix}$$

i.e. the first column (with which we multiply E_0) is separated from the remaining ones and the first row of $\mathbf{Q}$ is omitted. Then, Eq. (5) is rewritten in the following way:

$$\frac{\partial [F_1]}{\partial t} + \mathbf{Q}_1 [F_1] = -\mathbf{Q}_0 E_0(t) \tag{6}$$

Equation (6) describes a coupled, linear, inhomogeneous differential equation system (IDE), where E_0 (see Fig. 1) can be considered as source.

As well known from mathematics, the general solution of linear inhomogeneous differential equations is composed of the general homogeneous solution and a particular inhomogeneous one:

$$[F_1] = [F_h] + [F_i] \tag{7}$$

To obtain the homogeneous part, we must solve

$$\frac{\partial [F_h]}{\partial t} + \mathbf{Q}_1 [F_h] = [0] \tag{8}$$

As usual in the MoL, this system is diagonalized by transformation to principal axes:

$$\mathbf{Q}_1 = \mathbf{T}\boldsymbol{\Omega}\mathbf{T}^{-1} \tag{9}$$

where $\mathbf{T}$ and $\boldsymbol{\Omega}$ are the eigenvector/eigenvalue matrix of $\mathbf{Q}_1$. By transforming the fields according to

$$[F_\mathrm{h}] = \mathbf{T}[\overline{F}_\mathrm{h}] \tag{10}$$

we obtain

$$\frac{\partial[\overline{F}_\mathrm{h}]}{\partial t} + \boldsymbol{\Omega}[\overline{F}_\mathrm{h}] = [0] \tag{11}$$

with the solution:

$$[\overline{F}_\mathrm{h}(t)] = e^{-\boldsymbol{\Omega}t}[\overline{F}_\mathrm{h}(0)] \tag{12}$$

Next, we must find a particular solution of the inhomogeneous differential equation system. For this purpose, we transform $\mathbf{Q}_0$ and $[F_i]$ according to:

$$\overline{\mathbf{Q}}_0 = \mathbf{T}^{-1}\mathbf{Q}_0 \qquad [\overline{F}_\mathrm{i}] = \mathbf{T}^{-1}[F_\mathrm{i}]$$

Then, we obtain from Eq. (6):

$$\frac{\partial[\overline{F}_\mathrm{i}]}{\partial t} + \boldsymbol{\Omega}[\overline{F}_\mathrm{i}] = -\overline{\mathbf{Q}}_0 E_0(t) \tag{13}$$

As is known from mathematics, a particular solution can be constructed by variation of the constants, i.e. with the ansatz:

$$[\overline{F}_\mathrm{i}] = e^{-\boldsymbol{\Omega}t}[c(t)] \tag{14}$$

The first factor comes from the solution of the homogeneous differential equation and the second one has to be determined. Introducing Eq. (14) into Eq. (13) leads to

$$\frac{\partial[c(t)]}{\partial t} = -e^{\boldsymbol{\Omega}t}\overline{\mathbf{Q}}_0 E_0(t) \tag{15}$$

hence:

$$[c(t)] = -\int e^{\boldsymbol{\Omega}t}\overline{\mathbf{Q}}_0 E_0(t)\,dt \tag{16}$$

It should be mentioned that $\overline{\mathbf{Q}}_0$ just causes constant factors. Hence, analytic expressions can be given if the product between $e^{\boldsymbol{\Omega}t}$ and $E_0(t)$ can be integrated in closed form. Since the first factor ($e^{\boldsymbol{\Omega}t}$) only depends on the structure, the choice for $E_0(t)$ decides, if we are able to find such a closed form for the solution.

Usually, $E_0(t)$ is defined in time intervals; e.g. $E_0 = f_1(t)$ for $0 < t < \tau_1$, $f_2(t)$ for $\tau_1 < t < \tau_2$, etc. For this example, the fields must be continuous at $t = \tau_1, \tau_2 \ldots$

Then, the time dependent fields are computed with the following procedure: we begin with determining the eigenvectors/eigenvalues of the structure. It is important to note that this has to be done only once. After that, a particular solution of the inhomogeneous differential is constructed for the first time interval. Since we formulated an initial value problem, the fields at time $t = 0$ are known, e.g., $[F_\mathrm{in}(t=0)] = [0]$. With these initial values and the particular solution of the IDEs we are able to determine the homogeneous part for $t = 0$:

$$[F_\mathrm{h}^\mathrm{I}(t=0)] = [F_\mathrm{in}(t=0)] - [F_\mathrm{i}^\mathrm{I}(t=0)] \tag{17}$$

Then, the field distribution at the end of this time interval (i.e. for $t = \tau_1$) is determined analytically with Eqs. (12) and (14) and we can write for the whole fields:

$$[F^\mathrm{I}(t=\tau_1)] = [F_\mathrm{i}^\mathrm{I}(t=\tau_1)] + [F_\mathrm{h}^\mathrm{I}(t=\tau_1)] \tag{18}$$

(The superscript "I" corresponds to the first time interval.) At the beginning of the second time interval, again a solution of the inhomogeneous problem is constructed and the homogeneous part is used to ensure the continuity of the fields for $t = \tau_1$, resulting in:

$$[F_\mathrm{h}^\mathrm{II}(t=\tau_1)] = [F^\mathrm{I}(t=\tau_1)] - [F_\mathrm{i}^\mathrm{II}(t=\tau_1)] \tag{19}$$

This procedure (construction of a solution of the IDE, and determining the homogeneous part from the continuity-condition of the fields) is repeated for all intervals. The remaining computations to determine the field distribution at an arbitrary time are also done analytically, with very low numerical effort. For this purpose we have to store the homogenous and inhomogeneous solutions at the beginning of each time interval, i.e., $[F_\mathrm{h}^\mathrm{I}(0)]$, $[F_\mathrm{i}^\mathrm{I}(0)]$, $[F_\mathrm{h}^\mathrm{II}(\tau_1)]$, $[F_\mathrm{i}^\mathrm{II}(\tau_1)]\ldots$ and apply Eqs. (12 and 14). The storage itself is not very memory demanding since the homogeneous solutions are vectors, whereas the inhomogeneous ones are stored as matrices but with very few columns. For the example given in this paper (see Sect. 3) we have matrices with maximum seven columns (a sinusoidal wave was turned on/off with a polynomial of the sixth degree.)

Note: Since the geometric structure does not change with time, we could have written the Eq. (17–19) also for the transformed (overlined) fields.

2.2 Dispersive material

Dispersive material is introduced into the algorithm similarly to what is done in the FDTD (see e.g. Ziolkowski and Heyman, 2001; Ziolkowski, 2003). The material dependencies are written with the polarization and the magnetization:

$$\boldsymbol{D} = \varepsilon_0\boldsymbol{E} + \boldsymbol{P} \qquad \boldsymbol{B} = \mu_0(\boldsymbol{H} + \boldsymbol{M}) \tag{20}$$

In this paper, we use left handed material as example. The material parameters are described by the Drude model in the following way (see e.g. Ziolkowski, 2003):

$$\varepsilon(\omega) = \varepsilon_0\left(1 - \frac{\omega_\mathrm{pe}^2}{\omega(\omega - j\Gamma_\mathrm{e})}\right) \tag{21}$$

$$\mu(\omega) = \mu_0 \left(1 - \frac{\omega_{\mathrm{pm}}^2}{\omega(\omega - j\Gamma_{\mathrm{m}})}\right) \tag{22}$$

As before, we restrict ourselves in the following to the 1D case and introduce the following abbreviations:

$$E = E_x \qquad H = Z_0 H_y \qquad P = Z_0 P_x \qquad M = Z_0 M_y$$

Then, the material parameters in Eqs. (21–22) relate E and P resp. H and M in the following way:

$$-\omega^2 P + j\omega\Gamma_{\mathrm{e}} P = \omega_{\mathrm{pe}}^2 c_0 E \tag{23}$$

$$-\omega^2 M + j\omega\Gamma_{\mathrm{e}} M = \omega_{\mathrm{pe}}^2 H \tag{24}$$

With the help of an inverse Fourier-transform this can be rewritten to differential equations:

$$\frac{\partial^2 P}{\partial t^2} + \Gamma_{\mathrm{e}} \frac{\partial P}{\partial t} = \omega_{\mathrm{pe}}^2 c_0 E \tag{25}$$

$$\frac{\partial^2 M}{\partial t^2} + \Gamma_{\mathrm{m}} \frac{\partial M}{\partial t} = \omega_{\mathrm{pm}}^2 H \tag{26}$$

Since Eqs. (25–26) contain derivatives with respect to t we could apply them to the TD-MoL scheme. However, only time derivatives of P and M occur and we can simplify (in terms of numerical effort) these expressions by introducing electric and magnetic currents Ziolkowski and Heyman, 2001; Ziolkowski, 2003:

$$J = \frac{\partial P}{\partial t} \qquad K = \frac{\partial M}{\partial t}$$

Then, Eqs. (25–26) together with Maxwell's equations result in the following coupled differential equation system Ziolkowski and Heyman, 2001; Ziolkowski, 2003:

$$\frac{\partial E}{\partial t} = -c_0 \frac{\partial H}{\partial z} - c_0 J \tag{27}$$

$$\frac{\partial J}{\partial t} = -\Gamma_{\mathrm{e}} J + \omega_{\mathrm{pe}}^2 c_0 E \tag{28}$$

$$\frac{\partial H}{\partial t} = -c_0 \frac{\partial E}{\partial z} - K \tag{29}$$

$$\frac{\partial K}{\partial t} = -\Gamma_{\mathrm{m}} K + \omega_{\mathrm{pm}}^2 H \tag{30}$$

As can be seen, first derivatives with respect to t occur on the left hand side of these equations, whereas only derivatives with respect to z and the quantities themselves occur on the right hand side. Therefore, after discretization and combining the fields and currents in a vector

$$[F] = \begin{pmatrix} [E] \\ [J] \\ [t] \\ [K] \end{pmatrix}$$

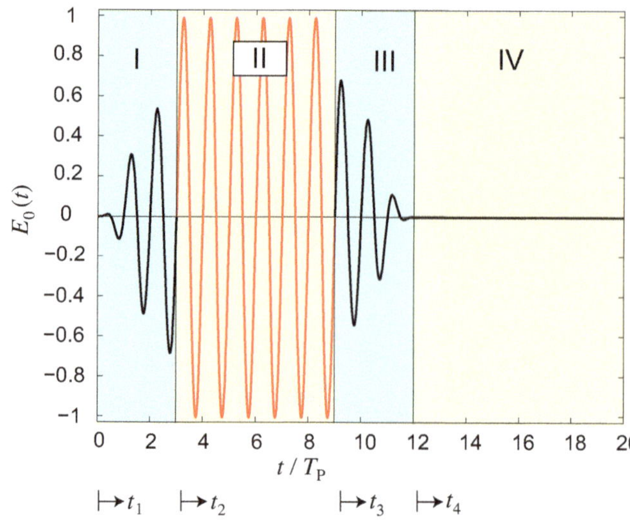

Fig. 2. Input time signal $E_0(t)$

we formally end up with the same system of coupled, linear, inhomogeneous differential Eq. (5) that we had in the dispersion free case. Obviously, $[F]$ and the operator Matrix $\mathbf{Q}$ differ here from this dispersion free case. However, for the further steps in the algorithm we may refer to Sect. 2.1 and do not need to repeat them.

3 Numerical results

3.1 Input field

To validate the algorithm, we examined various structures, where a sinusoidal wave was excited, which was smoothly turned on/off. In principle, the input field E_0 is shown in Fig. 2.

As can be seen, four intervals occur. Mathematically, the field in these intervals is given by the following expressions:

$$E_0(t) = \begin{cases} \mathrm{I}: \; A_0 \sum\limits_{k=0}^{N_1} a_k t_1^k \sin\omega_0 t_1 \\ \mathrm{II}: \; A_0 \sin\omega_0 t_2 \\ \mathrm{III}: \; A_0 \sum\limits_{k=0}^{N_2} b_k t_3^k \sin\omega_0 t_3 \\ \mathrm{IV}: 0 \end{cases} \tag{31}$$

For such input fields the integral in Eq. (16) can be given in closed form as is shown in the appendix.

In particular, the following parameters taken from Ziolkowski and Heyman, 2001; Ziolkowski, 2003 were used to turn on/off the sinusoidal wave:

$$\frac{E_{0\mathrm{on}}(t1)}{A_0} = \left(\frac{10t_1^3}{(m_{\mathrm{on}}T_{\mathrm{p}})^3} + \frac{-15t_1^4}{(m_{\mathrm{on}}T_{\mathrm{p}})^4} + \frac{6t_1^5}{(m_{\mathrm{on}}T_{\mathrm{p}})^5}\right) \sin\omega_0 t_1$$

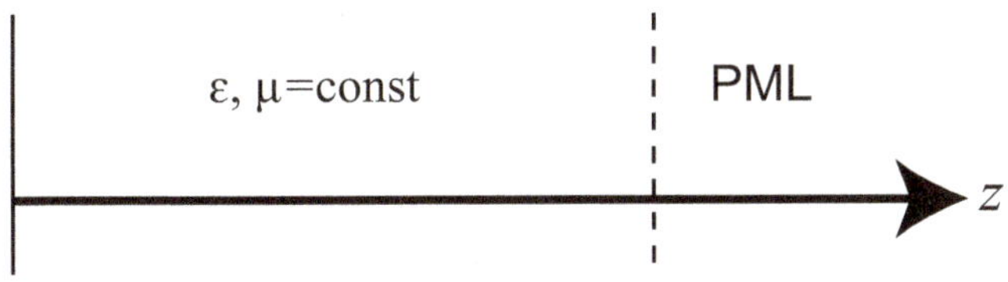

Fig. 3. Test structure with perfectly matched layers

$$\frac{E_{0\text{off}}(t_3)}{A_0} = \left(1 - \frac{10t_3^3}{(m_{\text{off}}T_\text{p})^3} + \frac{15t_3^4}{(m_{\text{off}}T_\text{p})^4} + - \frac{6t_3^5}{(m_{\text{off}}T_\text{p})^5}\right) \sin\omega_0 t_3$$

T_p stands for the cycle period and is related to the frequency via $\omega_0 T_p = 2\pi$. The number of periods to turn on/off the signal are indicated by $m_{\text{on,off}}$. In our studies we used $m_{\text{on}} = m_{\text{off}} = 2$, and a constant sinusoidal wave for $m_\text{s} = 4$ periods so that the local time variables $t_{1,\ldots,4}$ and the global one (t) are related in the following way:

$$t_1 = t \qquad t_2 = t - m_{\text{on}}T_p \qquad t_3 = t - (m_{\text{on}} + m_\text{s})T_p$$
$$t_4 = t - (m_{\text{on}} + m_\text{s} + m_{\text{off}})T_p$$

3.2 Examined structures

As first test, the structure shown in Fig. 3 was examined. The field was injected into the homogeneous section which is terminated with perfectly matched layers (PML). Numerical results are shown in Fig. 4. The electric field is presented at different moments in times. As expected, the field travels from left to right (i.e. in positive z–direction). Since there is no dispersion, the shape of the field remains unchanged.

Furthermore, it can be seen that the fields are damped by the PMLs and that no visible reflections occur, neither at the interface between the homogeneous structure and the PML nor at the end of the device. These results suggest that the algorithm (incl. PMLs) works in principle and has been implemented correctly .

A more complicated device was studied in the following. It consists of the concatenation of standard dielectric– (i.e. "right handed– (RHM)") and left handed–materials (LHM). In LHMs the material parameters ε_r and μ_r are negative in the interesting frequency regions. This leads to peculiar characteristics, e.g. the phase velocity is negative and waves travel "backwards". Due to physical reasons, LHMs must be strongly dispersive. In this paper, we will not discuss all these characteristic but would like to refer the interested reader e.g. to the books Eleftheriades and Balmain, 2005; Caloz and Itoh, 2006 and to the Ziolkowski-papers Ziolkowski and Heyman, 2001; Ziolkowski, 2003.

Since waves travel backwards in LHMs one might wonder if basic physical principles, like causality, are violated. Therefore, Ziolkowski used the FDTD and made time domain calculations Ziolkowski and Heyman, 2001; Ziolkowski, 2003. It was found that the front of the wave always propagates forward, i.e. that the causality principle is fulfilled. In this paper, we use the LHM structure as example to test our developed algorithm for dispersive materials.

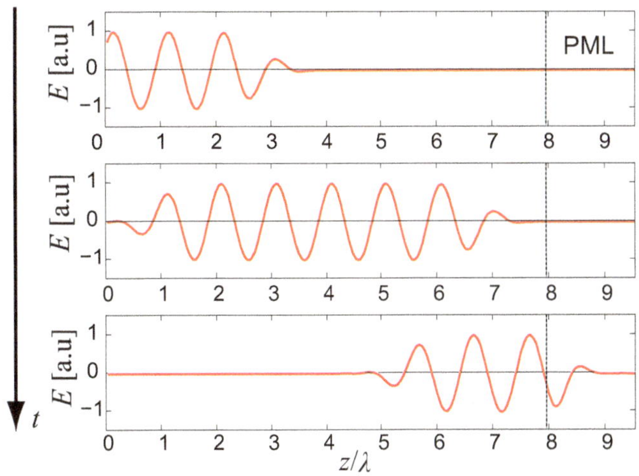

Fig. 4. Field determined at different times in the structure shown in Fig. 3

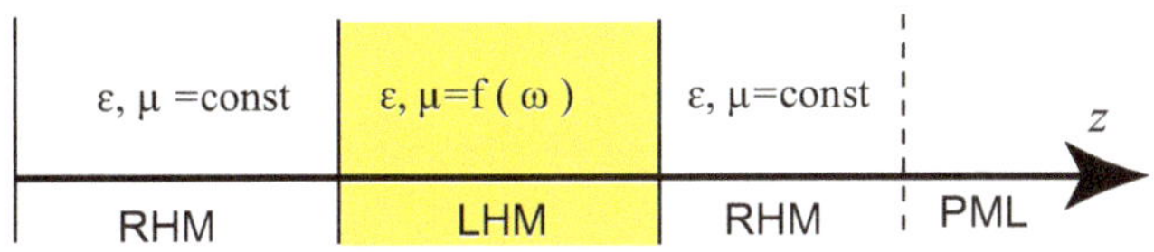

Fig. 5. Concatenation of right handed and left handed material

The structure under study is presented in Fig. 5. The left handed material is sandwiched between right handed material with $\varepsilon_\text{r} = \mu_\text{r} = 1$. Again the structure is truncated by PMLs. For the LHM we take permittivity and permeability from the Drude-Model Eqs. (21–22). Particularly, the following parameters from Ziolkowski and Heyman, 2001; Ziolkowski, 2003 were taken:

$$\omega_\text{p}^2 = 2\omega_0^2 \qquad \Gamma = 3.75 \times 10^{-4}\omega_\text{p}$$

Hence, for ω_0 (i.e. the frequency of the sine-function) the relative permittivity and permeability are:

$$\varepsilon_\text{r}(\omega_0) = \mu_\text{r}(\omega_0) \approx -1$$

The numerical results are shown in Fig. 6. Let us begin with a few remarks. Generally, the time progresses from the top graph to the bottom one. In each of the graphs two curves are sketched, which show the situation in a short time distance.

In the top graph the electric field has just reached the first RHM-LHM interface. So, the wave has travelled to the right in the standard RHM. This behaviour is also recognized in the beginning of the LHM.

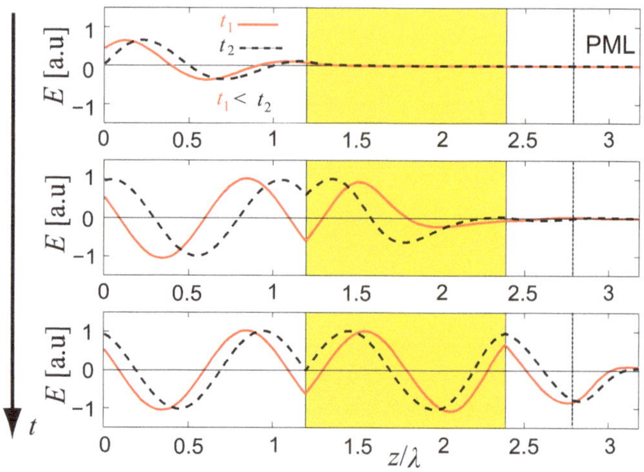

Fig. 6. Electric field distribution in the structure shown in Fig. 5 , snapshots at three different times

In the middle graph the wave front has arrived at the second (LHM-RHM) interface. As before, the wave front travels to the right. However, it can also be observed that at the LHM input, the phase of the wave is already propagating backwards i.e. to the left.

Finally, the bottom graph shows the steady state. Here we observe a positive phase velocity in the two right handed media, but a negative phase velocity in the LHM. We can also see that the PML region damps the fields and does not cause reflections at that interface. These results agree with the ones from the FDTD and can be interpreted in two ways:

1. The FDTD findings (Ziolkowski and Heyman, 2001; Ziolkowski, 2003) are confirmed. Left handed materials do not violate the causality principle, because the wavefront propagates in positive direction.

2. (more important) The TD-MoL is able to treat material dispersion.

4 Summary and conclusions

In this paper we presented a time domain Method of Lines. It could be shown that the algorithm works in principle, at least as long as we restrict ourselves to 1D-problems in space. Dispersive materials can be treated. A few potential advantages (compared to the FDTD) are:

- The solutions in time are given in an analytic form. Therefore, we can look at field distributions at arbitrary (particularly late) times immediately.

- In the FDTD fine grid sizes (in space) also require very short time-steps to keep the algorithm stable. Since analytic expressions are used in the TD-MoL, we may look at the fields at arbitrary times without stability problems and without any loss of accuracy.

However, besides these advantages a few problems must be named as well. In principle we could apply the developed method also to higher (2-D–3-D) problems. Unfortunately, the numerical effort rises significantly when higher dimensions are considered. Therefore, the topic of the current work deals with reducing this numerical effort to be able to treat 2-D, 3-D problems in the future.

Appendix A

The turning on/off the sinusoidal waves was described by the functions given in Eq. (31). For computing a particular solution of the inhomogeneous differential equation system (13) we must determine $[c(t)]$ with the integral given in Eq. (16). The integral contains the product of a diagonal matrix $(e^{\boldsymbol{\Omega}t})$ with a constant vector $(\overline{\boldsymbol{Q}}_0)$. Therefore, we end up with a system of decoupled scalar equations, and we must compute integrals of the form

$$\int t^k e^{\Omega t} \sin\omega_0 t\, dt \qquad \int e^{\Omega t} \sin\omega_0 t\, dt$$

For the first integral we obtain

$$\begin{aligned}\int t^k e^{\Omega t} \sin\omega_0 t\, dt &= \int t^k e^{\Omega t}\left(\frac{e^{j\omega t}-e^{j\omega t}}{2j}\right) dt \qquad \text{(A1)}\\ &= \frac{1}{2j}\int t^k \left(e^{(\Omega+j\omega)t}-e^{(\Omega-j\omega)t}\right) dt\\ &= \frac{t^k}{2j}\left(\frac{e^{(\Omega+j\omega)t}}{\Omega+j\omega}-\frac{e^{(\Omega-j\omega)t}}{\Omega-j\omega}\right)+\\ &\quad -\frac{k}{2j}\int t^{k-1}\left(\frac{e^{(\Omega+j\omega)t}}{\Omega+j\omega}-\frac{e^{(\Omega-j\omega)t}}{\Omega-j\omega}\right) dt\end{aligned}$$

As seen, the integral is solved recursively. The expression in the last row can be found in mathematical formularies. (Here we used Bronstein and Semendjajew (1979).)

To compute $\int e^{\Omega t} \sin\omega_0 t\, dt$ we can introduce $k = 0$ into Eq. (A1). Alternatively, we find from Bronstein and Semendjajew (1979):

$$\int e^{\Omega t} \sin\omega_0 t\, dt = \frac{e^{\Omega t}}{\Omega^2+\omega_0^2}\left(\Omega\cos\omega_0 t+\omega_0\sin\omega_0 t\right) \qquad \text{(A2)}$$

As can be shown very easily, the expressions in Eq. (A1) can be transformed into the one in Eq. (A2) in case $k = 0$.

References

Bronstein, I. and Semendjajew, K.: Taschenbuch der Mathematik, Verlag Harri Deutsch, Thun Frankfurt/Main, 7 Edn., 1979.

Caloz, C. and Itoh, T.: Electromagnetic Metamaterials, Wiley-Interscience, Hoboken, NJ, USA, 2006.

Eleftheriades, G. V. and Balmain, K. G., eds.: Negative-Refraction Metamaterials, Wiley-Interscience, Hoboken, NJ, USA, 2005.

Gerdes, J.: Hybrid Time–Frequency Domain Eigenmode Propagation Analysis of Optical Waveguides Based on the Method of Lines, in: ICTON Conf., 4, 324–327, Rome, Italy, 2007.

Nam, S., El–Ghazaly, S., Ling, H., and Itoh, T.: Time–domain method of lines, Electron. Lett., 24, 128–129, 1988a.

Nam, S., El–Ghazaly, S., Ling, H., and Itoh, T.: Time–domain method of lines applied to a partially filled waveguide, in: IEEE MTT-S Int. Symp. Dig., 2, 627–630, 1988b.

Nam, S., Ling, H., and Itoh, T.: Characterization of uniform microstrip line and its discontinuities using the time–domain method of lines, IEEE Trans. Microwave Theory Tech., 37, 2051–2057, 1989a.

Nam, S., Ling, H., and Itoh, T.: Time–domain method of lines applied to planar guided wave structures, IEEE Trans. Microwave Theory Tech., 37, 897–901, 1989b.

Nam, S., Ling, H., and Itoh, T.: Time-Domain Method of Lines Applied to the Uniform Microstrip Line and its Step Discontinuity, in: IEEE MTT-S Int. Symp. Dig., 3, 997–1000, 1989c.

Pregla, R.: MoL-BPM Method of Lines Based Beam Propagation Method, in: Methods for Modeling and Simulation of Guided-Wave Optoelectronic Devices (PIER 11), edited by Huang, W. P., Progress in Electromagnetic Research, pp. 51–102, EMW Publishing, Cambridge, Massachusetts, USA, 1995.

Pregla, R.: Analysis of Electromagnetic Fields and Waves - The Method of Lines, Wiley & Sons, Chichester, UK, 2008.

Pregla, R. and Pascher, W.: The Method of Lines, in: Numerical Techniques for Microwave and Millimeter Wave Passive Structures, edited by: Itoh, T., 381–446, J. Wiley Publ., New York, USA, 1989.

Schiesser, W. E.: The Numerical Method of Lines Integration of Partial Differential Equations, Academic Press, San Diego , USA, 1991.

Smith, D. R. and Kroll, N.: Negative refractive index in left-handed materials, Phys. Rev. Lett., 85, 2933–2936, 2000.

Taflove, A.: The Finite–Difference Time–Domain Method, Computational Electrodynamics, Artech house, inc, Norwood, MA, 1995.

Werner, D. H. and Mittra, R.: A new field scaling interpretation of Berenger's PML and its comparison to other PML formulations, Microw. and Opt. Tech. Lett, 16, 103–106, 1997.

Ziolkowski, R. W.: Pulsed and CW Gaussian beam interactions with double negative metamaterial slabs, Opt. Expr., 11, 662–681, 2003.

Ziolkowski, R. W. and Heyman, E.: Wave propagation in media having negative permittivity and permeability, Phys. Rev. E, 64, 056625, 1–15, 2001.

Multi-sensor Doppler radar for machine tool collision detection

T. J. Wächter[1]**, U. Siart**[1]**, T. F. Eibert**[1]**, and S. Bonerz**[2]

[1]Technische Universität München, Lehrstuhl für Hochfrequenztechnik, Arcisstr. 21, 80333 Munich, Germany
[2]Ott-Jakob Spanntechnik GmbH, Industriestr. 3–7, 87663 Lengenwang, Germany

Correspondence to: T. J. Wächter (thomas.waechter@tum.de)

Abstract. Machine damage due to tool collisions is a widespread issue in milling production. These collisions are typically caused by human errors. A solution for this problem is proposed based on a low-complexity 24 GHz continuous wave (CW) radar system. The developed monitoring system is able to detect moving objects by evaluating the Doppler shift. It combines incoherent information from several spatially distributed Doppler sensors and estimates the distance between an object and the sensors. The specially designed compact prototype contains up to five radar sensor modules and amplifiers yet fits into the limited available space. In this first approach we concentrate on the Doppler-based positioning of a single moving target. The recorded signals are preprocessed in order to remove noise and interference from the machinery hall. We conducted and processed system measurements with this prototype. The Doppler frequency estimation and the object position obtained after signal conditioning and processing with the developed algorithm were in good agreement with the reference coordinates provided by the machine's control unit.

1 Introduction

Collisions within the operating space of machine tools have many different causes. Main causes for collisions during human operation are wrong programming (wrong part, compensation or offsets), wrong machine setup (wrong tool, clamping or raw parts) and inattention or faulty operation by the operator himself.

The so called technological collisions concern crashes between tool and work piece only. The second group, the geometric collisions, involves further elements of the machine like the bearings, holder, spindle, axis or the driving shaft.

In the case of tool collisions, tremendous impact forces cause heavy damage to the work piece, the tool and to other elements of the drive train. The results are expensive damage, costs for production downtimes and high rejection rates. The goal is to guarantee high machine availability to maximize efficiency and output.

Up to now, no system is known which reliably provides information in order to avoid collisions with non-moving components and to actively protect the machine components against physical damage. There are some promising sensor based systems that can decouple the inner main motor spindle from the outer spindle stock, hence protecting the components from mechanical overload to reduce damage to a minimum. Other methods, without sensors, monitor the current flux through the motor and driving chain and activate when a certain threshold is exceeded. An elaboration on available safety systems can be found in Abele et al. (2012). All of these are reactive systems which are activated after a crash. With these, mitigation of damage is possible, but not their prevention.

Optical systems like laser scanners or (multi-)camera based systems are active observers. The disadvantage of these systems is that they are blinded due to cooling fluid, drizzle, dust and dirt from the milling treatment which pollute the lens or the sensor space. So their use is limited to phases where the metal processing is stopped or to other applications with non-cutting manufacture.

In this paper, the implementation of a prototype for a new possible solution is presented based on an autonomous active radar sensor safety system which operates in the 24 to 24.25 GHz ISM band. The utilized sensors are commercially available low-cost modules for short-range radar (SRR) applications, containing all front-end parts of a CW radar, including complex down-conversion. Figure 1 shows the prototype concept, here equipped with four modules in a

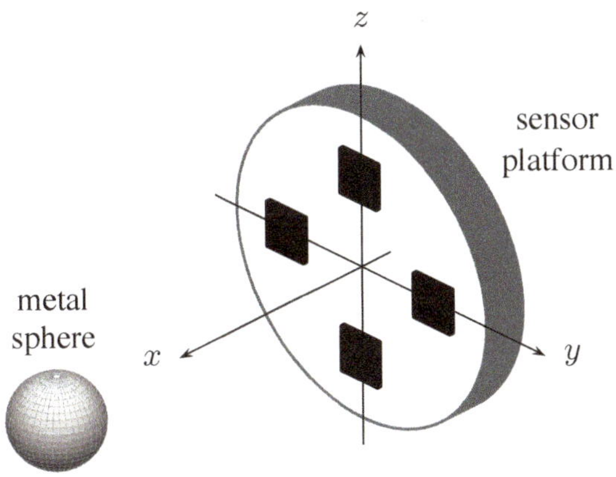

Figure 1. Concept of the toroidal holder with four modules in a symmetric arrangement and the internal coordinate system. A metal sphere serves as a measurement target.

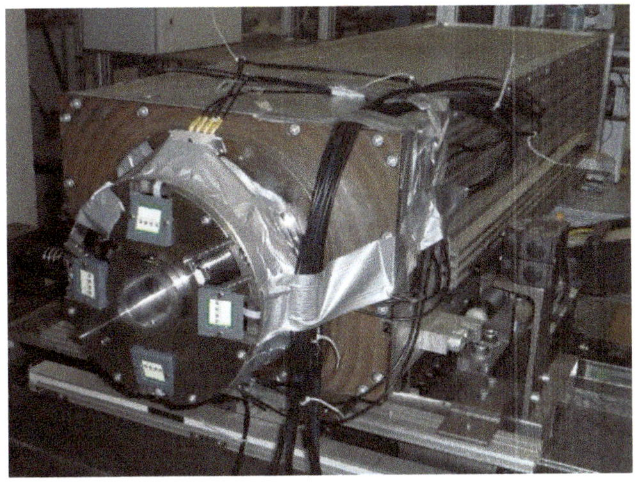

Figure 2. The implemented toroidal sensor platform equipped with four radars and integrated amplifiers embedded in an accessible, maneuverable test environment. Obtained signals are conducted via coaxial lines to the data storage and processing system.

symmetrical, perpendicular arrangement. The origin of the internal coordinate system is at the center point of the holder (sensor platform). Several modules are arranged around the milling machine tool to provide complete coverage of the three-dimensional surrounding space.

We designed a toroidal holder with integrated low-noise amplifiers (LNAs) which allows test scenarios with different antenna configurations involving up to five radar modules with variable radial distance, angle and positions in x direction. Figure 2 shows the implementation with four radar modules and the corresponding integrated LNAs embedded in an open test environment. Some important properties of this additional system and the CW concept are its independence from the acting machine tool, robustness against environmental influences and reliability due to its simplicity.

Microwave sensorics is more suited for this application than cameras and optical sensors. This is due to the superior properties of microwaves with the above-mentioned problems with dirt and moisture in the operating space. Microwave sensorics is also more suited than acoustical and ultrasonic sensors. The speed of sound (300 mm ms^{-1}) causes long signal delays compared to electromagnetic wave propagation.

Operation inside the enclosure of the machine and processing the diverse radar returns from there is one of the challenges for the signal processing. Multiple reflexions at the walls at arbitrary distances cause clutter effects and increase the noise level. Further challenges are the non-stationary environment, varying shapes and geometries of the milling tools as well as swarf from treatment in the operating zone.

In this first approach the system development concentrated on using CW radar only. This, on the one hand, is due to the robustness of the technology and the low cost arguments of the electronic hardware. On the other hand, the available space in any machine tool is limited. This only permits additional components of small size and low complexity. Furthermore, the large operating temperature range and the rapid fluctuation of temperature must be dealt with by the equipment to meet the high demands on linearity of frequency ramps in frequency modulated CW radar (FMCW) or pulsed radar to provide proper distance information. Also the use of the 24 GHz wideband technology with 5 GHz bandwidth as in automotive SRR was ceased as of June 2013 (Strohm et al., 2005). The necessary bandwidth for a specific range resolution Δr can be derived from $\Delta r = c_0/(2\Delta f)$, where c_0 is the speed of light and Δf is the absolute bandwidth. With the automotive SRR, a sufficient resolution of 3 cm is achievable if the full bandwidth of 5 GHz is be used.

A second figure of merit is the accuracy δr which is defined as (e.g., Skolnik, 1960; Barton, 1964),

$$\delta r = \tau_\text{p} \frac{c_0}{\sqrt{2\text{SNR}}}, \quad (1)$$

with the pulse rise time τ_p and the signal-to-noise ratio SNR. Just like the resolution, the accuracy depends on the bandwidth which determines $\tau_\text{p} \approx 0.35/\Delta f$. Both conditions, sufficient range resolution and accuracy, cannot be met in the 24 to 24.25 GHz ISM-band.

2 Designed multi-sensor Doppler radar network

From the viewpoint of hardware one of the limiting factors in this application is the available installation space in the machine tool. A block diagram of a single sensor module and the signal processing components is given in Fig. 3 showing the RF front-end, which includes the 24 GHz voltage controlled oscillator (VCO), a 90° hybrid coupler, the transmitting (Tx) and receiving (Rx) 4-patch antenna array and the IQ mixing element. Due to the homodyne mixing process the

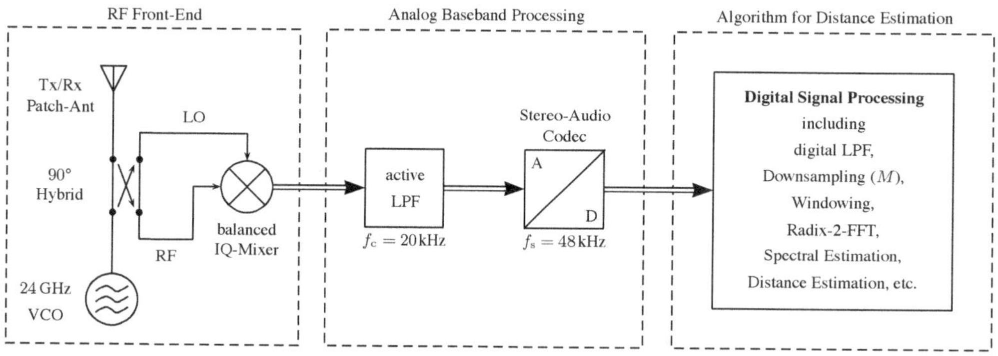

Figure 3. Block diagram of the radar system and signal processing components of a single module.

Table 1. Overview of several system parameters of the prototype.

system parameter	value
supply voltage	+5 V
power consumption	$\sim 4 \times 0.6$ W
center frequency	24.125 GHz
antenna gain	8.6 dBi
transmit power	+15 dBm (EIRP)
half-power beamwidth	80° (Azimuth)
	34° (Elevation)
IF bandwidth	3 Hz to 20 kHz
IF gain	40 dB
receiver sensitivity	−80 dBm
	($B = 20$ kHz, $R_{IF} = 1$ kΩ)
Doppler shift	160 Hz/(m s^{-1})
sampling frequency	48 kHz
ADC resolution	16 bit

output of the module is the downconverted complex Doppler signal. A frequency gap of 30 MHz between the carrier frequencies is introduced by tuning the VCOs to avoid crosstalk between the single modules of the radar network. Interfering down-conversion is then filtered out. The analog input stage consists of an active anti-aliasing low-pass filter (LPF) with a 3 dB cutoff frequency of 20 kHz, based on a low-noise operational amplifier. The input impedance is 1 kΩ. This configuration allows a receiver sensitivity of −80 dBm. A standard stereo-audio codec is used for converting the analog quadrature signal to a digital IQ data stream at a sampling frequency of $f_s = 48$ kHz. Some system parameters of the radar module and the input stage are given in Table 1, compare RFbeam Microwave GmbH (2011).

As the radar system acts in severe environmental conditions with fluid, drizzle, heat etc., hermetically sealed modules are necessary. The enclosures of the radar front-end in Fig. 2 are made from polyethylene and are closed and sealed for the final application. Additionally, they increase the flexibility for positioning and tilting of the modules. Coverage simulations based on deterministic ray-tracing showed that 15° tilting of the modules gives best coverage of the surrounding space around the tool (Azodi et al., 2013).

The input stage is built into a metal shielding enclosure. It is supposed to protect the circuit not only against electromagnetic interference but also against the harsh environmental conditions in machine tools. Experiments showed that interference from power supplies, switching electrical actuators and mechanical vibrations are there to effectively increase the noise level. The high-frequency interference can be suppressed after data acquisition by digital filtering as described in the next section. Low frequency components in the sub-kilohertz range remains as superimposed interfering signals.

3 Implemented digital signal processing stage

As the radar modules are based upon a homodyne architecture, the IQ mixer output appears in baseband centered around DC. The analog bandwidth of the active LPF is from 3 Hz to 20 kHz. This covers the needed range of relative velocities especially for slowly moving targets.

After digital conversion the signal is buffered and the DC bias is removed. Next a decimation stage follows. It consists of a digital LPF and subsequent downsampling by factor of M. Figure 4 illustrates the preprocessing steps between ADC and spectral frequency estimation. The digital LPF is a 4th order Chebyshev type I filter with 0.25 dB ripple and $f_c = 160$ Hz cutoff frequency. This cutoff equates to the Doppler shift due to the highest expected relative velocity of 1 m s^{-1}. The filter process reduces the noise contribution, hence the sensitivity level is decreased down to −101 dBm.

Each observation window has length NT seconds and is weighted with a window function after decimation. $T = 1/f_s$ is the sampling period and N is the number of sampling points. The decimation process does not change the observation period. The new sampling period is $T_M = MT = M/f_s$ and the number of sampling points is N/M. A Kaiser window with 80 dB sidelobe suppression has proven sufficient selectivity and low spectral leakage. The Kaiser window shape parameter value was chosen as $\alpha = 3.422$.

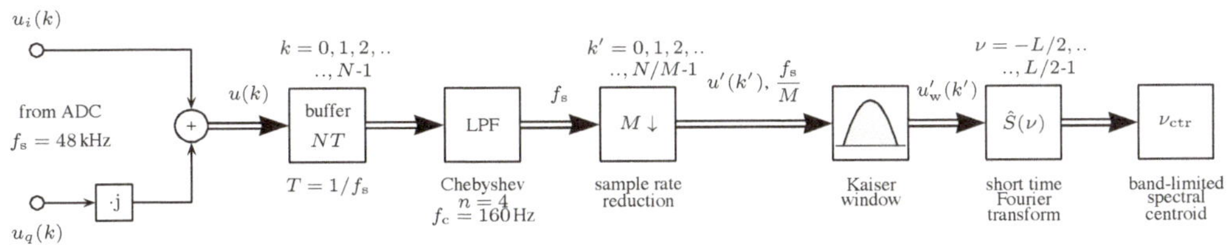

Figure 4. Preprocessing of complex signal from one single path.

Frequency resolution is determined by the duration of observation. The number of sampling points for each segment are kept to $N \sim 2^n$ for utilization of the efficient Radix-2-FFT algorithm. To extract Doppler shift and relative velocity, the short time Fourier spectrum is derived from each data segment by the estimator

$$\hat{S}(\nu) = \sum_{k'=0}^{L-1} u'_{\mathrm{w}}(k') \exp\left(-j2\pi \nu k'/L\right), \tag{2}$$

where k' is an integer and $u'_{\mathrm{w}}(k')$ is the preprocessed signal. The Fourier spectrum is calculated at discrete points ν, where $\nu = -L/2, \ldots, L/2-1$ and $L = N/M$.

Frequency estimation is based on computation of the bandwidth limited spectral centroid. The bandwidth for calculation is limited to a window which is identical to the LPF passband, excluding a small guard interval around DC. Exploiting the sign of the received Doppler shift provided by the IQ mixer architecture the integration limits can be set to the frequency indices $\nu_{\min} < \nu < \nu_{\max}$ where all indices are positive for positive Doppler shifts and negative for negative Doppler shifts. The spectral centroid is then calculated only in the frequency interval of interest by

$$\nu_{\mathrm{ctr}} = \frac{\sum\limits_{\nu=\nu_{\min}}^{\nu_{\max}} \nu \left|\hat{S}(\nu)\right|^2}{\sum\limits_{\nu=\nu_{\min}}^{\nu_{\max}} \left|\hat{S}(\nu)\right|^2}. \tag{3}$$

Limiting the summation in Eq. (3) reduces noise to a minimum and prevents erroneous biased estimation. In case of multi-target scenarios different methods must be utilized regarding more than one spectral maximum. But this case is not discussed in this paper since first investigations and performance tests assume that there is a single dominant target.

From ν_{ctr} the value of $\hat{f}_{\mathrm{d}} = \nu_{\mathrm{ctr}} f_{\mathrm{s}}/(LM)$ is calculated. An estimate for the relative velocity can be obtained with the well-known equation $\hat{v}_{\mathrm{rel}} = \hat{f}_{\mathrm{d}}\lambda/2$, where $\lambda = c_0/f_{\mathrm{c}}$ is the wavelength of the corresponding carrier frequency f_{c}.

Figure 5 shows a typical spectrogram acquired from the observation of the space in front of a sensor with one single dominant target. This example is based on real measurements performed with the prototype shown in Fig. 2. In this case the sensor platform is centrally arranged to the sphere and first approaches to it and afterwards departs from it. The spectrogram in Fig. 5 shows a distinct main frequency component.

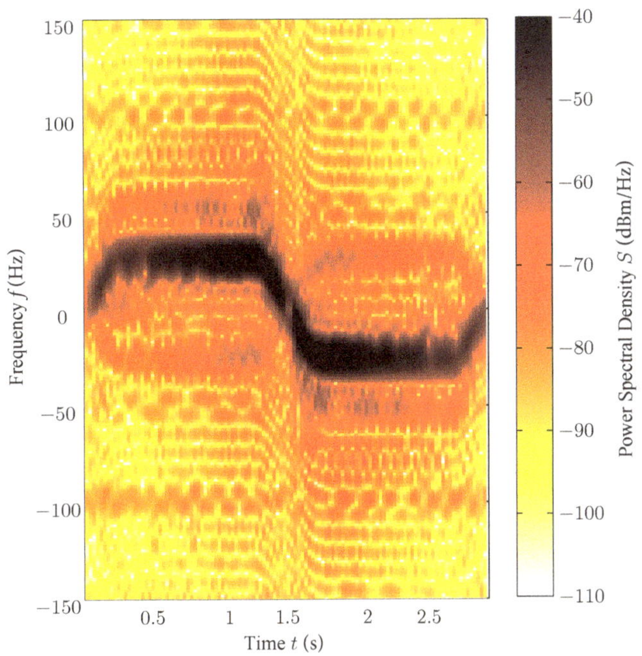

Figure 5. Doppler spectrogram for a closing and receding movement of the sensor platform. A metallic sphere serves as reflecting object.

Short phases of acceleration and deceleration at the beginning and the end and a turning point can be identified.

A critical point for unambiguous detection and tracking of objects are spurious frequency components that are observable in Fig. 5. These artifacts arise from IQ imbalance. The origins of this analog signal distortion are fabrication tolerances of the Schottky diodes and the transmission lines in the mixer stage. Shifts or distortions of the diode characteristic on the one hand and slightly unequal power division or phase deviation from ideal dividers on the other hand are the consequences. This causes variations between the output signals of the two mixing units which results in degraded signal quality. Like clutter, spurious peaks potentially increase the false alarm rate considerably. Especially, if solids with fluctuating radar cross section (RCS) have to be detected, e.g. cuboids or cylinders, image components will influence the algorithm performance of the decision unit when this special problem remains unconsidered. This issue is investigated in detail in future work.

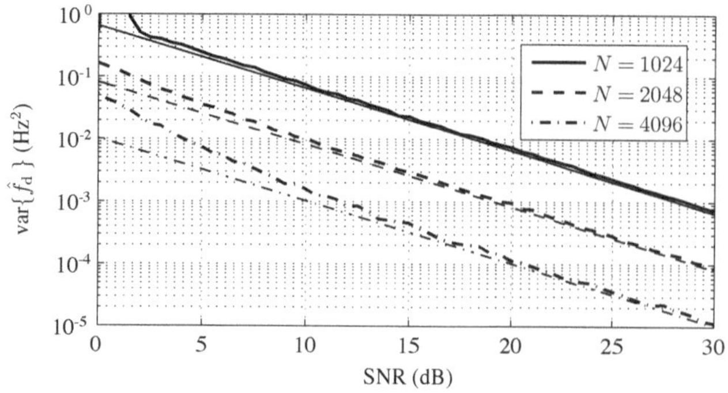

Figure 6. Frequency estimation simulation results for a single complex sinusoid with unknown amplitude and phase and the theoretical bounds for estimation variances. $f_s = 48\,\text{kHz}$, $f_d = 25.6\,\text{Hz}$.

4 Single target scenario

In a single target scenario with one dominant scatterer at position $\boldsymbol{r}_0 = (x_0,\, y_0,)^{\mathrm{T}}$ the observed discrete-time Doppler shift signal detected by the ith sensor is modeled by

$$u^i(k) = A^i \exp\left(\mathrm{j}(\Omega_{\mathrm{d}}^i k + \varphi^i)\right) + w^i(k), \tag{4}$$

where $\Omega_{\mathrm{d}} = \omega_{\mathrm{d}} T$ is the normalized Doppler frequency and w is band-limited measurement and processing noise. Measurements reveal that w is Gaussian white noise with the normal distribution

$$p(\boldsymbol{w}) \sim \mathcal{N}(0, \mathbf{C}), \tag{5}$$

where $\mathbf{C}$ is the noise covariance matrix.

In this application the parameters A, ω_{d} and φ are all unknown. In the following the index i is omitted. The Cramer-Rao-Bound (CRB) is a lower bound on the minimum achievable variances depending on the signal properties. In (Rife and Boorstyn, 1974) results for a discrete-time single complex sinusoid can be found. The CRB on the variance of the Doppler frequency estimate $\hat{f}_{\mathrm{d}} = \hat{\omega}_{\mathrm{d}}/(2\pi)$ is given by

$$\operatorname{var}\left\{\hat{f}_{\mathrm{d}}\right\} = \frac{12\sigma_N^2}{(2\pi)^2 A^2 T^2 N(N^2-1)}, \tag{6}$$

where σ_N^2 is the variance of the noise. The variance of $\hat{f}_{\mathrm{d}}$ depends on the $\mathrm{SNR} = A^2/\sigma_N^2$ and the observation time NT which is determined by a trade-off between accuracy and agility of signal processing algorithm. Simulation results for a noisy complex sinusoid with varying SNR and various numbers of sampling points N are illustrated in Fig. 6, together with the theoretical lower bounds given by Eq. (6).

For low SNR the frequency estimation is distorted and the variance rapidly increases. In Tab. 2 the values for three standard deviations 3σ of the estimated mean frequency are given. This range includes 99.7 % of all estimates.

The obtained Doppler frequency shift depends on the sensor-target geometry via the inner product of the distance

Table 2. Simulation results for the standard deviation: $f_s = 48\,\text{kHz}$, $f_d = 25.6\,\text{Hz}$, $\text{SNR} = 10\,\text{dB}$.

N	1024	2048	4096
3σ/Hz	0.8161	0.2999	0.1190
$3\sigma_{\mathrm{crb}}$/Hz	0.7662	0.2709	0.0958

vector $(\boldsymbol{r}_0 - \boldsymbol{s})$ and the velocity vector $\boldsymbol{v} = (v_x,\, v_y,\, v_z)^{\mathrm{T}}$. The equation for the general three-dimensional case with sensor at position $(s_x,\, s_y,\, s_z)^{\mathrm{T}}$ is

$$f_{\mathrm{d}} = \frac{2}{\lambda} \frac{(\boldsymbol{r}_0 - \boldsymbol{s})^{\mathrm{T}} \boldsymbol{v}}{\|\boldsymbol{r}_0 - \boldsymbol{s}\|}, \tag{7}$$

where $\|\cdot\|$ represents the standard Euclidean vector norm.

4.1 Data model

In this application a special kind of observation scenario arises. Several sensors are closely spaced on a maneuverable platform monitoring the space in front with one dominant scatterer. Figure 7 illustrates the scenario with sensor platform and object inside the closed machine. The platform typically moves with constant velocity on straight-lined trajectories.

Compilation of Eq. (7) for a radar network with several distributed sensors results in a system of nonlinear equations

$$\boldsymbol{f}_{\mathrm{d}} = \mathbf{A}(\boldsymbol{\xi}) + \boldsymbol{w}. \tag{8}$$

with the observation matrix $\mathbf{A}(\boldsymbol{\xi})$ which contains the the model of Eq. (7) and the parameter vector $\boldsymbol{\xi} = (x,\, y,\, z,\, v_x,\, v_y,\, v_z)^{\mathrm{T}}$ for the general three-dimensional case. The number of unknowns is reduced to four as the two-dimensional case is considered. To find the optimal $\boldsymbol{\xi}^*$ the constrained minimization problem (e.g., Bjoerk, 1996),

$$\begin{aligned} &\underset{\boldsymbol{\xi}}{\text{minimize}}\ F(\boldsymbol{\xi}) \\ &\text{such that } \boldsymbol{g}(\boldsymbol{\xi}) \le b_\ell, \quad \ell = 1, \ldots, p \end{aligned} \tag{9}$$

has to be solved, where $F(\boldsymbol{\xi})$ is the cost function and $\boldsymbol{g}(\boldsymbol{\xi})$ contains linear and nonlinear constraints for maximum distances between sensor and object and maximum velocities in all directions. To solve this nonlinear minimization problem and to find a solution for position $\boldsymbol{r}_0$ and velocity of the platform $\boldsymbol{v}$, Newton-type or Gauss-Newton method is applied. A fundamental problem is to find the global minimum. For cost functions and their Jacobian with several minima, the accuracy of the results strongly depend on the initial guess. More advanced methods are necessary which find the global minimum, e.g. grid search methods. These account for higher computing capacity at the beginning of a new trajectory. Then, the previously found solutions can be used recursively. The problem is investigated in detail in future works.

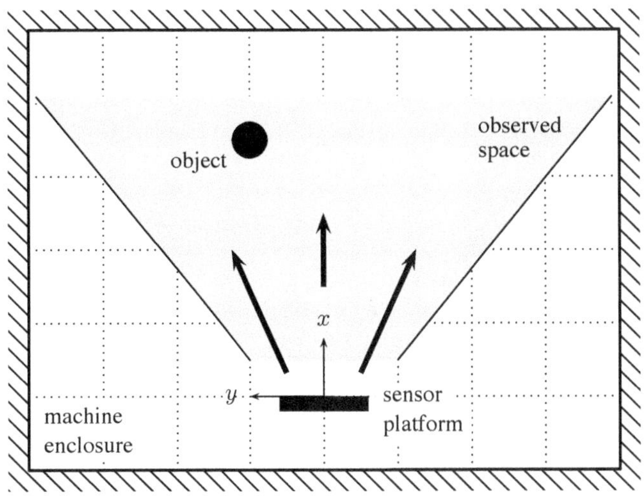

Figure 7. Illustration of a typical scenario of the application: One dominant scatterer at fixed position and the maneuverable sensor platform with radar modules. Arrows point out the typical rectilinear motion during manual operation.

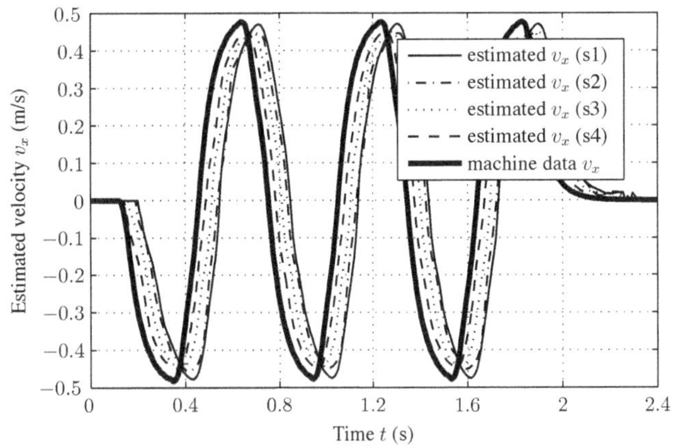

Figure 8. Estimated velocity from Doppler shift measurements for an axial movement of the sensor platform.

In the next section measurement data of a one-dimensional linear motion are analyzed to show the accuracy of the frequency estimation. The second example investigates a two-dimensional scenario. A nonlinear solver is used to obtain position estimates from synthetic and measurement data.

4.2 Measurement results for linear motion

This section provides the results of the algorithm applied to a set of measurements. The measurements are done with the in-house developed prototype depicted in Fig. 2. They clarify quality and accuracy of the algorithm for frequency estimation. The received signal streams are stored and available for off-line post processing. To obtain sufficient frequency resolution the processed block size is set to $N = 4096$ data points. This results in a resolution of $1/(NT) \approx 11.7$ Hz during the post processing. The decimation process has no influence on the value of NT. Variable update rates are achieved by overlapped data segments. Measurement and reference data for position and speed of the sensor platform with the same update rate of 4 ms are used for performance tests with the prototype. Reference data available from the machine tool control unit provides exact position and velocity data.

Results of the following scenario are presented: Comparable to Figs. 1 and 2, the sensor platform is equipped with four sensors. The movement of the sensor platform is bidirectionally (i.e. back and forth) on an axial path. A metal sphere with a diameter $d = 15$ cm serves as scattering object. Its position is fixed and aligned to the center of the platform. As the circumference of the sphere is much larger than the wavelength, the RCS of the sphere is in the optical region and equal the cross-section area $d^2\pi/4$ of the sphere.

Referred to the coordinate system in Fig. 1 the velocitiy components v_y and v_z of the platform are both equal to zero and the y and z coordinates remain constant. Figure 8 shows the obtained velocity from the internal machine's control unit (solid graph) and the estimated velocity v_x (dashed and dotted graphs) from the data of each of the four sensors. The small time shifts between the various sensor streams come from the data acquisition procedure. The exact value is known and can be compensated. At the beginning a receding movement is identified (negative sign) with continuously increasing velocity up to a deceleration phase around 0.35 to 0.45 s. Then, the turning point is reached and the movement with positive acceleration and velocity follows (positive sign). The movement pattern is repeated three times. Against the background of a dynamic scenario with ever-changing velocity, the estimates are in very good agreement with the true velocity in Fig. 8. In this scenario velocities of $0.47\,\mathrm{m\,s^{-1}}$ and accelerations up to $6.5\,\mathrm{m\,s^{-2}}$ are reached.

Minor discrepancy can be identified at the turning points where the velocities and change in position becomes small. This is due to the limited frequency resolution. Frequencies lower than 11.7 Hz (corresponding to $\sim 73.1\,\mathrm{mm\,s^{-1}}$ in this system) are actually mapped to zero due to lack of accuracy of the estimate. Improvements are achieved by using adaptive parallel analyzing of longer sequences and subsequent adjustment of the position.

4.3 Results for diagonal motion

As a further scenario a two-dimensional motion is investigated. Figure 9 shows the results for a diagonal approaching motion where v_x and v_y are nonzero. Using the coordinate system of Fig. 1, the starting position of the object (sphere, $d = 15$ cm) is at $(0.5\,\mathrm{m}, 0.5\,\mathrm{m}, 0)^{\mathrm{T}}$ and the constant velocity of the sensor platform is $(-0.707\,\mathrm{m\,s^{-1}}, -0.707\,\mathrm{m\,s^{-1}}, 0)^{\mathrm{T}}$. The position of the object is represented by circles. The results for synthetic data (+) and for real measurement data

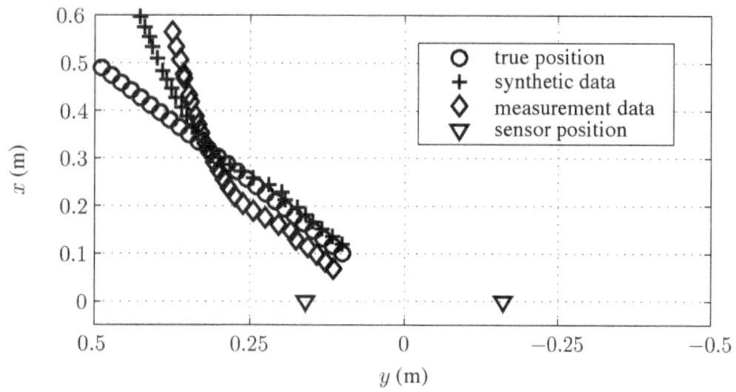

Figure 9. Target localization close to the sensor platform. The initial guess of the parameter vector for the iterative nonlinear solver is $(0.91\,\text{m}, 0.53\,\text{m}, 0, -0.66\,\text{m}\,\text{s}^{-1}, -0.33\,\text{m}\,\text{s}^{-1}, 0)^{\text{T}}$.

(◇) are very similar and close to the real position of the object.

The algorithm is based on the approach of Chan and Jardine (1990) which uses a special cost function involving the time derivative of the Doppler shift $\dot{f}_{\text{d}}$ (rate of change). For further details the reader is referred to the original paper.

In this example the absolute errors between real and estimated position are lower than 4 cm, what is already sufficient for collision warning in machine tools. Further investigations are in progress to improve reliability and flexibility of the algorithm to make it applicable for arbitrary linear motions.

5 Conclusions

This paper presented an implementation of a prototype system that aims at a new application of 24 GHz CW radar technology in machine tools of collision detection and avoidance. Important aspects for the mechanical design, the RF hardware, the components for the baseband processing and the elements of the developed algorithm for velocity estimation have been described. This first approach treats single-target scenarios in a two-dimensional space. For validation the case of linear motion has been investigated. Results from conducted measurements reveal very good agreement of velocity estimates with the reference data provided by the machine's control unit. The application of a nonlinear solver for position estimation in a two-dimensional motion scenario showed acceptably low position estimation errors. This prototype system serves as the basis for further investigations and developments towards a multi-sensor radar system for fast collision detection in machine tools. Due to the rather weak sensitivity of the used modules and the coarse Doppler resolution due to short observation windows the effect of multiple targets and multiple reflections has not been considered. It is expected that this simple approach is suitable for the majority of collision scenarios. However, specification of the actual false alarm and missed detection rates requires further studies. Further topics for future work are extension to three dimensions, the question of the minimum necessary number of sensors for uniqueness of the solution, and their optimum positioning around the tool. Although the observation time is very short in this application the effect of cross section fluctuations also needs further investigations.

Acknowledgements. The authors would like to thank M. Berger and D. Korff from the Institute of Production Management, Technology and Machine Tools (PTW) of the Technische Universität Darmstadt for their help and support during the execution of the measurements.

This work is funded by the Bavarian Ministry of Economic Affairs and Media, Energy and Technology within the economic development scheme "Mikrosystemtechnik Bayern" under grant BAY158/002.

Edited by: U. van Rienen

References

Abele, E., Brecher, C., Gsell, S. C., Hassis, A., and Korff, D.: Steps towards a protection system for machine tool main spindles against crash-caused damages, Springer J. Prod. Eng., 6.6, 631–642, doi:10.1007/s11740-012-0422-6, 2012.

Azodi, H., Siart, U., and Eibert, T. F.: A fast three-dimensional deterministic ray tracing coverage simulator for a 24 GHz anti-collision radar, Adv. Radio Sci., 11, 55–60, doi:10.5194/ars-11-55-2013, 2013.

Barton, D. K.: Radar System Analysis, 1. Ed., Prentice Hall, 1964.

Björck, Å.: Numerical Methods for Least Squares Problems, 1. Ed., Society for Industrial and Applied Mathematics (SIAM), doi:10.1137/1.9781611971484, 1996.

Chan, Y.-T. and Jardine, F. L.:, Target localization and tracking from Doppler-shift measurements, IEEE J. Oceanic Eng., 15, 251–257, doi:10.1109/48.107154, 1990.

RFbeam Microwave GmbH: K-LC2 Radar Transceiver – Datasheet, available at: http://www.rfbeam.ch (last access: 12 July 2013), 2011.

Rife, D. and Boorstyn, R. R.: Single tone parameter estimation from discrete-time observations, IEEE Trans. Inf. Theory, 20, 591–598, doi:10.1109/TIT.1974.1055282, 1974.

Skolnik, M. I.: Theoretical accuracy of radar measurements, IRE Trans. Aeronaut. Navig. Electron., ANE-7.4, 123–129, doi:10.1109/TANE3.1960.4201757, 1960.

Strohm, K. M., Bloecher, H.-L., Schneider, R., and Wenger, J., Development of future short range radar technology, European Radar Conference 2005, 165–168, doi:10.1109/EURAD.2005.1605591, 2005.

Permissions

All chapters in this book were first published in ARS, by Copernicus Publications; hereby published with permission under the Creative Commons Attribution License or equivalent. Every chapter published in this book has been scrutinized by our experts. Their significance has been extensively debated. The topics covered herein carry significant findings which will fuel the growth of the discipline. They may even be implemented as practical applications or may be referred to as a beginning point for another development.

The contributors of this book come from diverse backgrounds, making this book a truly international effort. This book will bring forth new frontiers with its revolutionizing research information and detailed analysis of the nascent developments around the world.

We would like to thank all the contributing authors for lending their expertise to make the book truly unique. They have played a crucial role in the development of this book. Without their invaluable contributions this book wouldn't have been possible. They have made vital efforts to compile up to date information on the varied aspects of this subject to make this book a valuable addition to the collection of many professionals and students.

This book was conceptualized with the vision of imparting up-to-date information and advanced data in this field. To ensure the same, a matchless editorial board was set up. Every individual on the board went through rigorous rounds of assessment to prove their worth. After which they invested a large part of their time researching and compiling the most relevant data for our readers.

The editorial board has been involved in producing this book since its inception. They have spent rigorous hours researching and exploring the diverse topics which have resulted in the successful publishing of this book. They have passed on their knowledge of decades through this book. To expedite this challenging task, the publisher supported the team at every step. A small team of assistant editors was also appointed to further simplify the editing procedure and attain best results for the readers.

Apart from the editorial board, the designing team has also invested a significant amount of their time in understanding the subject and creating the most relevant covers. They scrutinized every image to scout for the most suitable representation of the subject and create an appropriate cover for the book.

The publishing team has been an ardent support to the editorial, designing and production team. Their endless efforts to recruit the best for this project, has resulted in the accomplishment of this book. They are a veteran in the field of academics and their pool of knowledge is as vast as their experience in printing. Their expertise and guidance has proved useful at every step. Their uncompromising quality standards have made this book an exceptional effort. Their encouragement from time to time has been an inspiration for everyone.

The publisher and the editorial board hope that this book will prove to be a valuable piece of knowledge for researchers, students, practitioners and scholars across the globe.

List of Contributors

N. Fichtner
Institute for High Frequency Engineering, Technische Universität München, Arcisstr. 21, 80333 München, Germany

P. Russer
Institute for High Frequency Engineering, Technische Universitüt München, Arcisstr. 21, 80333 München, Germany

G. Saala
Institut für Hoch- und Höchstfrequenztechnik, Universität der Bundeswehr München, Werner-Heisenberg-Weg 39, 85577 Neubiberg, Germany

D. Müller
Institut für Hoch- und Höchstfrequenztechnik, Universität der Bundeswehr München, Werner-Heisenberg-Weg 39, 85577 Neubiberg, Germany

J. Hopf
Institut für Hoch- und Höchstfrequenztechnik, Universität der Bundeswehr München, Werner-Heisenberg-Weg 39, 85577 Neubiberg, Germany

S. Lindenmeier
Institut für Hoch- und Höchstfrequenztechnik, Universität der Bundeswehr München, Werner-Heisenberg-Weg 39, 85577 Neubiberg, Germany

T. Reichthalhammer
Fachgebiet Höchstfrequenztechnik, Technische Universität München, Munich, Germany

E. Biebl
Fachgebiet Höchstfrequenztechnik, Technische Universität München, Munich, Germany

M. A. Eberspächer
Lehrstuhl für Hochfrequenztechnik, Technische Universität München, Germany

T. F. Eibert
Lehrstuhl für Hochfrequenztechnik, Technische Universität München, Germany

B. Schaffer
Fachgebiet Höchstfrequenztechnik, Technische Universität München, Munich, Germany

G. Kalverkamp
Fachgebiet Höchstfrequenztechnik, Technische Universität München, Munich, Germany

M. Chaabane
Fachgebiet Höchstfrequenztechnik, Technische Universität München, Munich, Germany

E. M. Biebl
Fachgebiet Höchstfrequenztechnik, Technische Universität München, Munich, Germany

A. Bartsch
BMW Group Research and Technology, Munich, Germany

F. Fitzek
BMW Group Research and Technology, Munich, Germany

R. H. Rasshofer
BMW Group Research and Technology, Munich, Germany

M. A. Eberspächer
Lehrstuhl für Hochfrequenztechnik, Technische Universität München, Germany

M. Bauer
Lehrstuhl für Hochfrequenztechnik, Technische Universität München, Germany

T. F. Eibert
Lehrstuhl für Hochfrequenztechnik, Technische Universität München, Germany

S. Bertl
Fachgebiet Hochfrequente Felder und Schaltungen, Technische Universitt München, Arcisstr. 21, 80333 München, Germany

A. Kirschner
Fachgebiet Hochfrequente Felder und Schaltungen, Technische Universitt München, Arcisstr. 21, 80333 München, Germany

J. Detlefsen
Fachgebiet Hochfrequente Felder und Schaltungen, Technische Universitt München, Arcisstr. 21, 80333 München, Germany

R. Brem
Lehrstuhl für Hochfrequenztechnik, Technische Universität München, 80290 Munich, Germany

T. F. Eibert
Lehrstuhl für Hochfrequenztechnik, Technische Universität München, 80290 Munich, Germany

A. T. Ott
Lehrstuhl für Hochfrequenztechnik Technische Universität München 80290 Munich, Germany

C. J. Eisner
Lehrstuhl für Hochfrequenztechnik Technische Universität München 80290 Munich, Germany

M. D. Blech
Institut für Hochfrequenztechnik Universität Stuttgart, 70550 Stuttgart, Germany

T. F. Eibert
Lehrstuhl für Hochfrequenztechnik Technische Universität München 80290 Munich, Germany

R. H. Rasshofer
BMW Group Research and Technology, Hanauer Str. 46, 80992 Munich, Germany

M. Spies
Ingenieurbüro Spies, Seibersdorf 18, 86558 Hohenwart, Germany

H. Spies
Ingenieurbüro Spies, Seibersdorf 18, 86558 Hohenwart, Germany

H. Azodi
Lehrstuhl für Hochfrequenztechnik, Technische Universität München, Arcisstr. 21, 80333 Munich, Germany

U. Siart
Lehrstuhl für Hochfrequenztechnik, Technische Universität München, Arcisstr. 21, 80333 Munich, Germany

T. F. Eibert
Lehrstuhl für Hochfrequenztechnik, Technische Universität München, Arcisstr. 21, 80333 Munich, Germany

B. Cetin
Institute for Electromagnetic Fields and Microwave Electronics, ETH Zurich, Switzerland

H.-R. Benedickter
Institute for Electromagnetic Fields and Microwave Electronics, ETH Zurich, Switzerland

P. Leuchtmann
Institute for Electromagnetic Fields and Microwave Electronics, ETH Zurich, Switzerland

H. Brüns
Institut für Elektrotechnik und Informationstechnik, Christian-Albrechts-Universität zu Kiel, Germany

L. Klinkenbusch
Institut für Elektrotechnik und Informationstechnik, Christian-Albrechts-Universität zu Kiel, Germany

R. A. M. Mauermayer
Lehrstuhl für Hochfrequenztechnik, Technische Universität München, 80290 Munich, Germany

T. F. Eibert
Lehrstuhl für Hochfrequenztechnik, Technische Universität München, 80290 Munich, Germany

M. Stallein
Universität Paderborn, Theoretische Elektrotechnik, Warburger Strasse 100, 33098 Paderborn, Germany

B. Bandlow
Universität Paderborn, Theoretische Elektrotechnik, Warburger Strasse 100, 33098 Paderborn, Germany

R. Schuhmann
Universität Paderborn, Theoretische Elektrotechnik, Warburger Strasse 100, 33098 Paderborn, Germany

C. H. Schmidt
Department of Electrical Engineering, University of California, Los Angeles, USA

S. F. Razavi
Department of Electrical Engineering, University of California, Los Angeles, USA

T. F. Eibert
Lehrstuhl für Hochfrequenztechnik, Technische Universität München, Germany

Y. Rahmat-Samii
Department of Electrical Engineering, University of California, Los Angeles, USA

C. Morhart
Fachgebiet Höchstfrequenztechnik, Technische Universität München, Munich, Germany

E. M. Biebl
Fachgebiet Höchstfrequenztechnik, Technische Universität München, Munich, Germany

N. Burschäpers
Universität Paderborn, EIM, Fachgebiet Theoretische Elektrotechnik, Germany

S. Fiege
Universität Paderborn, EIM, Lehrstuhl für Mathematik und ihre Anwendungen, Germany

R. Schuhmann
Universität Paderborn, EIM, Fachgebiet Theoretische Elektrotechnik, Germany

A. Walther
Universität Paderborn, EIM, Lehrstuhl für Mathematik und ihre Anwendungen, Germany

M. Goppelt
Institute of Micowave Techniques, University of Ulm, 89081 Ulm, Germany

H.-L. Blöcher
Group Research & Advanced Engineering, Daimler AG, 89081 Ulm, Germany

W. Menzel
Institute of Micowave Techniques, University of Ulm, 89081 Ulm, Germany

F. Argin
Institut für Elektrotechnik und Informationstechnik, Christian-Albrechts-Universität zu Kiel, Germany

H. Ahrens
Institut für Elektrotechnik und Informationstechnik, Christian-Albrechts-Universität zu Kiel, Germany

L. Klinkenbusch
Institut für Elektrotechnik und Informationstechnik, Christian-Albrechts-Universität zu Kiel, Germany

U. Siart
Lehrstuhl für Hochfrequenztechnik, Technische Universität München, 80290 München, Germany

S. Adrian
Lehrstuhl für Hochfrequenztechnik, Technische Universität München, 80290 München, Germany

T. Eibert
Lehrstuhl für Hochfrequenztechnik, Technische Universität München, 80290 München, Germany

K. A. Yinusa
Lehrstuhl für Hochfrequenztechnik, Technische Universität München, Arcisstr. 21, 80333 Munich, Germany

C. H. Schmidt
Lehrstuhl für Hochfrequenztechnik, Technische Universität München, Arcisstr. 21, 80333 Munich, Germany

T. F. Eibert
Lehrstuhl für Hochfrequenztechnik, Technische Universität München, Arcisstr. 21, 80333 Munich, Germany

J. Gao
Graduate School of Computational Engineering, Technische Universität Darmstadt, Dolivostraße 15, 64293 Darmstadt, Germany
Institut für Theorie Elektromagnetischer Felder, Technische Universität Darmstadt, Schloßgartenstraße 8, 64289 Darmstadt, Germany

I. Munteanu
Graduate School of Computational Engineering, Technische Universität Darmstadt, Dolivostraße 15, 64293 Darmstadt, Germany
CST – Computer Simulation Technology AG, Bad Nauheimer Straße 19, 64289 Darmstadt, Germany

W. F. O. Müller
Institut für Theorie Elektromagnetischer Felder, Technische Universität Darmstadt, Schloßgartenstraße 8, 64289 Darmstadt, Germany

T. Weiland
Graduate School of Computational Engineering, Technische Universität Darmstadt, Dolivostraße 15, 64293 Darmstadt, Germany
Institut für Theorie Elektromagnetischer Felder, Technische Universität Darmstadt, Schloßgartenstraße 8, 64289 Darmstadt, Germany
CST – Computer Simulation Technology AG, Bad Nauheimer Straße 19, 64289 Darmstadt, Germany

H. Ahrens
Institut für Elektrotechnik und Informationstechnik, Christian-Albrechts-Universität zu Kiel, Germany

F. Argin
Institut für Elektrotechnik und Informationstechnik, Christian-Albrechts-Universität zu Kiel, Germany

L. Klinkenbusch
Institut für Elektrotechnik und Informationstechnik, Christian-Albrechts-Universität zu Kiel, Germany

C. V. Motrescu
Institut für Elektrotechnik und Informationstechnik, Christian-Albrechts-Universität zu Kiel, Germany

L. Klinkenbusch
Institut für Elektrotechnik und Informationstechnik, Christian-Albrechts-Universität zu Kiel, Germany

A. T. Ott
Lehrstuhl für Hochfrequenztechnik, Technische Universität München, 80290 Munich, Germany

M. Shalaby
Lehrstuhl für Hochfrequenztechnik, Technische Universität München, 80290 Munich, Germany

U. Siart
Lehrstuhl für Hochfrequenztechnik, Technische Universität München, 80290 Munich, Germany

E. Kaliyaperumal
Lehrstuhl für Hochfrequenztechnik, Technische Universität München, 80290 Munich, Germany

T. F. Eibert
Lehrstuhl für Hochfrequenztechnik, Technische Universität München, 80290 Munich, Germany

J. Engelbrecht
Professur für Kommunikationstechnik/Funksysteme, Hochschule für Technik und Wirtschaft Dresden, 01069 Dresden, Germany

R. Collmann
Professur für Kommunikationstechnik/Funksysteme, Hochschule für Technik und Wirtschaft Dresden, 01069 Dresden, Germany

K. Bavelis
Institut für Theorie Elektromagnetischer Felder, Technische Universität Darmstadt, Schloßgartenstr. 8, 64289 Darmstadt, Germany

E. Gjonaj
Institut für Theorie Elektromagnetischer Felder, Technische Universität Darmstadt, Schloßgartenstr. 8, 64289 Darmstadt, Germany

T. Weiland
Institut für Theorie Elektromagnetischer Felder, Technische Universität Darmstadt, Schloßgartenstr. 8, 64289 Darmstadt, Germany

R. Baltes
Chair of Electromagnetic Theory, Dept. of Physics and Mechatronics, Saarland University, Saarbrücken, Germany

J. Al Ahmar
Chair of Microintegration and Reliability, Dept. of Physics and Mechatronics, Saarland University, Saarbrücken, Germany

O. Farle
Chair of Electromagnetic Theory, Dept. of Physics and Mechatronics, Saarland University, Saarbrücken, Germany

R. Dyczij-Edlinger
Chair of Electromagnetic Theory, Dept. of Physics and Mechatronics, Saarland University, Saarbrücken, Germany

L. Li
Technische Universität München, Lehrstuhl für Hochfrequenztechnik, Arcisstrasse 21, 80333 Munich, Germany

K. Wang
Technische Universität München, Lehrstuhl für Hochfrequenztechnik, Arcisstrasse 21, 80333 Munich, Germany

H. Li
Politecnico di Torino, Electronic and Telecommunication, Corso Duca degli Abruzzi, 24, 10129 Torino, Italy

T. F. Eibert
Technische Universität München, Lehrstuhl für Hochfrequenztechnik, Arcisstrasse 21, 80333 Munich, Germany

O. Wiedenmann
Technische Universität München, Lehrstuhl für Hochfrequenztechnik, Arcisstr. 21, 80333 Munich, Germany

R. Ramakrishnan
Technische Universität München, Lehrstuhl für Umformtechnik und Gießereiwesen, Walther-Meißner-Straße 4, 85747 Garching, Germany

P. Saal
Technische Universität München, Lehrstuhl für Umformtechnik und Gießereiwesen, Walther-Meißner-Straße 4, 85747 Garching, Germany

E. Kılıç
Technische Universität München, Lehrstuhl für Hochfrequenztechnik, Arcisstr. 21, 80333 Munich, Germany

U. Siart
Technische Universität München, Lehrstuhl für Hochfrequenztechnik, Arcisstr. 21, 80333 Munich, Germany

T. F. Eibert
Technische Universität München, Lehrstuhl für Hochfrequenztechnik, Arcisstr. 21, 80333 Munich, Germany

W. Volk
Technische Universität München, Lehrstuhl für Umformtechnik und Gießereiwesen, Walther-Meißner-Straße 4, 85747 Garching, Germany

F. Fitzek
BMW Group Research and Technology, Munich, Germany

R. H. Rasshofer
BMW Group Research and Technology, Munich, Germany

E. M. Biebl
Technische Universität München – Fachgebiet Höchstfrequenztechnik, Munich, Germany

M. D. Blech
Institut für Hochfrequenztechnik, Universität Stuttgart, 70550 Stuttgart, Germany

A. T. Ott
Lehrstuhl für Hochfrequenztechnik, Technische Universität München, 80290 Munich, Germany

P. Neumeier
Abteilung Leistungsoptimierte Systeme, Fraunhofer-Institut für Integrierte Schaltungen IIS, 90411 Nürnberg, Germany

M. Möller
Institut für Hochfrequenztechnik, Universität Stuttgart, 70550 Stuttgart, Germany

T. F. Eibert
Lehrstuhl für Hochfrequenztechnik, Technische Universität München, 80290 Munich, Germany

C. Classen
University of Paderborn, Fachgebiet Theoretische Elektrotechnik, Germany

E. Gjonaj
TU Darmstadt, Institut Theorie Elektromagnetischer Felder, Germany

U. Römer
TU Darmstadt, Institut Theorie Elektromagnetischer Felder, Germany

R. Schuhman
University of Paderborn, Fachgebiet Theoretische Elektrotechnik, Germany

T. Weiland
TU Darmstadt, Institut Theorie Elektromagnetischer Felder, Germany

H.-L. Blöecher
Daimler AG, Group Research and Advanced Engineering, 89081 Ulm, Germany

M. Andres
University Ulm, Institute of Microwave Techniques, 89081 Ulm, Germany

C. Fischer
Daimler AG, Group Research and Advanced Engineering, 89081 Ulm, Germany

A. Sailer
University Ulm, Institute of Microwave Techniques, 89081 Ulm, Germany

M. Goppelt
University Ulm, Institute of Microwave Techniques, 89081 Ulm, Germany

J. Dickmann
Daimler AG, Group Research and Advanced Engineering, 89081 Ulm, Germany

M. Kijowski
Institute of Electrical and Information Engineering, Christian-Albrechts-Universität zu Kiel, Germany

L. Klinkenbusch
Institute of Electrical and Information Engineering, Christian-Albrechts-Universität zu Kiel, Germany

T. Banova
Technische Universität Darmstadt, Graduate School of Computational Engineering, Dolivostraße 15, 64293 Darmstadt, Germany
Technische Universität Darmstadt, Institut für Theorie Elektromagnetischer Felder (TEMF), Schlossgartenstraße 8, 64289 Darmstadt, Germany

W. Ackermann
Technische Universität Darmstadt, Institut für Theorie Elektromagnetischer Felder (TEMF), Schlossgartenstraße 8, 64289 Darmstadt, Germany

T. Weiland
Technische Universität Darmstadt, Institut für Theorie Elektromagnetischer Felder (TEMF), Schlossgartenstraße 8, 64289 Darmstadt, Germany

S. F. Helfert
FernUniversität in Hagen, Hagen, Germany

T. J. Wächter
Technische Universität München, Lehrstuhl für Hochfrequenztechnik, Arcisstr. 21, 80333 Munich, Germany

U. Siart
Technische Universität München, Lehrstuhl für Hochfrequenztechnik, Arcisstr. 21, 80333 Munich, Germany

T. F. Eibert
Technische Universität München, Lehrstuhl für Hochfrequenztechnik, Arcisstr. 21, 80333 Munich, Germany

S. Bonerz
Ott-Jakob Spanntechnik GmbH, Industriestr. 3–7, 87663 Lengenwang, Germany